AF411310

Green Economy and Sustainable Development

Green Economy and Sustainable Development

Editors

Sergey Zhironkin
Michal Cehlar

MDPI • Basel • Beijing • Wuhan • Barcelona • Belgrade • Manchester • Tokyo • Cluj • Tianjin

Editors
Sergey Zhironkin
Surface Mining Department
T.F. Gorbachev Kuzbass State
Technical University
Kemerovo
Russia

Michal Cehlar
Institute of Earth Sources
Faculty of Mining, Ecology,
Process Technologies and
Geotechnology
Technical University of Košice
Košice
Slovakia

Editorial Office
MDPI
St. Alban-Anlage 66
4052 Basel, Switzerland

This is a reprint of articles from the Special Issue published online in the open access journal *Energies* (ISSN 1996-1073) (available at: www.mdpi.com/journal/energies/special_issues/green_economy_sustainable_development).

For citation purposes, cite each article independently as indicated on the article page online and as indicated below:

LastName, A.A.; LastName, B.B.; LastName, C.C. Article Title. *Journal Name* **Year**, *Volume Number*, Page Range.

ISBN 978-3-0365-3223-3 (Hbk)
ISBN 978-3-0365-3222-6 (PDF)

Contents

Preface to "Green Economy and Sustainable Development"

Today sustainable development is associated with increasing the responsibility of business, governments and the entire society around the world for achieving a balance between current and future needs for subsoils, energy, traditional and new materials, and transport. Such responsibility equalizes environmental and social problems (poverty and malnutrition, inequalities in access to wealth and income distribution), which often have common roots. This, in turn, highlights the importance of interdisciplinary and multilateral research of sustainable development issues, the results of which can answer the questions about green economy perspectives.

The green economy—the newest way to obtain and use resources—is a product of the Fourth Industrial Revolution, and it concentrates many of the achievements of Industry 4.0. The related structural shifts in the economy are caused by the emergence of new industries of waste recycling, zero-emission energy production, absorption of greenhouse gases emissions, green urbanism, and post-mining. These shifts should be summed up in a parallel increase in productivity and labor safety, improved access to drinking water, food, energy, as well as in joining the efforts of national states and businesses in the fight against climate change, in replacing minerals with renewable resources.

Sergey Zhironkin, Michal Cehlar
Editors

 *energies*

Editorial

Green Economy and Sustainable Development: The Outlook

Sergey Zhironkin [1,2,3,*] and Michal Cehlár [1,4]

1   Institute of Trade and Economy, Siberian Federal University, 79 Svobodny av., 660041 Krasnoyarsk, Russia
2   School of Core Engineering Education, National Research Tomsk Polytechnic University, 30 Lenina st., 634050 Tomsk, Russia
3   Open Pit Mining Department, T.F. Gorbachev Kuzbass State Technical University, 28 Vesennya st., 650000 Kemerovo, Russia
4   Institute of Earth Sources, Faculty of Mining, Ecology, Process Technologies and Geotechnology, Technical University of Košice, Letná 9, 040 01 Košice, Slovakia; michal.cehlar@tuke.sk
*   Correspondence: zhironkinsa@kuzstu.ru

Citation: Zhironkin, S.; Cehlár, M. Green Economy and Sustainable Development: The Outlook. *Energies* **2022**, *15*, 1167. https://doi.org/10.3390/en15031167

Received: 31 January 2022
Accepted: 3 February 2022
Published: 4 February 2022

Publisher's Note: MDPI stays neutral with regard to jurisdictional claims in published maps and institutional affiliations.

Modern theories that make up the paradigm of sustainable development, and the best practices derived from them, are based on the consistency of individual and public needs, factors of economic growth and ecosystem conservation. The trend of green economy expansion is moving from a challenge facing modern society to the dominant area of scientific thinking, which is increasingly focused on solving the problems of reducing the anthropogenic impact on the environment, primarily on the climate. At the same time, the level of scientific and technological progress, and the quality of life that modern civilization has reached, require the preservation and increase in the specific consumption of various resources in the long term. Therefore, green economy technologies that ensure the transition to sustainable development are initially focused on the optimal and integrated use of non-renewable resources and the maximum involvement of renewable resources in the production of goods specific for a post-industrial era [1]. Based on this, the evolution of sustainable development methodology should be carried out in the system of innovative development of industrial technologies and their adaptation to the latest trends in energy and urbanism, ecology, finance and investment.

At the same time, sustainable development is associated with increasing the responsibility of business, governments and the entire society, around the world, for achieving a balance between current and future needs for subsoils, energy, traditional and new materials, and transport. Such responsibility equalizes environmental and social problems (poverty and malnutrition, inequalities in access to wealth and income distribution), which often have common roots. This, in turn, highlights the importance of interdisciplinary and multilateral research on sustainable development issues, the results of which can answer questions about green economy perspectives.

The green economy—the newest way to obtain and use resources—is a product of the Fourth Industrial Revolution and concentrates many of the achievements of Industry 4.0. The related structural shifts in the economy are caused by the emergence of new industries of waste recycling, zero-emission energy production, absorption of greenhouse gases emissions, green urbanism, and post-mining. These shifts should be matched by a parallel increase in productivity and labor safety, improved access to drinking water, food, energy, as well as in joining the efforts of national states and businesses in the fight against climate change, in replacing minerals with renewable resources.

However, a true transition to a green economy is possible only with the sustainable development of all industries and the saturation of both production and consumption with green technologies. Green production should be developed in basic industries (mining, energy, engineering, chemistry, transport), as well as in high-tech industries that set new horizons for environmentally oriented modernization. These processes, integral to sustainable development, are united in the "green growth" concept, the main trend of green economy development, which consists in increasing the production of so-called "sustainable goods and services", the main feature of which is recycling and zero emissions of toxic substances. At the same time, there is no consensus among academicians on the extent to which the growth trend of the green economy will coincide with traditionally driven growth (increase in the production of "high-carbon" energy, industries with large volumes of waste). Therefore, the creation of platforms for discussing the problems of sustainable development and the green economy is important for the publication of research papers on a wide range of related topics.

Thus, the discussions in the field of sustainable development, unfolding in the pages of scientific journals around the world, should provide the answers to a number of challenges facing the world economy.

The first challenge is to achieve the sustainable development goals set by the UN [2] (including ensuring universal access to affordable and clean energy sources, promoting decent work and economic growth, ensuring sustainable cities and communities, responsible consumption and production, et al.), without a simultaneous deterioration in the functioning of existing industries that concentrate investments, jobs, and tax payments. In other words, the balance between current and future production and consumption has not yet been found. The road is left wide open for unbalanced government policies towards a green economy, which could potentially cause a decline in investment activity and significantly slow down economic growth, especially in developing countries. Therefore, modern science faces the issue of ensuring the evolutionary development of green economy and regulating its expansion in such a way that the investments, jobs and consumer demand do not provoke national and global economic crises. It is important that the new "green" jobs are usually created in traditional industries and infrastructure, associated with the reduction in energy consumption, waste generation and emissions of toxic substances.

The second challenge is to ensure the investment attractiveness of the processes that form the green economy and the development of B2B, B2C, B2G, C2B, C2G, G2B, G2G relations in it, in which the development and implementation of standards for companies' environmental responsibility will not destroy the existing incentives for innovation. Setting eco-efficiency requirements for business can be an effective first step towards a green economy, sustainable energy production from alternative sources and sustainable urban development, without increasing pressure on the environment. The fact that these problems are reflected in a wide range of scientific publications inspires optimism for an increase in eco-efficient investments, the share of which reaches its maximum in the green economy.

The third challenge is the sustainable development of the energy sector and the expansion of its "green" segment producing zero emissions. The key trend in the development of modern industry is its ESG transformation, which combines the reduction of greenhouse gases emissions in energy production and transport (in the economy as a whole—a decrease in specific energy consumption), advances social responsibility and the efficiency of stakeholders' interactions, in solving sustainable development problems. As part of the ESG transformation, the development of green energy is aimed at the decarbonization of industry and cities, as a response to the growth of greenhouse gases emissions and the unfavorable climate changes caused by them. It implies increasing the share of carbon-free transport and cities during transition to the use of renewable energy sources (solar, wind, biomass, biogas, geothermal), as well as hydrogen fuel. It is important that the global research community does not demonstrate a technocratic approach to solving the problems of the green economy and energy, but fundamentally explores the possibilities of developing responsible ESG investment that takes into account environmental and social factors, when making decisions on the creation of new energy and industrial systems, as well as expanding urban agglomerations.

The fourth challenge to sustainable development is associated with the innovations in traditional industries that occupy an important place in the modern economy, in particular, mining. In the transition to a green economy, the most important step should be to ensure the main value of modern society—the labor safety and the protection of human lives. The technological level of the extractive industries sets the pace for reducing the negative anthropogenic impact on the ecosystem and, at the same time, improving the quality of life of the population in industrial clusters, which is in line with the UN sustainable development goals. Therefore, research into the issues of modernization of mining equipment, in order to improve labor safety and prevent man-made disasters, should take an important place in the mainstream of sustainable development.

Considering the importance of these challenges for sustainable development, the purpose of this Special Issue is to disseminate the results of cutting-edge research and broadcast the opinions of scientists from around the world, providing technological breakthroughs in green energy and urbanism, recycling and modernization of basic industries, conducting fundamental research on the economic problems of the transition to sustainable development.

Acting as the Guest Editors of this Special Issue, we got a chance to see the comprehensive understanding of sustainable development and the green economy emerging in the international scientific community. The positive response of researchers from different countries, supported by the number and high quality of articles published in this Special Issue, contributed to the expansion of horizons in sustainable development research, by combining economic and technological approaches. These approaches are based on the accumulated results of theoretical analysis and practical experience in the balanced use of non-renewable and renewable resources, the transition towards a circular economy and zero emissions. A consistent and multi-stage progress of green economy allows for the protection the world community from crises of overproduction and underconsumption, increasing the equity of access to the modern blessings of civilization, and goes beyond the rigid framework of academic discussions. The stable expansion of the green economy should be accompanied by an increase in the social well-being of not only current, but also future generations, which increases the responsibility of researchers, investors and governments. This is the main idea that this Special Issue promotes.

Articles collected in the Special Issue and selected for publication cover a wide range of sustainability and green economy issues, including green urban development and recycling, strategic production planning and market integration, zero-emission alternative energy sources, and Industry 4.0 technologies to reduce energy consumption, innovative modernization of extractive industries to improve labor safety. A special place is occupied by studies of the post-industrial transformation of industrial clusters.

Articles published in this Special Issue, as a part of Energies, are grouped in accordance with their thematic area, which corresponds to the aforementioned challenges for green economy expansion and sustainable development.

With regard to the first challenge (fundamental research in achieving sustainable development goals and stabilizing the green transformation of the economy), four articles address the sustainability of cities, planned product aging, energy markets integration, regional operational programs financing. Their provisions are related to solving the fundamental problems of sustainable development.

The article by Margeta et al. [3] provides an overview of the problems of energy and water supply in modern cities. The authors note that the high concentration of the world population in urbanistic areas causes an overconcentration of energy and water consumption in them, which makes it relevant to find ways to reduce the dependence of modern cities on their external sources, taking into account climatic features and vulnerability to natural disasters. The relevance of the presented study lies in the fact that the Seawater Steam Engine (SSE) technology proposed by the authors, which is fully consistent with the philosophy of sustainable development and based on the original "loop" concept, makes it possible to ensure the energy security of the city and its provision with critical resources, using methods and means of accumulation of electric and thermal energy, based on geothermal sources.

Niklewicz-Pijaczyńska et al. [4] presented a method for identifying the purchasing attitudes of durable goods' buyers, considering deliberate product aging. The authors analyze the negative externalities associated with the efforts of producers to program the replacement of goods by consumers, which generate significant private and social costs, and hinders sustainable development. The article presents a critique of the deliberate aging of products as inconsistent with the goals of sustainable development and contradictive to its model, and shows the results of an analysis of the factors of purchase of restitution goods, mainly, the income of buyers and the specifics of the goods. The main conclusion of the authors can be considered as the recognition of planned product aging as inappropriate for the tasks of preserving the environment by the majority of consumers, who positively assessed the transition to closed cycles of durable goods production.

In the article by Rybak et al. [5], the problem of achieving the Seventh Sustainable Development Goal (SDG7: affordable, reliable, sustainable and clean energy for all by 2030) is explored from the point of view of the integration of energy markets of European countries. The spatial information system proposed by the authors, based on the indicators used by Eurostat, was subjected to cluster and TSA analysis. As a result, it was found that in the course of the transformation of the energy systems of European countries, new unique clusters arise, and there is a need for additional indicators for assessing the production of affordable and clean energy. The author's findings, presented in the article, contribute to SDG7 achievement, in terms of realizing the potential of green energy to balance economic growth and increase health and well-being rates, as well as avoiding crises in energy production and consumption.

Mach et al. [6] reviewed the effectiveness of Regional Operational Programs, in the context of sustainable development of Polish provinces, by analyzing the correlation between public regulation and regional economic indicators that characterize sustainability (with an emphasis on employment and income indicators, housing construction). In the presented study, the authors used a method for determining the closeness of the relationship between public spending in the field of sustainable development and macroeconomic determinants, based on the temporal variability of the correlation and regression dependencies of the used determinants. Its use made it possible to determine the positive impact of Regional Operational Programs on regional entrepreneurial and socio-institutional capital. Its increase advances the regions on the path to sustainable development, which has a practical aspect for the effectiveness of EU funds disbursement.

With regard to the second challenge—the development of economic relations and processes that form a modern green economy—three articles have been published. They explore the problems of economic evaluation in waste disposal and recycling, increasing the efficiency in the use of material resources and optimizing their production, consumption and environmental regulation of the processes of the post-industrial economy. The scientific provisions outlined in these articles are of a practical nature and designed to help solve the short- and medium-term problems of the transition to the green economy.

The article by Stehlíková et al. [7] provides an approach to economic incentives for improving waste management (using the example of the Slovak Republic), taking into account the economic situation of households—the second waste producer after industry. In the course of the study, the authors used statistical methods, such as correlation analysis of economic indicators of household well-being and the quality of waste management, the results of which were verified using polynomial dependencies. The extension of this methodology to the regions of Slovakia, using cluster analysis, made it possible to determine the economic incentives for processing communal waste, to reduce their storage volumes and energy production.

Tausova et al. [8] analyzed the compliance of established practices and emerging trends in the use of material resources with the Environmental Policy Goals adopted in the European Union. The authors rightly note that the transition to a circular economy includes, in addition to new technologies for the design and consumption of materials, a reduction in the material and energy intensity of existing industrial production. The article provides a comparative analysis for the consumption of material resources in EU countries, using the cluster approach and statistical data processing by JMP software. The authors have established that the relationship between tax regulation on the use of natural resources and energy production, and changes in the material consumption of industry, is close to linear. The results of the study made it possible to identify the heterogeneity of EU countries in the effectiveness of the economic policy of reducing the use of natural material resources and switching to their recycling, which helps to improve the process of setting energy and environmental goals.

Raszka et al. [9], in their article, give a comprehensive assessment of the potential for the post-industrial development of cities and municipalities, taking into account social, economic and environmental criteria. In their study, the authors set the task of identifying strategic goals, the achievement of which means sustainable development acceleration, using a multidimensional analysis method for this, considering the distances between individual diagnostic variables. The application of the research method to the Wałbrzych Region (Poland) made it possible to rank the municipalities in terms of the social, economic and environmental potential of post-industrial development, which outlines the ways to improve regional environmental and economic policy. The authors paid special attention to the changes in the living conditions of the population and the structure of jobs, development of cultural service and tourism.

Eight articles are devoted to the third challenge to sustainable development—overcoming the green energy limitations and decreasing greenhouse gases emissions. The authors consider the issues of optimizing the use of green energy, taking into account the achieved standards of living, the dependence of carbon dioxide emissions on changes in energy consumption, the development of alternative solar energy, greening of traditional thermal engineering, the impact of diffusion of Industry 4.0 technologies on energy consumption, achievement of near-zero carbon dioxide emission in urban areas. The main conclusions and proposals presented in these articles are mostly practical and can be used both in strategic and operational management of the green economy processes.

Malinowski [10] analyzes the relationship between the quality of life in the EU and the development of alternative energy, focusing on the key role of green modernization of energy production, in ensuring a high standard of living for future generations. The article presents the results of empirical studies on the relationship between the development of green energy and living standards, based on the TOPSIS method and multi-criteria selection of variables, which include publicly available data from the EUROSTAT database. The results of the statistical analysis performed showed a relatively high value of the Spearman rank correlation coefficient, between quality of life indicators and the development of green energy, which confirms the idea of the inevitability of the transition to alternative energy sources in the mainstream of sustainable development as a condition for increasing the well-being of future generations.

Fong et al. [11] studied the dependence of the dynamics of energy consumption and emissions of carbon dioxide (the main greenhouse gas) as the main obstacle to sustainable development. The originality of the author's approach is the use of economic indicators (labor, capital and energy resources) to analyze carbon dioxide emissions in relation to GDP production. The statistical data obtained in the Guangdong-Hong Kong-Macau Greater Bay Area were used as the factual basis of the study, and analyzed in three stages (SBM-DEA model, SFA analysis). The results of the study made it possible to propose a number of recommendations for the development of state incentives for energy efficient and low-carbon production, and the provision of modern environmental protection standards.

In the article by Rybár et al. [12], the concept of a non-metallic flat plate solar collector is reproduced and a technique for evaluating its efficiency is shown. In accordance with the principles of the transition to a green economy, the development of green energy should be accompanied by a reduction in the material intensity of energy production from alternative sources, in particular, a reduction in the content of metals in solar equipment. In accordance with this, the authors proposed, and experimentally confirmed, the prospect of replacing the metal parts of the solar collector with a block of foam glass. The article presents a two-stage evaluation of the flat plate solar collector, a quantitative evaluation with a theoretical curve of the efficiency, followed by verification during experimental tests. As a result of experimental verification, the functionality of the concept of the flat plate solar collector was confirmed and problem areas were identified that require special attention when developing a prototype.

Bajno et al. [13], in their article, considered the possibilities of improving the traditional chimney system in reducing energy consumption, based on data obtained from the use of artificial neural networks in predicting temperature distributions in the building of chimney systems. The transition to the operation of energy-efficient buildings requires solving the problems of thermal energy losses, as soon as possible, in relation to a large number of existing buildings, as well as for facilities under construction. The authors conclude that improving the insulation of chimney systems is required to reduce energy consumption in older buildings, as well as avoiding the overheating of parts of rooms by regulating the amount of heat supplied to various parts of the building.

Beer et al. [14], in their article, presented the results of the development of the concept of a concentrated solar heater for segmented heat accumulators and an assessment of the prospects for its use in cooking, without burning fossil fuels. The traditional design of the earth oven, considered by the authors, is successfully supplemented with batteries that receive thermal energy from solar radiation in sufficient quantities. The original design proposed by the authors is quite simple and includes solar vacuum heat pipes, a solar radiation concentrator and heat accumulators. The design parameters were determined based on computational fluid dynamics and the transient simulation of selected operating situations and applied for three spatial points. As a result of the simulation, the possibility of safe and healthy cooking using heat accumulators was confirmed.

In the article by Khouri et al. [15], the prospects for innovative development of electricity production from solar energy, associated with the expansion of the range of materials used in the solar industry, are considered. The authors analyze the prospects for the use of the high purity polyvinyl butyral as a sealing material, based on the results of the molding process, homogenization, and analysis of physical and chemical characteristics during laboratory tests. The article pointed out that the applicability of polyvinyl butyral varies, depending on the location of the solar power plant, affecting UV radiation, atmospheric permeability, temperature ranges.

In an article by Wachnik et al. [16], the problem of the information gap in reducing the energy consumption in the industry, due to Industry 4.0 technologies diffusion, is considered. The authors advocate an absolute sense of the success of energy reduction industry-specific projects, associated with the development of IT, allowing for the optimization of decisions in the design of energy systems. Using the example of projects to minimize energy consumption at enterprises located in Poland, the authors presented the characteristics of information gaps in ICT projects in the industry, which are part of the methodology for reducing information risks in the energy sector and, ultimately, energy consumption.

Xu et al. [17] proposed the use of a three-stage planning method to achieve zero carbon dioxide emissions in the process of developing an integrated energy-planning scheme. The authors used a goal of the setting of regional integrated energy planning, which includes the definition of objects and the development of a strategy, as well as the formation of a methodology. The article proposes indicators for achieving planning goals, which are indices of the share of renewable energy in primary energy, the share of renewable energy in total consumption, and the reduction of carbon dioxide emissions, which are determined for each analyzed area for subsequent comparison. It is important that the results of the author's research can be translated, to improve integrated energy planning in various regions.

The fourth challenge facing sustainable development—the innovative modernization of traditional extractive industries to improve labor safety—is the subject of an article by Szurgacz et al. [18] The authors explore the safety issues of machines and workers in coal mines, associated with mechanical support in a longwall complex. In conditions of increasing output capacity of longwall, it is important to ensure the introduction of new technologies to protect the working space from the impact of rock mass. In accordance with that, the authors develop the mean of automation in the hydraulic control of the roof support, suggesting the introduction of a two-valve unit into the hydraulic system that provides automatic expansion of the support sections. The article presents the results of tests of a real installation of automated control of mechanical support in a longwall, showing a reduction in working time of the support operator, which is important for improving labor safety and productivity.

The articles published in the Special Issue, "Green Economy and Sustainable Development", highlight topics of the future prospects for reducing the anthropogenic impact on the environment, while maintaining the trend towards improving the well-being of mankind. In the near future, the economies of countries—the main producers of raw materials and energy—should take a massive step forward, in terms of reducing greenhouse gases emissions (to zero in the future), expanding materials recycling and responsible waste management. The contribution of science to accelerating the transition to sustainable development and green economy is to form a systemic response to such challenges as reducing the anthropogenic impact on the environment, ensuring access to alternative energy with zero emissions for everyone, creating jobs and attracting investments in the green economy, increasing environmental and labor safety in traditional industries, including the mining of minerals.

To help in solving the problems of sustainable development, the current Special Issue brings together the work of researchers from the world's leading centers of Earth science, on the economic assessment of the prospects for the development of green energy and green urbanism, strategic planning in sustainable development, innovative modernization of mining and thermal power, material engineering.

Along with this, in the modern world, there are many gaps that hinder the transition to sustainable development and determine the further advancement of scientific ideas. These include post-mining and restoration of biodiversity in industrial clusters, the promotion of new materials and technologies for recycling and renewable energy, the development of unmanned equipment on the Industry 4.0 technology platform, etc. We are confident that this Special Issue of "Energies", dedicated to sustainable development, will contribute to the consolidation and popularization of the ideas of research teams from many countries around the world, including China, Australia, Slovakia, Poland, Russia and Germany.

Author Contributions: The authors contributed equally to this work. All authors have read and agreed to the published version of the manuscript.

Funding: This research received no external funding.

Institutional Review Board Statement: Not applicable.

Informed Consent Statement: Not applicable.

Acknowledgments: This research was supported by TPU development program.

Conflicts of Interest: The authors declare no conflict of interest.

References

1. Zhironkin, S.; Gasanov, M.; Barysheva, G.; Gasanov, E.; Zhironkina, O.; Kayachev, G. Sustainable Development vs. Post-Industrial Transformation: Possibilities for Russia. *E3S Web Conf.* **2017**, *21*, 04002. [CrossRef]
2. United Nations. #Envision2030: 17 Goals to Transform the World for Persons with Disabilities. Available online: https://www.un.org/development/desa/disabilities/envision2030.html (accessed on 25 January 2022).
3. Margeta, K.; Glasnovic, Z.; Zabukovec Logar, N.; Tišma, S.; Farkaš, A. A Concept for Solving the Sustainability of Cities Worldwide. *Energies* **2022**, *15*, 616. [CrossRef]
4. Niklewicz-Pijaczyńska, M.; Stańczyk, E.; Gardocka-Jałowiec, A.; Gródek-Szostak, Z.; Niemczyk, A.; Szalonka, K.; Homa, M. A Strategy for Planned Product Aging in View of Sustainable Development Challenges. *Energies* **2021**, *14*, 7793. [CrossRef]
5. Rybak, A.; Rybak, A.; Kolev, S.D. Analysis of the EU-27 Countries Energy Markets Integration in Terms of the Sustainable Development SDG7 Implementation. *Energies* **2021**, *14*, 7079. [CrossRef]

6. Mach, Ł.; Bedrunka, K.; Dąbrowski, I.; Frącz, P. The Relationship between ROP Funds and Sustainable Development—A Case Study for Poland. *Energies* **2021**, *14*, 2677. [CrossRef]
7. Stehlíková, B.; Čulková, K.; Taušová, M.; Štrba, Ľ.; Mihaliková, E. Evaluation of Communal Waste in Slovakia from the View of Chosen Economic Indicators. *Energies* **2021**, *14*, 5052. [CrossRef]
8. Taušová, M.; Čulková, K.; Tauš, P.; Domaracká, L.; Seňová, A. Evaluation of the Effective Material Use from the View of EU Environmental Policy Goals. *Energies* **2021**, *14*, 4759. [CrossRef]
9. Raszka, B.; Dzieżyc, H.; Hełdak, M. Assessment of the Development Potential of Post-Industrial Areas in Terms of Social, Economic and Environmental Aspects: The Case of Wałbrzych Region (Poland). *Energies* **2021**, *14*, 4562. [CrossRef]
10. Malinowski, M. "Green Energy" and the Standard of Living of the EU Residents. *Energies* **2021**, *14*, 2186. [CrossRef]
11. Fong, W.; Sun, Y.; Chen, Y. Examining the Relationship between Energy Consumption and Unfavorable CO_2 Emissions on Sustainable Development by Going through Various Violated Factors and Stochastic Disturbance–Based on a Three-Stage SBM-DEA Model. *Energies* **2022**, *15*, 569. [CrossRef]
12. Rybár, R.; Beer, M.; Mudarri, T.; Zhironkin, S.; Bačová, K.; Dugas, J. Experimental Evaluation of an Innovative Non-Metallic Flat Plate Solar Collector. *Energies* **2021**, *14*, 6240. [CrossRef]
13. Bajno, D.; Bednarz, Ł.; Grzybowska, A. The Role and Place of Traditional Chimney System Solutions in Environmental Progress and in Reducing Energy Consumption. *Energies* **2021**, *14*, 4720. [CrossRef]
14. Beer, M.; Rybár, R.; Rybárová, J.; Seňová, A.; Ferencz, V. Numerical Analysis of Concentrated Solar Heaters for Segmented Heat Accumulators. *Energies* **2021**, *14*, 4350. [CrossRef]
15. Khouri, S.; Behun, M.; Knapcikova, L.; Behunova, A.; Sofranko, M.; Rosova, A. Characterization of Customized Encapsulant Polyvinyl Butyral Used in the Solar Industry and Its Impact on the Environment. *Energies* **2020**, *13*, 5391. [CrossRef]
16. Wachnik, B.; Kłodawski, M.; Kardas-Cinal, E. Reduction of the information gap problem in Industry 4.0 projects as a way to reduce energy consumption by the industrial sector. *Energies* **2022**, *15*, 1108. [CrossRef]
17. Xu, X.; Wang, Y.; Ruan, Y.; Wang, J.; Ge, K.; Zhang, Y.; Jin, H. Integrated Energy Planning for Near-Zero Carbon Emission Demonstration District in Urban Areas: A Case Study of Meishan District in Ningbo, China. *Energies* **2022**, *15*, 874. [CrossRef]
18. Szurgacz, D.; Borska, B.; Diederichs, R.; Zhironkin, S. Development of a Hydraulic System for the Automatic Expansion of Powered Roof Support. *Energies* **2022**, *15*, 680. [CrossRef]

Article

Reduction of the Information Gap Problem in Industry 4.0 Projects as a Way to Reduce Energy Consumption by the Industrial Sector

Bartosz Wachnik [1,*], Michał Kłodawski [2] and Ewa Kardas-Cinal [2]

1 Faculty of Mechanical and Industrial Engineering, Warsaw University of Technology, Narbutta 85, 02-524 Warsaw, Poland
2 Faculty of Transport, Warsaw University of Technology, Koszykowa 75, 00-662 Warsaw, Poland; michal.klodawski@pw.edu.pl (M.K.); ewa.kardascinal@pw.edu.pl (E.K.-C.)
* Correspondence: bartek@wachnik.eu

Abstract: Reducing energy consumption should be treated as crucial for contemporary information and communication technology (ICT) projects under conditions of Industry 4.0. This research proposes a wider look at the factors influencing the success of ICT industry projects, considering not only technological and procedural conditions or implementation methods but also information and competency resources, thus allowing for correct decisions to be taken during project implementation. The article analyzes the information gap in Industry 4.0 projects completed in enterprises based in Poland, following the concept of sustainable development and minimization of energy consumption. The research was completed between 2018 and 2021 in medium enterprises, and the result is a qualitative characteristic of the information gap in ICT projects from the client's perspective. The research can help develop a complete methodology for Industry 4.0 ICT projects to limit the level of uncertainty and risk while reducing energy consumption.

Keywords: energy consumption; information gap; Industry 4.0; IT project; Green I

Citation: Wachnik, B.; Kłodawski, M.; Kardas-Cinal, E. Reduction of the Information Gap Problem in Industry 4.0 Projects as a Way to Reduce Energy Consumption by the Industrial Sector. *Energies* **2022**, *15*, 1108. https://doi.org/10.3390/en15031108

Academic Editors: Sergey Zhironkin and Michal Cehlar

Received: 3 December 2021
Accepted: 20 January 2022
Published: 2 February 2022

Publisher's Note: MDPI stays neutral with regard to jurisdictional claims in published maps and institutional affiliations.

1. Introduction

Currently, the industrial sector is the largest energy consumer, and its buildings have recorded a very dynamic energy consumption increase. According to the International Energy Agency, industry is responsible for 37% of global energy consumption, a trend that is forecast to continue [1]. Changes will be possible only because of the fourth industrial revolution, which provides access to Big Data and the possibility of using it in factory operations.

Industry 4.0 leads in the digitization and networking of production as well as the related transformation of business models and strategies. It includes digitally networked production as well as assembly, maintenance and marketing. What distinguishes Industry 4.0 is its decentralized components, which make independent decisions based on digital information and feed the production system. In the future, not only autonomous but self-learning systems are expected. Thus, Industry 4.0 is based on four pillars: the cyber-physical systems (CPS), the Internet of Things, the Internet of Services, and the smart factory [2].

CPSs are physical and engineering systems in which operations are monitored, coordinated, controlled and integrated by a computational and communication core [3]. A CPS describes the combination of information and software components with mechanical and electronic parts that communicate through a data infrastructure such as the Internet. One characteristic is high complexity.

The Internet of Things is a concept that interconnects many objects into a single network. It concerns not only computers, smartphones or tablets, but also household appliances and televisions as well as industrial automation such as production lines, machines and robots.

The Internet of Services comprises modular production stations that can be flexibly modified and extended. Services are offered and bundled into value-added services by different providers, and then communicated to users and consumers through a variety of selling channels [4].

The concept of a smart factory refers to the application of modern robotics and digital systems in the efficient management of industrial production. The coordination of an enormous amount of data from sensors embedded in machines and software for internal and external processes increases production efficiency and reduces the need for constant staff supervision. The most advanced factories of this type can adapt production volume to fluctuations in demand, react to sudden changes in the supply chain, and correct irregularities to avoid production downtime by using Big Data, the Industrial Internet of Things (IIoT), and simulation techniques such as digital twins or digital shadows.

Industry 4.0 offers almost unlimited possibilities and is characterized by increased production energy efficiency, which results in maintaining energy security, lower energy expenses, and environmental protection. Therefore, the development and implementation of various types of Industry 4.0 projects are extremely important for modern economies. It should be noted, however, that very often the success of such projects depends on the appropriate flow and availability of information both from the investor and the project contractor. The lack of necessary information (information gap) in the investor-contractor relationship hinders decision making and delays or prevents project implementation.

The purpose of this paper is to identify informational components in the customer's perspective of Industry 4.0 projects to reduce energy consumption and promote sustainable growth.

The final product, which is the result of the Industry 4.0 ICT project, consists of a number of information components, which are not separate items of trade, but appear on the market as integral parts of the services provided under the project. From the customer's perspective, information on each component is necessary for the implementation of an IT project. If the client has information included in the components, the project could be carried out without involving the supplier. The scope of the client's component information significantly influences the effectiveness of the project. In Industry 4.0 project management, a noteworthy issue is how to manage the information gap between customer and supplier.

The remainder of this paper is organized as follows. Section 3 provides a formulation of research goals and methods for developing the information gap theory, including the presentation of projects selected for the study. The authors' intention is to achieve the following research goals:

G1: The identification of information components over the life cycle of the project from the client's perspective.

G2: A characteristic of information gaps to identify information components.

Section 4 presents results with information components and a discussion on inheriting information gaps. In Sections 5 and 6, the conclusions are summarized.

2. The Development of Industry 4.0 Considering Sustainable Growth and Reduction of Energy Consumption—Literature Review

2.1. Energy Consumption Review

One of the most important issues in modern economies is how to manage the demand and supply of energy. In Poland in 2019, transport accounted for the largest amount of national energy consumption—33%. Households came in second (26.3%), and industry was third (23.9%) [5]. The situation was similar across the EU: transport, 31.3%; households, 26.9%; industry, 24.6%; and services, 13.9%, [5]. This indicated that improving industrial energy performance can make an important contribution toward achieving significant improvements to local and global energy efficiency. It also reinforced the fact that industrial energy efficiency has to be considered when designing new enterprises and searching for technological, organizational and IT solutions in existing enterprises.

The problem of how to reduce energy consumption in various industries is of interest to many researchers. Publications [6,7] have reviewed research on energy savings, carbon reduction, and improved energy efficiency in the Chinese cement industry, and the potential for reduced energy consumption and CO_2 emissions was discussed in [8]. The same issue was analyzed for India and presented by the authors in [9], while [10] presents a comparison of energy consumption in cement production using different technologies. In addition, the authors presented the possibility of using a tool to evaluate and compare energy consumption in cement production plants based on specific variables and production parameters.

Other examples pertaining to the reduction of energy consumption include the analysis and evaluation of the benefits of efficient energy use; proposals such as electricity recovery [11] to improve efficiency; presentation of theoretical and practical aspects of energy saving in the chemical [12], paper [13] and wood industries based on a Swedish example [14] and the Chinese [15] and Polish metallurgical industries [16]. In [17] the authors presented a forecast of energy consumption connected with food production and an analysis and evaluation of alternative political solutions aiming at reducing energy consumption for food production in the medium and long term.

The implementation of new means to reduce industry energy intensity often requires a systemic approach and top-down energy policy support. Many studies have been conducted in the EU to assess the effectiveness of specific industry-focused energy efficiency measures. Many researchers have paid attention to the type of policy, its accessibility to the industry and its effectiveness [18–23].

The industry-reducing energy consumption issue is inextricably linked with the Green IT concept, defined in [24] as "the optimal use of information and communication technology (ICT) for managing the environmental sustainability of enterprise operations and the supply chain, as well as that of its products, services, and resources, throughout their life cycles". This includes energy-saving IT equipment as well as the optimization of resource requirements for workstations and cooling servers, and the sustainable life cycle of IT equipment from purchase to recycling. The different approaches to defining and understanding Green IT and the associated risks are presented in [25–28]. Energy savings in information management and processing systems can lead to significant energy savings in an area of the industry that today is almost entirely computerized.

2.2. Sustainable Development and Energy-Consuming

According to Robert Solow [29], "the recipe for growth" in principle does not differ between countries. We can distinguish two basic types of growth:

- "brute force growth" [30], based on a quantitative increase of investment (more work and capital equals more product); and
- "intelligent growth", based on qualitative changes (e.g., technological development) or institutional changes [30] the key factor of which is increased efficiency.

The term "sustainable development" was introduced by Hans Carl von Carlowitz and referred to a form of forest management where each felled tree was replaced with a new seedling [31]. Concepts of sustainable development are present in different areas of human activity, like business [32–34] or environmentally positive systems. The basic task of sustainable development is to maximize the net benefits of economic growth while protecting and ensuring that natural resources are renewed over the long term [35]. The Industry 4.0 concept constitutes an intelligent form of development aimed at increasing efficiency through technological, institutional, and social progress and minimizing energy consumption. ICT projects in Industry 4.0 are an important area of sustainable growth research. Thanks to ICT solutions, it is possible to support the implementation of sustainable development concepts in the following areas:

1. Ecology—IT solutions allow for a switch from an energy-intensive economy to a model that considers the protection of the environment. This type of solution may concern

a. Limiting the energy intensity of ICT (e.g., cloud data migration); and
b. Limiting the energy intensity of industry solutions through ICT (e.g., using artificial intelligence)

In this area, we need to consider the results of research into information ecology.

2. Society—allowing direct communication between employees and contractors and taking action in real time; and
3. Economy—leading to benefits in a way that does not eliminate the social and ecological benefits of stakeholders.

Considering these areas, the solutions described as Green IT or Sustainable IT [36–38], that are used for designing, producing, and using IT systems [39] are being introduced. According to Andreopoulou [40], Green IT is a design, construction and information diffusion technique designed to achieve optimal environmental governance and optimize and improve the organization's operational processes.

Harnessing [25] offers another definition of Green IT: "the study and practice of designing, manufacturing, using, and disposing of computers, servers, and associated subsystems—such as monitors, printers, storage devices, and networking and communications systems—efficiently and effectively with minimal or no impact on the environment. Green IT also strives to achieve economic viability and improved system performance and use while abiding by our social and ethical responsibilities". Green IT is an element of sustainable development and Industry 4.0 concerning the configuration of information processing systems that will soon gain dominance, especially in industry, such as the Internet of Things or blockchain [41]. It concerns the quality of decisions made as a function of knowledge about the mechanism of future investment and may translate into aspects of sustainable development such as energy consumption [42–44].

To sum up, the completion of Industry 4.0 projects under conditions of sustainable growth bring not only economic but also social and ecological benefits. Completing this type of project requires knowledge and skills linked to the implementation of Industry 4.0 projects as well as competence in sustainable growth conditions. Consequently, this research regards project management in Industry 4.0 not only under conditions of sustainable growth, but also under conditions of rapidly increasing uncertainty and risk.

This research concerns medium enterprises in 2018 based in Poland, which was the first country from the former communist bloc to be classified as a developed country [45].

2.3. Project Management in Industry 4.0

An important part of Industry 4.0 projects is high-tech projects in the area of business process digitization. According to Cao and Zhang [46], many project implementation methods in current use (e.g., ITIL), will be replaced by new methods that will accommodate artificial intelligence [47], agility, a lower level of formality (especially in communication and project documentation), information asymmetry between the supplier and the client, and the possibility of strong organizational turbulences, both internal and external. The nature of Industry 4.0 project implementation is both technologically and organizationally complex, necessitating flexible project completion. Consequently, project management is expected to have an increasingly more adaptive character and require the use of different, often contradictory methods, that will lead to a combination of established routine and the use of hard skills, especially digital skills. Currently, Industry 4.0 projects are completed under conditions of uncertainty [48,49] in a highly turbulent environment. The external and internal conditions of a project stimulate the development of agile and adaptive responses that lead to further changes in structures and operating methods, resulting in a better adjustment to change.

Research indicates that it would be appropriate to include the concept of uncertainty in the management methods of Industry 4.0 projects that involve sustainable growth and reduced energy consumption and to consider the information gap theory to support decision-making.

One of the examples may be an attempt to include the hypothesis-driven development (HDD) concept to increase the effectiveness of the Agile and Waterfall implementation methods [50–52]. It is an exploratory-adaptive approach that assumes a gradual identification of uncertainties and information gaps in individual project sprints, resulting in a group of hypotheses regarding functional and technological requirements. The HDD concept assumes a method of defining functional and technological requirements where instead of the user perspective, the requirements are based on three factors.

1. A hypothesis that can refer to information components that may be characterized by the users' information gap;
2. A business result, along with benefits, when the hypothesis is proven to be true; and
3. Conditions that allow for the hypothesis to be fulfilled.

An example of decision-making support in large and advanced projects under conditions of uncertainty is Robust Decision Making (RDM) [53]. It combines an analysis of two factors: first, decision-making based on technological and organizational assumptions of project completion and predicted scenarios (which may result from information gaps of the components) and second, exploratory modelling that uses stress tests in extremely difficult conditions caused by uncertainty from the research object or its environment. The RDM concept may include guidelines regarding energy consumption reduction and sustainable growth for the specific project and the industry, especially Key Performance Indicator (KPI) criteria and process requirements.

Yet another example is Dynamics Adaptive Planning (DAP), which supports activities linked to designing a plan of adaptive possibilities as the conditions change and the knowledge resulting from uncertain events is gained. DAP includes a characteristic of a process-monitoring system within the project along with a definition of tasks that need to be undertaken to achieve goals.

To sum up, the high level of uncertainty within an organization and its turbulent environment necessitates the use of new Industry 4.0 project completion concepts that entail reduced energy consumption. To achieve this goal, the authors propose the use of the information gap theory to support decision-making under conditions of uncertainty.

2.4. Information Gap in Industry 4.0 Projects

According to Ben-Haim [54], the information gap theory can support business decision-making under deep and dynamically changing uncertainty, which may result from technological and functional requirements of Industry 4.0 projects that include sustainable growth and reduced energy consumption. This study was based on the definition and characteristics of the information gap devised by Oleński [55] and Wachnik [56].

In IT projects, the range, character and dynamic of the information gap on both the client and the supplier strongly determine the degree of uncertainty in decision-making at each stage of the project life cycle.

As part of a market transaction of an Industry 4.0 project, we distinguished a cluster of contracts for the purchase of software licenses, hardware, and IT services aimed at creating a unique configuration. A project consists of many information components throughout its life cycle. These pieces of information constituting an information component are not separate subjects of trade, but appear on the market as integral parts of the services provided [55,57]. From the perspective of the client and the supplier, information consisting of components, information, and knowledge about each component are necessary to satisfy their partially conflicting goals. Hence, by having information about the components, the client and the supplier can shape mutual relations more successfully. The structure of information components changes at different stages and phases of the IT project life cycle, which results from the learning curve. The goal of this study is to identify information components from the customer's perspective to reduce energy consumption and sustainable growth, and the information gap between the client and the supplier is a noteworthy issue. Information gap management for individual components consists in decreasing it on one side and decreasing or increasing it on the other side of an economic relation: in this case, an IT project. According to Babik [57], information needs are shaped by two factors: the kind of task being solved and the person's knowledge and experience. Among these needs, two subsets can be distinguished: information necessary to solve a task but available to the system user, and necessary information that is not directly available.

Obtaining information necessary to close the gap entails costs. The bigger the gap, the higher the cost, which increases the more we attempt to close it. The types of activities undertaken to obtain information, as discussed in the literature [58,59], indicate varied approaches among information users to the solved problems, which define their information needs. Hence, information service processes offered to the users need a diversified approach for each of them. This means that it is necessary to include the user in a dynamic model, to organize effective servicing of information needs. Particularly noteworthy are the models and standards linked to the Information Literacy (IL) concept, designed in the 1990s: The Big6 [58] and SCONUL [59].

3. Research Methods

3.1. Research Goals

The research goals concern the information gap in ICT projects in Industry 4.0 completed in enterprises implementing sustainable growth and seeking to reduce energy consumption.

G1: The identification of information components in the entire life cycle of the researched project from the client's perspective.

G2: A characteristic of information gaps for the identified information components.

3.2. Research Method

An analysis was performed based on the results of a study of four purposefully selected enterprises. The case study method made it possible to identify and analyse the gap in an Industry 4.0 ICT project, which is discussed more extensively further on. The research performed led to conclusions and that will allow these projects to be more effective and efficient under competitive global conditions.

The research in this article is interdisciplinary, contributing to the development of the information gap theory and business informatics in enterprises developing in the conditions of sustainable growth and limiting energy consumption [60,61]. As a result of thus formulated research goals, the direct observation method was selected. The research method was selected based upon

1. A desire to perform a qualitative description of the entire life cycle of an ICT project spanning many years;
2. The relevant characteristics of the projects resulting from the characteristics of Industry 4.0 projects and sustainably developing enterprises;
3. The interdisciplinary character of the ICT projects;
4. The observation of the information gap areas that cannot be studied using interviews during the project life cycle;
5. The possibility of tracking the dynamics of the information gap during the project life cycle.

The research method phases are presented in Table 1.

The selection of cases was deliberate and made on the basis of five basic criteria: data availability, vividness of the case, diversity in multiple case studies, critical phenomenon and a metaphor that directs the researcher to a specific direction of the studied phenomenon. The first was the purely pragmatic question of the availability of data, which led to the most incisive description of those enterprises that were especially pertinent to the research question. Projects had to meet four substantive criteria to be selected:

1. Be concerned with ICT solutions belonging to the Industry 4.0 project group;
2. Be implemented in enterprises that paid particular attention to sustainable growth;

3. Be avant-garde; i.e., employees had no experience in implementing this type of projects; and
4. Be implemented in a relatively large diversification of industries.

Table 1. Research method phases.

Stage 1	Formulating the research question
Stage 2	Selection of cases
Stage 3	Development of data-collection tools
Stage 4	Fieldwork
Stage 5	Data analysis
Stage 6	Formulating generalizations
Stage 7	Confrontation with the literature
Stage 8	Study conclusion—generalization

The second criterion was the vividness of the case, which illustrated the properties being studied in an extreme form which, however, allowed for an unambiguous interpretation of them. The third concerned diversity by requiring that many cases be investigated in a manner that presents different circumstances or contradictory situations.

The number of cases ranged from 4 to 10 to allow for the analysis of phenomena that had a different course or took place in different industries. This enabled the formulation of generalizations largely free from circumstantial factors or industry. The selection then consisted in creating appropriate cases (e.g., a project with high technology, a mature market and a local–global enterprise.

The fourth was the critical phenomenon—either extreme or counter to generally accepted opinion—which allowed generalizations to be formed.

The fifth concerned a metaphor which directed the researcher's attention to a specific course of the phenomenon under study or allowed for the assumption of a specific research position. For instance, the lifecycle metaphor required the selection of cases that allowed for the observation of its emergence, development, maturity, decline and disappearance. The projects, completed between 2018 and 2021, are characterized below (Table 2).

The study, study involved observations of the phenomenon and the events as well as co-participation: receiving interpretations from the participants, making contact with the participants, assuming the position of an outsider, or not participating in the work of the observation participants but remaining in the background. As part of direct observation, the study focused on the functioning of the project group, technical and organizational conditions, and project management methods. Direct observation is characterized by a high level of difficulty and layering of all observed phenomena, resulting in a feeling of overload (selective observation). To minimize the impact of subjectivity on the evaluation during observation, a passive approach was applied, i.e., not actively participating in the project, taking on a managerial position, or not being a project group member.

During the research completion

1. The observed knew that the study was being carried out by a researcher from the Warsaw University of Technology conducting studies in IT project management;
2. The strong requirement for privacy and confidentiality by the enterprises where the research was carried out was respected. The research description did not name the enterprises and the context described was at times slightly modified while ensuring credibility;
3. The observed were treated as subjects;
4. The majority of data characterizing the project over its life cycle was accessible.

Table 2. Characteristic of the studied projects. Source: Own study.

	A	B	C	D	E
Enterprise description	Furniture production	Food manufacturing	Transport services	Production of car components	Financial services
Project description	Infrastructure transfer to Cloud	Using Big Data to identify business-marketing data	Using IoT to monitor the work of the transport fleet	Implementation of an AI-based system identifying defective products	Digitization of the invoice documents introduced into the financial module of the ERP system
Benefits resulting from the project that has an impact on sustainable development	Limiting the use of Resources	Making better decisions	Extending the life cycle of the transport fleet	Limiting the number of defective components being released into the market	Lower paper usage
Number of participants	50	40	120	5	25
Studied project duration (phase/completion time)	F 1—3 months F 2—1 month F 3—2 years	F 1—6 months F 2—6 months F 3—1 year	F 1—6 months F 2—2 months F 3—2 years	F 1—6 months F 2—1 month F 3—2 years	F 1—3 months F 2—3 months F 3—2 years

The study was conducted mainly through collecting and analyzing information, and evaluating the completed project through participation in project meetings, private conversations with project participants, and access to project documents.

4. Results

4.1. Information Components

The identification of information components was completed over the entire cycle of the IT projects. An analysis of the methods—Sure Step (recommended by Microsoft for MIS, i.e., ERP and CRM [62]), and Accelerated SAP (recommended by SAP AG for SAP implementations [63] included in literature, and other project completion standards (PMI, PRINCE 2 [64–69]) and models (ITIL [70])—allowed for a generalization of an IT project life cycle consisting of 3 stages.

Stage one, i.e., the preparation stage, comprised two phases.

1. Pre-implementation—The creation of the problem domain model, a user-needs analysis and definition of the system's functional requirements, analysis of the organization's IT infrastructure, project group definition, identification of significant risk factors, ex ante economic analysis of the investment and preliminary definition of the implementation project;

2. System and supplier selection—The preparation of a potential suppliers' list, creation of Request for Proposal (RFP) forms, analysis and evaluation of offers according to established criteria, substantive and trade negotiations, and contract formulation. The result of the first, preparatory stage was a selection of a system and a supplier; the completion of Phase 2 was a multi-dimensional task where many organizational, legal, social, and technical factors needed to be considered.

Stage two, the completion of an IT implementation project comprised five phases.

1. Initiation—An implementation planning session as an initial meeting and a technological project consisting of the installation and configuration of components in the hardware layer, system software and application software layer;

2. Analysis—Training for key users and a functional analysis that included analytical workshops and the designing of a theoretical prototype;

3. Design—Customization of the project;

4. Implementation—Preliminary data migration and acceptance testing for the completed customization along with tuning, developing workplace instructions and training for key users; and

5. Go-live—System go-live and post-go-live support during system stabilization.

Stage three, IT system operation, comprises two phases.

1. Post-implementation analysis and identification of operational needs—covering tasks linked to an ex-post analysis of the implemented IT project and an identification of needs linked to system operation;
2. Selection of an appropriate supplier of post-implementation services linked to system operation—covering the preparation of a potential suppliers' list, creation of RFP forms, analysis and evaluation of offers according to established criteria, substantive and trade negotiations, and creation of Service Level Agreements (SLAs) [71].

Tables 3–5 present information components for the project life cycle phases where users identified information gaps. Each component was marked as C.X.Y., where X was the phase where it first appeared, and Y was the component number. The tables used color gradation for the individual components to mark the impact of the information gaps on the effectiveness of IT project completion: (1. dark grey—high impact; 2. light grey—moderate impact; 3. white—low impact). The study used a definition of IT project effectiveness that compared the results and the expected results for the following criteria: time, business range, project completion cost, and functional-technological requirements. The projects were a successes (fully met all the criteria) or partial successes, (failed to meet at most two criteria).

Table 6 outlines how the project effectiveness criteria were met.

Thirty-one components where the respondents identified information gaps throughout the entire project life cycle were identified. In Stage 1, the preparation stage, the components were characterized by the high impact of the information gap on the project's success, mainly regarding the definition of the TCO for the project life cycle and the range of functional and technological requirements resulting from the sustainable growth concept (including energy consumption reduction), the so-called Green IT. Stage 2 had the same component characterizations but mostly regarding the risk and uncertainty management methods, range of functional and technological requirements resulting from reduced energy consumption and the sustainable growth concept, methods of maintaining implementation costs according to plan and the knowledge and experience of the project group members, which allowed them to make the organization more efficient. In Stage 3, the operation stage, one component characterized by high impact from the information gap was identified: the range of functional and technological requirements resulting from the sustainable growth and reduced energy consumption concept. The project group members identified the strongest impact of information gaps on project success in three main areas:

1. Managing the TCO of an IT system,
2. Implementing functional and technological requirements linked to Green IT, and
3. Managing risk and uncertainty in an IT project.

According to the study, project managers indicated information gaps in the 31 information components, which could constitute a source of risk factors and uncertainty throughout the project life cycle. A theoretical and empirical analysis conducted during the study indicated that information gaps were independent of the class of information systems. The study showed that information gaps for the components depended upon:

1. The character of the project resulting from the energy consumption reduction and sustainable growth concept,
2. The functional-technological character of the customization based on the technical possibilities of the parameterization and software,
3. The level of investment in the knowledge of the project group regarding the project and the concept of reduced energy consumption (sustainable growth) during the first stage, and
4. The level of trust between among members of the project group.

Table 3. The structure of information components on the client's side during Stage 1 of the IT implementation projects. Source: Own study.

Stage 1	Phase 1	Competence in estimating the TCO in the entire project life cycle C.1.1	Functional and technological requirements range resulting from the sustainable growth concept C.1.2	The structure of the IT implementation project group and the role and responsibility of its members during the project completion C.1.3	A precise definition of the technological and functional requirements for the system C.1.4	Project completion method C.1.5		
Stage 2	Phase 2	Security of data access C.2.6	Methods of charging the license fee C.2.7	The right to modify the completed customization, inc. the right to source code modification C.2.8	Integration with other IT systems C.2.9	Project schedule C.2.10	Project completion method C.1.5	Future SI development of the producer and future license price policy C.2.11	Functional and technological requirements

Table 4. The structure of information components on the client's side during Stage 2 of the IT implementation projects. Source: Own study.

Stage 2	Phase 3	Project documentation and completion methods C.3.13	Methods of risk and project uncertainty management C.3.14	Project schedule C.2.10	The role and responsibility of the project group members during the implementation C.3.15	Project completion method C.1.5	–
Stage 2	Phase 4	Knowledge and experience of the project group members that can make the organization more efficient, resulting from analogous IT projects C.4.16	Methods of maintaining implementation costs according to plan C.4.17	Functional and technological requirements range resulting from the sustainable growth concept C.2.13	Methods of collecting functional requirements by the project group members during the functional analysis C.4.18	Methods of functional and technological knowledge transfer C.4.19	Methods of risk and project uncertainty management C.3.14
Stage 2	Phase 5	Internal testing methods for the completed customization C.5.20	Documentation methods of the completed customization C.5.21	The impact of customization method selection on the TCO of the IS C.5.22	Functional and technological requirements range resulting from the sustainable growth concept C.2.13	Competence in the area of communication in a project based on remote working C.5.23	Methods of risk and project uncertainty management C.3.14
Stage 2	Phase 6	Data migration methods C.6.24	Knowledge transfer methods C.6.25	Functional and technological requirements range resulting from the sustainable growth concept C.2.13	Methods of system tuning completion after receiving the results of acceptance testing C.6.26	Competence in the area of communication in a project based on remote working C.5.23	Methods of acceptance testing completion C.6.27
Stage 2	Phase 7	System go-live method C.7.28	Methods of risk and project uncertainty management C.3.14	Project schedule C.2.10	Competence in the area of communication in a project based on remote working C.5.23	Functional and technological requirements range resulting from the sustainable growth concept C.2.13	–

Table 5. The structure of information components on the client's side during Stage 3 of the IT implementation projects. Source: Own study.

Stage 3	Phase 8	Identification of needs linked to the system's functional development C.8.29	Economic evaluation (ex-post) of the IT project completion C.8.30	Methods of risk and project uncertainty management C.3.14	Project schedule C.2.10	Functional and technological requirements range resulting from the sustainable growth concept C.2.13
Stage 3	Phase 9	Range of development tasks C.9.30	Operational task range C.9.31	Methods of risk and project uncertainty management C.3.14	Project schedule C.2.10	Functional and technological requirements range resulting from the sustainable growth concept C.2.13

Table 6. Project effectiveness criteria. Source: Own study.

	A	B	C	D	E
Enterprise type	Furniture production	Food manufacturing	Transport services	Production of car components	Financial services
Project type	Cloud migration of infrastructure	The use of Big Data for the identification of business-marketing data	The use of IoT to monitor the work of the transport fleet	Implementation of an AI-based system to identify poor quality products	Digitization of invoice documents introduced into the financial module of the ERP system
Time	☑	☑	○	☑	☑
Business range	☑	○	☑	☑	☑
Project completion cost	☑	○	○	☑	☑
Functional-technological requirements	☑	☑	☑	☑	☑

☑Criterion met　　　○—Criterion not met

4.2. Inheriting Information Gaps

The study identified a phenomenon accompanying the information gap, "information gap inheritance" [56,57] for the components indicated. The inheritance for each of the three stages is presented in Figures 1 and 2. A characteristic of the information gap, its impact on IT project completion efficiency, was identified, and it affected information gaps in subsequent information components. Analogously, as in Tables 3–5, color gradation was used (1. dark grey—high impact; 2. light grey—moderate impact; 3. white—low impact) to illustrate how information gaps affected IT project completion. Inheriting information gaps means that they became more widespread and ingrained, increasing uncertainty. The impact of the information gap on IT project effectiveness and the inheritance of the information gap resulted in the decreasing resilience of the information gap party—who is often the decision-maker—and directly affected their decision-making methods. Figures 1 and 2 present sequences of the components for which information gaps greatly or moderately affected IT project completion.

The inheritance and the inclusivity of information gap characteristics in Stage 1 caused the uncertainty in Stage 2 and 3 to increase. The domino effect meant that the resilience of the internal stakeholder to information uncertainty in Stages 2 and 3 decreased. Research showed that in Stage 1, the sequence of inheriting information gaps referred to components that concerned mostly three issues:

1. Agreeing on advantageous trade conditions of license purchases of the software and external outsourcing services, both for Stage 2 and 3 of the project life cycle,
2. Recommendations resulting from reduced energy consumption and the sustainable growth concept (Green IT) regarding Industry 4.0 project completion,
3. The security of business activity in influencing development possibilities of the IT system and security of data access.

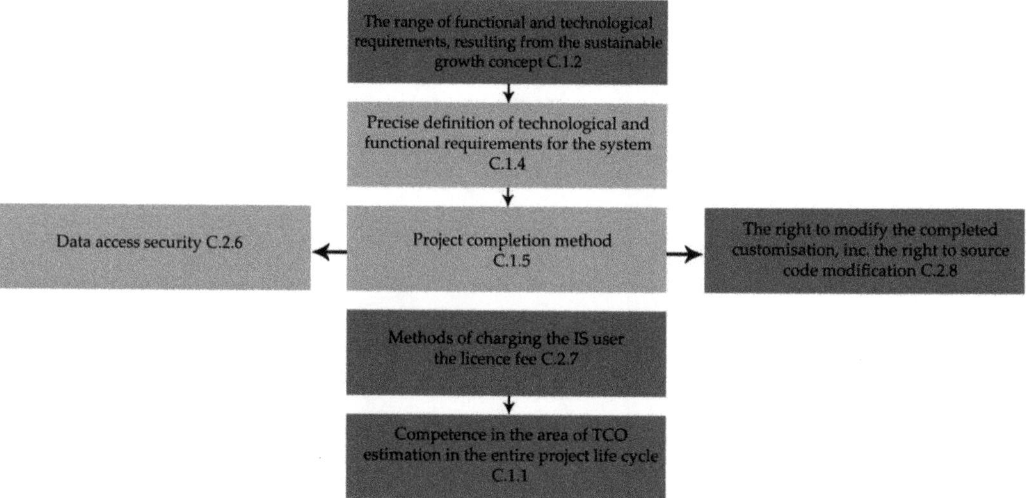

Figure 1. Inheriting information gaps for the relevant components in Stage 1. Source: Own study.

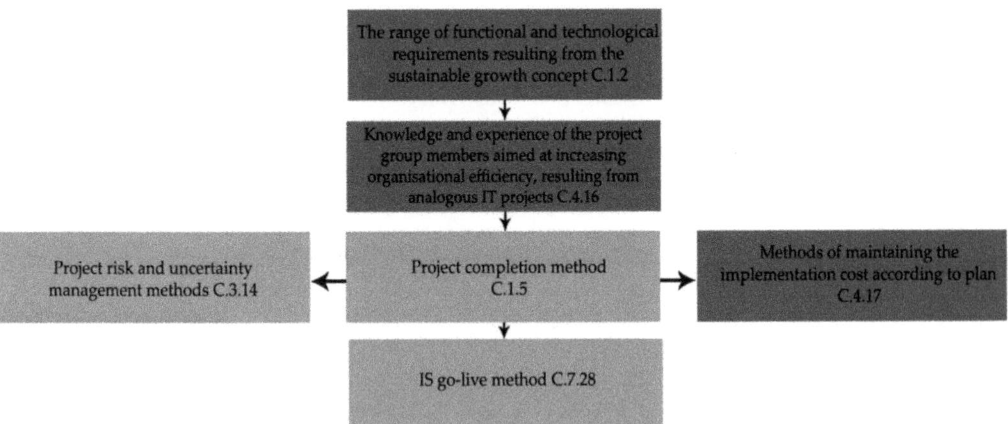

Figure 2. Inheritance of information gaps for the significant components in Stage 2. Source: Own study.

The research showed that in Stage 2, the sequence of inheriting information gaps referred to the components largely relating to two issues:

1. The knowledge and experience of project group members regarding the concept of sustainable growth and
2. Risk and uncertainty management following the concept of sustainable growth.

In Stage 3, the sequence of inheriting information gaps referred to two components: the range of functional and technological requirements resulting from the sustainable growth concept (C.1.2) and the methods of managing project risk and uncertainty (C.3.14).

To sum up the phenomenon of information gap inheritance throughout the entire project life cycle, the information gaps with the highest impact on the project success, concerned the components that characterized the knowledge and experience of the project group members in the sustainable growth concept as well as knowledge and experience supporting the possibility of achieving a trade relationship with an external supplier, thereby obtaining a correct TCO for the project life cycle. The cost of closing the information gap for the indicated components in Stage 1 led to project success in Stage 2.

The components organized according to the sequence of inheritance confirmed the main development determinants of the Green IT concept as presented in the literature [72–74].

5. Discussion

Research regarding the information gap theory mainly concerned non-probabilistic models of decision-making that allowed for the establishment of priorities or choices under conditions of uncertainty [75]. The completion of Industry 4.0 projects in organizations following the sustainable growth concept (Green IT) under conditions of Volatility, Uncertainty, Complexity, and Ambiguity (VUCA) [76] required particular knowledge and skills form decision-makers to manage risk and uncertainty. To increase the effectiveness of Industry 4.0 projects that support sustainable growth, we can use the knowledge regarding information gaps, which helps create and evaluate alternative operational what-if scenarios in

1. Structured issues (solving resource problems, scheduling turbulence, or choosing an optimal type of services or license in the context of the project's TCO) and
2. More abstract issues, based on Big Data, scientific theories, empirical knowledge, as well as contextual knowledge and understanding, i.e., selecting a model structure, completing a prognosis, or formulating a policy.

The completion of two research goals presented in Section 3.1 of this article provided insight into the development of Industry 4.0 projects based on the sustainable growth concept.

The completion of the first research goal allowed for the identification of 31 information components affected by the information gap of the project manager. The components concerned the three stages of the project life cycle. For each component, the strength of the information gap on the project completion was identified, and the information components were used to create operational what-if scenarios that were included in implementing Green IT projects in Industry 4.0. It is important to note that the components characterized by the highest or moderate strength of the impact of the information gap on the effectiveness of project completion over its life cycle mainly concerned

1. The shaping of beneficial economic relations between the supplier and the recipient of services,
2. Maintaining the TCO of the implemented IT system on the planned level,
3. Defining the functional and business requirements of the system and confronting them with the Green IT concept, and
4. Managing risk and uncertainty.

The information components affected by information gaps supported the hypothesis definition in functionality and technology as part of HDD, as a method for increasing the effectiveness of the Agile and Waterfall methods.

The completion of the second research goal allowed the identification of the sequence of information gap inheritance during the three stages. Embedding the information gap in the individual components increased the uncertainty of a given project task or goal both vertically and horizontally, which may have led to a domino effect, making the stakeholder more prone to uncertainty. The sequencing of the inheritance of the components affected by the information gap can be useful in the RDM concept supporting decision-making in exploratory modeling. The sequencing of inheritance may allow the researching the subject through stress tests in unfavorable conditions resulting from uncertain events, caused by both internal and external factors. Additionally, sequencing the inheritance of information gaps may help define constraints in individual phases and conditions for success, using the Dynamics Adaptive Planning model presented in the literature [75]. Having access to information regarding the sequencing of information gaps and their impact on project success, we can—as part of the exploratory model—perform a sensitivity analysis that will allow us to choose the best option.

The research results were in line with the current research trend concerning the information gap theory [75,77–80]. The catalog of information components along with their in-depth analysis constituted a fresh contribution to the applications of information gap theory. The results were mainly helpful in non-probabilistic decision-making models. Identifying and analyzing the information components that contain information gaps helped prioritize or alter choices when making decisions under conditions of deep uncertainty in ICT project management.

6. Conclusions

We need to stress that the sustainable development concept—which in the area of ICT represents Green IT in Industry 4.0 projects—aims to provide decision-makers with a wider perspective on IT investments. Rather than underlining short-term economic and non-economic gains, it indicates benefits resulting from sustainable growth such as ecological and social responsibility, and an ethical approach. The use of the information gap theory allowed for the development of methods that limited uncertainty in Green IT project completion.

The research suggested, however, that Green IT increased additional costs linked to project completion that decision-makers have to face: investing in the knowledge and skills of their employees and technology, as well as adapting project completion methods to limit information gaps over the project life cycle. Additional project costs distributed, throughout the life cycle as part of the TCO are linked to reducing the information gap.

It should be emphasized that the research is characterized by the following limitations:

1. The surveyed enterprises operated in Poland;
2. The projects were implemented in a mixed manner by both specialized suppliers and customers, but the information gap was only studied from the perspective of the client's project manager; and
3. The projects were implemented in medium enterprises and the SME group

The authors would like to explain more two main limitations of this research. The first is linked to limiting the project sample to medium enterprises conducting activity in Poland. The study included both local Polish enterprises and divisions of international companies. Limiting the study to Poland-based enterprises also meant that the results applied to a developed country, while in developing countries the map of information components and the inheritance structure could be different.

The second limitation is linked to the fact that the information gap was only studied from the perspective of the client's project manager. Future research could concern the information gap from the supplier side, which could represent the information gap between the two key project group members more thoroughly.

To sum up, the study indicated that the identification and reduction of the information gap in Industry 4.0 projects supporting reduced energy consumption and sustainable growth helps reduce uncertainty and increases the likelihood of project success at the same time. The reduction of the information gap, however, entails additional costs for the entrepreneurs.

Author Contributions: Conceptualization, B.W. and M.K.; methodology, B.W.; validation, B.W., M.K. and E.K.-C.; formal analysis, B.W., M.K. and E.K.-C.; investigation, B.W.; resources, B.W., M.K. and E.K.-C.; data curation, B.W., M.K. and E.K.-C.; writing—original draft preparation, B.W.; writing—review and editing, M.K. and E.K.-C.; visualization, B.W., M.K.; supervision, E.K.-C.; project administration, B.W. All authors have read and agreed to the published version of the manuscript.

Funding: This research received no external funding.

Data Availability Statement: Not applicable.

Conflicts of Interest: The authors declare no conflict of interest.

References

1. International Energy Outlook. 2019. Available online: https://www.eia.gov/outlooks/ieo/pdf/IEO2021_Narrative.pdf (accessed on 1 December 2021).
2. Hermann, M.; Pentek, T.; Otto, B. *Design Principles for Industry 4.0 Scenarios: A Literature Review*; Working Paper No. 01, 2015; Technische Universität Dortmund: Dortmund, Germany, 2015.
3. Rajkumar, R.; Lee, I.; Sha, L.; Stankovic, J. Cyber-physical systems: The next computing revolution. In Proceedings of the 47th Design Automation Conference (DAC '10), Anaheim, CA, USA, 13–18 June 2010; Association for Computing Machinery: New York, NY, USA, 2010; pp. 731–736. [CrossRef]
4. Buxmann, P.; Hess, T.; Ruggaber, R. Internet of Services. *Bus. Inf. Syst. Eng.* **2009**, *1*, 341. [CrossRef]
5. Eurostat. [NRG_BAL_S]. Available online: http://eurostat.ec.europa.eu (accessed on 1 December 2021).
6. Madlool, N.A.; Saidur, R.; Rahim, N.A.; Kamalisarvestani, M. An overview of energy savings measures for cement industries. *Renew. Sustain. Energy Rev.* **2013**, *19*, 18–29. [CrossRef]
7. Madlool, N.A.; Saidur, R.; Hossain, M.S.; Rahim, N.A. A critical review on energy use and savings in the cement industries. *Renew. Sustain. Energy Rev.* **2011**, *15*, 2042–2060. [CrossRef]
8. Ke, J.; Zheng, N.; Fridley, D.; Price, L.; Zhou, N. Potential energy savings and CO_2 emissions reduction of China's cement industry. *Energy Policy* **2012**, *45*, 739–751. [CrossRef]
9. Thirugnanasambandam, M.; Hasanuzzaman, M.; Saidur, R.; Ali, M.B.; Rajakarunakaran, S.; Devaraj, D.; Rahim, N.A. Analysis of electrical motors load factors and energy savings in an Indian cement industry. *Energy* **2011**, *36*, 4307–4314. [CrossRef]
10. Galitsky, C.; Price, L.; Zhou, N.; Fuqiu, Z.; Huawen, X.; Xuemin, Z.; Lan, W. *Guidebook for Using the Tool BEST Cement: Benchmarking and Energy Savings Tool for the Cement Industry*; (No. LBNL-1989E); Lawrence Berkeley National Lab. (LBNL): Berkeley, CA, USA, 2008.

11. Czosnyka, M.; Wnukowska, B.; Karbowa, K. Energy efficiency as a base of the energy economy in a modern industrial plants. In *Proceedings of the 2018 Progress in Applied Electrical Engineering (PAEE), Koscielisko, Poland, 18–22 June 2018*; IEEE: New York, NY, USA, 2018; pp. 1–6. [CrossRef]

12. Leites, I.L.; Sama, D.A.; Lior, N. The theory and practice of energy saving in the chemical industry: Some methods for reducing thermodynamic irreversibility in chemical technology processes. *Energy* **2003**, *28*, 55–97. [CrossRef]

13. Bujak, J. Energy savings and heat efficiency in the paper industry: A case study of a corrugated board machine. *Energy* **2008**, *33*, 1597–1608. [CrossRef]

14. Johnsson, S.; Andersson, E.; Thollander, P.; Karlsson, M. Energy savings and greenhouse gas mitigation potential in the Swedish wood industry. *Energy* **2019**, *187*, 115919. [CrossRef]

15. Shao, Y. Analysis of energy savings potential of China's nonferrous metals industry. *Resour. Conserv. Recycl.* **2017**, *117*, 25–33. [CrossRef]

16. Wolniak, R.; Saniuk, S.; Grabowska, S.; Gajdzik, B. Identification of energy efficiency trends in the context of the development of industry 4.0 using the polish steel sector as an example. *Energies* **2020**, *13*, 2867. [CrossRef]

17. Xu, Y.; Szmerekovsky, J. System dynamic modeling of energy savings in the US food industry. *J. Clean. Prod.* **2017**, *165*, 13–26. [CrossRef]

18. Diaz, N.; Ninomiya, K.; Noble, J.; Dornfeld, D. Environmental impact characterization of milling and implications for potential energy savings in industry. *Procedia CIRP* **2012**, *1*, 518–523. [CrossRef]

19. Price, L. Voluntary Agreements for Energy Efficiency or GHG Emissions Reduction in Industry: An Assessment of Programs Around the World. Presented at the Proceedings of the 2005 ACEEE Summer Study on Energy Efficiency in Industry, New York, NY, USA, June 2005. Available online: https://www.osti.gov/servlets/purl/881402 (accessed on 1 December 2021).

20. Modig, G. Evaluation of the Industrial Energy Efficiency Network in Norway. Report Prepared within the AID-EE Project, Ecofys. Available online: www.aid-ee.org/documents (accessed on 1 June 2017).

21. Schüle, R. Evaluation of Energy Concepts for Trade and Industry Sectors (ECTIS) in North Rhine-Westphalia. ECOFYS/Active Implementation of the European Directive on Energy Efficiency. 2006. Available online: https://www.academia.edu/50 048010/Evaluation_of_Energy_Concepts_for_Trade_and_Industry_Sectors_Ectis_in_North_Rhine_Westphalia (accessed on 1 December 2021).

22. Rezessy, S.; Bertoldi, P. Top down energy efficiency indicators in the scope of Directive 2006/32/EC. In Proceedings of the ODYSSEE—MURE Workshop, Madrid, Spain, 21–22 June 2010.

23. SenterNovem. *Long-Term Agreement on Energy Efficiency in the Netherlands LTA3. Results of 2009*; SenterNovem: The Hague, The Netherlands, 2010.

24. Mingay, S. Green IT: The New Industry Shock Wave. Gartner Inc., 2007. Available online: http://download.microsoft.com/download/E/F/9/EF9672A8-592C-4FA2-A3BF-528E93DF44EA/VirtualizationPublicSafety_GreenITWhitepaper.pdf (accessed on 1 December 2021).

25. Murugesan, S. Harnessing green IT: Principles and practices. *IT Prof.* **2008**, *10*, 24–33. [CrossRef]

26. Murugesan, S. Making IT green. *IT Prof.* **2010**, *12*, 4–5. [CrossRef]

27. Chou, D.C.; Chou, A.Y. Awareness of Green IT and its value model. *Comput. Stand. Interfaces* **2012**, *34*, 447–451. [CrossRef]

28. Chou, D.C. Risk identification in Green IT practice. *Comput. Stand. Interfaces* **2013**, *35*, 231–237. [CrossRef]

29. Solow, R. A contribution to the theory of economic growth. *Q. J. Econ.* **1956**, *70*, 65–94. [CrossRef]

30. Baumol, W.J.; Litan, R.E.; Schramm, C.J. *Good Capitalism, Bad Capitalism, and the Economics of Growth and Prosperity*; Yale University Press: New Haven, CT, USA, 2009; ISBN 978-0-300-15832-8.

31. Mazur-Wierzbicka, E. Koncepcja zrównoważonego rozwoju jako podstawa gospodarowania środowiskiem przyrodniczym. In *Funkcjonowanie Gospodarki Polskiej w Warunkach Integracji i Globalizacji*; Kopycińska, W.D., Ed.; Wydawnictwo Uniwersytetu Szczecińskiego: Szczecin, Poland, 2005.

32. Rudyk, T.; Szczepański, E.; Jacyna, M. Safety factor in the sustainable fleet management model. *Arch. Transp.* **2019**, *49*, 103–114. [CrossRef]

33. Jacyna, M.; Wasiak, M.; Lewczuk, K.; Chamier-Gliszczyński, N.; Dąbrowski, T. Decision problems in developing proecological transport system. *Rocz. Ochr. Srodowiska* **2018**, *20*, 1007–1025.

34. Pyza, D.; Jacyna-Gołda, I.; Gołda, P.; Gołębiowski, P. Alternative fuels and their impact on reducing pollution of the natural environment. *Rocz. Ochr. Srodowiska* **2018**, *20*, 819–836.

35. Pearce, D.W.; Turner, R.K. *Economics of Natural Resources and the Environment*; Harvester Wheatsheaf: London, UK, 1990; ISBN 0-7450-0225-0.

36. Molla, A.; Cooper, V. Green IT readiness: A framework and preliminary proof of concept. *Australas. J. Inf. Syst.* **2009**, *16*, 5–23. [CrossRef]

37. Howard, G.R.; Lubbe, S. Synthesis of Green IS Frameworks for Achieving Strong Environmental Sustainability in Organisations. In Proceedings of the South African Institute for Computer Scientists and Information Technologists Conference, Pretoria, South Africa, 1–3 October 2012. [CrossRef]

38. Izdebski, M.; Jacyna-Gołda, I.; Gołębiowski, P.; Pyza, D.; Żak, J. Decision problems in designing database architecture for the assessment of logistics services. *Zesz. Nauk. Transp. Politech. Śląska* **2020**, *108*, 53–71. [CrossRef]

39. Penzenstadler, B. *Infusing Green: Requirements Engineering for Green in and through Software Systems*; University of California: Irvine, CA, USA, 2015; Available online: http://ceur-ws.org/Vol-1216/paper8.pdf (accessed on 16 April 2021).
40. Andreopoulou, Z.S. Green informatics: ICT for green and sustainability. *J. Agric. Inform.* **2012**, *3*, 1–8. [CrossRef]
41. Lewczuk, K.; Kłodawski, M. Logistics information processing systems on the threshold of IoT. *Zesz. Nauk. Transp. Politech. Śląska* **2020**, *107*, 85–94. [CrossRef]
42. Lewczuk, K.; Kłodawski, M.; Gepner, P. Energy consumption in a distributional warehouse: A practical case study for different warehouse technologies. *Energies* **2021**, *14*, 2709. [CrossRef]
43. Cieśla, M.; Sobota, A.; Jacyna, M. Multi-Criteria decision making process in metropolitan transport means selection based on the sharing mobility idea. *Sustainability* **2020**, *12*, 7231. [CrossRef]
44. Stypułkowski, K.; Gołda, P.; Lewczuk, K.; Tomaszewska, J. Monitoring system for railway infrastructure elements based on thermal imaging analysis. *Sensors* **2021**, *21*, 3819. [CrossRef]
45. Radu, S. Poland Graduates to Developed Status. U.S. News, 2 October 2018. Available online: https://www.usnews.com/news/best-countries/articles/2018-10-02/poland-reclassified-as-a-developed-economy-by-the-ftse (accessed on 16 April 2021).
46. Cao, J.Q.; Zhang, S.H. ITIL Incident management process reengineering in industry 4.0 environments. In Proceedings of the 2nd International Conference on Advances in Mechanical Engineering and Industrial Informatics (AMEII 2016), Hangzhou, China, 9–10 April 2016; Volume 73, pp. 1011–1016. [CrossRef]
47. Izdebski, M.; Jacyna-Gołda, I.; Wasiak, M.; Jachimowski, R.; Kłodawski, M.; Pyza, D.; Żak, J. The application of the genetic algorithm to multi-criteria warehouses location problems on the logistics network. *Transport* **2018**, *33*, 741–750. [CrossRef]
48. Walker, W.E.; Marchau, V.A.W.J.; Swanson, D. Addressing deep uncertainty using adaptive policies: Introduction to section 2. *Technol. Soc. Change* **2010**, *77*, 917–923. [CrossRef]
49. Jacyna, M.; Semenov, I. Models of vehicle service system supply under information uncertainty. *Eksploat. I Niezawodn. Maint. Reliab.* **2020**, *22*, 694–704. [CrossRef]
50. Bland, D.J.; Osterwalder, A. *Testing Business Ideas: Field Guide for Rapid Experimentation*; John Wiley & Sons, Inc.: Hoboken, NJ, USA, 2019.
51. Geissdoerfer, M.; Bocken, N.M.P.; Hultink, E.J. Design thinking to enhance the sustainable business modelling process—A workshop based on a value mapping process. *J. Clean. Prod.* **2016**, *135*, 1218–1232. [CrossRef]
52. Cosenz, F.; Noto, G. A dynamic business modelling approach to design and experiment new business venture strategies. *Long Range Plann.* **2018**, *51*, 127–140. [CrossRef]
53. Marchau, V.A.W.J.; Walker, W.E.; Bloemen, P.J.T.M.; Popper, S.W. *Decision Making under Deep Uncertainty from Theory to Practice*; Springer International Publishing: Berlin/Heidelberg, Germany, 2019; ISBN 978-3-030-05252-2.
54. Ben-Haim, Y. *Info-Gap Decision Theory: Decisions under Severe Uncertainty*; Academic Press: London, UK, 2006; ISBN 978-149-330-098-3.
55. Oleński, J. *Ekonomika Informacji. Metody*; PWE: Warsaw, Poland, 2003; p. 20.
56. Wachnik, B. *Luka Informacyjna w Przedsięwzięciach Informatycznych. Problemy i Rozwiązania*; Polskie Wydawnictwo Ekonomiczne: Warszawa, Poland, 2020; ISBN 978-83-208-2416-2.
57. Babik, W. Zarządzanie Informacją we Współczesnych Systemach Informacyjno-Wyszukiwawczych—Nowe Wyzwania Współczesności. *Zag. Inf. Nauk.* **2000**, *1*, 51–63.
58. The Big6. Available online: http://www.big6.com (accessed on 16 April 2021).
59. The Society of College, National and University Libraries (SCONUL). Available online: http://www.sconul.ac.uk (accessed on 16 April 2021).
60. Ciesielska, M.; Wolnia-Bostrom, K.; Ohlander, M. Obserwacja. In *Badania Jakościowe. Metody i Narzędzia*; Jemielniak, D., Ed.; Wydawnictwo Naukowe PWN: Warszawa, Poland, 2012; Volume 2, pp. 41–67.
61. Czakon, W. *Podstawy Metodologii Badań w Naukach o Zarządzaniu*; Oficyna Wolters Kluwer Business: Warszawa, Poland, 2011.
62. Microsoft. Available online: https://www.microsoft.com (accessed on 16 April 2021).
63. SAP. Available online: http://www.sap.com (accessed on 16 April 2021).
64. Bradley, K. *Podstawy Metodyki PRINCE II CRM S.A.*; Centrum Rozwiązań Menadżerskich: Warszawa, Poland, 2002.
65. Esteves, J.; Bohorquez, V. An Updated ERP Systems Annotated Bibliography: 2001–2005. *Commun. Assoc. Inf. Syst.* **2007**, *19*, 1–59.
66. Esteves, J.; Pastor, J.A. Organizational and technological critical success factors behavior along the ERP implementation phases. In Proceedings of the 6th International Conference on Enterprise Information Systems, Porto, Portugal, 14–17 April 2004; Seruca, I., Filipe, J., Hammoudi, S., Cordeiro, J., Eds.; pp. 45–53. [CrossRef]
67. Chang, S.I. ERP Life cycle implementation, management and support. Implications for practice and research. In Proceedings of the 37th Annual Hawaii International Conference on System Sciences, Hawaii, HI, USA, 5–8 January 2004. [CrossRef]
68. Nguyen, T.H.; Sherif, J.S.; Newby, M. Strategies for Successful CRM Implementation. *Inf. Manag. Comput. Secur.* **2007**, *15*, 102–115. [CrossRef]
69. Best Practice Solutions. Available online: https://www.axelos.com/best-practice-solutions/prince2 (accessed on 16 April 2021).
70. Itil. Available online: https://www.itlibrary.org/ (accessed on 16 April 2021).
71. Auksztol, J. *Outsourcing Informatyczny w Teorii i Praktyce Zarządzania*; Wydawnictwo Uniwersytetu Gdańskiego: Gdańsk, Poland, 2008.
72. Deng, Q.; Ji, S. Organizational Green IT adoption: Concept and evidence. *Sustainability* **2015**, *7*, 16737–16755. [CrossRef]

73. Bayo-Moriones, A.; Lera-López, F. A firm-level analysis of determinants of ICT adoption in Spain. *Technovation* **2007**, *27*, 352–366. [CrossRef]
74. Greenpeace International. Make IT Green: Cloud Computing and Its Contribution to Climate Change. Greenpeace International: Amsterdam, The Netherlands, 30 March 2010. Available online: http://www.greenpeace.org/international/Global/international/planet-2/report/2010/3/makeit-green-cloud-computing.pdf (accessed on 27 February 2016).
75. Wall, T.A.; Walker, W.E.; Marchau, V.A.W.J.; Bertolini, L. Dynamic adaptive approach to transportation-infrastructure planning for climate change: San-Francisco-Bay-Area case study. *J. Infrastruct. Syst.* **2015**, *21*, 05015004. [CrossRef]
76. U.S. Army Heritage and Education Center. Who First Originated the Term VUCA (Volatility, Uncertainty, Complexity and Ambiguity)? Available online: https://usawc.libanswers.com/faq/84869 (accessed on 27 August 2021).
77. Ben-Haim, Y. *Info-Gap Economics. An Operational Introduction*; Palgrave Macmillan: London, UK, 2010.
78. Ben-Haim, Y. What Strategic Planners Need to Know, Workshop on Strategic Uncertainty in National Security, Samuel Neaman Institute, Technion, 26 June 2018. Available online: https://info-gap.technion.ac.il/files/2018/06/stpl007.pdf (accessed on 7 September 2020).
79. Ben-Haim, Y. Robust satisficing and the probability of survival. *Int. J. Syst. Sci.* **2014**, *45*, 3–19. [CrossRef]
80. Regev, S.; Shtub, A.; Ben-Haim, Y. Managing project risks as knowledge gaps. *Proj. Manag. J.* **2006**, *37*, 17–25. [CrossRef]

Article

Integrated Energy Planning for Near-Zero Carbon Emission Demonstration District in Urban Areas: A Case Study of Meishan District in Ningbo, China

Xiaoyan Xu [1], Ying Wang [2], Yingjun Ruan [2,*], Jian Wang [2,*], Kailiang Ge [3], Yongming Zhang [4] and Haikui Jin [1]

1 Tongji Architectural Design (Group) Co., Ltd., Shanghai 200092, China; 53xxy@tjad.cn (X.X.); 53jhk@tjad.cn (H.J.)
2 College of Mechanical and Energy Engineering, Tongji University, Shanghai 201804, China; 53wy@tjad.cn
3 State Grid Ningbo Electric Power Supply Company, Ningbo 315016, China; gekl@163.com
4 Sino-German College of Applied Sciences, Tongji University, Shanghai 201804, China; zhangyongming@tongji.edu.cn
* Correspondence: ruanyj@tongji.edu.cn (Y.R.); wangjiantjad@tongji.edu.cn (J.W.); Tel.: +86-021-6598-9750 (Y.R.); +86-021-3537-5002 (J.W.)

Abstract: Reasonable regional integrated energy planning is an important prerequisite for the construction of a Near-Zero Carbon Emission Demonstration District (NCEDD). An integrated energy planning scheme that is based on a three-step planning method with the objective of achieving an NCEDD is proposed in this paper. First, the planning objectives should be determined. After that, the planning strategies should be established. Finally, the planning approaches should be proposed according to the previously determined objectives and strategies. A case study considering the integrated energy planning of the Meishan International Near-Zero Carbon Emission Demonstration District (MINCEDD) is investigated to explain the planning method. In addition, the planning results, which are indicated as indexes, are explained, analyzed, and compared to the ones of other districts. The indexes include a proportion of renewable energy to primary energy (73% by 2030 and 108% by 2050), a proportion of renewable power to total power consumption (98% by 2030 and 111% by 2050), and CO_2 emission reduction rates (70% by 2030 and 100% by 2050) and are more advanced than other districts in China. This planning scheme and method can provide a reference for the integrated energy planning of NCEDDs in developed urban areas.

Keywords: integrated energy planning; near-zero carbon emission; near-zero carbon emission demonstration district; renewable energy; urban district planning methods

Citation: Xu, X.; Wang, Y.; Ruan, Y.; Wang, J.; Ge, K.; Zhang, Y.; Jin, H. Integrated Energy Planning for Near-Zero Carbon Emission Demonstration District in Urban Areas: A Case Study of Meishan District in Ningbo, China. *Energies* **2022**, *15*, 874. https://doi.org/10.3390/en15030874

Academic Editors: Michal Cehlar and Sergey Zhironkin

Received: 29 November 2021
Accepted: 19 January 2022
Published: 25 January 2022

1. Introduction

Strategic objectives were proposed during the 2015 United Nations Climate Conference in Paris which aimed to control the average global temperature within 2 °C of pre-industrial levels and make substantial effort to control temperature increases within 1.5 °C to achieve a peak of global greenhouse gas emissions (GHGs) as soon as possible, and achieve net-zero GHGs during the second half of this century [1].

Therefore, a series of policies have been proposed by countries to achieve carbon neutrality by 2050. For example, the European Union (EU) proposed that it would be carbon neutral by 2050 and hence released *"The European Green Deal"* in 2019 during the United Nations Climate Change Conference, which details key policies and core technologies in essential areas that are to be developed and the release of eight major goals [1,2]. The U.S. House of Representatives published *"The Congressional Action Plan for a Clean Energy Economy and a Healthy, Resilient, and Just America"* to promote the U.S. to achieve net-zero carbon emissions by 2050 [3]. Japan issued *"The Green Growth Strategy"*, which introduced the country's goal to be carbon neutral by 2050 and proposed specific development goals and essential development tasks for core industries [4].

In order to contribute to the net-zero GHGs, China announced that peak CO_2 emissions will be achieved by 2030 and that the nation will become as carbon neutral as possible by 2060 [5]. Additionally, tens of Chinese policies related to the reduction in carbon emissions, renewable energy, the electricity market, energy storage, etc., have been promulgated by various management departments from 2020 to 2021. One of these policies is called *"Opinions of the Central Committee and the State Council on the complete, accurate and comprehensive implementation of the new development concept to achieve carbon peaking and carbon neutral"*. The importance of energy planning is stressed by this policy. It is indicated that the carbon-neutral district is reaching the carbon emission limit rather than completely zero carbon emission. In the meantime, the requirements for both the supply and demand sides, as well as for multiple fields, such as energy, construction, industry, and transportation, need to be coordinated and put forward to achieve carbon neutrality [6]. Moreover, the Top Ten Actions for Carbon Peaking were announced in another Chinese policy called the *"Carbon Peaking Action Plan by 2030"*, which aim to promote peak carbon emissions by 2030 [7].

Cities and provinces in China are projected to reach peak CO_2 emission between 2020 and 2030, while some relatively developed cities, such as Beijing, Zhenjiang, Shenzhen, etc., were projected to reach peak CO_2 emissions before 2020 [8].

The definition of CO_2 emission is clarified by IPCC and is defined is the sum of the GHGs that are converted into equivalent CO_2 emissions by the indexes of global warming potential (GWP), global temperature potential (GTP), radiative forcing equivalence potential, etc. [9]. However, there are various definitions of CO_2 emissions when defining emission targets related to carbon neutrality among countries [10]. In most western and in Asian countries, such as Demark, Slovakia, Chile, Switzerland, Portugal, Canada, the United Kingdom (UK), Singapore, Germany, Sweden, France, Finland, Japan, etc., CO_2 emissions are defined as GHGs. CO_2, CH_4, and N_2O emissions are defined as CO_2 emissions in Bhutan and Uruguay. However, the definition in China and in many other countries is not clear. In this paper, CO_2 emissions refer to the GHGs in equivalent amounts of CO_2 based on GWP, including the emissions from fossil fuel combustions caused by transportation, electricity generation, manufacturing, building operations, etc.

Cai et al. [11] indicate that increases in carbon emissions are mainly driven by the population. In addition, over 90% of the total CO_2 emission in China are from urban areas [8]. This means that reducing CO_2 emissions in urban areas is significant for the total carbon reduction in the entire country, and conducting near-zero carbon emission programs is one of the current strategies in China that is being implemented to reduce carbon emissions in urban areas.

The Near-Zero Carbon Emission Demonstration District (NCEDD) project was first proposed in the China's 13th Five-Year Plan. This project is one of the 12 projects of the 13th Five-Year Plan and has caused widespread concern in the community. Moreover, the 13th Five-Year Plan proposed that GHG control measures in mature areas, which are subject to restricted and prohibited development, ecological functional areas, industrial mining areas and towns, etc., should be selected to carry out the NCEDD project and that 50 demonstration projects should be built by 2020.

In order to carry out the NCEDD, researchers, such as the IPCC, have mainly discussed near-zero carbon emissions in terms of its connotations. The IPCC indicates that carbon neutrality or net-zero CO_2 emissions are achieved when annual CO_2 emissions are considered to be balanced by decarburization technologies [9]. Near-zero carbon emissions are defined as being achieved when the net carbon emissions are approaching zero in a specific area. The relationship among near-zero carbon emissions, low carbon emissions, and zero carbon emissions is shown in Figure 1. Zero carbon emissions is the final objective for low-carbon community construction, while low carbon emissions and zero carbon emissions are necessary in the process to achieve zero carbon emissions. The net carbon emissions of a specific district can be decreased to zero gradually by reducing carbon emissions from the supply, contributing to zero carbon energy, etc. Moreover, near-zero carbon emissions refer to contexts that are approaching zero carbon emission situations.

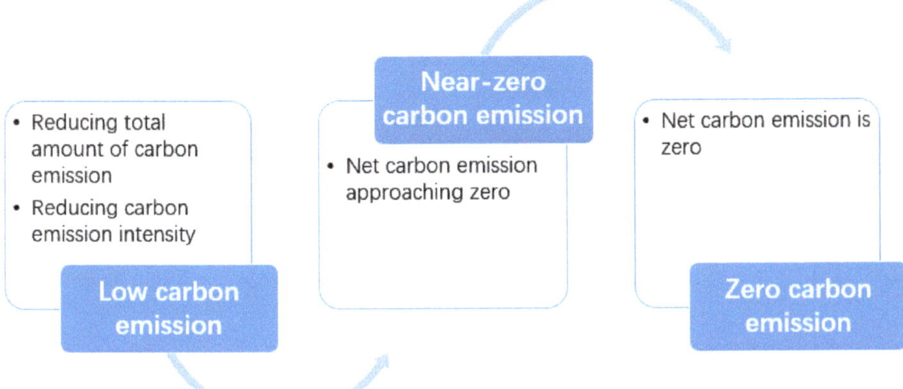

Figure 1. The relationship among near-zero carbon emissions, low carbon emissions, and zero carbon emissions.

The definition of an NCEDD can be explained based on the meaning of near-zero carbon emissions. Demonstration districts represent typical districts that are developed and constructed in advance. The development mode can play a demonstration role, and the experience can be replicated and promoted. Li et al. [12] define NCEDDs with near-zero carbon emission as follows: A typical district where the objective net carbon emissions approaching zero is achieved using approaches that reduce source carbon emissions, increase carbon sequestration, or contribute zero carbon energy, etc., within the boundaries of the county's (county-level city/district) administrative divisions and where the carbon-emission-reducing methods are replicable and can be promoted. Furthermore, two quantitative indexes are proposed for NCEDDs and are as follows: (1) It is possible for the total net carbon emissions to reach levels closer to zero by offsetting source carbon emissions and carbon sequestration or by offsetting source carbon emissions and by contributing zero carbon energy; (2) absolute carbon emissions reach their peak value. These quantitative indexes are also important integrated energy planning objectives for NCEDDs.

Liu [13] concludes four types of NCEDDs. The first type is 100% renewable energy demonstration districts, which are applicable to less developed areas with plentiful renewable energy resources. The second type is carbon-neutral demonstration districts, which are applicable for scenic areas with large amounts of forests and pastoral areas. The third type of NCEDD is partial demonstration in the industry, for example, near-zero carbon emission buildings. The fourth type of NCEDD is low carbon emission upgrades, which is further decarburized on the basis of current low carbon emission pilot projects. The first and second types are constrained by regional and resource issues, and they are not able to be promoted in developed urban areas. The third and fourth types only consider near-zero carbon emissions in a narrow sense. It is necessary to determine how to build NCEDDs in developed urban areas.

Reasonable district-integrated energy planning is an important part of constructing urban NCEDDs. Recently, the commonly used methods for energy planning are demand-side energy planning and supply-side energy planning. In Chinese urban planning systems, urban power planning, gas planning, and centralized cooling and heating planning mostly belong to supply-side energy planning, i.e., top-down planning. However, demand-side energy planning starts from the demand side, pays attention to the effective integration and reasonable application of various energy resources in urban areas, and is a bottom-up process. In recent years, demand-side energy planning has been utilized more and more often in district-integrated energy planning.

Recently, there has been increasing amounts of research into demand-side planning, a large amount of which has been conducted by Long [14–17], Jelić [18], Wang [19], and Yu [20] et al. Most of these researches proposed reducing energy demand, utilizing renewable energy sources instead of fossil fuels, and capturing and storing carbon (CCS). These are solutions also indicated by Khalilpour [21] as specific technical approaches for the reduction in carbon emissions. However, developed urban areas lack carbon sequestration resources, such as forests and pastoral areas, which makes it even more difficult for them to achieve near-zero carbon emissions. Apart from reducing energy demand and applying renewable energy, Dobbelsteen et al. [22] proposed seeking energy synergy rather than CCS in urban energy planning. In addition, energy cascade utilization, the reuse of waste heating, energy storage, etc., are ways in which energy synergy can be implemented, which can provide heating and cooling network connections at different temperature levels.

In existing research, there is generally no clear planning method for district energy planning and few implementation conditions for energy schemes [23–27]. There have been numerous limitations in the research that has been carried out recently. The Shenzhen Academy of Building Research Co., Ltd. [23] conducted a green and ecological district in Ya'an by energy demand forecasting and energy resources analysis. The renewable energy and heat pump systems are planned to take place after evaluation. Cheng et al. [24] proposed four available energy station schemes with integrated energy utilization and energy micro-grid in a northern district. Tang et al. [25] developed energy schemes for a district in Guilin by determining planning objectives and analyzing energy demand and energy resources. The Tianjin Architectural Design Institute et al. [26] utilized a planning method that consisted of project investigation, demand forecasting, the proposal of energy schemes, the evaluation and comparison of energy schemes, and the determination of recommended schemes. The energy schemes for the university park in Tianjin were proposed and determined by the HVAC, electrical, water supply, and drainage energy consumption sectors. Although this method is relatively clear, it is only suitable for building sector districts rather than industrial and traffic sectors. Zhejiang Zheneng Energy Service Co., Ltd. et al. [27] proposed an integrated energy planning scheme for a district called Xiangbao in Ningbo. The distributed energy stations are planned for buildings and industries. Additionally, special planning for solar energy, electricity, water source, and utility tunnels is proposed based on local conditions. However, there are limitations for the above researches in terms of the absence of clear planning methods, planning objectives, and integrated and implementable energy schemes. The limitations of the above mentioned researches are listed in Table 1.

Table 1. The limitations of recent research into district energy planning.

Planned District	Ya'an	A Northern District	Guilin	Tianjin	Xiaongbao
Absence of clear planning method	•	•	•	•	N/A
Absence of planning objectives	•	•	N/A	•	•
Absence of demand reducing schemes	•	•	•	•	•
Absence of quantitative analysis, layout, and application scenarios of energy schemes	•	•	•	N/A	N/A
Absence of renewable energy utilzation schemes	N/A	N/A	N/A	•	N/A
Absence of grid or storage schemes	•	N/A	•	•	N/A

"•" means the condition that is invloved in the district planning.

In this paper, a case study of the Meishan International Near-Zero Carbon Emission Demonstration District (MINCEDD) was investigated in order to propose an integrated energy planning scheme for this district. Moreover, a three-step planning method for the planning of NCEDDs was proposed and includes planning objectives, approaches, and strategies. After that, the planning results were analyzed and compared with respect to the indexes of renewable energy and CO_2 reduction.

2. Methods and Algorithms

2.1. The Three-Step Planning Method

A three-step planning method for NCEDD development is proposed that determines the planning objectives, planning strategies, and planning approaches. Firstly, the planning objectives are determined by focusing on targets such as CO_2 reduction, renewable energy utilization, etc. After that, the planning strategies are formed as the overall principles of the planning approaches, which consist of reducing energy demand, applying renewable energy based on local resource conditions, and improving integrated energy efficiency through means such as thoroughly utilizing waste heat on the demand side. Finally, the planning approaches are proposed based on the objectives and on the local conditions affecting access to natural resources, current and planned building construction, traffic and industry, etc., and can be categorized into the demand-side, supply-side, grid-side, and storage-side approaches, according to the energy application segment.

2.1.1. Determining the Planning Objectives

The objectives should be determined to satisfy the NCEDD target as well as to be appropriate the local conditions of the district. Therefore, the research team investigated local energy supply companies, local governments, local factories, etc. to thoroughly understand the current conditions for renewable energy utilization and potential, the electricity supply, industrial energy consumption, and waste energy potential, etc.

The research team evaluated and researched the energy utilization potential and advantage of the district by analyzing the current energy supply, energy consumption, and renewable and waste energy potential. After that, the NCEDD objectives were determined with respect to the indexes of renewable energy utilization and CO_2 emission reduction.

2.1.2. Establishing the Planning Strategies

In order to the achieve the objectives that were previously determined, the planning strategies were established from the aspects of reducing energy demand, improving energy efficiency, and utilizing both waste heat and renewable energy. The strategies must not only be established according to the objectives but should also be adapted to the local conditions of the planned district.

2.1.3. Proposing the Planning Approaches

Planning approaches are based on "demand-supply-grid-storage", which means reducing energy demand on the demand side, utilizing renewable energy on the supply side, and improving energy interconnection and peak load shifting on the grid and storage sides so as to achieve the objectives of the MINCEDD.

2.2. The Calculation Algorithms of Planning Indexes

After applying the three-step planning method, including the objectives, strategies, and approaches, indexes with respect to renewable energy utilization and CO_2 emission reduction are proposed and summarized in Table 2. The indexes were evaluated by analyzing and comparing other similar planned districts. The advancement and feasibility of the planning scheme based on the three-step planning method can be proven after the evaluation.

2.2.1. Proportion of Renewable Energy to Primary Energy

In order to explain the definition of the RE/PE index, the algorithm of this index is introduced followed by its explanation in this sub-section.

Algorithm: Total renewable energy consumption (tce) ÷ Total primary energy consumption (tce) × 100%

Table 2. The summary of major planning indexes.

Indexes	Objectives of 2030	Planning Results of 2030	Objectives of 2050	Planning Results of 2050
Proportion of renewable energy to primary energy (RE/PE)	$\geq$70%	73%	$\geq$90%	108%
Proportion of renewable power to total power consumption (RP/PC)	$\geq$90%	98%	100%	111%
CO_2 emission reduction (ton)	$\geq$750,000	780,000	$\geq$1,100,000	1,210,000
CO_2 emission reduction rate	–	70%	–	100%

Explanation: Renewable energy refers to the energy rather than fossil fuels and includes wind energy, solar energy, water energy, biomass, geothermal energy, ocean energy, etc. According to the second provision of the *"China National Renewable Energy Law"* [28], renewable energy power mainly consists of solar PV, wind power, biomass power, etc. Primary energy refers to the energy resources that exist in nature without processing and conversion, also known as natural energy, such as coal, oil, natural gas, hydropower, etc.

Total energy consumption is the total amount of annual energy consumed on the demand side in the planned district. RE/PE is the amount of total annual consumed renewable energy divided by the total annual consumed primary energy in the planned district.

2.2.2. Proportion of Renewable Power to Total Power Consumption

In order to explain the definition of the RP/PC index, the algorithm for this index is introduced followed by its explanation in this sub-section.

Algorithm: Renewable energy power generation (kWh) ÷ District total power consumption (kWh) × 100%

Explanation: Power generation refers to energy generation on the supply side. The meaning of renewable power is the power generated from the renewable energy that is explained in Section 2.2.1. RP/PC is the total annual renewable power generated divided by the total annual power consumed in the planned district.

2.2.3. CO_2 Emission Reduction

In order to explain the definition of the CO_2 emission reduction index, the algorithm of this index is introduced followed by its explanation in this sub-section.

Algorithm: (CO_2 emission caused by conventional energy supply − CO_2 emission caused by near-zero carbon emission energy supply) ÷ CO_2 emission caused by conventional energy supply × 100%

Explanation: Conventional energy is the primary energy that has been mass-produced and widely used, such as coal, oil, natural gas, etc. Zero-carbon emission energy is the energy that is generated and consumed without creating carbon emissions, such as solar energy, wind energy, ocean energy, geothermal energy, etc. The near-zero carbon emission energy supply refers to the combination of conventional and zero-carbon emission energy supply in planned districts, which leads to the overall near-zero carbon emission effect.

The three-step planning method process is shown in Figure 2.

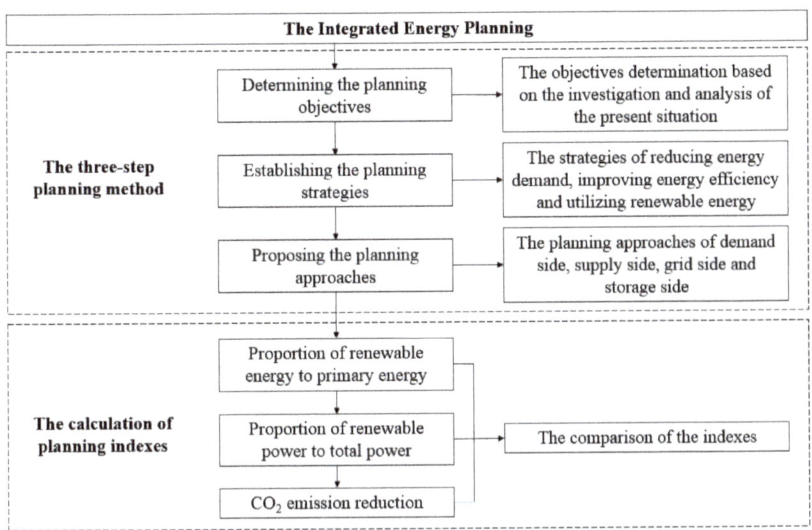

Figure 2. The three-step planning method process.

3. Present Situation of the Planning District

3.1. The Planning Area and Planning Period

The MINCEDD planning area, which is located in Ningbo, Zhejiang province, is 333 km^2 and includes Meishan Island in the Beilun area, Chunxiao Street, and Baifeng Town. The planning area is shown in Figure 3.

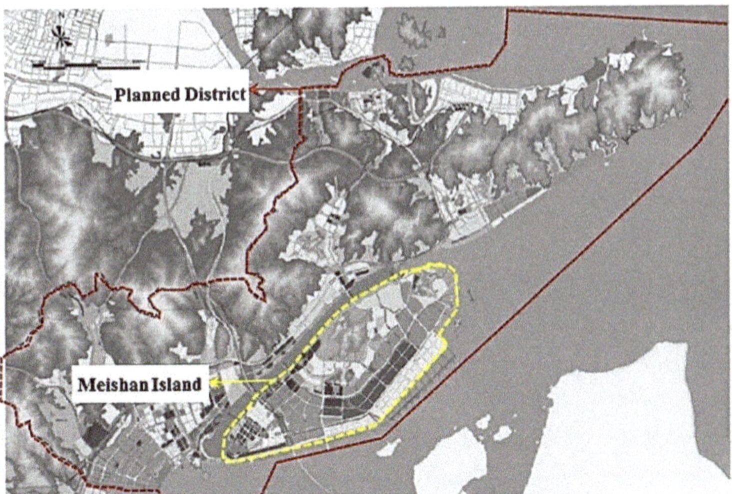

Figure 3. The MINCEDD planning area.

According to the regulatory plans, the main planning area comprises six core areas that make up a total land area of 43.49 km^2 and includes Financial Town, the Menhu Area, the Industry Area, the Bonded Area, the East Area, and the Li'ao Area (as is shown in Figure 4).

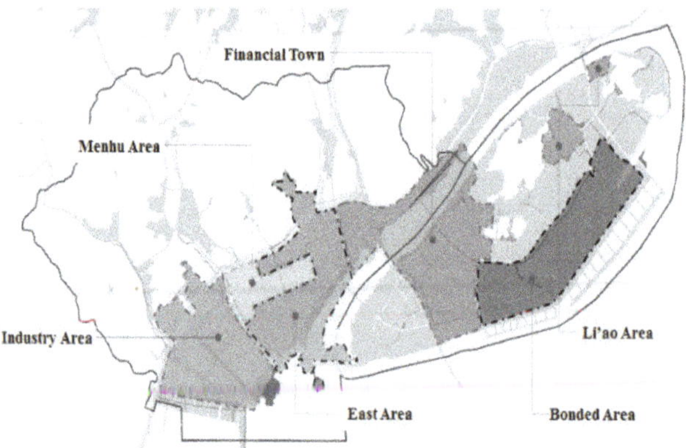

Figure 4. The planning areas.

The planning period is from 2020 to 2050, with mid-term planning taking place from 2020 to 2030 and long-term planning taking place from 2030 to 2050.

3.2. Local Energy Resources

The MINCEDD belongs to the subtropical monsoon climate zone (see Figure 5) [29] and contains abundant renewable resources and clean energy. In order to appropriately propose energy planning schemes based on local conditions, energy resources, including solar energy, wind energy, biomass energy, liquefied natural gas (LNG) cold energy, and hydrogen energy, are each analyzed and evaluated in the following.

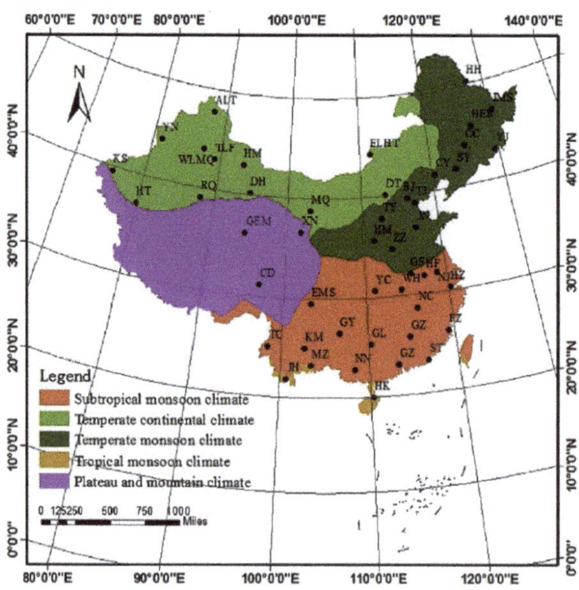

Figure 5. Climate zone distribution of China.

3.2.1. Solar Resource

Solar energy is one of the most competitive sources of renewable energy due to its abundance and clean operation. Total annual solar radiation of MINCEDD is greater than 4008 MJ/m^2/a [30].

3.2.2. Wind Resource

Wind resources are relatively rich in the MINCEDD [31]. Wind resources are simulated using computational fluid dynamics (CFD) numerical software. The simulation results show that in the winter and summer, the dominant wind directions are northwest and southeast, respectively. In addition, the dominant wind speed is about 7.48 m/s in winter and 7.95 m/s in summer and occurs at the height of 80 m. The wind speed distribution map at different heights in the planning area is shown in Figure 6.

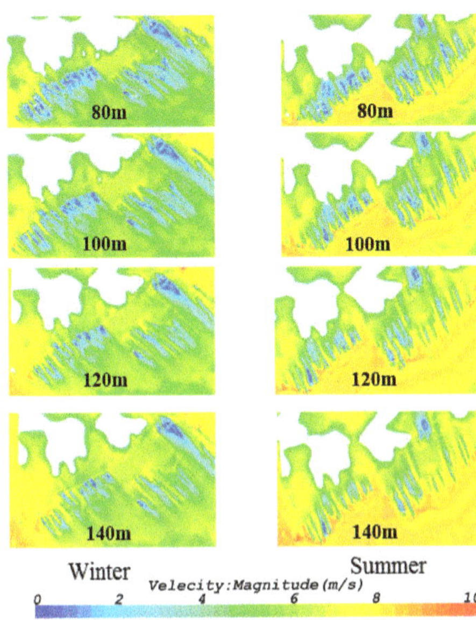

Figure 6. Wind speed chart at different heights.

3.2.3. Hydrogen Energy Resource

Ningbo is a city with a large amount of industrial by-product hydrogen, which is mostly expulsed from the Daxie Industrial Park. According to TrendBank statistics, Ningbo can supply about 72 kt/yr of by-product hydrogen, which indicates the large planning potential of by-product hydrogen from the industry sector. Moreover, the purity of industrial by-product hydrogen from Donghua Energy and from Wanhua Chlor-Alkali is 99.99% based on site investigations. Existing hydrogen by-product production is detailed in Table 3.

Table 3. Industrial by-product hydrogen in Ningbo area.

Company Name	Hydrogen Production (ton/a)
Zhenhai Refinery	144,000
Sinochem	117,000
Zhenyang Chemical Industry	8100
Siming Chemical	7200
Beilun Area	9400
Daxie Petrochemical	53,000
Wanhua Chemical	70,000
Wanhua Chlor-Alkali	15,000
Donghua Energy	20,000
Total	443,700

3.2.4. LNG Cold Energy Resource

An LNG receiving station that belongs to Zhonghai Ningbo Liquefied Natural Gas Co., Ltd. is located at the northeast end of the planning area. The external transmission capacity of the LNG receiving station is 7 million tons per year. Currently, only 400 ktons of LNG per year is used for cold energy air separation operations, while most of the remaining cold energy is released into the sea water. Hence, there is large application potential in which the remaining cold energy can be used. The LNG receiving station location is shown in Figure 7.

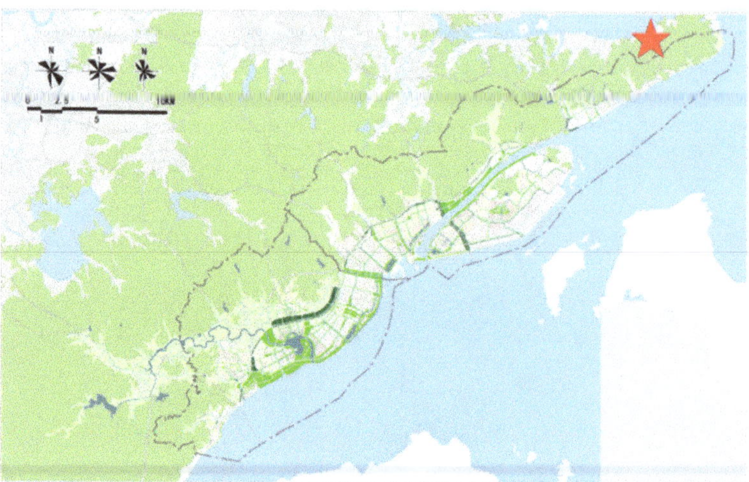

Figure 7. Geographical location of the LNG receiving station.

3.2.5. Other Resources

The biomass resources and low-grade heat sources in the planning area were investigated and evaluated. The existing biomass power plants cover all biomass resources in the planned district and part of the biomass resources beyond the MINCEDD in Ningbo (including domestic waste). There is additional capacity that can be used by the MINCEDD biomass power plant in the future even though the capacity of the plant will maintain its status due to garbage odor concerns in the selected location. In addition, a sewage treatment plant with a daily treatment capacity of 35,000 tons will be built in the near future. According to the progress being made in the construction of the sewage plant and the surrounding users, it is proposed that a small-scale sewage source heat pump system can be developed by the MINCEDD in the future. Apart from biomass and sewage source, low-grade heat sources, groundwater sources, soil sources, and seawater sources are not suitable for exploitation due to the environmental and policy constraints.

3.3. Current Energy Consumption

The total energy consumption in the MINCEDD was about 318 ktce in 2017 and includes 106.9 ktce of building consumption, 177.3 ktce of industry consumption, and 34.2 ktce of traffic consumption. The building, industry, and traffic sectors account for 33.5%, 55.7%, and 10.8% of the total energy consumption, respectively.

In addition, there are 235.8 ktce of primary energy that are consumed by electricity, which is 74.1% of the total primary energy consumption. The natural gas consumption is 27.5 ktce and fossil fuel (incl. coal and oil) consumption is 55 ktce, representing 8.6% and 17.3% of total primary energy, respectively. Figure 8 illustrates the energy consumption and percentages of different energy sectors and energy sources [32].

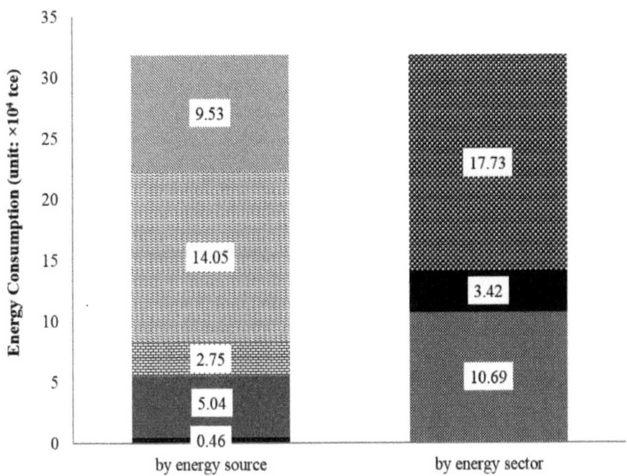

Figure 8. The composition of primary energy consumption.

4. The Integrated Energy Planning Scheme

4.1. The Planning Objectives

The primary energy consumption is 318 ktce in 2017, and the energy intensity per unit GDP is 0.15 tce/CNY. The integrated energy planning objectives for the MINCEDD were determined by taking NCEDD as a guide, planning based on the local population and GDP, integrating local energy resources and energy demand characteristics, and using the quantitative scenario analysis method [33]. The planning objectives are listed in Table 4.

Table 4. The planning objectives and index compliance.

Indexes	Objectives of 2030	Objectives of 2050
RE/PE	≥70%	≥90%
RP/PC	≥90%	100%
CO_2 emission reduction (ton)	≥750,000	≥1,100,000

4.2. The Planning Strategies

4.2.1. Reducing Energy Demand

The core of achieving the objective of near-zero carbon emissions is reducing energy consumption on the demand side and increasing the sources of the energy supply. Energy demands consist of building demand, industry demand, and traffic demand.

Building demand can be reduced by passive measures, which are as follows: (1) improving building envelope performance to reduce winter heating and summer cooling and (2) introducing natural ventilation to reduce the demand for cooling and heating during the transition seasons.

Industry demand can be reduced by industrial restructuring, which is achieved by restricting and gradually eliminating high-energy-demand industries (i.e., energy consumption per unit output value higher than 0.07 kgce/CNY) and introducing high output with low-energy-consumption industries (i.e., energy consumption per unit output value lower than 0.015 kgce/CNY).

Traffic demand is reduced by transportation energy transformation and by changing the travel mode, for instance, restricting cars with high emissions, encouraging public and green travel, etc.

4.2.2. Improving Energy Efficiency and Utilizing Waste Heat

1. Improving energy efficiency of building sector:

In buildings, energy consumption mainly takes place through lighting, sockets, air conditioning, power system, and special equipment.

In terms of lighting, a building's energy consumption can be reduced by using LED lights, installing intelligent control, controlling the lighting by zones, etc.

The amount of energy consumed by a building in terms of air conditioning can be reduced by improving the energy efficiency of a system and equipment, such as heating/cooling units, boilers, pumps, fans, air handling units, etc.

The amount of energy that a building consumes in terms of its power system consists of the energy that is used by water pumps, elevators, etc. The energy efficiency measures for water pumps refers to the selection of high-efficiency and variable frequency equipment. The energy efficiency measures for elevators are variable frequency, group controlling, etc.

Waste heat can be applied in the following ways. (1) Heat recovery on the air system side: Using a heat exchanger to recover the heat from exhaust air or fresh air so that the fresh air can be pre-heated or cooled and can reduce the energy for fresh air. (2) Units can recover condensation heat: The heat that is generated from cooling condensation water in the summer is recovered for domestic hot water heating to decrease the hot water heating energy, which has a significant energy-saving effect for buildings with a high demand for domestic hot water such as in-patient buildings and hotels. (3) Elevator energy feedback: The energy that is generated during elevator operation is given back to the power grid or to the micro power grid for reuse, which can save electricity consumption by about 30%.

2. Improving energy efficiency of industry sector:

First, the energy efficiencies of general equipment, such as boilers, motors, transformers, and air compressors, should be improved.

Second, heat recovery and cascade utilization are the major approaches for reusing industrial waste heat. This means the recovered heat is used for industrial processes or for heating the areas in the building. Most equipment, such as prime motors, heating furnaces, air compressors, annealing furnaces, VOC handling units, boilers, etc., only consume small amounts of heating energy. Recovering the remaining industrial heating energy for centralized heating can save primary energy, improve economic benefit, and reduce pollution.

4.2.3. Utilizing Renewable Energy Adapted to Local Conditions

For urban areas, utilizing renewable energy that can be adapted to local conditions is the last guaranteed method through which near-zero carbon emissions can be achieved and is based on decreasing energy demand, improving energy efficiency, and using waste heat. This means providing zero-carbon emission energy supply, which is close to or is more than fossil fuel consumption.

Renewable energy consists of solar energy, wind energy, ocean energy, geothermal energy, etc. Distributed renewable energy is the best way to use renewable energy sources. This can be achieved, for example, by installing a solar photovoltaic system, wind turbines, etc., for buildings or in communities. High-percentage renewable power is consumed by cascade grid connection and energy storage to maintain the stability of the power grid.

4.3. The Planning Approaches

4.3.1. Demand Side

1. Building Sector

The point and surface combination method is used for building sector energy demand reduction. From the point of view of the retrofitting objectives, energy conservation measures can be proposed for existing buildings, and design and construction requirements of energy conservation are proposed for new buildings. Therefore, all of the buildings will be constructed by conforming to the national ultra-low energy or near-zero energy building standards.

All existing buildings in MINCEDD will be retrofitted with energy conservation measures by 2030. Standards of ultra-low energy building are required for new buildings so that all new buildings conform to the current national standard *"Technical standard for nearly zero energy buildings"* [34] by 2030. Buildings built during 2030–2050 conform to the claims for the near-zero energy building standards determined by the current national standard *"Technical standard for nearly zero energy buildings"* [34]. The specific building performance parameters are listed in Table 5.

Table 5. Energy conservation measures for buildings.

	Residential Buildings		Public Buildings	
Construction Year	**2030**	**2050**	**2030**	**2050**
Building Envelope Performance				
Roof (W/m^2·K)	0.5	0.3	0.45	0.3
External Wall (W/m^2·K)	0.6	0.4	0.7	0.4
External Window (W/m^2·K)	1.8	1.4	1.8	1.4
Solar heat gain coefficient (Winter)	$\geq$0.40	$\geq$0.40	$\geq$0.40	$\geq$0.40
Solar heat gain coefficient (Summer)	$\leq$0.30	$\leq$0.30	$\leq$0.15	$\leq$0.15
Equipment Energy Efficiency				
Unitary air conditioner (EER)	4.0	4.5	4.0	4.5
Chiller (COP)	6.0	6.25	6.0	6.25
Multi-connected split air conditioner (IPLV(C))	5.0	6.0	5.0	6.0
Air-cooled heat pump (heating EER)	2.0	2.3	2.0	2.3
Air-cooled heat pump (cooling EER)	3.2	3.4	3.2	3.4
Gas boiler (D $\leq$ 2.0/Q $\leq$ 1.4)	92%	94%	92%	94%
Gas boiler (D > 2.0/Q > 1.4)	94%	96%	94%	96%
Pump [1]	Level 2	Level 1	Level 2	Level 1

[1] The pump energy efficiency is determined based on the related current national standard for pump.

In terms of surfaces, replicable projects for future community construction in accordance with *"Pilot Work Plan for Future Community Construction in Zhejiang Province"* are planned. The planning schemes for low-carbon communities are listed in Table 6.

Table 6. Low-carbon communities planning scheme.

Type of Community	2020–2025		2026–2030		2031–2040	2041–2050
	Quantity	Building Area (m^2)	Quantity	Building Area (m^2)	Building Area (m^2)	
Hydrogen community	/	/	3	400,000		fully promoted
Direct current community	3	300,000	10–12	1,000,000–1,500,000	7,000,000–9,000,000	
All electric community	3	300,000	10–12	1,000,000–1,500,000		
Zero energy community	5	500,000	10–15	1,000,000–2,000,000		

After applying the above building energy conservation measures, the building energy demand forecasting results are listed in Tables 7 and 8. Scenario 1 refers to the conventional scenario for 2030, i.e., energy increases under the current rate. Scenarios 2 and 3 refer to the near-zero carbon emission scenarios for 2030 and 2050, respectively.

Table 7. Annual power consumption per building area for Scenarios 1–3 (unit: kWh/m^2).

Scenario	Residential Building	Shopping Mall + Office	School
Scenario 1	30	150	70
Scenario 2	21	75	50
Scenario 3	20	74	47

Table 8. Annual power consumption for Scenarios 1–3 (unit: GWh).

Scenario	Residential Building	Shopping Mall + Office	School	Current Condition	Total
Scenario 1	127	686	92	319	1224
Scenario 2	82	318	61	319	780
Scenario 3	130	522	95	319	1066

2.   Traffic Sector

Green transportation and low-carbon travelling are recommended to reduce the amount of energy that is consumed by the traffic sector. In terms of "light-storage-charging" technology, there are seven power charging stations and five hydrogen refueling stations that are planned for green transportation. The vehicles will consume 100% of clean energy for public utilities, such as buses, taxi, and delivery vehicles, by 2030 and for private cars by 2050.

3.   Industry Sector

The industry energy conservation and low-carbon-emission objectives can be achieved by improving equipment efficiency and by integrating energy utilization and industrial restructuring. The introduction of industries with energy consumption per unit output below 0.20 tce/CNY into districts is encouraged. The evaluation and development recommendations for industrial structures are listed in Table 9.

4.3.2. Supply Side

In order to achieve the target proportion of renewable energy and renewable power, renewable energy is the major source of energy on the supply side. The scheme for the supply side focuses on photovoltaic and wind power generation and the full utilization of LNG cold energy and hydrogen energy. The renewable energy capacity is planned according to demand forecast results, consuming maximum renewable power, avoiding reverse power transmission, and decreasing peak–valley difference fluctuations. The layout of the energy scheme for the supply side is shown in Figure 9.

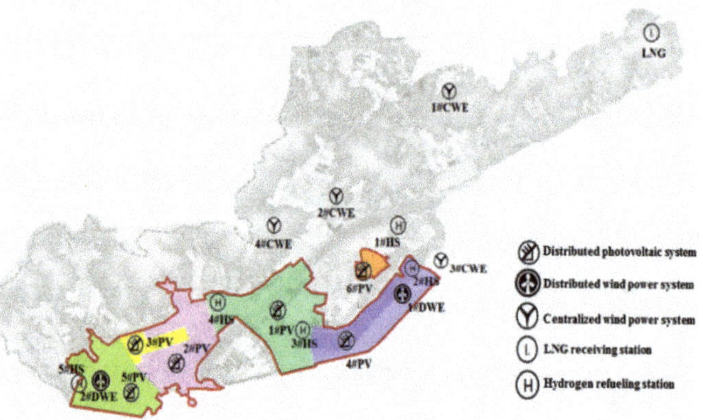

Figure 9. Source-side energy plan layout.

Table 9. The evaluation and development recommendations of industrial structure.

Items	EC	IO	EC/O	Evaluated Points					Evaluation				Development Recommendation
	(%)	(%)	(kgce/CNY)	EC	Output	EC/O	Total	Evaluation	P	D	R	E	
Car manufacturer	29.30	52.40	0.020	71	52	91	214	100.00	√				Assembly orientated
General equipment	9.90	10.40	0.025	90	10	89	190	88.45		√			
Rubber plastic	5.30	3.30	0.040	95	3	83	181	84.24		√			
Metalwork	2.20	2.10	0.030	98	2	87	187	87.15		√			Reducing plating and casting manufacturing
Railways and traffic	1.40	2.40	0.020	99	2	91	192	89.69		√			
Buildings	12.20	7.00	0.045	88	7	80	175	81.73		√			Developing prefabricated buildings and green and low-energy buildings
Other industries	13.30	13.50	0.030	87	14	87	187	87.29		√			
Gas production	18.90	7.20	0.070	81	7	70	158	73.63			√		Recovering cooling energy
Textile industry	5.70	1.50	0.110	94	2	52	148	69.02			√		Energy consumption control and waste heat recovery
Paper making	1.80	0.20	0.230	98	0	0	98	45.89				√	Restriction of high-pollution and -demand industries
Total	100.00	100.00		900	100	730	1730	807.09					
Peak	29.30	52.40	0.230	99	52	91	214	100.00					

Abbreviation interpretation: EC—energy consumption; IO—industrial output; EC/O—energy consumption per unit output; P—prior; D—develop; R—restrict; E—eliminate.

1. Solar Energy

To maximize the photovoltaic (PV) installation, it is suggested to make full use of a building's roof, façade, and structure to design building-integrated photovoltaics (BIPV). The maximum area of factory roofs ($\geq$60% of roof area) and available civil building roofs ($\geq$20% of roof area) are utilized to install solar photovoltaic systems. The capacity of installed solar photovoltaic systems will increase to 600 MW by 2030 and will increase to 642 MW by 2050. The planning scheme for photovoltaic systems for different types of buildings is detailed in Tables 10 and 11.

Table 10. PV layout plans for different types of buildings by 2030 (40 MW in existence).

Building Type	Roof Area (m^2)	Roof Usage Rate (%)	PV Area (m^2)	PV Capacity (MW)	Power Generation (GWh)
Existing industry	3,220,000	60	1,930,000	1,930,000	193
Residential	300,000	20	60,000	60,000	6
Commercial Office	910,000	20	183,000	183,000	18
Industry	4,070,000	60	2,441,000	2,441,000	244
Education	190,000	20	38,000	38,000	4
Warehouse	2,260,000	20	1,357,000	1,357,000	136
Total	10,950,000	/	6,009,000	6,009,000	601

Table 11. PV layout plans for different types of buildings by 2050 (40 MW in existence).

Building Type	Roof Area (m^2)	Roof Usage Rate (%)	PV Area (m^2)	PV Capacity (MW)	Power Generation (GWh)
Existing industry	3,217,000	60	1,930,200	193.0	193
Residential	332,000	20	66,500	6.6	7
Commercial Office	1,006,000	20	201,300	20.1	20
Industry	4,475,000	60	2,684,800	268.5	268
Education	207,000	20	41,300	4.1	4
Warehouse	2,487,000	20	1,492,200	149.2	149
Total	11,724,000	–	6,416,000	641.5	641

2. Wind Energy

Distributed onshore wind power and offshore wind power will be developed first. Then, centralized onshore wind power will be installed during the period of 2030~2050. The distributed wind power will be installed near factories and warehouses and will be more than 300 m away from residential areas. Offshore wind power will be installed away from channels and in the areas near the seacoast that are not sheltered. In addition, the installation of onshore wind power generation facilities on the ridge, where the wind resources are superior, is suggested. The wind capacities for different planning periods are shown in Table 12.

Table 12. The wind power capacity.

Type	Total Capacity of 2030 (MW)	Total Capacity of 2050 (MW)
Distributed wind power	66	131
Offshore wind power	258	337
Onshore wind power	95 (current condition)	320
Total	419	788

3. Hydrogen energy

It is recommended that hydrogen energy is obtained from industrial by-products and hydrogen electrolysis devices that are powered by surplus renewable power.

There are five hydrogen refueling stations with a capacity of 500 kg/d that are planned according to integrated energy supply stations for the traffic sector. Two of them will be built in the integrated stations by 2030 in order to satisfy the hydrogen demands of buses, delivery vehicles, and transportation vehicles. The other three hydrogen refueling stations will be built by 2050 in order to satisfy the increasing demand for the use of hydrogen in the traffic sector in the future. Furthermore, the main applications are the construction of hydrogen refueling stations, household hydrogen fuel cells, hydrogen fuel vehicles, and trucks. Hydrogen will be partially mixed with natural gas and will used for cooking and heating in buildings in the long term. The overall utilization plan for hydrogen energy is shown in Figure 10.

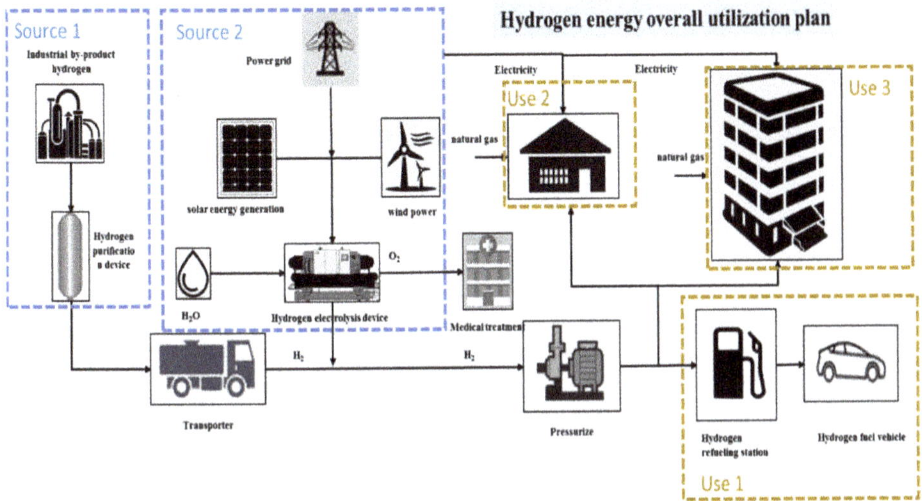

Figure 10. The overall utilization plan for hydrogen energy.

4. LNG cold energy

There is currently an LNG receiving station with a capacity of 7 Mton in the MINCEDD. Cold energy can be released and utilized after the LNG is gasified.

A cold energy cascade utilization industrial area that includes air separation, CO_2 liquefaction and dry ice manufacturing, medical refrigeration, and a data center is planned near the receiving station. Simultaneously, the natural gas that is produced after the gasification of LNG can be used for office buildings or small-scale commercial tri-generation system fuel cells in industrial parks and in surrounding areas. The layout proposed in the diagram of the industrial park is in Figure 11, and the LNG cold energy cascade utilization process is shown in Figure 12.

4.3.3. Grid Side

The port, industry, city, transportation, urban energy grid, green transport charging grid, and port microgrid will be planned, all of which compose the green energy grid that is required for energy dispatch.

1. Urban Energy Grid

In the MINCEDD, renewable energy mainly includes wind energy and solar energy. The scheme through which renewable energy will be connected to the grid is illustrated in Figure 13.

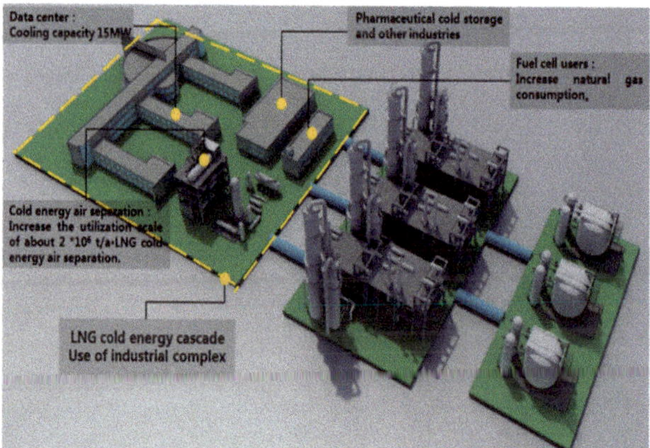

Figure 11. Schematic diagram of LNG cold energy industry layout.

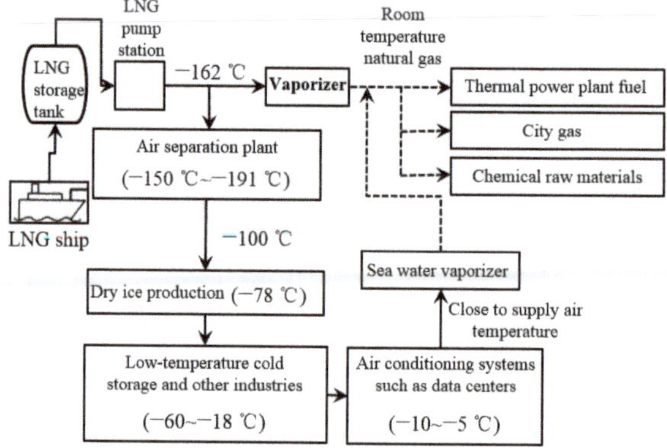

Figure 12. LNG cold energy cascade utilization flow chart.

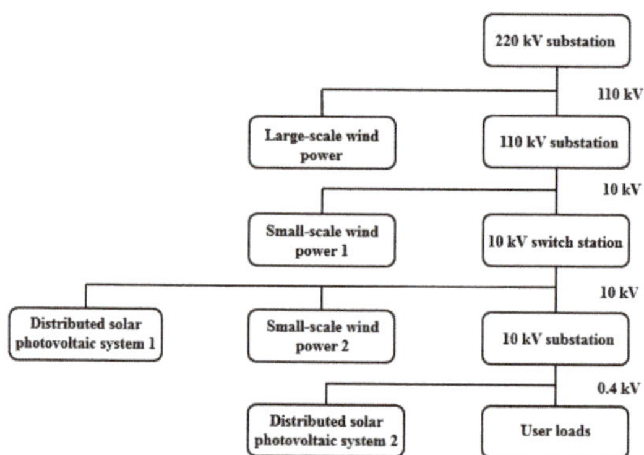

Figure 13. The scheme through which renewable energy will be connected to the grid.

The voltage that is necessary for solar PV to connect is 0.4 kV. The voltage of the rest power is increased to 10 kV in order to continue consumption when the consuming ability under 0.4 kV is insufficient. Small and large amounts of wind power are preferentially connected to the 10 kV and 110 kV substations, respectively. This means that small and large amounts of wind power can be connected to the secondary side of the 220 kV substation, the secondary side of the 110 kV substation, and the 10 kV switching station, which is based on a different scale. PV facilities are widely distributed in the planning area and are suitable for connecting to the 0.4 kV voltage class, which is mainly intended for self-use.

2. Green Transport Charging Grid

The charging infrastructures for electric vehicles (EV) and hydrogen energy vehicles are EV charging stations and hydrogen refueling stations, respectively, whose layouts constitute the green transport charging grid in Figure 14.

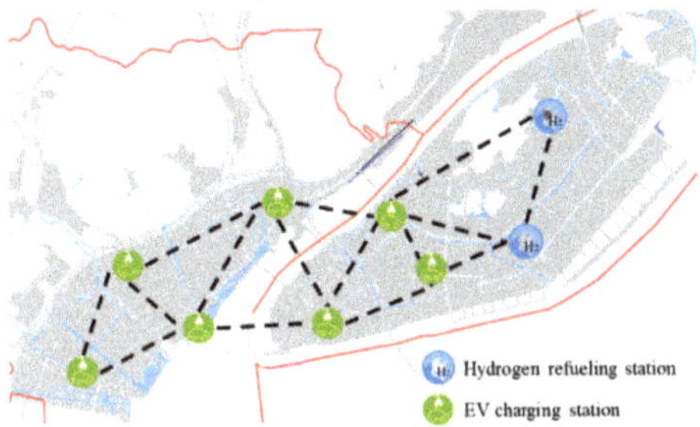

Figure 14. The layout of charging infrastructures.

According to the principle of optimal land utilization and the use characteristics of EVs, it is suggested that charging stations should be built together with existing public service facilities. PV devices can be installed on the ceiling of the charging station to increase the renewable energy use.

A hydrogen refueling station will be built in the Li'ao area, which is located on the main roads and is in a location that is convenient for hydrogen vehicle users. Additionally, a hydrogen refueling station will be built in the port area during long-term planning.

3. Port Microgrid

The concept of an onshore power supply is regarded as an effective solution to make ports free of greenhouse gas emissions, air pollutants, vibrations, and noise pollution.

The Meishan port is located in the southeast of the zero carbon emission demonstration area. The port has 10 berths. Shore power facilities will be built successively for each berth and will be able to output 50 Hz and 60 Hz alternating currents (AC) with a voltage class of 6.6 kV.

An AC/DC hybrid microgrid will be built in the port area for shore-to-ship power, which can improve the utilization of renewable energy distributed in the port area. The microgrid is shown in Figure 15.

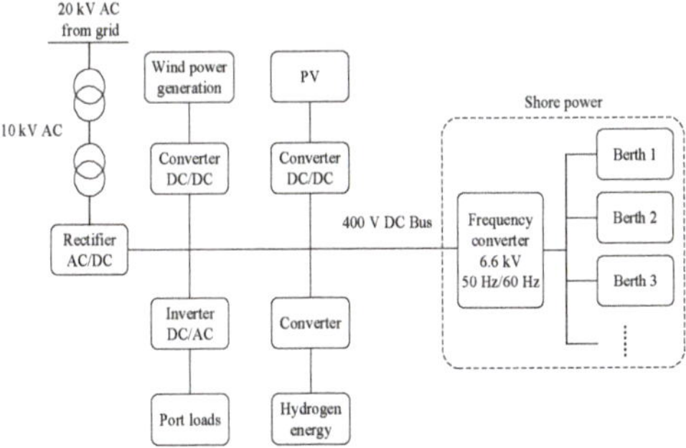

Figure 15. The hybrid micro-grid in the port.

In addition, hydrogen energy applications will be explored in transport ships and in logistics vehicles.

4.3.4. Storage Side

According to different types of energy and demands in near-zero carbon emission demonstration districts, multiple energy storage facilities are planned, which will consist of electrical energy storage (EES), thermal energy storage (TES), hydrogen storage, and compressed air energy storage (CAES) facilities. EES and TES can handle energy balances from the supply and demand sides in time series. Hydrogen storage and CAES facilities are oriented to the actual needs of industrial production.

1. Electrical Energy Storage

EES can transform electrical energy from power grids into a storable form and can ensure that it can be converted back into electrical energy if needed.

In the MINCEDD, there is no suitable terrain to construct pumped hydro storage facilities. Therefore, battery energy storage is widely utilized in this plan. There are three major ways to apply battery energy storage, which are as follows:

○   PV with small-scale battery energy storage

In order to maximize solar energy use and reduce the peak–valley difference, residential, commercial, and industrial users should be encouraged to install PV with small-scale battery energy storage for self-use with the 0.4 kV access voltage class. A battery storage device operating from 0.1 kWh to 0.2 kWh can be allocated for each kilowatt PV device. Users will be able to obtain a one-time subsidy for battery energy storage devices.

○   Medium-scale power storage facilities on the user side

Users with a large peak–valley difference, such as those in offices and commercial buildings, should be encouraged to install medium-scale power storage facilities on the user side, and these users will receive one-time subsidy for the installation of these power storage facilities and discounted electricity prices.

○   Large-scale centralized power storage facilities

Around the substation, centralized power storage facilities will be built to reduce load fluctuation and to improve the stability of the power distribution grid.

2. Thermal Energy Storage

TES refers to the temporary storage of thermal energy and is able to increase the utilization of thermal energy equipment. For the cold energy demand in industrial areas, cold energy storage facilities will be built to store cold energy during the valley–load period and will release it during the peak–load period, which can restrain the fluctuation that take place during load cooling and will allow industrial and residential demand to be met.

3. Hydrogen Storage

Hydrogen can be stored directly for use in fuel cells or can be transported to users. In the planned district, hydrogen storage facilities will be built for fuel cell vehicles and residential functions.

4. Compressed Air Energy Storage

In CAES systems, a compressor driven by off-peak or low-cost electricity is used to store energy by compressing air into an air reservoir. Since the self-discharge that is generated by CAES is very low, it can act as a long-term storage device. In addition, the investment cost for CAES is obviously low. In the industrial area, CAES facilities will be planned and constructed for the automobile production industry.

5. The Layout of Energy Storage Facilities

Based on land-use load types, the layout of energy storage facilities is planned as depicted in Figure 16.

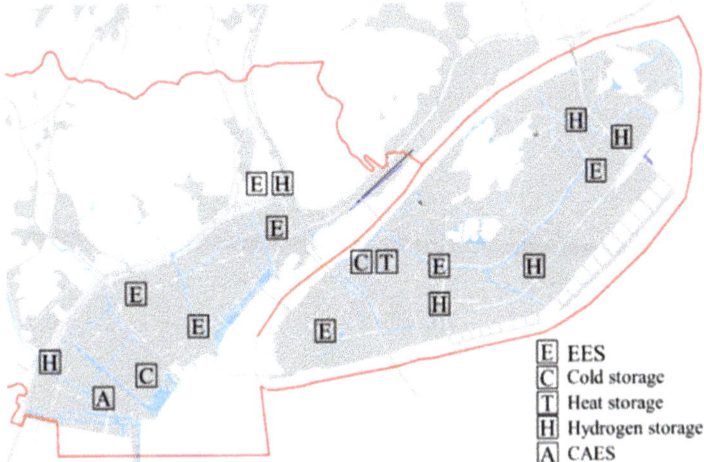

Figure 16. The layout of energy storage facilities.

According to the load forecast and the capacity of renewable energy sources that will be installed for the planning year (2030), the total capacity of the energy storage facilities will need to be about 300 MWh (electricity) in order to maximize the renewable energy power generation consumption, avoid the generation of renewable energy power generation, and reduce the fluctuations in the peak–valley differences. The capacities of the different storage types are listed in Table 13.

Table 13. The capacities of different storage types.

Storage Types	Capacity
PV + Small-scale Battery Energy Storage	60 MWh
Medium-scale Power Storage	80 MWh
Large-scale Centralized Power Storage	20 MWh
Heat Storage	30 MWh (heating)
Cold Storage	60 MWh (cooling)
Hydrogen Storage	2500 kg
CAES	30 MWh

4.4. The Planning Safeguard Measures

The MINCEDD planning scheme was proposed for the planning period from 2020 to 2050. The planning safeguard measures are under development by local governemnts and organizations currently. However, in order to achieve the planned scheme, safeguard measures in terms of technical measures and policies and subsidies are proposed in this article. Additionally, it is suggested that the local governemnts and organizations should make safeguard measures based on these measures.

4.4.1. Technical Measures

An integrated energy management platform is planned to be constructed to provide technical support for the realization of the planning scheme. As a district integrated energy service provider, it is considered to build a comprehensive energy monitoring and management platform based on 5G mobile internet. This platform is used as the hardware foundation, which can be combined with big data analysis to promote the implementation of energy conservation management and user behavior for energy conservation in the MINCEDD.

The platform is connected to the users to collect energy demand data and to the district energy supply system to collect the energy supply data from renewable energy systems, gas distributed energy supply systems, waste heat utilization systems, and energy storage systems. Hence, the comprehensive management and dispatch of energy demand and supply can be achieved through the platform. There are four modules for the operation of the platform, which are listed in Table 14.

Table 14. The integrated energy management platform modules.

Module Name	Functions
User energy management module	Energy consumption analysis, energy efficiency evaluation, user power management
Energy supply management module	Energy index analysis, energy supply quality analysis, energy supply optimization analysis
Energy transaction management module	Energy trading management, invoice management, expense management
Special value-added service module	User energy saving consultation and benefit analysis, big data analysis

4.4.2. Policies and Subsidies

The energy planning indexes that consist of renewable and clean energy capacity, energy conservation parameters for building, industrial and traffic sectors, grid capacity, and storage capacity can be included into land grant clauses during land bidding. After that, the indexes are committed to be implemented by developers or constructors during the construction period, whereby the planning indexes can be achieved.

Apart from the land grant clauses, policies and subsidies are formulated to ensure that the planning approaches are carried out (see Table 15).

Table 15. The policies and subsidies for the planning scheme implementation.

Planning Approach Category	Policies and Subsidies	Related Department
Solar PV	Subsidies for distributed PV	The management committee of the MINCEDD
Wind power	Subsidies for distributed wind power	The management committee, the development and reform committee, and the land and resources bureau of the MINCEDD
LNG energy	Subsidies and measures for the introduction of LNG energy utilization industries	The development and reform committee of the MINCEDD
Hydrogen energy	Subsidies for hydrogen production from renewable energy	The development and reform committee of the MINCEDD
Building energy conservation	Subsidies for ultra-low energy buildings	The management committee of the MINCEDD
Low carbon communities	Subsidies for the equipment of direct current low carbon communities, and hydrogen fuel cells	The management committee, the development and reform committee, and the power grid companies of the MINCEDD
Industry energy conservation	Measures for the introduction of high-tech or low-carbon green industries	The management committee, and the economic and information committee of the MINCEDD
Green traffic	Subsidies for electric vehicle charging facilities; preferential electricity price for public charging facilities	The management committee, and power grid companies of the MINCEDD
Power distribution grid	Measures for renewable power accessing to power distribution grid	The development and reform committee, and the power grid companies of the MINCEDD
Smart green port	Subsidies or preferential electricity price for the electricity of the ports	The management committee, the development and reform committee, and the port office of the MINCEDD

5. Planning Evaluation

5.1. Indexes Analysis

5.1.1. Proportion of Renewable Energy to Primary Energy

The authors of [32] propose that the proportion of non-fossil-fuel-based energy consumption increase to 20% by 2020. The current shares of renewable energy sources are 39.3 MW for solar PV, 94.5 MW for onshore wind power, and 32 MW for biomass. During the planning period, the capacities of solar PV and wind power will be increased to 642 MW and 788 MW by 2050. However, the biomass capacity will be kept the same due to the selected location and garbage odor concerns in that area. The total amount of renewable energy that will be used by 2030 is 460 ktce and 890 ktce by 2050. The authors of [32] estimate that the primary energy consumption will be 625 ktce by 2030 and 650 ktce by 2050. Therefore, the proportions of renewable energy to total primary energy will be 73% by 2030 and 108% by 2050, which are slightly higher than the planning objectives. Table 16 displays the current and planned renewable energy conditions.

Table 16. Current and planned renewable energy conditions.

Projects		Current Condition		by 2030		by 2050	
		Installed Capacity (MW)	Annual Energy Saving (ktce)	Installed Capacity (MW)	Annual Energy Saving (ktce)	Installed Capacity (MW)	Annual Energy Saving (ktce)
Solar PV		39.3	10	601	146	642	156
	Distributed	0	0	66	32	131	63
Wind power	Offshore	0	0	258	171	337	222
	Onshore	94.5	62	94.5	62	319.5	211
Biomass		32	48	32	48	32	48
Total		165.8	120	1052	459	1462	700

5.1.2. Proportion of Renewable Power to Total Power Consumption

Renewable power can be calculated by multiplying the installed capacity by the amount of power generated per unit capacity that is estimated according to past applications. In 2017, the total amounts of renewable power being consumed were 39.3 MW for solar PV, 94.5 MW for wind power, and 32 MW for biomass. The total amount of existing and planned renewable energy used by 2050 is 600 MW for solar PV, 419 MW for wind power, and 32 MW for biomass, with 1883 GWh of electricity being generated by 2030. The installed capacity of solar PV, wind power, and biomass is expected to be 642 MW, 788 MW, and 32 MW, respectively, by 2050. This can generate 2873 GWh of electricity by 2050. Furthermore, the MINCEDD is expected to consume 1928 GWh and 2581 GWh of power by 2030 and 2050, respectively. Hence, the RP/PC of the MINCEDD is expected to be 98% by 2030 and 111% by 2050. This means that the renewable power generated by 2030 can cover almost all of the electricity consumption in the district. Moreover, as the installed capacity of the renewable power is expected to increase and the electricity consumption is expected to decrease by 2050, the amount of renewable power that is generated will be greater than the electricity consumed. Therefore, the index of RP/RC is over 100%. The surplus of renewable power can be supplied to the national power grid for users outside the MINCEDD. Renewable energy generation is listed in Table 17.

Table 17. Current and planned renewable energy generation.

Projects		Current Condition		by 2030		by 2050	
		Installed Capacity (MW)	Annual Amount (GWh)	Installed Capacity (MW)	Annual Amount (GWh)	Installed Capacity (MW)	Annual Amount (GWh)
Solar PV		39.3	39	601	600	642	640
	Distributed	0	0	66	130	131	260
Wind power	Offshore	0	0	258	700	337	910
	Onshore	94.5	255	94.5	255	319.5	865
Biomass		32	198	32	198	32	198
Total		165.8	492	1052	1883	1462	2873

5.1.3. CO_2 Emission Reduction

The 13th Five-Year National Energy Plan and the 13th Five-Year Plan for Low Carbon Development in the Zhejiang province indicated that CO_2 emissions will be reduced by 18% by 2020 compared to 2015 [35]. The reductions in CO_2 emissions can be divided into supply side reductions and demand side reductions. Supply side reductions are caused by the application of renewable energy, hydrogen energy, LNG cascade utilization, and geothermal energy, and demand side reductions are caused by industrial waste energy and distributed energy station applications.

Compared to the current conditions, CO_2 emissions on the supply side are expected to be reduced by 0.46 Mton and 0.77 Mton by 2030 and 2050, respectively. Additionally, on the demand side, reductions of 0.32 Mton by 2030 and 0.44 Mton by 2050 are expected. The authors of [32] estimated that CO_2 emissions from the MINCEDD will be 1.08 Mton by 2030 and 1.20 Mton by 2050, both of which are under the benchmark scenario in which the district is developed without CO_2 reduction measures. The total CO_2 reduction rate is expected to approach 70% and 100%, respectively, by 2030 and 2050. This indicates that the CO_2 emissions in the MINCEDD are expected to be nearly zero by 2050, leading to the MINCEDD being in a near-zero carbon district. The CO_2 reductions for the supply side and demand side are listed in Table 18.

Table 18. The CO_2 emission reductions of the source and demand side.

Projects	Current Condition		by 2030		by 2050	
	Installed Capacity (MW)	Annual CO_2 Reduction (ton)	Installed Capacity (MW)	Annual CO_2 Reduction (ton)	Installed Capacity (MW)	Annual CO_2 Reduction (ton)
Solar PV	39.3	11,000	601	177,000	642	188,000
Wind power — Distributed	/	/	66	38,000	131	77,000
Wind power — Offshore	/	/	258	206,000	337	268,000
Wind power — Onshore	94.5	75,000	94.5	75,000	319.5	255,000
Biomass	32	58,000	32	58,000	32	58,000
Hydrogen energy	/	/	12.6 MW fuel cell + 2 hydrogen stations	12,000	30 MW fuel cell + 5 hydrogen stations	27,000
LNG cascade utilization	/	/	800,000 ton LNG/yr	41,000	800,000 ton LNG/yr	40,000
Geothermal energy	/	/	42.75	600	78.84	1100
Distributed energy stations sourced by natural gas	/	/	• fuel cell: 4.5 MW; • centrifugal cooling: 16,000 RT; • absorption cooling: 1000 RT; • boilers: 46 t/h	300	• fuel cell: 4.5 MW; • centrifugal cooling: 16,000 RT; • absorption cooling: 1000 RT; • boilers: 46 t/h	300
Total source side reduction	/	145,000	/	608,000	/	915,000
Demand side reduction	The measures of building energy conservations, industrial waste heat utilization, etc. are implemented on the demand side. The demand side CO_2 reduction is estimated as 320,000 ton by 2030 and 440,000 ton by 2050.					
Total reduction	/	145,000	/	928,000	/	1,355,000

5.2. Indexes Comparison

The MINCEDD indexes are compared to the similar indexes of the international project and other districts outside of Zhejiang. The results show that the MINCEDD indexes are all more advanced than other indexes.

5.2.1. Indexes Comparison with the International Project

The *"Ningbo Meishan Near-Zero Carbon Emission Demonstration Zone Construction Planning and International Cooperation Research Project Technical Report"* [32] is a preliminary international project that details the energy planning scheme for the MINCEDD. The MINCEDD energy plan that is introduced in this paper is developed based on the four aspects of the demand side, supply side, grid side, and storage side. The results show that the RE/PE is expected to reach 73% and that the RP/PC is expected to reach 98% by 2030. The MINCEDD indexes are more advanced than those determined in the international project (see Table 19).

Table 19. Comparison of the MINCEDD indexes and the international project.

Index	MINCEDD		The International Project	
	2030	2050	2030	2050
RE/PE	73%	108%	71.4%	93.8%
RP/PC	98%	111%	90%	100%

5.2.2. Indexes Comparison with Demonstration Districts outside Zhejiang

The MINCEDD was one of the five demonstration project plans for China State Grid in 2019, and the remaining four demonstration projects are the Beidaihe Integrated Energy Demonstration District (IEDD), the Zhengding IEDD, the Lankao IEDD, and the Guzhenkou IEDD [36].

The RE/PE for the Beiudaihe IEDD was 20% for 2020 and is 20% for the Zhending IEDD by 2030. In addition, the CO_2 emission reduction rate is expected to reach 15% by 2030. The RE/PE and the RP/PC for the Lankao IEDD were greater than 60% and 90%, respectively, in 2021. There is no specific energy-related index that has been identified for the Guzhenkou IEDD because the energy plan is focused on an intelligent power grid.

Compared to the indexes above, the indexes of RE/PE and RP/PC of the MINCEDD are higher than that of both the Beidaihe IEDD and Zhending IEDD. The RP/PC of the MINCEDD is expected to be at relatively the same level as that of the Lankao IEDD by 2030. In summary, the indexes of the MINCEDD are relatively more advanced than those of other demonstration districts outside of Zhejiang. The detailed indexes are listed in Table 20.

Table 20. The comparison of the MINCEDD indexes and those of other districts outside of Zhejiang.

District	MINCEDD		Beidaihe IEDD [37]	Zhengding IEDD [38]		Lankao IEDD [39]
Year	2030	2050	2020	2020	2030	2021
RE/PE	73%	108%	≥20%	3%	20%	≥60%
RP/PC	98%	111%	/	/	/	≥90%
CO_2 emission reduction rate	70%	100%	/	10%	15%	/

6. Conclusions and Recommendations

Energy consumption in developed urban areas is relatively high. Moreover, the renewable energy and carbon sequestration resources in these areas are limited. Integrated energy planning methods and schemes for NCEDDs in urban areas have been proposed in this paper to provide a reference for urban energy planning with near-zero carbon emission objectives.

A three-step planning method was proposed for the integrated energy planning of NCEDDs in urban areas that allows objectives to be determined, planning strategies to be established, and planning approaches to be proposed. Planning strategies include reducing energy demand; improving the energy efficiency of building, industry, and traffic sectors; and utilizing renewable energy sources that have been adapted to local conditions. Approaches are proposed according to these strategies, which encompass reducing energy demand and increasing renewable energy applications on the demand side and supply side, respectively, as well as improving energy interconnection and peak–valley differences in power levels.

The integrated energy planning for the MINCEDD was investigated as a case study to explain the planning method and scheme. The CO_2 emission reduction objectives in the MINCEDD are 0.75 Mton and 1.1 Mton by 2030 and 2050, respectively. The planned results show that annual CO_2 emissions will be reduced by 0.78 Mton by 2030 and 1.21 Mton by 2050 through the implementation of approaches that are related to the supply, demand, grid, and storage points of view. Furthermore, the CO_2 emission rates are expected to approach 70% and 100% by 2030 and 2050, respectively. Compared to the other districts with an integrated energy planning scheme, the renewable energy utilization and CO_2 emission reduction performances are relatively advanced in the MINCEDD.

In addition to the approaches proposed in this case study, other innovative measures can be implemented according to local conditions of the planned district. For example, energy buses (Ebus) (i.e., fifth-generation district energy network) are recommended for future integrated energy planning in urban areas, which provides opportunities for sharing heating and cooling energy under ultra-low temperature conditions compared to those of fourth-generation energy systems. Moreover, these measures can improve system efficiency by capturing low-grade heat sources and waste heat, which is an advanced heat recovery and energy synergy method that can be implemented in various building types [40].

Moreover, this method is only being used a few cases at present, and its feasibility needs to be verified and improved through continuous use. Therefore, this method should be used reasonably according to the characteristics of the project and should not be copied completely when planning demonstration districts. Although this method can ensure the rationality of the energy plan to a certain extent, it cannot guarantee the implementation of these plans. Therefore, when planning a demonstration district, it is necessary to clarify the implementation factors of the planned energy schemes, including the scenarios, location, stages, capacities, energy efficiencies, etc., of the applications of energy systems. In addition, it is necessary to implement safeguard measures to ensure the implementation of the plan. The safeguard measures can be policy requirements; financial subsidy support, the establishment of management platforms, publicity and guidance; the development of integrated energy service operation models; innovative energy business models; etc. Apart from the technical measures, the three-step planning method and the integrated energy planning scheme should be carried out with policies and incentives issued by the government and organizations to ensure its implementation. Future research into the implementation of this three-step planning method for other NCEDD should be conducted to demonstrate the rationality, feasibility, and limitations of this conducted research. It is suggested that this NCEDD planning case study be used as a reference in future projects.

Author Contributions: Writing—original draft preparation, X.X., Y.W., Y.Z. and H.J.; writing—review and editing, X.X.; supervision, Y.R. and J.W.; project administration, Y.R. and K.G. All authors have read and agreed to the published version of the manuscript.

Funding: This research was funded by National Natural Science Foundation of China (No. 51978482).

Data Availability Statement: Not applicable.

Conflicts of Interest: The authors declare no conflict of interest.

Nomenclature

NCEDD	near-zero carbon emission demonstration district
GHGs	greenhouse gas emissions
CCS	carbon capture and storage
LNG	liquefied natural gas
CFD	computational fluid dynamics
GDP	gross domestic product
LED	light emitting diode
VOC	volatile organic compounds
EER	energy efficiency ratio
COP	coefficient of performance
IPLV(C)	integrated part load value (cooling)
D	nominal evaporation of boiler, t/h
Q	nominal heating output of boiler, MW
PV	photovoltaics
BIPV	building integrated photovoltaics
EV	electric vehicle
AC	alternating current
EES	electrical energy storage
TES	thermal energy storage
CAES	compressed air energy storage
Tce	ton of standard coal equivalent
kgce	kilogram of standard coal equivalent
CNY	Chinese Yuan

References

1. United Nations. The Paris Agreement. 2015. Available online: https://www.un.org/ga/search/view_doc.asp?symbol=FCCC/CP/2015/L.9/Rev.1 (accessed on 21 January 2021).
2. Zheng, J. Analysis on the Impact of EU Green New Deal and Green Agreement. *Environ. Sustain. Dev.* **2020**, *2*, 40–42.
3. Tang, Y.; Chen, W. The Impact of Biden's Climate and Energy Policy Proposition on China and its Suggestions and Countermeasures. *World Sci.-Tech. R&D* **2021**, *43*, 605–615.
4. Li, H.; Cui, G. Main Points and Enlightenment of Japan's Green Growth Strategy. *China Electr. Equip. Ind.* **2021**, *6*, 52–55.
5. Xi, J. Speech at the General Debate of the 75th United Nations General Assembly. *China Daily*, 24 September 2020. Available online: http://www.chinadaily.com.cn/a/202009/24/WS5f6c08aca31024ad0ba7b776.html (accessed on 21 January 2021).
6. China CCTV. The Central Committee of the Communist Party of China and the State Council issued the "Opinions on the Complete, Accurate and Comprehensive Implementation of the New Development Concept to Achieve Carbon Peak and Carbon Neutrality". *Shanghai Energy Conserv.* **2021**, *10*, 1054.
7. Wang, Z.; Xue, L. The Release of "Carbon Peak Action Plan by 2030". *Resour. Guide* **2021**, *11*, 7.
8. He, C.; Jiang, K.; Chen, S.; Jiang, W.; Liu, J. Zero CO_2 Emissions for an Ultra-Large City by 2050: Case Study for Beijing. *Curr. Opin. Environ. Sustain.* **2019**, *36*, 141–155. [CrossRef]
9. IPCC. *Special Report on Global Warming of 1.5 °C*; Cambridge University Press: Cambridge, UK, 2018.
10. Deng, X.; Xie, J.; Teng, F. What is Carbon Neutrality. *Clim. Chang. Res.* **2021**, *1*, 107–113.
11. Cai, B.; Li, W.; Dhakal, S.; Wang, J. Source Data Supported High Resolution Carbon Emissions Inventory for Urban Areas of the Beijing-Tianjin-Hebei Region: Spatial Patterns, Decomposition and Policy Implications. *J. Environ. Manag.* **2017**, *206*, 786–799. [CrossRef]
12. Li, Y.; Sun, L.; Zhuang, G. Analysis on the Connotation and Construction Path of Near-Zero Carbon Emission Demonstration Zone. *Bus. Econ.* **2017**, *36*, 21–25.
13. Liu, C. China Demonstration Project Construction Path and Supporting Measures in Near-zero Carbon Emission Zone. *China Econ. Trade Guide* **2017**, *36*, 58–60.
14. Long, W. Demand-Side Urban Energy Planning. *HV&AC* **2015**, *45*, 60–66.
15. Long, W.; Liu, K. Some Key Points in Urban Demand Side Energy Planning. *HV&AC* **2017**, *47*, 2–9.
16. Long, W. Demand Side Energy Planning of Green Industry Park. *Shanghai Energy Conserv.* **2016**, *10*, 533–539.
17. Long, W. Demand-Side Energy Planning Complied with Supply Side Structural Reform-Preface of Book "Demand-Side Community Energy Planning and Energy Micro-Net Technologies". *HV&AC* **2016**, *46*, 135–138.
18. Jelić, M.; Batić, M.; Tomašević, N. Demand-Side Flexibility Impact on Prosumer Energy System Planning. *Energies* **2021**, *14*, 7076. [CrossRef]
19. Wang, J.; Zeng, P.; Liu, J.; Li, Y. Coordinated Planning of Multi-Energy Systems Considering Demand Side Response. In *Energy Reports, Proceedings of the 2020 7th International Conference on Power and Energy Systems Engineering (CPESE 2020), Fukuoka, Japan, 26–29 September 2020*; Elsevier: Fukuoka, Japan, 2020.
20. Yu, H.; Huang, Z.; Pan, Y.; Chen, W.; Jing, R.; Li, Y.; Li, Y.; Lin, M.; Liu, H.; Liu, J.; et al. *Guidelines for the Implementation of Urban Demand Side Energy Planning*; China Construction Industry Press: Beijing, China, 2018; p. 375. Available online: http://ss.zhizhen.com/detail_38502727e7500f266958a49170b691e38c0b18b32cfbbeb31921b0a3ea25510134114c969f2eae5cea2a197fb58a956c1ecadaf343b123bdcbb66c56596b1631d3553196fe37f46855337573c23a06c5? (accessed on 28 November 2021).
21. Khalilpour, K.R. The Transition from X% to 100% Renewable Future: Perspective and Prospective, Polygeneration with Polystorage for Chemical and Energy Hubs. *Acad. Press* **2019**, 513–549. Available online: http://ss.zhizhen.com/detail_38502727e7500f265bcd780c2f14aa8e2db5e5bc5fdc4f821921b0a3ea255101c944b624736f9e85f6b33ab1fdc5048fccc22aa85d4f7f2889125aed905f86484f16e318dea09d1ddf7462043f5301cf? (accessed on 28 November 2021).
22. Dobbelsteen, A.; Roggema, R.; Tillie, N.; Broersma, S.; Fremouw, M.; Martin, C.L. *Urban Energy Transition*; Elsevier: Amsterdam, The Netherlands. [CrossRef]
23. Shenzhen Academy of Building Research (Shenzhen, China). Special Energy Plan for Ya'an Green Ecological City. 2013, *unpublished report*. Available online: http://www.upssz.net.cn/newsinfo_803_997.html (accessed on 28 November 2021).
24. Cheng, L.; Zhang, J.; Huang, R.; Wang, C.; Tian, H. Case Analysis of Multi-Scenario Planning Based on Multi-Energy Complementation for Integrated Energy System. *Electr. Power Autom. Equip.* **2017**, *37*, 282–287.
25. Tang, M.; Xu, Z. Taking the Energy Planning of the Central District of Lingui New District in Guilin as an Example Discuss the Main Route of Demand Side Energy Planning. In Proceedings of the 2017 Annual Conference of Shanghai Institute of Refrigeration, Shanghai, China, 18 December 2017; China Academic Journal Electronic Publishing House: Shanghai, China, 2017.
26. Tianjin Architectural Design Institute (Tianjin, China). Energy Planning of Tianjin Health Industry Park University of Traditional Chinese Medicine. 2019, *unpublished work*.
27. Zhejiang Zheneng Energy Service (Zhejiang, China); Tongji Architectural Design Group (Shanghai, China). Integrated Energy Planning for Xiangbao District in Ningbo. 2018, *unpublished work*.
28. Renewable Energy Law of the People's Republic of China. 2006. Available online: http://bfqk.5read.com/pdf/67/%E4%B8%AD%E5%9B%BD%E3%80%8A%E5%8F%AF%E5%86%8D%E7%94%9F%E8%83%BD%E6%BA%90%E6%B3%95%E3%80%8B%E8%BF%91%E6%97%A5%E5%87%BA%E5%8F%B0.pdf (accessed on 28 November 2021).

29. Zhou, Z.; Wang, L.; Lin, A.; Zhang, M.; Niu, Z. Innovative Trend Analysis of Solar Radiation in China during 1962–2015. *Renew. Energy* **2017**, *119*, 675–689. [CrossRef]

30. Zhang, Y.; Ren, J.; Pu, Y.; Wang, P. Solar Energy Potential Assessment: A Framework to Integrate Geographic, Technological, and Economic Indices for a Potential Analysis. *Renew. Energy* **2019**, *149*, 577–586. [CrossRef]

31. Wang, Q.; Kwan, M.-P.; Fan, J.; Zhou, K.; Wang, Y.-F. A Study on the Spatial Distribution of the Renewable Energy Industries in China and Their Driving Factors. *Renew. Energy* **2019**, *139*, 161–175. [CrossRef]

32. Rocky Mountain Institute; National Center for Climate Change Strategy Research and International Cooperation (Beijing, China). Construction Planning and International Cooperation Project Report of Ningbo Meishan Near-Zero Carbon Emission Demonstration Zone. 2019, *unpublished work*.

33. Zhang, C.; Su, B.; Zhou, K.; Yang, S. Decomposition Analysis of China's CO_2 Emissions (2000–2016) and Scenario Analysis of Its Carbon Intensity Targets in 2020 and 2030. *Sci. Total Environ.* **2019**, *668*, 432–442. [CrossRef]

34. China National Standardization Committee. *Technical Standard for Nearly Zero Energy Buildings GB/T 51350-2019*; China Construction Industry Press: Beijing, China, 2019.

35. Development and Reform Commission of Zhejiang Province. The 13th Five-Year Plan for Low-Carbon Development in Zhejiang Province. 2016. Available online: http://www.zj.gov.cn/art/2016/5/20/art_1229540818_4666989.html (accessed on 28 November 2021).

36. The General Office of the People's Government of Zhejiang Province. Action Plan for Building a National Clean Energy Demonstration Province in Zhejiang Province (2018–2020). Available online: http://www.zj.gov.cn/art/2018/9/25/art_12290193 65_61718.html (accessed on 28 November 2021).

37. He, Z. The Start of Construction of Hebei Beidaihe Ubiquitous Power Internet of Things in Comprehensive Demonstration Park Project. *National Grid News*, 14 November 2019.

38. Kang, W.; Wu, Y. Hebei Zhengding Builds County-Level Energy Internet Comprehensive Demonstration Zone. *National Grid News*, 8 August 2019.

39. Wang, D. Preliminary Study on Henan Electric Power Lankao Energy Internet Comprehensive Demonstration Project. *China Electric Power News*, 11 November 2019.

40. Revesz, A.; Jones, P.; Dunham, C.; Davies, G.; Marques, C.; Matabuena, R.; Scott, J.; Maidment, G. Developing Novel 5th Generation District Energy Networks. *Energy* **2020**, *201*, 117–389. [CrossRef]

energies

Article

Development of a Hydraulic System for the Automatic Expansion of Powered Roof Support

Dawid Szurgacz [1,2], Beata Borska [3], Ryszard Diederichs [1] and Sergey Zhironkin [4,5,6,*]

1　Center of Hydraulics DOH Ltd., ul. Konstytucji 147, 41-906 Bytom, Poland; dawidszurgacz@vp.pl (D.S.); rdider@poczta.onet.pl (R.D.)
2　Polska Grupa Górnicza S.A., ul. Powstańców 30, 40-039 Katowice, Poland
3　KWK Ruda Ruch Halemba, ul. Halembska 160, 41-717 Ruda Śląska, Poland; borskab@gmail.com
4　Department of Trade and Marketing, Siberian Federal University, 79 Svobodny Ave., 660041 Krasnoyarsk, Russia
5　Department of Open Pit Mining, T.F. Gorbachev Kuzbass State Technical University, 28 Vesennya Str., 650000 Kemerovo, Russia
6　School of Core Engineering Education, National Research Tomsk Polytechnic University, 30 Lenina Str., 634050 Tomsk, Russia
*　Correspondence: zhironkinsa@kuzstu.ru

Abstract: A mechanical support in a longwall complex requires proper supervision and control of operating parameters because it is responsible for the safety of machines and workers. Its main functions are to protect the excavation against the adverse effects of rock mass and to move it along with the operational progress of the shearer and the scraper conveyor. In the era of Industry 4.0 as well as green economy and sustainable development, the development of machines and devices included in the longwall complex is an important area of invention. When it comes to the mechanical support included in the longwall complex, the authors propose changes to the hydraulic control system. The main purpose of the work was to develop a system that can automatically expand the legs of the powered support. Tests were carried out under real conditions while equipping the support section with a prototypical hydraulic system and with a system monitoring the correct operation of the prototype. This article presents the development of a prototypical installation under real conditions for testing on a longwall. The innovative solution proposed by the authors consisted of introducing a prototypical double valve block into the hydraulic system of the housing, which enables the function of automatic section expansion. The authors proposed a solution aimed at shortening the work time of the powered-support-section operator. The research results made it possible to evaluate the usefulness of the proposed solution. This type of solution can be considered technical support in the process of coal mining in an underground mine. Analysis presented in this paper, along with results from research conducted and its conclusions, can be of practical help to enclosure users to improve reliability and achieve optimal performance.

Keywords: powered roof support; hydraulic leg; hydraulic system; work automation; tests under real conditions

check for
updates

Citation: Szurgacz, D.; Borska, B.; Diederichs, R.; Zhironkin, S. Development of a Hydraulic System for the Automatic Expansion of Powered Roof Support. *Energies* **2022**, *15*, 680. https://doi.org/10.3390/en15030680

Academic Editor: Helena M. Ramos

Received: 17 December 2021
Accepted: 13 January 2022
Published: 18 January 2022

Publisher's Note: MDPI stays neutral with regard to jurisdictional claims in published maps and institutional affiliations.

1. Introduction

Continuous technological progress and economic development in the era of the fourth industrial revolution, referred to as Industry 4.0, require companies to constantly change and implement innovative technical solutions [1–3]. This issue also applies to the mining industry of solid, liquid and gaseous minerals, which is expected to intensify production while minimizing negative impacts on the environment [4–6]. This determines the conduct of research and the search for new solutions—both in terms of improving the reliability of machines [7–10] and optimizing production processes [11–17] as well as improving work safety [18–22] and caring for the environment [23].

Sustainable development is also observed in hard-coal mining, which deals with the intensification of the mining process and an increase in efficiency. This situation is accompanied by the deterioration of mining and geological conditions, an increase in natural hazards and a generation of new operational problems, among others related to the selection of progressively lower layers of coal beds (over 1000 m). It all adds up to ever-higher requirements for the level of reliability of machines, in particular longwall systems, while needing to maintain the required level of safety [24–27].

In order to meet the previously mentioned requirements and constantly improve the production efficiency and safety of the crew, it is necessary to search for new solutions in the field of coal-bed selection [28–33], improve working conditions and minimize hazards [34,35] as well as conduct research aimed at improving the reliability of machines [36,37]. A key element influencing the increase in efficiency of the mining process and the increase in safety is automation of the longwall complex [38–43].

The powered support (Figure 1) is one of the key elements of longwall systems, next to the extracting machine (shearer or plow) and the scraper conveyor. Its main task is to control the roof and protect the crew working along the wall and other machines against the fall of the roof rocks and against the chunks of coal falling off the face of the longwall [44–47]. The support moves behind the front of the wall and, at the same time, is responsible for moving the gutters of the wall scraper conveyor. In order for the support to perform its functions properly, it is important to select it correctly, taking into account the mining and geological conditions and parameters of the wall. Recent research in the development of supports is presented in the works [48–57].

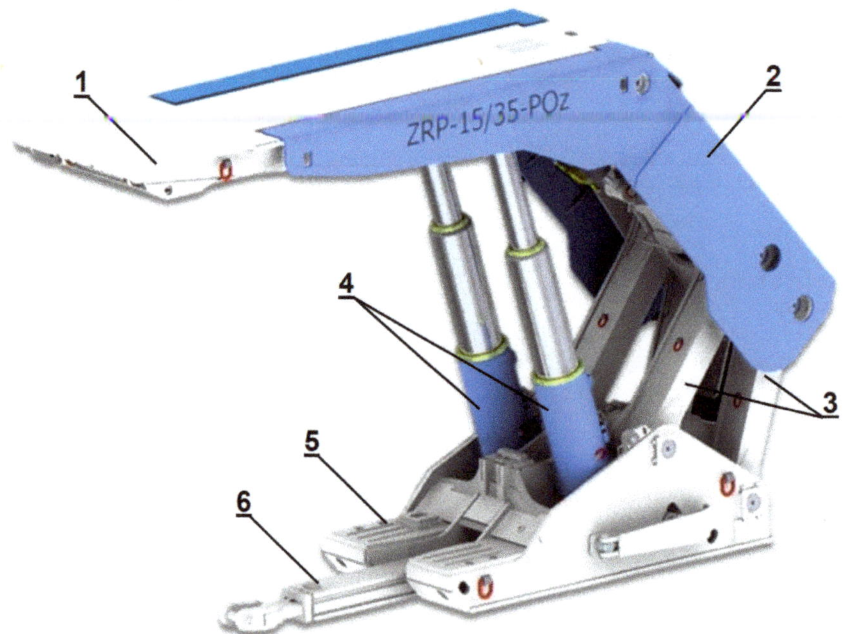

Figure 1. Section of the powered roof support: 1, canopy; 2, shield support; 3, lemniscate tie rods; 4, leg; 5, floor base; 6, shifting system.

Currently, leg and protection supports with a lemniscate canopy guiding mechanism are commonly used. In this type of support, the canopy is pivotally connected to the sill piece by means of a shield system as well as by means of front and rear connectors, called lemniscates. This type of structure makes it possible for the end of the canopy to move along the lemniscate curve, almost perpendicularly to the floor, thanks to which the roof is protected up to the sidewall [41].

The main element of the hydraulic system of powered supports are hydraulic legs, which are expanded between the canopy and the sill piece. Their parameters largely determine the quality of the support of the section. An additional actuator connected with the canopy and the shield system stabilizes the section and allows adjustment of the angle of the canopy inclination in relation to the sill piece. Other elements of power hydraulics are the shifter, actuators of the tilting and extending canopy, actuators of the face of the wall cover and transition area, the sill piece lifting actuator and corrective actuators [41].

Each actuator is controlled by a hydraulic distributor. The distributors are connected with hydraulic locks, consisting of controlled check valves and pressure limiting valves, which also have a safety function against overloads resulting from impact of rock mass. Each section of the powered support can be cut off with a shut-off valve, and a check valve installed in front of the drain line protects the section against pressure [41].

The support operation cycle can be divided into the following phases: drawing off, moving the sections, expanding the support and the apparatus for repositioning the conveyor. All functions are performed by means of hydraulic actuator in conjunction with the control hydraulics (Figure 2). Individual phases of the work cycle may be implemented in an adjacent, pilot or electro-hydraulic control system.

(a)

(b)

Figure 2. *Cont.*

(c)

Figure 2. Work cycle for the powered roof-support section: (**a**) withdrawing the powered roof-support section, (**b**) moving the powered roof-support section to a new location, (**c**) spragging the powered roof-support section.

This paper focuses on the development of the automatic expansion of the powered support system. An innovative solution was proposed by introducing a prototypical double valve block into the hydraulic system, which enables the function of automatic expansion of the housing. In this regard, the prototypical system was tested and analyzed under real conditions. The purpose of this study was to determine the usefulness of a prototypical system with the automatic-expansion function. The proposed changes to the hydraulic system of the mechanical casing are described in Section 2. Results from the tests, together with analysis of powered roof-support operation before application of the new solution and after introducing changes in the hydraulic system, are presented in Section 3. Section 4 discusses effects from the new solution.

2. Materials and Methods

For proper operation of the support and ensured good conditions for maintaining the roof of the excavation, selection of the appropriate support, which entails choosing the force with which the support acts on the roof, is an important factor. Depending on the phase of the support operation cycle, there are three types of load-carrying capacities: initial, nominal and working [52].

Initial load capacity is obtained when the support is expanded, and it depends on the supply pressure in the main supply line in the wall. After the support takes over the pressure of the roof rocks, the section's load capacity changes to the working-load capacity. In contrast, the nominal load capacity is the force at which the leg reveals its compliance, which depends on the opening pressure of the safety valves in the leg valve block [52].

2.1. Carrying-Load-Capacity Calculation Model

Taking into account the above definitions, the carrying-load capacity of a support leg can be described with the following relations [52]:

- Initial load capacity

$$F_w = \frac{\pi d^2}{4} \cdot p_{zas}, \ N \tag{1}$$

where
d—leg working diameter (mm);
p_{zas}—supply pressure (pressure in the supply line) (MPa);
N—the SI unit of force (newton).

- Working load

$$F_r = \frac{\pi d^2}{4} \cdot p_{rob}, \; N \tag{2}$$

where

p_{rob}—leg working diameter (MPa);

- Nominal load capacity

$$F_n = \frac{\pi d^2}{4} \cdot p_{nom}, \; N \tag{3}$$

where

p_{nom}—nominal pressure (safety valve opening pressure) (MPa).

It should be noted in the case of double-telescopic legs, during the transition of the work of the leg from the 1st to the 2nd degree, there will be a decrease in its load-carrying capacity, resulting from the smaller diameter of the 2nd-stage cylinder. The value of the working-load capacity is between the initial load capacity and the nominal load capacity. The nominal load capacity is the maximum capacity the support can achieve.

$$F_w \leq F_r \leq F_n \tag{4}$$

Selection of the appropriate load capacity is of great importance for controlling the roof properly. If the initial load capacity is too high, the roof rocks may be destroyed. On the other hand, if the nominal load capacity is too high, the effect of the coal-bed stress relief caused by the pressure of the roof may disappear. Coal-bed stress relief has a direct impact on the mining process, as it causes the initial crushing of the unmined coal near the face of the wall, which means the miner uses less energy and increases the proportion of coarse-grained classes.

2.2. Layout Concept for Automatic Expansion of Legs

A special hydraulic system was tested (Figure 3), which was equipped with a double valve block with an automatic-expansion function. This system consisted of a threshold valve (2c) with a check valve (2d) and was connected to the supply line (15) through a second check valve (6) and a shut-off valve (10). The threshold valve (2c), which was located in the block, had an opening pressure setting of 9 MPa. This meant the automatic pressure function did not work below this number. When the pressure of 9 MPa was exceeded during leg expansion, the automatic expansion system was turned on. It ensured the expansion of the leg to the required initial equal-load capacity to the maximum value of the pressure in the supply line despite the fact that the expansion function was interrupted by the operator.

The use of a double valve block (2) in the hydraulic system for the automatic expansion of the support legs is a necessary condition because it prevents a pressure drop in the space under the piston of a leg in the case of internal leakage. Its purpose is to ensure initial- and working-load capacities are attained. Additionally, this system was equipped with an excess flow valve (5) in the piston space. This valve is designed to protect against damage to the hydraulic lines connecting the valve bank with the space over the piston of the leg. It is effective protection in the case of internal leakage and increase in pressure in the space over the piston to a value exceeding the strength of the connecting pipes as well as when the expansion-control function is performed in the section of a single leg. The automatic expansion system of an additional check valve (6), which was connected in series with this system, was implemented. It aimed to eliminate the possibility of liquid flowing back to the main supply line in the event of contamination of the check valve (2d).

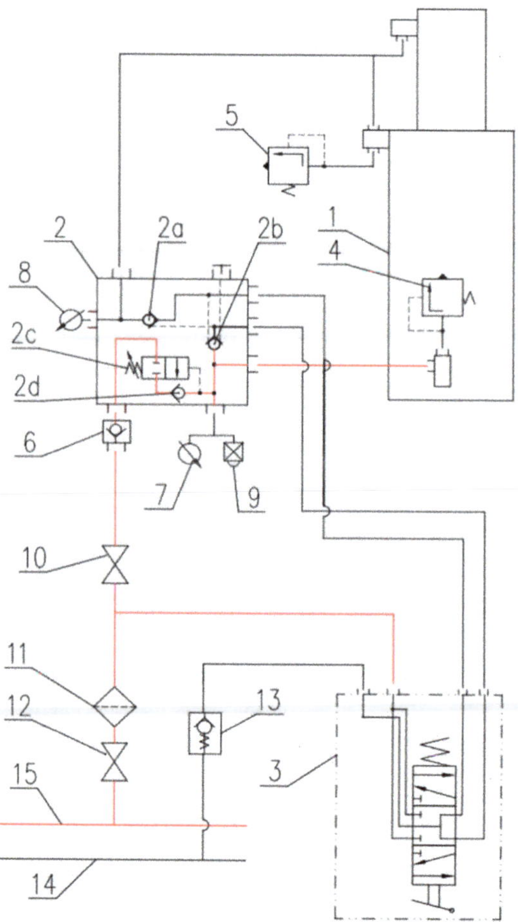

Figure 3. Conception of the hydraulic system for the automatic expansion of powered roof support: 1, leg; 2, valve block with double function of expanding the leg; 3, four-way distributor; 4, and; 5, safety valve; 7, and; 8, manometer; 9, pressure indicator; 10, and; 12, shut-off valve; 11, filter; 13, check valve; 14, runoff line; 15, power bus.

In the new solution proposed by the authors, the system automatically equalizes the pressure in the plunger space of the legs to the required value. In the event of pressure fluctuations in the supply line, the new system will automatically react when the pressure in the line increases and will then compensate the pressure in the piston cavity of the leg to achieve the required initial and operating supports. In addition, the new system automatically reacts to any pressure drops that may occur between the working cycles of a mechanically powered roof support, e.g., due to internal leaks. If a drop in pressure is detected in the subpiston space of the leg, the system automatically repressurizes to ensure stable operation of the section. The pressure in the legs is equalized automatically, without operator intervention—unlike the classic solution. In a classical system, the section operator would have to continuously monitor the pressure in the racks and the supply line to provide the required support. If the section operator does not notice the pressure drop, the section will be operating with too little support.

3. Results

The tests were carried out in a longwall, where an automatic-expansion system was installed in the powered support section. In order to obtain the test results, the proposed solution was equipped with a wireless pressure-monitoring system. Pressure sensors in the piston space were installed in the stands of the powered support. Pressure was measured in a continuous system with a sampling rate of 1 s. Thanks to this, it was possible to constantly monitor the operation of the powered support equipped with the hydraulic system for automatic leg expansion. Tests and analysis were conducted on the basis of the pressure monitoring records in the powered support legs for selected sections of the powered support. Analysis was carried out for two specific stages of longwall exploitation:

- Before the assembly of the prototypical system of automatic leg expansion;
- After installing the automatic leg expansion.

3.1. Analysis of the Expansion of a Powered Support Section in a Longwall

The presented diagram (Figure 4) shows a clear drop in pressure in the legs after the completion of the expansion operation of the powered support. The presented pressure drop in the initial period of the expansion of the powered support results from the local conditions of interoperability between the support and the rock mass as well as from the influence of the adjacent sections in the longwall and the characteristics of the safety valves used in the hydraulic system. This pressure drop can change from 2 to 5 MPa in a matter of minutes. The analyzed charts show a large pressure difference between the expanded sections, amounting to over 10 MPa. A similar measurement is visible in the previous and next expansion cycles. The pressure difference is the result of interrupting the expansion of the support before obtaining the required initial load-bearing capacity. In the diagram (Figure 4), the pressure difference between the legs of sections 43, 44 and 45 is 20 MPa.

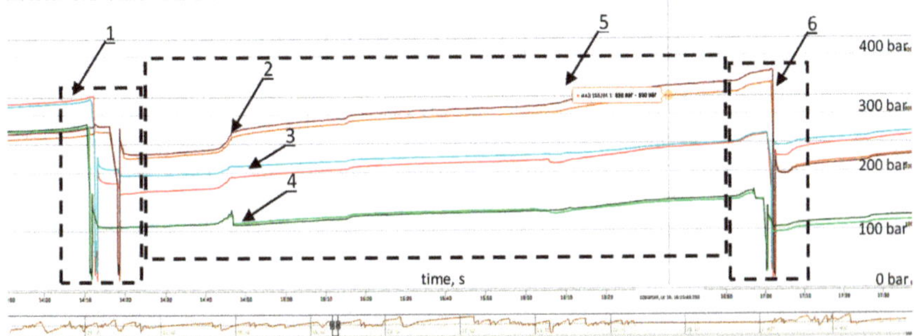

Figure 4. Measurement of the pressure difference in the legs of the powered support: 1, work performed by the support (leg withdrawal, readjustment, expanding; 2, powered support no. 43; 3, powered support no. 44; 4, powered support no. 45; 5, powered support operation at working-load capacity; 6, support operation.

3.2. Analysis of the Measurements for the Expansion of the Powered Support before the Installation of the Automatic System

The presented Figure 5 includes the measurement of the powered support operation prior to the installation of the automatic leg-expansion system for supports no. 43, 44 and 45. The use of a wireless pressure-monitoring system made it possible to obtain the measurements. The measurement shows the minimum value of the initial load capacity as a result of which the pressure difference between supports no. 43, 44 and 45 is slight, up to 4 MPa.

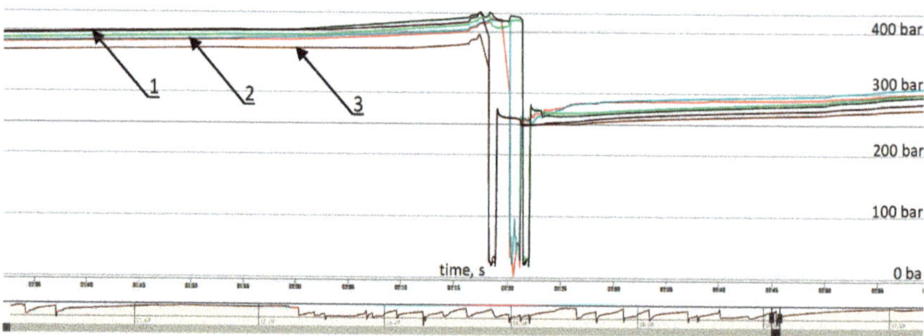

Figure 5. Measurement of the pressure difference in the legs of the powered support during its operation: 1, powered support in longwall no. 43; 2, powered support in longwall no. 44; 3, powered support in longwall no. 45.

Eliminating the pressure difference (5 MPa) is quite difficult because it results from the nature of the pressure change in the main supply line. The pressure change is illustrated in detail in Figure 6. This diagram clearly shows that the value of the final pressure in the legs during expansion depends on the instantaneous pressure value in the supply line.

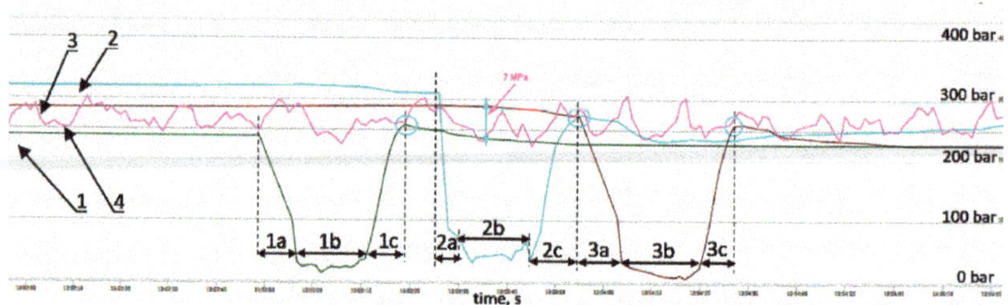

Figure 6. The course of the registered pressure drop and working time for the support legs: 1, powered support no. 45; 1a, leg withdrawal time 8 s; 1b, adjustment time 14.5 s; 1c, expansion time 8 s; 2, powered support no. 44; 2a, prop withdrawal time 7 s; 2b, adjustment time 12.5 s; 2c, expansion time 10.5 s; 3, powered support no. 43; 3a, leg withdrawal time 9.5 s; 3b, adjustment time 16 s; 3c, expansion time 7.5 s; 4, main supply line.

Figure 7 shows the expansion of the support at a time when the pressure in the main power line is below the value to which the mechanical support should be expanded. Using the pressure-monitoring light signal, the support operator receives information whether the pressure in the main line will increase, allowing the expansion of the support to the required value of initial load capacity. However, this is done at the expense of extending the support's expansion time, which may affect the effectiveness of the longwall exploitation.

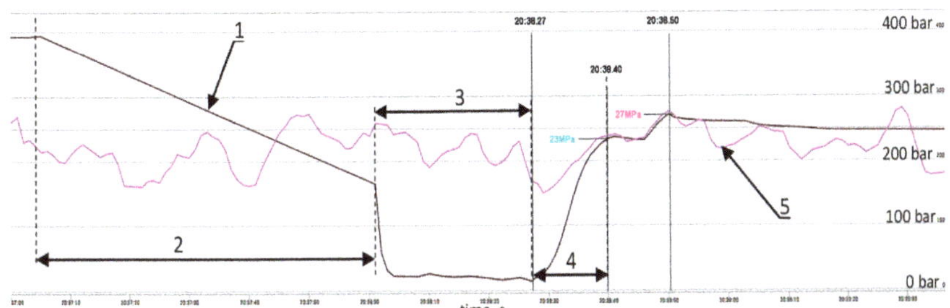

Figure 7. The course of the pressure value in the legs: 1, powered support no. 43; 2, leg withdrawal time 56 s; 3, adjustment time 26 s; 4, expansion time 23 s; 5, supply pressure.

3.3. Analysis of the Measurements for the Expansion of the Powered Support before the Installation of the Automatic System

Figure 8 shows the pressure course in the legs for powered support no. 43, 44 and 45 in the longwall after assembling the prototypical system in order to verify its correct operation under real conditions. Analyzing the measurement, it can be concluded that powered support no. 44 reaches the maximum value of the main-line pressure above 25 MPa shortly after expansion. The pressure created in the legs of support no. 43 and 45, after their proper expansion, decreases below 25 MPa. A pressure difference of 5 MPa is maintained between support no. 44, 43 and 45 during the work cycle performed by the powered support (i.e., leg withdrawal, adjustment and expanding). Figures 9 and 10 show in detail that the analyzed sections of the powered support were expanded to nearly the same value of initial pressure. The expansion of powered support section no. 44 to the maximum value of the main supply-line pressure results mainly from the correct operation of the automatic expansion system.

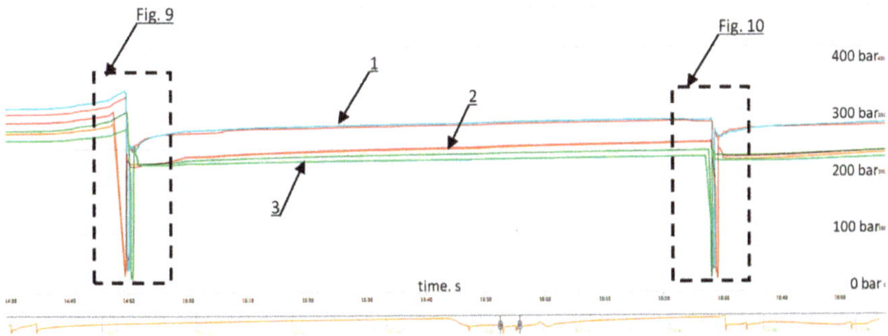

Figure 8. The course of the pressure value after the installation of the automatic expansion system for the selected powered support: 1, powered support in longwall no. 43; 2, powered support in longwall no. 44; 3, powered support in longwall no. 45.

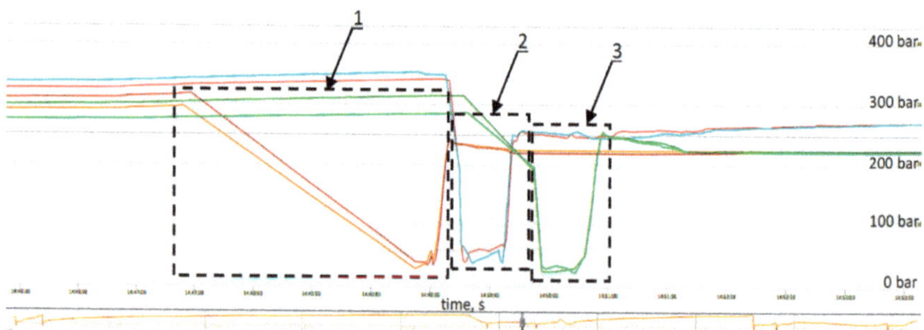

Figure 9. The pressure value progress after installing the automatic expansion system for the selected powered support: 1, working time of powered support no. 43 is 131 s; 2, working time of powered support no. 44 is 32 s; 3, working time of powered support no. 45 is 34 s.

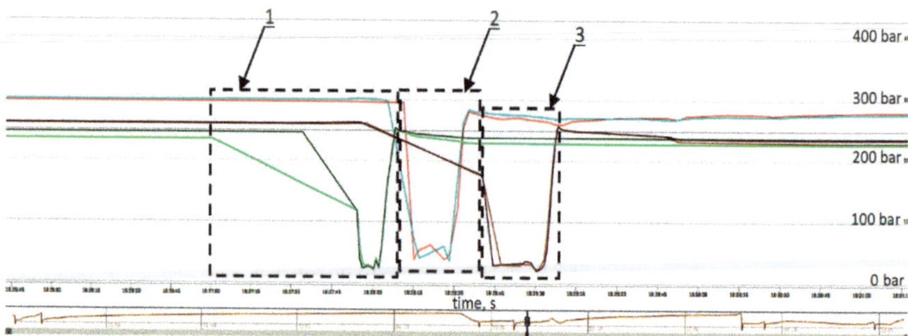

Figure 10. The pressure value progress after installing the automatic expansion system for the selected powered support: 1, working time of powered support no. 43 is 67 s; 2, working time of powered support no. 44 is 27 s; 3, working time of powered support no. 45 is 28 s.

4. Discussion

The research and analysis performed for the operation of a powered longwall support with the use of the automatic leg-expanding system showed that the powered support can be expanded in a longwall excavation with large differences in initial load-bearing capacity. The differences mostly result from pressure fluctuations in the main supply line and the pressure dropping shortly after expanding the support. The pressure drop in the hydraulic system after expanding the sections in the longwall excavation causes uneven support of the roof along the length of the wall. This affects the improper cooperation of the powered support with the rock mass.

The expansion of a powered longwall support with minimal initial load-bearing capacity significantly influences the expansion time of the powered support. As observed in Figure 7, it is approximately 30 s. There are differences in the support expansion between powered supports no. 43, 44 and 45, resulting from pressure fluctuations in the main power-supply line and the typical pressure drop in the support legs occurring shortly after expansion of the powered support in the longwall.

Analysis of the prototypical system of automatic expansion of the support legs enables the expansion of the support legs to the maximum initial load capacity despite temporary pressure drops in the power-supply line, which is displayed in Figures 5–7. The prototypical system also eliminates the pressure drop in the support legs shortly after expanding the support (Figures 5 and 7). The use of a prototypical hydraulic system for automatic expansion of the support causes small differences in the initial load capacity between supports no. 43, 44 and 45. It ensures stable operation of the powered support section in the longwall.

The prototypical automatic expansion system can significantly reduce the expansion time of the powered longwall support by reducing the level of the minimum expansion pressure. It is displayed in Figure 7, where the pressure threshold that should be exceeded during expansion is kept at the level of 25 MPa.

By equipping the entire longwall system with a powered support with an automatic expansion system, the threshold of the minimum expansion pressure of the support could be set to 20 MPa. Then, further expansion of the powered support to the maximum initial load capacity above 25 MPa would be performed by the automatic expansion system. This is shown in Figure 4, where the support's expansion time was reduced to about 20 s.

Table 1 summarizes the cycle times of operation for 11 selected sections of a powered support equipped with an automatic leg-expansion system and a support without the system.

Table 1. Working-cycle times of sections (1*) for a powered support equipped with an automatic leg-expansion system and a support without the system—divided into the leg withdrawal (2*), readjustment (3*) and section-expansion phase (4*).

Section Number	For a System with Automatic Leg Expansion				For a System without Automatic Leg Expansion			
	2* (s)	3* (s)	4* (s)	1* (s)	2* (s)	3* (s)	4* (s)	1* (s)
1	19	12	6	37	14	14	19	47
15	11	12	8	31	15	13	19	47
30	10	17	5	32	11	18	23	52
45	9	16	5	30	14	12	15	41
60	12	21	5	38	14	11	18	43
75	16	17	8	41	10	13	21	44
90	11	21	8	40	10	10	24	44
105	13	13	8	34	10	17	24	51
120	14	20	8	42	12	19	25	56
135	16	24	8	48	16	13	20	49
150	21	19	8	48	20	20	21	61

The use of the automatic expansion system for legs shortens the section's working time in its last phase, i.e., during expansion. Figure 11 compares the section's operation time for a powered support equipped with an automatic leg expansion system and for a support without the system. In this paper, the authors focused on real-world testing to validate the performance of a prototypical dual valve block. The next stage will be a statistical analysis of the obtained test results in order to optimize the section shifting time and thus improve the efficiency of powered roof-support operation [58,59].

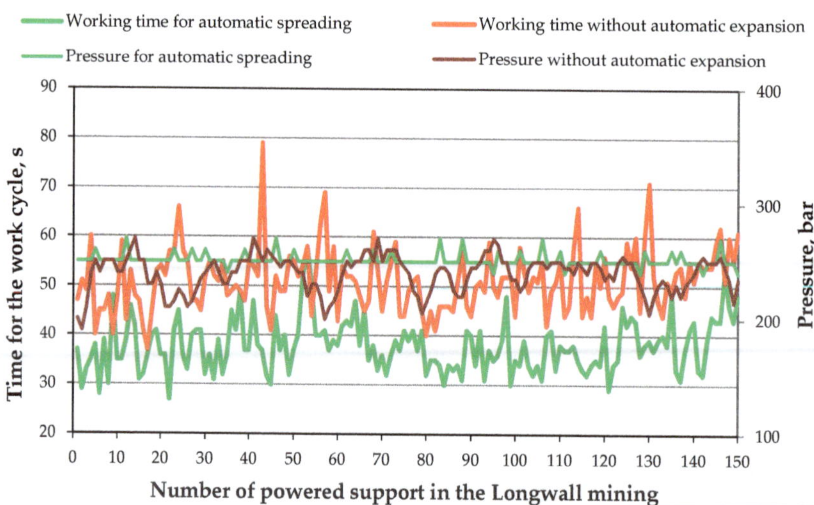

Figure 11. Compares the section operation time for a powered support equipped with an automatic leg-expansion system and for a support without the system.

5. Conclusions

This paper compares the operation of the powered support section before and after the introduction of the automatic leg-expansion system. Research was carried out on one longwall, analyzing the work of the same sections, and thus the same parameters of the longwall and similar mining and geological conditions were maintained.

As this research showed, the introduction of a prototypical automatic leg-expansion system eliminated the problem of pressure drops in the legs immediately after expanding and the problem of obtaining the required initial load-bearing capacity, which was noticed in the case of a traditional control system.

The new solution for the control system, proposed by the authors, made it possible to obtain stable operation of the section, which is important for the correct control of the roof and for ensuring safety. This system is resistant to pressure fluctuations in the main supply line and to internal leaks, thanks to which the successive sections obtained similar load-bearing-capacity values. At the same time, the proposed solution reduces the leg-expansion time, and ultimately, in relation to the operation of the entire mechanical complex, it may shorten the time of the mining cycle.

Results of these tests and practical conclusions can be helpful for users and housing manufacturers in order to improve the reliability of a mechanical support in a longwall complex, which can improve the safety and efficiency of the entire mechanical complex. The next stage could be an integrated study of the entire longwall complex. Future research will examine how a change in the hydraulic system of the casing affects the mining-cycle time. Moreover, in the future, based on the experience gained and the research results obtained, tests of the proposed hydraulic system can be carried out for various types of powered support.

Author Contributions: Conceptualization, D.S. and B.B.; methodology, D.S. and B.B.; software, B.B.; validation, D.S., B.B. and S.Z.; formal analysis, R.D.; investigation, D.S.; resources, R.D.; data curation, S.Z.; writing—original draft preparation, D.S, B.B. and R.D.; writing—review and editing, D.S. and B.B.; visualization, R.D.; supervision, D.S.; project administration, D.S, B.B. and S.Z.; funding acquisition, D.S. and S.Z. All authors have read and agreed to the published version of the manuscript.

Funding: This research received no external funding.

Institutional Review Board Statement: The study was conducted according to the guidelines of the Declaration.

Informed Consent Statement: Not applicable.

Data Availability Statement: Not applicable.

Conflicts of Interest: The authors declare no conflict of interest.

References

1. Król, R.; Kisielewski, W. Research of loading carrying idlers used in belt conveyor-practical applications. *Diagnostyka* **2014**, *15*, 67–74.
2. Grzesiek, A.; Zimroz, R.; Śliwiński, P.; Gomolla, N.; Wyłomańska, A. A Method for Structure Breaking Point Detection in Engine Oil Pressure Data. *Energies* **2021**, *14*, 5496. [CrossRef]
3. Bazaluk, O.; Velychkovych, A.; Ropyak, L.; Pashechko, M.; Pryhorovska, T.; Lozynskyi, V. Influence of Heavy Weight Drill Pipe Material and Drill Bit Manufacturing Errors on Stress State of Steel Blades. *Energies* **2021**, *14*, 4198. [CrossRef]
4. Szurgacz, D.; Zhironkin, S.; Vöth, S.; Pokorný, J.; Spearing, A.J.S.; Cehlár, M.; Stempniak, M.; Sobik, L. Thermal Imaging Study to Determine the Operational Condition of a Conveyor Belt Drive System Structure. *Energies* **2021**, *14*, 3258. [CrossRef]
5. Bortnowski, P.; Gładysiewicz, L.; Król, R.; Ozdoba, M. Energy Efficiency Analysis of Copper Ore Ball Mill Drive Systems. *Energies* **2021**, *14*, 1786. [CrossRef]
6. Bajda, M.; Hardygóra, M. Analysis of Reasons for Reduced Strength of Multiply Conveyor Belt Splices. *Energies* **2021**, *14*, 1512. [CrossRef]
7. Gładysiewicz, L.; Król, R.; Kisielewski, W.; Kaszuba, D. Experimental determination of belt conveyors artificial friction coefficient. *Acta Montan. Slovaca* **2017**, *22*, 206–214.
8. Kawalec, W.; Suchorab, N.; Konieczna-Fuławka, M.; Król, R. Specific energy consumption of a belt conveyor system in a continuous surface mine. *Energies* **2020**, *13*, 5214. [CrossRef]
9. Wajs, J.; Trybała, P.; Górniak-Zimroz, J.; Krupa-Kurzynowska, J.; Kasza, D. Modern Solution for Fast and Accurate Inventorization of Open-Pit Mines by the Active Remote Sensing Technique—Case Study of Mikoszów Granite Mine (Lower Silesia, SW Poland). *Energies* **2021**, *14*, 6853. [CrossRef]
10. Patyk, M.; Bodziony, P.; Krysa, Z. A Multiple Criteria Decision Making Method to Weight the Sustainability Criteria of Equipment Selection for Surface Mining. *Energies* **2021**, *14*, 3066. [CrossRef]
11. Wodecki, J.; Góralczyk, M.; Krot, P.; Ziętek, B.; Szrek, J.; Worsa-Kozak, M.; Zimroz, R.; Śliwiński, P.; Czajkowski, A. Process Monitoring in Heavy Duty Drilling Rigs—Data Acquisition System and Cycle Identification Algorithms. *Energies* **2020**, *13*, 6748. [CrossRef]
12. Borkowski, P.J. Comminution of Copper Ores with the Use of a High-Pressure Water Jet. *Energies* **2020**, *13*, 6274. [CrossRef]
13. Bazaluk, O.; Slabyi, O.; Vekeryk, V.; Velychkovych, A.; Ropyak, L.; Lozynskyi, V. A Technology of Hydrocarbon Fluid Production Intensification by Productive Stratum Drainage Zone Reaming. *Energies* **2021**, *14*, 3514. [CrossRef]
14. Góralczyk, M.; Krot, P.; Zimroz, R.; Ogonowski, S. Increasing Energy Efficiency and Productivity of the Comminution Process in Tumbling Mills by Indirect Measurements of Internal Dynamics—An Overview. *Energies* **2020**, *13*, 6735. [CrossRef]
15. Doroszuk, B.; Król, R. Conveyor belt wear caused by material acceleration in transfer stations. *Min. Sci.* **2019**, *26*, 189–201. [CrossRef]
16. Kawalec, W.; Błażej, R.; Konieczna, M.; Król, R. Laboratory Tests on e-pellets effectiveness for ore tracking. *Min. Sci.* **2018**, *25*, 7–18. [CrossRef]
17. Baiul, K.; Khudyakov, A.; Vashchenko, S.; Krot, P.V.; Solodka, N. The experimental study of compaction parameters and elastic after-effect of fine fraction raw materials. *Min. Sci.* **2020**, *27*, 7–18. [CrossRef]
18. Adach-Pawelus, K.; Pawelus, D. Influence of Driving Direction on the Stability of a Group of Headings Located in a Field of High Horizontal Stresses in the Polish Underground Copper Mines. *Energies* **2021**, *14*, 5955. [CrossRef]
19. Janus, J.; Krawczyk, J. Measurement and Simulation of Flow in a Section of a Mine Gallery. *Energies* **2021**, *14*, 4894. [CrossRef]
20. Zimroz, P.; Trybała, P.; Wróblewski, A.; Góralczyk, M.; Szrek, J.; Wójcik, A.; Zimroz, R. Application of UAV in Search and Rescue Actions in Underground Mine—A Specific Sound Detection in Noisy Acoustic Signal. *Energies* **2021**, *14*, 3725. [CrossRef]
21. Ziętek, B.; Banasiewicz, A.; Zimroz, R.; Szrek, J.; Gola, S. A Portable Environmental Data-Monitoring System for Air Hazard Evaluation in Deep Underground Mines. *Energies* **2020**, *13*, 6331. [CrossRef]
22. Uth, F.; Polnik, B.; Kurpiel, W.; Baltes, R.; Kriegsch, P.; Clause, E. An innovate person detection system based on thermal imaging cameras dedicate for underground belt conveyors. *Min. Sci.* **2019**, *26*, 263–276. [CrossRef]
23. Huang, P.; Spearing, S.; Ju, F.; Jessu, K.V.; Wang, Z.; Ning, P. Control Effects of Five Common Solid Waste Backfilling Materials on in Situ Strata of Gob. *Energies* **2019**, *12*, 154. [CrossRef]
24. Krauze, K.; Mucha, K.; Wydro, T.; Pieczora, E. Functional and Operational Requirements to Be Fulfilled by Conical Picks Regarding Their Wear Rate and Investment Costs. *Energies* **2021**, *14*, 3696. [CrossRef]

25. Kotwica, K.; Stopka, G.; Kalita, M.; Bałaga, D.; Siegmund, M. Impact of Geometry of Toothed Segments of the Innovative KOMTRACK Longwall Shearer Haulage System on Load and Slip during the Travel of a Track Wheel. *Energies* **2021**, *14*, 2720. [CrossRef]

26. Prostański, D. Empirical Models of Zones Protecting Against Coal Dust Explosion. *Arch. Min. Sci.* **2017**, *62*, 611–619. [CrossRef]

27. Ji, Y.; Zhang, Y.; Huang, Z.; Shao, Z.; Gao, Y. Theoretical analysis of support stability in large dip angle coal seam mined with fully-mechanized top coal caving. *Min. Sci.* **2020**, *27*, 73–87. [CrossRef]

28. Qiao, S.; Zhang, Z.; Zhu, Z.; Zhang, K. Influence of cutting angle on mechanical properties of rock cutting by conical pick based on finite element analysis. *J. Min. Sci.* **2021**, *28*, 161–173. [CrossRef]

29. Wang, J.; Wang, Z. Systematic principles of surrounding rock control in longwall mining within thick coal seams. *Int. J. Min. Sci. Tech.* **2019**, *29*, 591–598. [CrossRef]

30. Kumar, R.; Singh, A.K.; Mishra, A.K.; Singh, R. Underground mining of thick coal seams. *Int. J. Min. Sci. Tech.* **2015**, *25*, 885–896. [CrossRef]

31. Mo, S.; Tutuk, K.; Saydam, S. Management of floor heave at Bulga Underground Operations—A case study. *Int. J. Min. Sci. Tech.* **2019**, *29*, 73–78. [CrossRef]

32. Hu, S.; Ma, L.; Guo, J.; Yang, P. Support-surrounding rock relationship and top-coal movement laws in large dip angle fully-mechanized caving face. *Int. J. Min. Sci. Technol.* **2018**, *28*, 533–539.

33. Jixiong, Z.; Spearing, A.J.S.; Xiexing, M.; Shuai, G.; Qiang, S. Green coal mining technique integrating mining-dressing-gas draining-backfilling-mining. *Int. J. Min. Sci. Technol.* **2017**, *27*, 17–27.

34. Juganda, A.; Strebinger, C.; Brune, J.F.; Bogin, G.E. Discrete modeling of a longwall coal mine gob for CFD simulation. *Int. J. Min. Sci. Technol.* **2020**, *30*, 463–469. [CrossRef]

35. Ji, Y.; Ren, T.; Wynne, P.; Wan, Z.; Zhaoyang, M.; Wang, Z. A comparative study of dust control practices in Chinese and Australian longwall coal mines. *Int. J. Min. Sci. Technol.* **2016**, *25*, 687–706. [CrossRef]

36. Woźniak, D.; Hardygóra, M. Method for laboratory testing rubber penetration of steel cords in conveyor belts. *Min. Sci.* **2020**, *27*, 105–117. [CrossRef]

37. Bajda, M.; Błażej, R.; Hardygóra, M. Optimizing splice geometry in multiply conveyor belts with respect to stress in adhesive bonds. *Min. Sci.* **2018**, *25*, 195–206. [CrossRef]

38. Peng, S.S.; Feng, D.; Cheng, J.; Yang, L. Automation in U.S. longwall coal mining: A state-of-the-art review. *Int. J. Min. Sci. Technol.* **2019**, *29*, 151–159. [CrossRef]

39. Ralston, J.C.; Hargrave, C.O.; Dunn, M.T. Longwall automation: Trends, challenges and opportunities. *Int. J. Min. Sci. Technol.* **2017**, *27*, 733–739. [CrossRef]

40. Ralston, J.C.; Reid, D.C.; Dunn, M.T.; Hainsworth, D.W. Longwall automation: Delivering enabling technology to achieve safer and more productive underground mining. *Int. J. Min. Sci. Technol.* **2015**, *25*, 865–876. [CrossRef]

41. Szurgacz, D.; Zhironkin, S.; Cehlár, M.; Vöth, S.; Spearing, S.; Liqiang, M. A Step-by-Step Procedure for Tests and Assessment of the Automatic Operation of a Powered Roof Support. *Energies* **2021**, *14*, 697. [CrossRef]

42. Klishin, V.I.; Klishin, S.V. Coal Extraction from Thick Flat and Steep Beds. *J. Min. Sci.* **2010**, *46*, 149–159. [CrossRef]

43. Buyalich, G.; Buyalich, K.; Byakov, M. Factors Determining the Size of Sealing Clearance in Hydraulic Legs of Powered Supports. *E3S Web Conf.* **2017**, *21*, 3018. [CrossRef]

44. Buyalich, G.; Byakov, M.; Buyalich, K. Factors Determining Operation of Lip Seal in the Sealed Gap of the Hydraulic Props of Powered Supports. *E3S Web Conf.* **2017**, *41*, 1045. [CrossRef]

45. Buyalich, G.; Byakov, M.; Buyalich, K.; Shtenin, E. Development of Powered Support Hydraulic Legs with Improved Performance. *E3S Web Conf.* **2019**, *105*, 3025. [CrossRef]

46. Stoiński, K.; Mika, M. Dynamics of Hydraulic Leg of Powered Longwall Support. *J. Min. Sci.* **2003**, *39*, 72–77. [CrossRef]

47. Świątek, J.; Janoszek, T.; Cichy, T.; Stoiński, K. Computational Fluid Dynamics Simulations for Investigation of the Damage Causes in Safety Elements of Powered Roof Supports—A Case Study. *Energies* **2021**, *14*, 1027. [CrossRef]

48. Gil, J.; Kołodziej, M.; Szurgacz, D.; Stoiński, K. Introduction of standardization of powered roof supports to increase production efficiency of Polska Grupa Górnicza SA. *Min. Inform. Autom. Electr. Eng.* **2019**, *56*, 33–38.

49. Rajwa, S.; Janoszek, T.T.; Prusek, S.S. Influence of canopy ratio of powered roof support on longwall working stability–A case study. *Int. J. Min. Sci. Technol.* **2019**, *29*, 591–598. [CrossRef]

50. Wang, X.; Xu, J.; Zhu, W.; Li, Y. Roof pre-blasting to prevent support crushing and water inrush accidents. *Int. J. Min. Sci. Technol.* **2012**, *22*, 379–384.

51. Frith, R.C. A holistic examination of the load rating design of longwall shields after more than half a century of mechanised longwall mining. *Int. J. Min. Sci. Technol.* **2015**, *26*, 199–208. [CrossRef]

52. Szurgacz, D. Dynamic Analysis for the Hydraulic Leg Power of a Powered Roof Support. *Energies* **2021**, *14*, 5715. [CrossRef]

53. Wan, L.; Zhang, S.; Meng, Z.; Xie, Y. Analysis of the protection performance of face guard for large mining height hydraulic support. *Shock. Vib.* **2021**, *2021*, 6631017. [CrossRef]

54. Rajwa, S.; Janoszek, T.; Prusek, S. Model tests of the effect of active roof support on the working stability of a longwall. *Comput. Geotech.* **2020**, *118*, 103302. [CrossRef]

55. Klishin, V.I.; Fryanov, V.N.; Pavlova, L.D.; Nikitenko, S.M.; Malakhov, Y.V. Rock mass-multifunction mobile roof support interaction in mining. *J. Min. Sci.* **2021**, *57*, 361–369. [CrossRef]

56. Rudzki, P.; Krot, P. Dynamics control of powered hydraulic roof supports in the undergroung longwall mining complex. *IOP Conf. Ser. Earth Environ. Sci.* **2021**, *942*, 012014. [CrossRef]
57. Gabov, V.; Babyr, N.; Zadkov, D. Mathematical modelling of operation of the hydraulic support system of the powered support sections with impulse-free continuous regulation of its resistance to the roof rock lowering. *IOP Conf. Ser. Mater. Sci. Eng.* **2021**, *1064*, 012045. [CrossRef]
58. Burgan, H.I.; Aksoy, H. Monthly flow duration curve model for ungauged river basins. *Water* **2020**, *12*, 338. [CrossRef]
59. Burgan, H.I.; Aksoy, H. Daily flow duration curve model for ungauged intermittent subbasins of gauged rivers. *J. Hydrol.* **2022**, *604*, 127429. [CrossRef]

 *energies*

Article

A Concept for Solving the Sustainability of Cities Worldwide

Karmen Margeta [1,*], Zvonimir Glasnovic [2,†], Nataša Zabukovec Logar [3,4], Sanja Tišma [1] and Anamarija Farkaš [1]

1   The Institute for Development and International Relations, Lj. F. Vukotinovića 2, 10000 Zagreb, Croatia; sanja.tisma@irmo.hr (S.T.); afarkas@irmo.hr (A.F.)

2   Faculty of Chemical Engineering and Tachnology, University of Zagreb, Savska cesta 16, 10000 Zagreb, Croatia; zvonglas@fkit.hr

3   National Institute of Chemistry, Hajdrihova ulica 19, 1000 Ljubljana, Slovenia; natasa.zabukovec@ki.si

4   Faculty of Science, University of Nova Gorica, Vipavska 13, 5000 Nova Gorica, Slovenia

*   Correspondence: karmen.margeta@irmo.hr or karmen.margeta@gmail.com

†   Retired, independent scientist, current address: zvonimir.glasnovic@gmail.com.

Abstract: Considering that more than half of the world's population today lives in cities and consumes about 80% of the world's energy and that there is a problem with drinking water supply, this paper presents a way to solve the problem of the sustainability of cities by enabling their complete independence from external sources of energy and drinking water. The proposed solution entails the use of Seawater Steam Engine (SSE) technology to supply cities with electricity, thermal energy and drinking water. The system would involve the seasonal storage of electricity and thermal energy, supported by geothermal heat pumps. The strategy of the distribution network would be based on the original concept of the "loop". In cities that do not have enough space, SSE collectors would be placed above the lower parts of the city like "canopies". The city of Zagreb (Croatia) was selected as a case study due to its size, climate and vulnerability to natural disasters. The results show that Zagreb could become sustainable in 30 years with the allocation of less than 2% of GDP and could become a paradigm of sustainability for cities worldwide. This paper encourages the development of the "Philosophy of Sustainability" because the stated goals cannot be achieved without a change in consciousness.

Keywords: sustainable city; climate change; Seawater Steam Engine; renewable energy sources; drinking water; heating and cooling; electricity

check for updates

Citation: Margeta, K.; Glasnovic, Z.; Zabukovec Logar, N.; Tišma, S.; Farkaš, A. A Concept for Solving the Sustainability of Cities Worldwide. *Energies* **2022**, *15*, 616. https://doi.org/10.3390/en15020616

Academic Editors: Beata Zofia Filipiak, Sergey Zhironkin and Michal Cehlar

Received: 29 October 2021
Accepted: 13 January 2022
Published: 16 January 2022

Publisher's Note: MDPI stays neutral with regard to jurisdictional claims in published maps and institutional affiliations.

1. Introduction

The IPCC report (2021), Summary for Policymakers [1], once again shows that the UN has no successful response to climate change as global temperatures continue to rise and extreme weather conditions become more frequent. In this sense, the authors of this paper also doubt the reality of the IPCC's new forecasts/scenarios of slow global temperature growth that extends until 2100. That is, nowhere in this IPCC report (as in most previous reports) is climate breakdown predicted. This could be a realistic possibility, not because the geological history of the Earth lacks higher concentrations of CO_2 and global temperatures, but because the climate system does not have the "elasticity" to withstand such a sudden rise in temperature in such a short period of time. On the other hand, even assuming that a slow temperature increase of 3.3–5.7 °C will occur until 2100 (according to the worst IPCC scenario "3.3–5.7 °C under SSP5-8.5" [1]), this increase will have dramatic consequences on human health and lives and biodiversity. A temperature of 48.8 °C was measured in Sicily (Italy) on 11.08.2021 [2], and an increase in global temperature in 2020 of "only" 1.2 ($\pm$0.1) °C [3] was observed. These figures show that temperature extremes, which will (in the event that the year 2100 is "peacefully welcomed"—that is, without climate breakdown) obviously become more frequent and, with higher temperatures and longer durations, could significantly increase human mortality [4]. Therefore, the authors believe that the appeal by over 230 editors of scientific journals in the field of medicine should be taken extremely seriously: "Call for emergency action to limit global temperature increases, restore biodiversity, and protect health", published on 6 September 2021 [5], citing: "Indeed, no temperature rise is 'safe'. In the past 20 years, heat related mortality among people aged over 65 has increased by more than 50%. Higher temperatures have brought increased dehydration and renal function loss, dermatological malignancies, tropical infections, adverse mental health outcomes, pregnancy complications, allergies, and cardiovascular and pulmonary morbidity and mortality. Harms disproportionately affect the most vulnerable, including children, older populations, ethnic minorities, poorer communities, and those with underlying health problems". Thus, the situation is alarming, and decisive action needs to be taken to halt the further rise in global temperature.

Alternative solutions listed in the IPCC report (2021) that would reduce CO_2 emissions, such as the so-called "Most strong-mitigation scenarios" that use "geoengineering" or "climate engineering" methods, which are mostly related to so-called "Carbon Dioxide Removal" (CDR) and "Solar Radiation Modification" (SRM), can hardly be expected to have sufficiently large effects in the radical reduction of CO_2.

It is unrealistic to expect that CDR methods could remove such large amounts of CO_2 from the atmosphere, especially not at the rate at which CO_2 is emitted into the atmosphere, so it is logical to ask: What is the purpose of developing CDR methods to remove CO_2 from the atmosphere when much larger amounts of CO_2 are still emitted into the atmosphere at a much faster rate, and can such methods even overcome the climate breakdown, especially since the removal and emission of CO_2 are not correlated? Such methods of removing CO_2 from the atmosphere while CO_2 emissions into the atmosphere increase could have a countereffect, which is to further encourage the fossil fuel industry to continue emitting CO_2, which would be in line with the substantial aid given to this industry in the form of subsidies; thus, the arguments advocating such methods (under the auspices that they will reduce global temperature) seem extremely unconvincing. Therefore, these trends are opposite to the proclaimed "mitigation". They could only add to the possible healing of the atmosphere in a carbon-neutral era.

Furthermore, SRM methods, i.e., methods that would reduce solar radiation, are even more problematic, because no one can predict the short-term effects of such interventions, nor their long-term effects and consequences on the climate and ecosystem. These could also be very dangerous experiments with unpredictable implications.

The authors are of the opinion that we should not continue to hope for salvation from climate change by geoengineering methods, because the efficiencies of the application of these technological solutions have not yet been determined. Decisive actions should be taken, primarily in the form of new policies, but also in the form of new, economically viable and environmentally friendly technologies that allow a more realistic assessment of whether they can produce results in the expected time period [6].

In attempts to solve the problem of climate change, the proposed economic methods and measures have a significant impact. They were first incorporated into the Kyoto Protocol [7] and proposed by prominent world economists, such as Stern [8], who proposed three measures as a solution: (1) pricing carbon, (2) supporting innovation and (3) acting to remove barriers to energy efficiency; Nordhaus (Nobel laureate for economics in 2008 "for integrating climate change into long-run macroeconomic analysis"), who proposed a global taxing policy, convinced that taxes are proven instruments that will deliver results; and Romer (also Nobel laureate for economics in 2008 "for integrating technological innovations into long-run macroeconomic analysis"), who sought to encourage new ideas and long-term development. These measures were incorporated into the Paris Agreement [9] (whose "key architect" is the eminent French economist Laurence Tubiana), and all methods and measures seek to sustain economic growth. However, it is clear that permanent economic growth (with further growth of the world population) is not possible with limited natural resources. Therefore, it will be necessary to establish different economic theories and methods by which human civilization will be harmonized with the environment in order to achieve sustainability.

It is clear that the IPCC, among other things, uncritically adopts economic models that project a reduction in CO_2 emissions by about 50% by 2030 (which is already contrary to the trends of further temperature increase and CO_2 emissions). The UN does not actually have a solution to tackle climate change, and such reports [1] (which are based on projections instead of concrete actions) are just wasting valuable time. The first step in solving the problem of climate change should be an official admission by the UN itself that it is not able to solve such a complex problem with current tools and approaches.

The proof that the UN is not able to solve this problem was expressed through the media by the UN Secretary-General himself, with the statement "We're losing the battle" (Independent, 17 May 2019) or "We're losing the race" (Guardian, 18 September 2019), which is a definitive recognition that the UN is aware of the problem. This is why the Secretary-General also publicly seeks "meaningful plans", i.e., cit. "I expect there will be an announcement and unveiling of a number of meaningful plans on dramatically reducing emissions over the next decade, and on reaching carbon neutrality by 2050" (BBC News, 22 September 2019). In response to the appeal of the UN Secretary-General, the first three authors of this paper have provided such a meaningful plan and technological solutions [6] that would radically reduce emissions by 2050 to achieve carbon neutrality.

However, despite the aforementioned weaknesses of the UN in terms of insufficient knowledge and capacity to tackle climate change, there are clearly no instruments to force countries (parties) to abide by any international climate agreement, as these countries are largely governed by interests that are opposed to stopping climate change.

The proof of this is the BloombergNEF report [10], which showed that the 20 strongest economies in the G20 (which are responsible for almost $\frac{3}{4}$ of CO_2 emissions) have directly supported fossil fuels with as much as USD 3.3 trillion since 2015 (since the adoption of the Paris Agreement), which clearly indicates the hypocrisy in the policy for preventing climate change, i.e., the fact that some of these countries are publicly advocating for preventing climate change and transitioning to "green economies" while, on the other hand, significantly and directly supporting fossil fuels (the question arises as to what the term "mitigation" means in general, i.e., what does "reduction" relate to if energy production from fossil fuels, and thus CO_2 emissions, increases?). In this sense, the Paris Agreement becomes a meaningless document, because, apart from setting the wrong targets (global temperature cannot be stopped through our will because the Earth is not a boiler that can be turned off when the global temperature reaches 2 °C, which is explained in detail in [6]), there are no sanctions for countries (parties) that do not adhere to this agreement.

The responsibility of the UN, whose action should be more decisive and courageous in addressing problems and pursuing more responsible policies, is great, but it is equal to, if not greater than, that of the G20 countries because stopping climate change is a matter of human survival.

All of these failures of the UN more than clearly indicate the fact that humanity today, in 2021, is not only far from achieving sustainability but also facing a climate breakdown and its own disappearance and that it is necessary to urgently abandon the inappropriate objectives of the Paris Agreement and strategies to address the biggest problem facing humanity, which, in its essence, comes down to adaptation and mitigation; these strategies are even quantified in The Value of Sustainable Urbanization [11] as follows: 75% climate change mitigation (51% RES, 13% green buildings, 6% urban transport, 5% energy efficiency) and 25% climate change adaptation (10% coastal protection, 6% flood control, 5% drought/agriculture, 4% heat/greening). These estimates are utterly arbitrary and unrealistic, so such plans will make it virtually impossible to defend cities from climate change. In addition, supporting fossil fuels should be abandoned (the authors are aware that this will be the most difficult to achieve, but in a previous paper [6], they proposed a solution to this problem based on reorientation so that economies would not suffer damage) and completely different goals and strategies should be set as soon as possible, as well as sanctions for irresponsible parties. Decisive actions should be taken that could reverse these trends in the direction of achieving sustainability and stopping climate change, because we must not forget that the climate system has a huge inertia and that it will be horribly difficult to reverse these negative flows.

This paper presents an original concept for realizing the sustainability of cities worldwide using innovative technology (Seawater Steam Engine) for energy and water production and "distribution loops" for energy and water distribution in the city itself. The proposed concept can maintain a balance between the economy, society and the environment as fundamental components of sustainable development. The city of Zagreb (Croatia) was selected as a case study on the basis of three criteria: its size, geographical position and climate and vulnerability to natural disasters, and it could become a paradigm for other cities worldwide.

The term sustainability is used quite imprecisely in the literature, especially when it comes to sustainable communities ("sustainable cities", "green cities", "smart cities" and "sustainable smart cities"), which are graded from lower to higher sustainability. However, there are also opinions that a city can be either only sustainable or not sustainable at all, and it is very important to determine what is actually meant by the sustainability of cities, which currently account for as much as 78% of the world energy consumption [12] and whose population tends to grow, which contributes to the increase in CO_2 emissions.

On the other hand, cities are ranked by sustainability according to the following criteria: "Based on energy, transportation infrastructure, affordability, pollution, air quality, CO_2 emissions and % of green space" [13]. Each city is given a percentage of sustainability for some of these indicators, which is another indication that there is actually no city in the world that is sustainable. In this regard, the authors find this ranking problematic, because sustainability is not compatible with competition, but it should be an imposed obligation/command from above ("top-down" principle); otherwise, there will be no success.

For the overall sustainability of Earth, it is insufficient that only one or some cities are fully sustainable while others remain far from it. There is no sense in countries competing on sustainability because the climate is common to all cities and countries, and therefore, the same/common criteria that lead in the direction of sustainability should be established for all of them. Given that not all cities and countries are equally developed, the criterion of fairness on the way to achieving sustainability should certainly be respected [6].

Therefore, to achieve overall sustainability on Earth, which is *conditio sine qua non* of human survival, it is crucial to stop CO_2 emissions from cities because they are also the largest consumers of energy. As cities without drinking water (as another key criterion for achieving sustainability) cannot be sustainable, in this paper, the sustainability of cities will mean their complete independence from external sources of energy and drinking water as key factors for their sustainable survival.

In line with the topic of the sustainability of cities (which are also the biggest polluters), this paper also intends to:

(1) Point out to the UN that a formal approach to finding a new solution/path (which was publicly pointed out by the Secretary-General by seeking a meaningful plan (BBC News, 22 September 2019) is needed and clearly and directly address obstacles [10] encountered in attempts to address climate change;

(2) Draw the UN's attention to the fact that the growing morbidity and excessive deaths of people from climate change [5] could lead to a situation where the question of responsibility for this is raised in the future and that this is the reason for urgent policy change towards climate change;

(3) Provide assistance to the UN (i.e., UNFCCC and IPCC, which, on their website, implicitly call for assistance: "Engage with the IPCC" [14]). The authors of this paper firmly believe that the solutions described in [6] are, at this historical moment, the only correct way for the salvation of humankind.

2. Methodology

In order to achieve the sustainability of a city, which, in the case of this paper, means that it is continuously, throughout the year, provided with energy and drinking water through significant RES systems (solar, wind) that are interminable, it is necessary to provide storage of both energy and drinking water. In the case of using solar energy, energy storage should balance summer surpluses with winter energy shortages (for the northern hemisphere, and for the south, it is of course the other way around), which means that energy storage must be seasonal.

As cities are relatively large consumers of energy, seasonal energy storage, in today's era of technological development, can obviously be economically implemented only with pump-storage hydroelectric (PSH) technology, because it can store even the largest amounts of energy and is a mature technology with very high efficiencies (75–92%) and relatively low prices (USD 0.434/W [15]). The situation is similar with drinking water if it is obtained with the help of solar energy, because the storage must be relatively large in order to supply cities with drinking water during the winter.

Present-day technologies produce energy and drinking water separately from the RES system. Specifically, RES systems provide energy as their output, and if it is to be used for the production of drinking water (e.g., from seawater), then this energy is used to drive desalination systems. Therefore, it is necessary to separately install the capacities of RES systems that produce energy and those of RES systems that drive the desalination system. In the case of solar systems, this means that more collectors are needed than in the case when both resources (energy and water) are produced by the same collector surface.

In the case of PV or wind systems, only electricity is produced as their output, which can then be delivered to cities, or desalination devices can be powered by this electricity; however, in that case, these RES systems cannot also provide heat/cooling energy to cities.

Due to all of the above reasons, it was necessary to develop a completely new technology, Seawater Steam Engine, which can simultaneously produce energy (electricity and heat for heating and cooling) and drinking water from seawater or other unclean water sources. Thus, from the input solar energy, SSE technology can provide as many as four products as its output, which puts it far ahead of other technologies known today.

In addition, this technology is mainly made of currently available raw materials such as iron (steel), which implies a relatively low cost, and due to its simplicity of construction, it could be equally available to developed and less developed countries. From the above, the potential for application is evident: i.e., SSE technology could play a key role in stopping climate change.

For these reasons, this paper shows, for the first time, how SSE technology can make a city sustainable (completely independent of external energy sources and drinking water). The proposed system differs from previous RES technologies, which most often produce electricity (which is the easiest to transmit over long distances) that is delivered to the power system and from there to individual cities. There is no storage of this energy from the RES system, and the power system experiences constant shocks, depending on the ability of the RES system to produce electricity. Moreover, in this way, cities receive only electricity indirectly (through the electricity system), while other forms of energy (heat, i.e., heating and cooling) and drinking water cannot be obtained.

This paper describes a completely different principle of achieving the sustainability of cities than all previous ones, both in the production of energy and drinking water and in its distribution (using loops), which changes the relationship in all three aspects of sustainability.

2.1. Distribution of the World Population in Cities

From the estimate of the distribution of the world population in cities (2020) [16], shown in Figure 1, it can be seen that 56.2% of the population lives in cities, of which small cities (5.3%), smallest cities (3.7%) and other urban areas (23.2%) together make up 32.2% of the population and medium cities (12.1%), large cities (4.3%) and megacities (7.6%) together make up 24% of the world's population; that is, more than 30% of the population lives in cities with less than 1 million inhabitants, and over 20% lives in larger cities. The number of cities with less than 1 million inhabitants, according to [17], starts in 558th place, which means that out of 10,000 cities in the world, 94.4% of cities (in number) have a population below 1 million. Of the total population, 37% live in coastal communities by the sea [18], which is important information in terms of the possibility of using seawater as a resource for the application of new technologies that can supply these cities with drinking water.

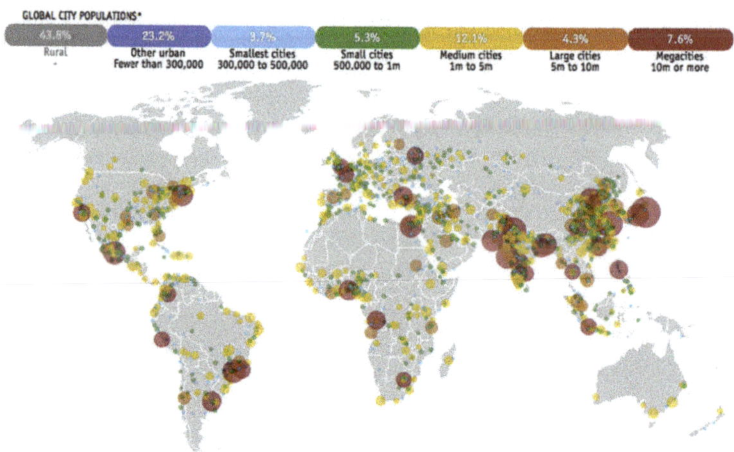

Figure 1. Population of cities worldwide, distribution estimated for 2020 [16]. Reprint with permission 5217100788999; 2021, Elsevier.

Certain statements made by the UN [11], namely, "In all regions, local governments and their organizations are contributing to the advancement of sustainable urbanization by fostering climate change mitigation actions, urban resilience, alternative economic models and social inclusion policies" and "Nearly 10,000 cities and local governments have set emissions reduction targets with accompanying policies and programs to meet those targets", are in fact a recognition that "sustainable urbanization" is of a more declarative nature, with minor effects that are far from being able to protect the climate. What is the purpose of "contributing to the advancement of sustainable urbanization" if these "contributions" are far smaller than the "increase of environmental pollution" by CO_2 emissions from these same cities and how long should the climate system "wait" (so that it does not breakdown) until the current minor contributions outgrow the pollution and completely nullify it? Therefore, the same issue arises again, which is that the UN does not have the instruments to make the world's cities sustainable, but everything remains on a voluntary basis, and today, every city "contributes as much as it can".

Regarding the definition of sustainable communities, the Institute for Sustainable Community states: "Throughout the world, people want the same things: access to clean air and water; economic opportunities; a safe and healthy place to raise their kids; shelter; lifelong learning; a sense of community; and the ability to have a say in the decisions that affect their lives" [19]. However, this definition is extremely vague because it is not about assessing the quality of people's lives but about the factors that would make a city sustainable, and as such, it would no longer emit CO_2 because it is the key category of sustainability.

2.2. Organization of the Transformation Process of Cities towards Sustainability

The transformation of today's cities towards achieving sustainability is an extremely complex task with many variables and limitations, and therefore, a systems engineering approach is of special importance for solving it from the very beginning, i.e., from setting a policy whose task is to set objectives.

The objective is defined as the direction in which changes are to be made. In this case, it means to make a city completely sustainable, i.e., independent of external energy sources and drinking water (for which the terms "self-supply", "stand-alone" and "grid-off" systems are used in energetics). The task of systems engineering is to achieve this objective, which, in this case, is the most demanding engineering task in general, because it needs to align natural resources (which are stochastic) with the needs of citizens of a city, which are also stochastic. Thus, engineers will be faced with limited resources in solving this problem, which, in addition to natural resources, comprise money, labor, construction time, the existing construction of a city, possibility of construction in certain locations, etc. Furthermore, engineers will also be faced with various perturbations. For this reason, the notion of systems engineering includes the art and science of choosing from a large number of possible alternatives, a set of actions (alternative or alternatives) that best meet the set objectives of decision makers. The terms art and science indicate that it is necessary to include engineering experience and engineering sciences (calculation and analyses) in an appropriately organized manner in order to best or optimally determine the required solution.

This means that it is especially important to follow the steps of systems analysis (as a segment of systems engineering, consisting of the following steps: setting objectives, determining key performance indexes for evaluating the achievement of goals, formulating alternatives, formulating a solution selection model, selecting the "best" solution and performing feedback analysis). This procedure is very complex in this case and requires the engagement of a large number of associates of different professions, which may lead to a loss of perceiving the integrity of the problem.

Therefore, objectives should be set by mayors (or their administrations) as politicians, with evaluation criteria set by the main engineer, who develops alternatives with a group of planners; designers and management engineers, as systems engineers, should develop models and make calculations, while the feedback analysis of the results should achieve proper communication between all experts and ensure the integrity of problem solving.

However, with all of the complexities and difficulties in solving the problem of the sustainability of cities, this objective offers great local benefits, because these cities would significantly improve their economies. The construction of such large infrastructure facilities and works in cities with many new technologies would strongly encourage the development of industry in these cities, with a multitude of innovations, and such strong development would change the attitudes of citizens towards sustainability; therefore, all of this would be reflected in the overall economic, social and cultural development of these cities.

The social aspect primarily refers to the fact that sustainable cities should not live life with the same rhythm as before, which is independent of nature, but should live in harmony with nature or subordinate their rhythm of life to nature, all in order to enable the consumption of energy in the summer when the most solar energy is available, and in the winter, excessive energy consumption should be avoided (e.g., various events—sports, cultural and other events—should be held mainly in the summer, while during the winter, these events should be reduced and/or have lower energy consumption).

The environmental aspect primarily refers to the placement of the fields of solar collectors, which should minimize damage to the environment. In this sense, the construction of relatively large PSH reservoirs could have very positive effects on social life, as they could be multifunctional lakes that could be used for both recreation and sports.

2.3. Seawater Steam Engine—Technology for the Realization of Sustainable Cities

As stated in [6,20–22], Seawater Steam Engine (SSE) technology uses three natural sources—RES energy, seawater or water from other unclean water sources (rivers, lakes, etc.) and gravity (thus, three natural resources)—in order to produce electricity and heat (for heating and cooling) and drinking water (which could also be referred to as "quadri-generation"), which would be provided continuously throughout the year. Seasonally stored thermal energy would also be supplied by geothermal heat pumps, which means that the fourth natural resource, i.e., Earth, would be used.

Therefore, with this production potential, this technology would achieve continuity of supply to consumers throughout the year, which is the key to sustainability.

The overall sustainability of Earth can be achieved only by nullifying anthropogenic CO_2 emissions, which is possible if we primarily nullify anthropogenic CO_2 emissions in cities (that make up 78% of the world's energy), which is the subject of this paper. Therefore, in this paper, the sustainability of cities is observed through the lens of the complete independence of cities from external sources of energy and drinking water; i.e., this energy and drinking water must be provided by the RES system. Other aspects of sustainability (listed in, e.g., [10,19]) are not of interest for this paper, because the primary issue is to stop anthropogenic CO_2 emissions and thus prevent climate breakdown. Simply put, overall sustainability on Earth will be achieved if all cities in the world are made sustainable as soon as possible.

Therefore, this paper, unlike all previous approaches to urban sustainability (where relatively small and insufficient steps towards sustainability have been made and have so far yielded no results globally, because global temperatures continue to rise), describes a completely original concept of a sustainable city as a basic unit ("brick"), which would achieve overall sustainability on Earth, as shown in Figure 2.

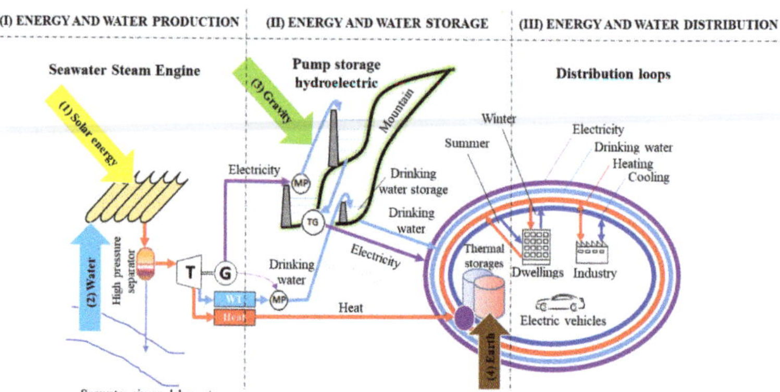

Figure 2. An original concept of a sustainable city: **(I)** energy and drinking water production by SSE technology; **(II)** storage of electricity (by pump-storage hydroelectric), thermal energy (for heating and cooling) and drinking water; **(III)** distribution of electricity, thermal energy (heating and cooling) and drinking water by distribution loops.

This concept consists of three parts: (I) energy and water production by SSE technology, (II) energy and water storage (pump-storage hydroelectric for electricity storage, thermal energy storage for heating and cooling and water reservoir for drinking water storage) and (III) energy and water distribution by distribution loops.

Solar energy can be used as input energy into the SSE system, as well as other renewable sources (wind, biomass, hydro, etc.), which significantly expands the scope of application of this technology (integrated SSE system) in locations with relatively little solar energy.

Seawater and water from rivers, lakes and other sources of unclean water can be used as input water. This water is heated and evaporates in the pipes of parabolic trough collectors (concentrated solar power), after which it enters a high-pressure separator, where steam is separated from the concentrated water, which is then returned to the impure water source [6].

Steam from the high-pressure separator is discharged to turbines (T), where it is converted into mechanical work and generates electricity in generators (G). This energy drives motors and pumps (MP) of pump-storage hydroelectric (PSH), which pumps water from the lower to the upper water reservoir, which serves as seasonal energy storage that balances summer surpluses and winter shortages of solar energy and thus ensures a continuous supply of electricity throughout the year. It is produced in turbines and generators (TG) and is delivered to the city electricity network.

Condensed steam, i.e., water from the turbine (T), is drained to the water treatment plant (WT), after which it is pumped (using MP) to the drinking water storage that supplies the city with drinking water. This reservoir must also be a seasonal tank to ensure a continuous supply of drinking water to the city.

Thermal energy obtained in the SSE system is discharged into the city thermal network, made in the form of loops for reliability. This network distributes heat and cooling energy (obtained by absorption cooling) to the city in the way that it is radiantly distributed from the main loops to all consumers. Seasonal heat storage would also be used, which would ensure the continuous supply of heat to the city for space heating and hot water and an "ice bank" for storing cooling energy, thus ensuring the continuity of cooling during the summer months.

As an additional heat source in these heat storage systems, geothermal heat pumps (fourth natural resource, i.e., Earth) would be used, whose coefficient of performance (COP) for large systems reaches a value of up to 10. In this way, all areas in the city can be continuously provided with heating and hot water during the winter and continuous cooling during the summer months. Since industrial plants themselves produce heat during the production process, they have the possibility of constant heat exchange with the heating and cooling network.

The distribution of electric and thermal (heating and cooling) energy and drinking water through loops would ensure high security of supply, because if a failure occurs in one part of the loop—high-voltage networks, heating systems, cooling systems and drinking water supply systems—the supply of the city would automatically be ensured on the other side of the loop, whereby all of these systems (complete with transformer and pumping stations) would be automatically monitored and managed. Such cities would basically be "Sustainable smart cities".

In sustainable cities, it is necessary that all vehicles be CO_2-free, i.e., electrically powered, for which the SSE system would provide enough electricity.

The first step in designing an SSE system for a city is its sizing, i.e., determining its nominal power, which, on the one hand, depends on the available solar radiation and, on the other hand, on the energy consumption of a city. Considering that these are stochastic quantities (possibilities for energy production from solar radiation and energy consumption of the city, which need to be harmonized) and the energy is stored seasonally using PSH technology, the power (size) of the SSE system must match the energy consumption throughout the year from the available solar energy. This can be realized by optimizing the system, which was carried out in [20]. For this purpose, a simulation–optimization model based on dynamic programming was developed, because it is a multistage decision-making process in terms of time (discretization to "i" time steps), where the selected objective function is to determine the optimal power of the SSE generator $P_{el(NOM)}$.

Based on the known nominal power, the area of the SSE generator A_{coll} is determined, and when that area is known, then the amount of produced thermal $Q_{(i)}$ and electric E_{el} energy and the amount of drinking water V_{DW} can be determined from it.

Formulas for the calculation of key parameters of the SSE system—nominal power of the SSE generator, produced heat and electricity and drinking water [20,23,24]—are as follows:

$$P_{el(NOM)(i)} = \frac{\rho g H_{TE(i)}}{\eta_{OS}\eta_{SE}\eta_{PSI}R_{coll(i)}E_{S(i)}}V_{ART(i)} \tag{1}$$

$$Q_i = A_{coll} \cdot F \cdot \overline{\eta}_{opt} \cdot \overline{\Phi} \cdot E_{S(coll)(i)} \qquad (2)$$

$$E_{el(i)} = \eta_{Q-EL} \cdot f_m \cdot A_{coll} \cdot F \cdot \overline{\eta}_{opt} \cdot \overline{\Phi} \cdot E_{S(coll)(i)} \qquad (3)$$

$$V_{DW} = \frac{1}{\rho_{H_2O}} \cdot \frac{P_{el(NOM)}}{\eta_{ME} \cdot \eta_{TUR} \cdot \Delta h} \sum_{i=1}^{N} f_{t-DIR(i)} \cdot T_{S(i)} \qquad (4)$$

where $P_{el(NOM)}$ is the nominal electric power of the SSE generator; ρ is water density (ρ = 1000 kg/m^3); g is the gravitational constant (g = 9.81 m/s^2); $H_{TE(i)}$ is the average total head (m); η_{OS} is the efficiency of the open thermodynamic system; η_{SE} is the efficiency of the collector field of SSE power plants; η_{PSI} is the efficiency of the pumping system and inverter; R_{coll} is a conversion factor for converting mean daily radiation on a horizontal plane to mean radiation on the aperture of tracking parabolic collectors; $E_{S(i)}$ is the mean daily solar radiation on a horizontal plane at Earth's surface; $V_{ART(i)}$ is the mean daily value of artificial water inflow that can be pumped by the SSE generator from the sea into the upper reservoir; $Q_{(i)}$ is the thermal energy production (kWh); A_{coll} is the aperture area of parabolic collectors (m^2); F is the heat removal factor; η_{opt} is long-term average optical collector efficiency; Φ is the long-term average utilization factor of solar energy based on the Hottel–Whillier concept; $E_{S(coll)}$ is the average daily value of the collected solar energy; E_{el} is electric energy production (kWh); η_{Q-EL} is the average value of conversion efficiency of thermal into electric energy; f_m is the load matching factor to the characteristics of the SSE generator; V_{DW} is drinking water production (m^3); η_{ME} is the conversion efficiency of mechanical turbine power into electric generator power; η_{TUR} is isentropic (inner) turbine efficiency; Δh is the difference between the enthalpy of vapor at the inlet and outlet of the turbine; $f_{t-DIR(i)}$ is a factor that describes the proportion of the number of hours of direct radiation exceeding 250 W/m^2; $T_{S(i)}$ is insolation (duration of sunshine); I is the time stage (increment) related to the dynamic programming; N is the number of time stages (for a time stage of 1 day, N = 365).

3. Results

3.1. Why Was Zagreb Chosen as the Paradigm of Sustainability for All Other Cities in the World?

In terms of selecting a city for the case study, which could be a paradigm for all other cities in the world, the authors used three criteria:

- City size,
- Geographical position and climate,
- Vulnerability to natural disasters.

According to data on the sizes of cities, out of a total of 10,000 cities in the world, the city of Zagreb, according to [17], is in 836th place with a stated population of 684,878. However, it is obvious that these data are not up to date because, according to the aforementioned data, Zagreb actually had 806,341 inhabitants in 2019, which would still place it in the category of "small cities" with populations of 500,000 to 1 million. Thus, the city of Zagreb, due to its size (slightly less than 1 million inhabitants), can be a good model for other cities with a population below 1 million, which make up 90% of the total number of cities in the world (where 32.2% of the world's population lives), offering the significant potential to make them all sustainable.

Since it is stated in [25], cit.: "We found that only less than 1/8 of the human population lives south of the equator while around 50% of the dwell population within the area between 20° N and 40° N", it was logical to choose a city that would be on the northern edge of the belt due to climatic conditions, i.e., irradiated amounts of solar radiation, because it means that the problem of the sustainability of these cities (which have more solar energy) could be solved with SSE technology that uses solar radiation.

Zagreb is located at 45° north latitude (N), with a direct component of solar radiation of 1081 kWh/m^2/a (Table 1) and, in this sense ("of the dwell population within the area between 20° N and 40° N"), has a favorable location, despite the fact that the stated values of solar radiation for concentrated solar power (CSP) systems are considered uneconomical because they require larger collector areas. As the intensity of solar radiation per unit area of the collector is practically the same (1000 W/m^2) at these locations and only the duration of solar radiation (insolation) changes, CSP systems could be successfully used at locations with smaller amounts of solar radiation, but it is important to put the issue of sustainability above the status of the economy, because sustainability has no price. Therefore, the opinion of the authors of this paper is that the application of CSP, i.e., SSE technology with concentrator collectors, should be extended to locations up to about 1000 kWh/m^2, where energy storage, which would bridge the winter energy shortages and enable their continuous supply, should be higher, i.e., with hydropower (as the largest type of energy storage facility today), with the potential that the entire world's electricity system becomes completely sustainable [26].

Table 1. Input data for: global G_h, diffusion D_h and direct B_n solar radiation per month in a year for reference location (Ref.) of the island Vis (Croatia) and the city of Zagreb (Croatia).

Location	Vis–Ref. (Croatia)			Zagreb (Croatia)		
	G_h	D_h	B_n	G_h	D_h	B_n
Month	[kWh/m^2]			[kWh/m^2]		
January	54	25	87	31	19	43
February	77	31	78	56	32	59
March	125	50	117	91	47	88
April	158	71	136	121	59	105
May	203	77	171	171	83	144
June	222	73	167	175	88	132
July	224	74	222	181	79	157
August	198	64	168	156	75	134
September	150	46	167	98	59	74
October	111	34	111	68	34	79
November	65	24	74	33	21	37
December	50	22	80	24	16	28
Total	1632	590	1578	1203	612	1081

Therefore, in terms of location and climate conditions, or irradiated solar energy, the way to realize the sustainability of Zagreb with SSE technology could be more difficult and demanding than in most other cities in the world, which means that, in this sense, Zagreb could be a good model for considering this problem.

When addressing the problem of sustainability, the vulnerability of a certain city in terms of natural disasters, in which climate change has an increasing impact, carries special weight. It is not enough just to make a city sustainable, but it should also be preserved, which means that it is necessary to make it as resilient and prepared as possible for natural disasters. Zagreb is located in a seismically active zone, with the possibility of an earthquake of 6.5 on the Richter scale. Floods can also occur in Zagreb due to extreme weather events, which are becoming more frequent, and no one can predict the level of such floods, which makes the city vulnerable to natural disasters. Therefore, the measures that Zagreb would take must also account for this adverse environmental impact.

In this sense, the upper water storage of the SSE system could be used to regulate high waters at the entrance of the Sava River into the city and defend the city from floods with an original strategy by building another large reservoir (lake). In this reservoir, water would be pumped during high waters (pumping energy would be obtained by pumping water from the upper water storage of the SSE system at a rate that would optimally control this system so that large waters are later used for energy production, which should not be a problem to perform for Zagreb as a sustainable smart city), which could be applied to many other cities in the world.

Another reason for choosing the city of Zagreb is the experiences of the authors, who have already worked on the problems of energy efficiency and RES systems [26–30] and achieving the sustainability of Zagreb [31]; therefore, they can perceive all relevant issues of such a complex intervention in a particular case.

3.2. Input Data

In order to verify the success of solving the problem of the sustainability of cities with SSE technology, the case of the city of Zagreb was selected, with 806,341 inhabitants (2019), an area of 641 km^2 and a population density of 1258 inhabitants/km^2, according to public data. In the western part of the city, there is sufficient space for the accommodation of SSE technologies, and the terrain configuration with altitude differences (Medvednica Mountain) is suitable for the accommodation of hydro storage.

Gross domestic product (GDP) for the city of Zagreb in 2018 amounted to EUR 17,544 million [32], which is an important parameter for assessing the economic possibilities that the city be made sustainable "on its own".

Zagreb is located in a moderately continental climate, with the values of solar radiation shown in Table 1, which were obtained from the Meteonorm program 7.3.4. [33], while the data for the island of Vis (Croatia) were calculated in a previous paper [34]. The island of Vis (Croatia) was chosen because an optimization model was developed for it, which calculated all characteristic parameters of the SSE system [20] and can be applied in this case as well.

For parabolic trough collectors (as used by SSE technology), only the component of direct solar radiation B_n (kWh/m^{2}a) on the collector surface accompanying the sun is relevant.

Table 2 shows the data on the structure of energy consumption in the city of Zagreb [35] and individual consumption expressed in (GWh) (far right column).

Table 2. Structure of energy consumption of the city of Zagreb in 2015 * [35].

Measuring Unit: TJ	Industry	Transportation	Households	Services	Agriculture	TOTAL (GWh)
Coal	0.0	0.0	2.9	0.0	0.0	0.8
Natural gas	982.6	117.6	7314.4	2816.4	17.3	3124.5
Firewood	4.5	0.0	1922.4	45.9	0.0	548.0
Solar energy	0.0	0.0	86.3	0.0	0.0	24.0
Geothermal energy	0.0	0.0	0.0	25.3	0.0	7.0
Biofuel	0.0	202.5	0.0	0.0	0.0	56.3
Other biomass and waste	0.0	0.0	141.6	11.9	0.0	42.6
Liquefied petroleum gas	37.5	393.9	136.0	225.1	0.0	220.1
Motor gasoline	31.2	3585.0	0.0	0.0	4.5	1005.8
Petroleum	0.0	0.0	0.0	0.0	0.0	0.0
Diesel fuel	670.5	8097.8	0.0	0.0	123.9	2470.1
Extra light heating oil	128.1	0.0	187.9	738.9	8.5	295.4
Heating oil	48.2	0.0	0.0	72.3	0.0	33.5
Electric energy	1523.9	331.9	3856.0	5161.7	4.0	3021.5
Steam and hot water	1723.2	0.0	3565.4	861.2	0.0	1708.3
TOTAL	5149.7	12,728.7	17,212.9	9958.7	158.2	12,557.8

* According to data from 2015, the population of Zagreb was 799,565, while in 2019, that number was only 1% higher and amounted to 806,341 [36]. Such a small difference could not have significantly affected the structure of energy consumption.

According to [29], the annual water consumption of the city of Zagreb for 2020 amounted to 54,683,000 m^3 of water, which is about 55 hm^3 of water per year.

3.3. Output Data

Based on the above input data and reference values of the SSE system for the island of Vis in Croatia (where the optimal nominal power of the SSE generator was determined by a simulation–optimization model based on dynamic programming, because it is a multistage decision-making process in a temporal sense, with a nonlinear objective function and constraints [20]), which are calculated by Equations (1)–(4), the results for the city of Zagreb were obtained by linear correlation (comparison with the results obtained for the island of Vis specifically; due to lower values of solar radiation, Zagreb will need higher unit power and a larger collector area compared to Vis, and due to significantly higher energy consumption in Zagreb, the SSE system will need proportionally higher power in relation to Vis), and the results are shown in Table 3.

Table 3. Calculation results of relevant parameters of SSE system for the city of Zagreb.

	Characteristic Parameters	**VIS (Croatia)**	**ZAGREB (Croatia)**
1 *	B_n (kWh/m^2/a)	1578	1081
2 *	Population	3637	806,341
3 *	$E_{el(consumption)}$ (GWh/a)	20	3022
4 *	$V_{(consumption)}$ (hm^3/a)	0.45	55
5	$P_{el(NOM)}$ (MW)	52	11,529
6	A_{coll} (km^2)	0.37	39
7	$Q_{(production)}$ (GWh/a)	179	18,509
8	$E_{el(production)}$ (GWh/a)	63	6478
9	$V_{(DW)(production)}$ (hm^3)	0.48	73
10	V_{PSH} (hm^3)	7	1057
11	W_{DW} (hm^3)	0.08	13

* Input data from Tables 1 and 2.

In the first row of Table 3, the values of direct solar radiation B_n (kWh/m^2/a) for the island of Vis [34] and the city of Zagreb [25] are given, which shows that in Zagreb, these values of direct solar radiation Bn are about 30% lower, which results in about 30% of the required higher unit values. The second row shows the number of inhabitants, and the third row shows total electricity consumption $E_{el(consumption)}$ (GWh/a) (data for Zagreb are for 2015—taken from Table 2), while the fourth row shows total water consumption $V_{(consumption)}$ (hm^3/a) for Zagreb in 2020, which changed very little compared to 2015.

The fifth row of Table 3 contains the values of the nominal power of the SSE generator $P_{el(NOM)}$ (MW) (Equation (1)), which for Zagreb, should amount to 11,529 MW, while the required area of the collector field A_{coll} (km^2) for this power is given in row 6 and would amount to 39 km^2. Rows 7, 8 and 9 show the values of produced thermal energy $Q_{(production)}$ (GWh/a) = 18,509 GWh/a (Equation (2)) and electricity $E_{el(production)}$ (GWh/a) = 6478 GWh/a (Equation (3)) and drinking water $V_{(DW)(production)}$ (hm^3) = 73 hm^3 (Equation (4)), respectively. The 10th row of Table 3 shows the volume of hydro storage V_{PSH} (hm^3) (1057 hm^3, which can be situated in several lakes about 100 m deep, located on the mountain "Medvednica", along which the construction of the SSE system is planned, which would ensure continuous energy supply throughout the year. The 11th row shows the required volume of drinking water storage W_{DW} (hm^3) (13 hm^3—also located on the mountain "Medvednica") that is necessary to supply the city with drinking water in periods when there is not enough solar radiation.

The calculation results for relevant parameters of the SSE system for the city of Zagreb (Table 3) show that the SSE system could fully meet the needs of Zagreb for electricity because they amount to 3021.5 GWh/a (Table 2), as well as the needs for drinking water (it can produce 73 hm^3/a, and needs are 55 hm^3/a). In addition, the system produces a large amount of heat (18,509 GWh/a), of which 1/3 is converted into electricity, while the other 2/3 of energy could meet the needs of the city for heating and cooling (about 6000 GWh/a), which are now satisfied using: coal, natural gas, firewood, solar energy, geothermal energy, biofuels, other biomass and waste, liquefied petroleum gas, extra light fuel oil, fuel oil, steam and hot water (6060 GWh/a in total), as can be seen in Table 2.

The energy currently used for transportation in Zagreb for motor gasoline and diesel fuel is 3476 GWh/a. Considering that the SSE system could produce 6478 GWh/a of electricity, of which Zagreb needs 3021 GWh, there is still 3457 GWh/a of electricity left, which coincides with the energy required for the transportation sector. Therefore, if Zagreb switched to electric cars, its energy needs in the transportation sector could be fully met, and the city of Zagreb could become the first sustainable city in the world to fully meet all of its needs for electricity and heat, including its transportation sector (with implications for preserving clean air), and meet its needs for drinking water, thus becoming the first city in the world to be completely independent of external sources of electricity and heat (heating/cooling) energy and drinking water.

Figure 3 shows the situation of SSE and the distribution energy system (heating and cooling). The collector field of parabolic trough collectors is positioned in the western part of the city (where there is enough space for its accommodation) along the Sava River, from which water can be pumped. With the height difference that exists on Medvednica Mountain, where the upper water reservoir can be located, summer surpluses and winter shortages of solar energy could be balanced, thus ensuring a continuous supply of electricity to the city throughout the year. The existing "Zaprešić Lake" or the Sava River could be used as the lower reservoir.

Figure 3. Layout of the collector fields of SSE technology (which uses Sava River as input water) and distribution loops of heating and cooling for the city of Zagreb.

For the purpose of continuously supplying the city with drinking water, the appropriate "drinking water storage" would be located on Medvednica Mountain.

The distribution network of heating and cooling energy, as well as the supply of electricity and drinking water, would operate in a novel way, with a loop within the city (Figure 3). This would achieve high reliability of supply (from which other parts of the city would be radiantly supplied). In the eastern part of the city, it would reach the location "Borongaj", which could accommodate: the control center and management center "Sustainable Smart City of Zagreb", "Institute of Sustainable Technologies" and the associated "Study of Sustainable Technologies", the first of their kind in the world, so that they can further develop this innovative technology and its application and, in this sense, as well as in terms of philosophy of sustainability, educate future generations that could disseminate knowledge and technology to other cities around the world, all in order to make them sustainable as soon as possible.

Since the Borongaj location has sufficient usable area, it could be used for the possible addition of solar photovoltaic collectors to increase the reliability of electricity supply to consumers, thus increasing the reliability of the city's supply with heat and cooling energy and drinking water, because supply reliability of sustainable cities is crucial.

3.4. Investment

Considering that this is a completely new SSE technology, as well as an original method of distributing heat (heating and cooling), electricity and drinking water (loop), only a rough estimate can be made for the necessary investments on the basis of similar systems that have been developed and on the basis of the authors' design experiences.

The investment consists of SSE collector fields (through which seawater or water from other unclean water sources passes), power block, water treatment technology, PSH technology, thermal energy storages, ice banks, distribution loops (35 km long) and adaptations to existing systems.

The total investment in the SSE system can be estimated on the basis of the costs for solar parabolic trough collectors, reported in [37], which states the price of 177 USD/m^2 for solar parabolic collectors for 2025. The price includes: site preparation, collector structure (incl. assembly), pylons and foundations, drives, mirrors, receivers, cabling, HTF system (fluid) and HTF system (excl. fluid).

Thus, for a collector area of 39 km^2 (Table 2, row 6):

$$177 \text{ USD/m}^2 \times 39,000,000 \text{ m}^2 = 6,903,000,000 \text{ USD}$$

Investment costs for the power block and water treatment technology should be added to this value, so the total investment could amount to about USD 7 billion.

The investment in the accumulation system can be estimated from the data from [15], which states the amount of 0.434 USD/W for PSH systems. The capacity of hydro storage is also evaluated linearly by correlation and is 703 MW, which amounts to an investment of:

$$703,000,000 \text{ W} \times 0.434 \text{ USD/W} = \text{USD } 305,000,000$$

Since another large reservoir is required for high water regulation and a drinking water reservoir needs to be built, the entire reservoir system, complete with engines, pumps, hydro turbines and generators, could be estimated at about USD 1 billion.

The investment in the distribution network system is difficult to estimate because it is an original concept of the distribution of heat (heating and cooling), electricity and drinking water, which would be in the form of a loop (ring) in the city, with a total length of 35 km. In addition, the heat distribution system would consist of seasonal thermal energy storage (TES) and ice banks, which, according to the experience from the sustainable city Borongaj (Zagreb, Croatia) [38], can be estimated at about 300 TES of 80,000 m^3 of water and about 6000 probes (sonde) of geothermal heat pumps, 100 m long. The whole system of distribution and storage of thermal energy (heating and cooling) with geothermal heat pumps to ensure the continuity and reliability of supply to consumers throughout the year, complete with pumping stations and substations, could be assessed to about USD 100 million per kilometer of the loop from which energy and drinking water would continue to be radiantly supplied to all parts of the city. This means that this investment could be assessed as follows:

$$35 \text{ km} \times \text{USD } 100 \text{ million} = \text{USD } 3.5 \text{ billion}$$

Assuming that the adaptation and construction of new lines in the old network and its connection to the new "loop system" are estimated to amount to USD 1 billion, the total investment could be estimated with a value of:

$$7 + 1 + 3.5 + 1 = \text{USD } 12.5 \text{ billion}$$

Assuming that this investment will be realized by 2025, after 30 years (2055), the average annual investment would be around USD 0.4 billion/year.

The GDP for the city of Zagreb in 2018 was EUR 17,544 million [32], with the exchange rate of EUR 1 = USD 1.18 (on 13 September 2021), which amounts to USD 20,702 million, i.e., USD 20.7 billion.

Thus, the value of USD 0.4 billion/year accounts for 1.93% of GDP, or an investment of about:

$$2\% \text{ of GDP per year}$$

which clearly shows that the city of Zagreb has the economic strength to implement this system.

This rough economic analysis of only investment costs does not include benefits, meaning that from the construction of the energy and drinking water supply system (from 2025), Zagreb would not have any costs related to fossil fuels. Furthermore, with parallel investment in the energy efficiency of buildings in Zagreb, energy consumption would be significantly reduced, so the entire SSE system and distribution system could be smaller, which means a smaller investment. This is why it is very important to plan properly and strictly adhere to the systems engineering approach and find the optimal solution and dynamics for the construction of all systems that would make the city of Zagreb sustainable, which means independence of external energy sources and drinking water.

4. Discussion

4.1. Various Approaches to Combating Climate Change

The conclusions of COP26 [39] once again show the failure of UN policy because they list the same methods of combating climate change (II. Adaptation and IV. Mitigation) as listed in the Paris Agreement, which, to date, have yielded no results. This is best illustrated in citations 20 and 21, item IV. Mitigation, Glasgow Climate Pact, which reads:

> "20. *Reaffirms the Paris Agreement temperature goal of holding the increase in the global average temperature to well below 2 °C above pre-industrial levels and pursuing efforts to limit the temperature increase to 1.5 °C above pre-industrial levels;*
>
> 21. *Recognizes that the impacts of climate change will be much lower at the temperature increase of 1.5 °C compared with 2 °C and resolves to pursue efforts to limit the temperature increase to 1.5 °C".*

Therefore, despite the fact that after 6 full years since the adoption of this agreement, no results have been achieved in the direction of reducing climate change (on the contrary, it is increasing, with more extreme events and with increasing global temperature), COP26 "reaffirms the Paris Agreement temperature goal".

This can only indicate that the UN either does not really notice the weaknesses of the Paris Agreement or is simply not ready to recognize them. Given that the former is unlikely (because the climate situation is getting worse and worse), it is clear that the UN is not yet ready to acknowledge the mistakes of the Paris Agreement, regardless of the reason.

In 2021, the UN came to an extremely absurd conclusion, i.e., to stick to an agreement that has not yielded results for 6 years and that is obviously unenforceable, while still insisting on its maintenance (COP26) and thus preventing any other policy from being established. Thus, the Paris Agreement (which was supposed to bring a solution) has actually become the main obstacle to solving the problem of climate change.

For this reason, it is necessary to completely change the global policy on combating climate change to an approach that is not reduced to "as much as any party can" and "carbon pricing". Instead, the strategy should be to reorient the high-carbon economy into a green economy by adopting a new technology that would be a powerful weapon in the fight against climate change, that could produce both energy and drinking water and that would be equally accessible to both developed and underdeveloped countries. Such a policy was developed and called the Climate New Deal by the first three authors of this paper, and it is explained in more detail in [6].

However, this completely new policy also requires completely new strategies in the fight against climate change, which would not be economically based (such as the Kyoto Protocol and the Paris Agreement with "carbon pricing") but driven by engineering-based strategy (Section 2.3), and which would be based on making all cities (as well as the largest energy consumers) completely sustainable, which means being completely independent of external sources of energy and drinking water. Thus, overall sustainability could be achieved by making all cities in the world sustainable so that each city, out of a total of 10,000 cities, would become a "brick" in building a fully sustainable civilization.

This also means that all cities would have to change their previous policies and strategies to achieve sustainability, which have been based on showing certain percentages indicating how much sustainability they have achieved in a particular area.

This paper shows how the new Climate New Deal policy can be implemented in cities with up to 1 million inhabitants with new SSE technology, but also in cities with more than 1 million inhabitants (using new strategies for mixing collectors—canopies), and solutions for megacities have also been proposed. Thus, this work can become a guide for mayors of all cities in the world on how to make their cities sustainable.

Therefore, this paper shifts the burden of combating climate change to cities, and if all of them would strictly implement the strategy of achieving sustainability, overall sustainability could be achieved.

The essential difference in policies to combat climate change is that they would no longer set percentages for reducing CO_2 emissions by a certain year (2030 or 2050) or seek to reduce the use of a fossil fuel (e.g., coal) or aim for climate neutrality, because past experiences in the fulfillment of agreements have shown that such tasks are unlikely to be carried out; rather, according to the authors of this paper, a new, much more ambitious policy ("Climate New Deal") should be sought, with the strategy aimed at achieving the sustainability of all cities in the world by 2050.

4.2. Solving Sustainability Problems for Cities with a Population of over 1 Million

Regarding the problem of sustainability, cities with a population of over 1 million (medium, large and megacities) in locations with over 1000 kWh/m^2/a of direct solar radiation generally do not have space for the collector systems of SSE technology, and distances from the city to the SSE system could be too large, in which case the collector layout strategies must obviously be altered. The roofs of buildings, currently used for photovoltaic and thermal collectors, are insufficient surfaces to achieve more serious energy production. Therefore, with such limitations, radical new solutions clearly need to be resorted to. The authors of this paper propose to place so-called "canopies" over low buildings, i.e., constructions with SSE collectors, which could achieve the required production of all types of energy and drinking water and would be in the vicinity of consumers. It is clear that the construction of structures with collector systems within the city would change the image of the city, but these structures could be made so as not to obscure the sun above a certain percentage in winter, and in summer, due to the shade they would produce, less cooling energy for the city would be required. Simply put, there would be entire neighborhoods covered by SSE collectors, so cities would have "canopies" with multiple functions (futuristic trends in solar architecture). In this way (by covering SSE collectors and placing them over low parts of the city), the SSE system would be in the city itself and would supply it with the necessary energy and drinking water in the same way (or according to the same concept) as proposed for the city of Zagreb. These cities could achieve sustainability, as the SSE system could continuously supply them with energy and drinking water throughout the year.

On the other hand, if cities with more than 1 million inhabitants are located in areas with relatively little solar energy (below 1000 kWh/m^2/a of direct solar radiation), then, clearly, different strategies must be applied. In this case (in this transitional period, e.g., in the next 30 years), there is no sustainable solution other than to use the energy of the RES system produced outside these cities (hydro, wind, biomass, etc.) and then transmit it to them in the form of electricity, which should become the main energy supply of such cities. As the transmission of electricity is far more efficient relative to heat transfer, electricity would be converted into thermal energy in the city itself in the SSE system (heating and cooling), after which all other processes would be the same as in SSE systems that directly use solar energy or would be reduced to them. However, this also raises the question of whether it will be possible to produce enough RES energy for these cities to be sustainable and whether this transition to RES systems will be possible in a certain period of time (the next 30 years) in order to prevent further climate change. Given that cities can only be sustainable or not sustainable, i.e., they cannot be "unsustainable-sustainable cities" because that would be an oxymoron, such cities would have to change and undergo a radical transformation in terms of depopulation policy whereby their populations are reduced to the limits required to achieve sustainability in parallel with the process of increasing energy participation from RES systems.

The authors are aware of the criticisms that such strategies for solving the problem of sustainability of large cities could provoke. However, the following truth is hard to deny:

> *If fossil fuels are responsible for the big cities of today, seriously threatening the climate system, then it is obvious that the transition to RES systems would lead in the other direction (reversal of trends), i.e., in the direction of the depopulation of such large cities and stopping climate change.*

Therefore, these processes should be reversed so that the situation with the climate system (global warming) is also reversed in the direction of its mitigation and global temperature drop.

If we do not want to give up such unsustainable cities and still want to prevent climate breakdown, there is only one solution left: in the upcoming period (30 years), these large cities must more intensively use nuclear energy, which does not emit CO_2. However, there are major ethical dilemmas because nuclear energy is not acceptable to people due to fears of accidents and the production and disposal of radioactive waste, and it is not a renewable energy source. However, stuck between the threat of climate breakdown, strong (actually the greatest) resistance of big cities to be sustainable and fears of nuclear power, humanity will have to choose, and the authors of this paper propose the transformation of big cities towards sustainability, which is the underlying focus of this paper.

4.3. The Issue of Energy Efficiency in Cities

Since buildings in the world consume about 35% of energy and are responsible for 38% of CO_2 emissions [40] and given that older buildings predominate in many cities, consuming more energy, in order to achieve sustainability, it is logical to take measures to significantly reduce their consumption.

In this sense, a good example is the city of Zagreb, which has a relatively large number of old buildings whose consumption exceeds 250 kWh/m^2/a, so the total energy consumption of the city buildings (both old and new) makes up almost half of the overall energy consumption. This also means that there is a very high potential to primarily reduce energy consumption in buildings by applying energy efficiency measures to the level of passive standard (15 kWh/m^{2}a), which would significantly reduce the total energy consumption in buildings [41,42] and significantly affect the size of the SSE system. Thus, by a parallel reduction in energy consumption in buildings and the construction of the SSE system, the SSE system could be significantly smaller.

4.4. Natural Resources for Implementation of the SSE Technology and Planning of Sustainable Cities

The advantage of SSE technology is that it is relatively simple and mainly uses available sources of raw materials (iron), so it is equally available to developed and underdeveloped countries. This is of particular importance because countries/companies that have invested heavily in the development of RES technologies will certainly not be willing to donate them to less developed countries

The question that could also be asked is why water for some cities would be obtained from SSE technology when it can be obtained by purifying water from rivers flowing through them. The answer is that rivers currently are polluted by different pollutants, i.e., wastewater from settlements (sewers) and industry, and microbiologically (viruses, bacteria, fungi and parasites) and chemically contaminated (heavy metals, nitrates, nitrites, detergents, pesticides, ammonia, etc.). Furthermore, their contamination with antibiotics is a growing problem, as well as the possible problem of radioactive waste (if the river operates to cool a nuclear power plant). These all result in the fact that fewer and fewer cities can safely use drinking water from their rivers or other natural water sources, because not all pollutants can be safely addressed and removed by using available water treatment methods. Therefore, obtaining drinking water with SSE technology (evaporation and subsequent condensation and additional treatment) could be the safest way to purify it.

Planning for sustainable cities is the most important and meaningful job because the overall success of climate protection depends on it. Partial solutions, such as those used in cities today (bicycle paths, electric cars, increasing green areas, placing photovoltaic collectors on roofs, painting facades and streets white, "greening buildings", etc.) might have a positive impact from a sociological point of view. However, they will not make these cities sustainable, nor make them resistant to increasing climatic extremes (floods, fires, droughts, etc.); rather, success in the direction of sustainability requires serious interventions and responsible management structures in cities that will be able to recognize the strategies for achieving sustainability, such as the system proposed in this paper.

4.5. Climate Change and Maintaining Sustainability

In the dynamics of the transformation of cities towards sustainability, it is necessary to emphasize another problem that relates to the fact that, with climate change, the input energy parameters of certain renewable energy sources also change (in addition to possible changes in water resources in terms of its absence if, for example, it is a lake or a small river). These primarily refer to the sun and wind, but also to hydropower and ocean energy (waves, tides, salinity gradient energy, and ocean temperature differences), so care should be taken to keep the city sustainable, not only in the present but also in the upcoming variable climatic conditions as society gradually changes/adapts to them.

When selecting a particular renewable energy source, the long-term data series provided by hydrometeorological institutes for a particular location in the city to be made sustainable will not be good enough but should be limited to shorter periods of time (e.g., 5 years). Special attention should be paid to trends (increases or decreases in solar radiation, wind energy, etc.). With greater climate change, it is logical that deviations from the budget for energy production from the SSE (RES) system at a particular location will be greater. In this sense, it is best to rely on the expertise of design engineers (who can perceive all relevant aspects of the problem and make decisions within given constraints), who will determine the reserves in the capacity of the SSE (RES) system in order to keep cities permanently sustainable (it is generally much more difficult to plan SSE/RES systems than fossil fuel energy systems, which are easily manageable—coal, oil and gas are added to power plants according to consumer needs), which also means that SSE (RES) systems can permanently and reliably supply energy and drinking water. The installed equipment of the RES system has its own lifespan, but it will be periodically maintained and changed, and the old equipment will be disposed of sustainably (which is extremely important, especially if PV collectors are used), as well as equipment for water reservoirs (e.g., geo foils), all in order to permanently maintain the sustainability of cities [43].

4.6. New Philosophy: Philosophy of Sustainability

Given the experience with the COVID-19 pandemic, where people, despite the dangers to their health and lives, offered great resistance to lifestyle changes, it is logical to expect that people will offer even greater resistance to more serious lifestyle changes aimed at sustainability. People do not want to change their current way of life or their worldview, and despite so much informatization, they do not have convincing answers from authorities to new questions, so they resort to skepticism. Therefore, it is necessary to change people's consciousness, which is one of the most difficult processes where a new philosophy could play a decisive role. All other disciplines originate from philosophy, and today, they are so extensive that it is difficult to perceive them, so they need to be comprehended from the philosophical aspect. All relevant aspects of achieving sustainability should be critically observed, and in that sense, a new way of thinking (a different consciousness of people) will be articulated.

Although the engineering profession should play a key role in achieving sustainability, because the whole process is very limited in time, the authors believe that this will not be possible without introducing a new philosophy as a parallel process of changing consciousness, which should provide answers to all questions surrounding the harmonization of civilization with nature and set new paradigms for achieving sustainability (e.g., related to cities, so they can continue to maintain and/or grow in the future only if the environment allows it). For this reason, the authors encourage the development of a new philosophical direction, "Philosophy of Sustainability".

5. Conclusions

This paper answers the question of how to systematically solve the problem of sustainability in 10,000 cities, as many as there are in the world today, whose number has increased significantly in the 5000-year-long period since the first industrial revolution and expanded in the second (because the energy of fossil fuels could be transmitted through electricity over long distances and brought into cities, where factories and workforces were situated), with an increasing number of inhabitants, all the way to megalopolises. Such cities have become the largest consumers of energy (78% of total energy consumption) and thus the largest polluters and the main cause of climate change.

To this end, this paper primarily criticizes the achievements related to climate agreements to date, because goals evidently will not be achieved in the future, as evidenced by rising global temperatures (which, in 2020, increased to as much as 1.2 ($\pm$0.1) °C compared to the pre-industrial period [3]). Even the latest appeal by over 230 editors of scientific journals in the field of medicine, i.e., the "call for emergency action to limit global temperature increases, restore biodiversity, and protect health" [4], clearly warns that a further rise in global temperature will disrupt biodiversity, human health and lives (described in Section 1).

In contrast to these approaches, which obviously do not yield results, the authors of this paper offer a completely different solution to achieve sustainability, emphasizing the need for a new ("top-down") climate policy, as already presented by the first three authors [6], which is based on the reorientation of the high-carbon economy into a green economy and the need for a radically new technology that should produce both energy (heat for heating and cooling and electricity) and drinking water. Of particular importance, the solution should be equally accessible to both developed and less developed countries because climate change is a common problem facing all of humanity.

In addition, the authors had to look critically at the ways in which cities today are trying to be made sustainable, which boil down to achieving certain percentages of sustainability in certain areas of sustainability. This shows that, at present, no city is completely sustainable: i.e., no city is completely independent of external sources of energy and drinking water, and this cannot be achieved using the previous approaches, so overall sustainability will not be achievable with the previous strategies. This leads to the conclusion that it is necessary to change strategies in order to achieve overall sustainability, which comes down to making all cities in the world sustainable, applying a systematic engineering approach to each of these cities, so in this way, they would be "bricks" that would build the overall sustainability of the Earth ("bottom-up").

Therefore, stopping climate change would be implemented by a new "top-down" global policy ("Climate New Deal" [6]) and a "bottom-up" strategy to make all cities in the world sustainable.

This paper pays special attention to this new strategy for achieving sustainability, which envisages the application of new Seawater Steam Engine technology in cities. The proposed SSE system could continuously supply cities throughout the year with electricity and heat (heating and cooling) and drinking water using four natural resources—solar energy (or some other renewable energy source), seawater (or other unclean water sources), gravitational potential energy and the Earth's thermal energy (Section 2.3).

This paper also presents an original concept of a sustainable city (using Zagreb, which has slightly less than 1 million inhabitants, as a case study), which can guide decision makers (mayors) of cities in the transition of today's cities towards sustainability. Thus, it is an original concept of energy and drinking water production with SSE technology, an original application of energy storage and an original concept of energy distribution through loops, which significantly increases the reliability of the supply of these resources (Section 2.3). The results for the example of the city of Zagreb show that the proposed system of production, storage and distribution of energy and drinking water can fully meet all needs of the city for energy and drinking water throughout the year, which is key to achieving sustainability. Furthermore, the paper presents an estimate of investment, which, for the example of the city of Zagreb (whose GDP was 22,695 EUR/inhabitant in 2018 [32]), shows that to achieve sustainability requires investments of about 2% of GDP in the next 30 years, which shows that Zagreb also has the economic strength to finance such a system (Section 3.4).

The paper also proposes a radically new strategy for solving the problem of collector accommodation in smaller cities (less than 1 million inhabitants), but also in larger cities, by installing SSE collector systems as new "canopies over cities" (Section 4.2). For cities with over 1000 kWh/m^2 of solar energy, despite the fact that such systems are considered uneconomical for this level of solar radiation, this paper emphasizes that in achieving sustainability, the question of the price should not be raised because sustainability has no price.

For cities that do not have enough solar energy, it is suggested that they continue to be supplied with energy outside of the city, which would of course be from the RES system.

The paper also addresses the problem of nuclear energy (Section 4.2), highlighting that humanity is stuck between a possible climate breakdown, the resistance of cities to achieving sustainability and the fear of nuclear energy (which of course does not emit CO_2), and in these conditions, decisions on how to solve the problem of big cities need to be made. The authors' suggestion is, of course, that energy be brought to cities without sufficient solar energy through other RES systems.

In addition to the aforesaid issues, the authors address another very sensitive topic: if cities cannot be made sustainable, their depopulation should be addressed; i.e., the number of inhabitants should be reduced to the limits required for their sustainability (Section 4.2).

Aware of the problem that without a change in consciousness, people will not be able to achieve the sustainability of cities and thus global sustainability (because the success of the whole action depends on all of humanity), the authors encourage the development of a new philosophical direction: "Philosophy of Sustainability" (Section 4.6).

This paper is therefore also an open appeal to the UN to revise the Paris Agreement after a public debate at the UN climate conference (COP28) in 2023 (when the first verification of the achievement of the goals of the Paris Agreement should be made) and adopt a different policy that could reverse the negative flows of global temperature increase, i.e., reduce global temperature.

Contemporary civilization puts the importance of living in cities ahead of sustainability and therefore continues to use fossil fuels (even subsidizing them) that have enabled the emergence of these cities and this way of life. However, it is not possible to achieve sustainability while cities continue to survive on the consumption of fossil fuels. Thus, humanity is facing a serious dilemma: i.e., either it will make all present-day cities in the world (as the largest consumers of energy) completely sustainable, or it will face the breakdown of the climate system and disappear.

The choice is very difficult because people generally do not want to give up this contemporary way of life. Nevertheless, the authors of this paper hope that the international community, i.e., those who make decisions on its behalf, will understand the seriousness of the situation and have enough empathy, both towards current generations (given that climate breakdown could happen relatively quickly because climate change does not occur linearly; rather, it is accelerating) and towards those yet to come and make the right decision.

Author Contributions: K.M., Z.G. and N.Z.L. conceived the presented idea, created the concept of the paper and original strategies for solving the problem of sustainability of all cities in the world, made calculations for the Sustainable City of Zagreb and wrote the manuscript. S.T. and A.F. supervised the manuscript content related to the economy and provided critical feedback. All authors have read and agreed to the published version of the manuscript.

Funding: This research received no external funding.

Institutional Review Board Statement: Not applicable.

Informed Consent Statement: Not applicable.

Data Availability Statement: All data presented in this study are included within the article.

Conflicts of Interest: The authors declare no conflict of interest.

References

1. IPCC. 2021: Summary for Policymakers. In *Climate Change 2021: The Physical Science Basis. Contribution of Working Group I to the Sixth Assessment Report of the Intergovernmental Panel on Climate Change*; Masson-Delmotte, V., Zhai, P., Pirani, A., Connors, S.L., Péan, C., Berger, S., Caud, N., Chen, Y., Goldfarb, L., Gomis, M.I., et al., Eds.; Cambridge University Press: Cambridge, UK, 2021; in press. Available online: https://www.ipcc.ch/report/ar6/wg1/downloads/report/IPCC_AR6_WGI_SPM.pdf (accessed on 5 September 2021).
2. World Meteorological Organization (WMO). Mediterranean Gripped by Extreme Heat, with New Reported Temperature Record. 2021. Available online: https://public.wmo.int/en/media/news/mediterranean-gripped-extreme-heat-new-reported-temperature-record (accessed on 27 August 2021).
3. World Meteorological Organization (WMO). 2020 Was One of Three Warmest Years on Record. 2021. Available online: https://public.wmo.int/en/media/press-release/2020-was-one-of-three-warmest-years-record (accessed on 27 August 2021).
4. Mora, C.; Dousset, B.; Caldwell, I.R.; Powell, F.E.; Geronimo, R.C.; Bielecki, C.R.; Counsell, C.W.W.; Dietrich, B.S.; Johnston, E.T.; Louis, L.V.; et al. Global risk of deadly heat. *Nat. Clim. Chang.* **2017**, *7*, 501–506. Available online: https://www.nature.com/articles/nclimate3322#citeas (accessed on 27 August 2021). [CrossRef]
5. BMJ. "Over 200 Health Journals Call on World Leaders to Address 'Catastrophic Harm to Health' from Climate Change: Wealthy Nations Must Do Much More, Much Faster." ScienceDaily, 6 September 2021. Available online: www.sciencedaily.com/releases/2021/09/210906091017.htm (accessed on 27 August 2021).
6. Glasnovic, Z.; Margeta, K.; Logar, N.Z. Humanity Can Still Stop Climate Change by Implementing a New International Climate Agreement and Applying Radical New Technology. *Energies* **2020**, *13*, 6703. [CrossRef]
7. United Nations. *Kyoto Protocol*; United Nations: Tokyo, Japan, 1997. Available online: https://unfccc.int/kyoto-protocol-html-version (accessed on 5 September 2021).

8. Stern, N. *Stern Review on the Economics of Climate Change*; Government of the United Kingdom: London, UK, 2006. Available online: http://mudancasclimaticas.cptec.inpe.br/~{}\{\}rmclima/pdfs/destaques/sternreview_report_complete.pdf (accessed on 5 September 2021).
9. United Nations. *Paris Agreement*; United Nations: Paris, France, 2015. Available online: https://unfccc.int/files/essential_background/convention/application/pdf/english_paris_agreement.pdf (accessed on 5 September 2021).
10. BloombergNEF. Climate Policy Factbook, Three Priority Areas for Climate Action. 2021. Available online: https://assets.bbhub.io/professional/sites/24/BNEF-Climate-Policy-Factbook_FINAL.pdf (accessed on 15 September 2021).
11. UN-Habitat. World Cities Report 2020, The Value of Sustainable Urbanization. 2020. Available online: https://unhabitat.org/sites/default/files/2020/10/wcr_2020_report.pdf (accessed on 15 September 2021).
12. United Nations. Climate Action, Cities and Pollution. 2021. Available online: https://www.un.org/en/climatechange/climate-solutions/cities-pollution (accessed on 15 September 2021).
13. Cassells, K. The World's Most Sustainable Cities. Uswitch, 14 May 2021. Available online: https://www.uswitch.com/gas-electricity/most-sustainable-cities/ (accessed on 15 September 2021).
14. IPCC. Engage with the IPCC. 2017. Available online: https://www.ipcc.ch/about/engage_with_the_ipcc/ (accessed on 5 September 2021).
15. Deane, J.P.; Ó Gallachóir, B.P.; McKeogh, E.J. Techno-economic review of existing and new pumped hydro energy storage plant. *Renew Sustain. Energy Rev.* **2010**, *14*, 1293–1302. [CrossRef]
16. García-Ayllon, S. Rapid development as a factor of imbalance in urban growth of cities in Latin America: A perspective based on territorial indicators. *Habitat Int.* **2016**, *58*, 127–142. [CrossRef]
17. World Population Review, World City Populations 2021. Available online: https://worldpopulationreview.com/world-cities (accessed on 15 September 2021).
18. United Nations. Ocean Conference, Factsheet: People and Oceans, New York, 5–9 June 2017. Available online: https://oceanconference.un.org/documents (accessed on 15 September 2021).
19. Institute for Sustainable Community. What Is Sustainable Community. Available online: https://sustain.org/about/what-is-a-sustainable-community/ (accessed on 15 September 2021).
20. Glasnovic, Z.; Margeta, K.; Premec, K. Could Key Engine, as a new open-source for RES technology development, start the third industrial revolution? *Renew. Sust. Energ. Rev.* **2016**, *57*, 1194–1209. [CrossRef]
21. Margeta, K.; Glasnovic, Z. Seawater Steam Engine-The most powerful tecnology for building sustainable communities and stopping the climate change. In Proceedings of the II International Energy & Environmental Summit—2017, Dubai, United Arab Emirates, 18–20 November 2017. [CrossRef]
22. Glasnovic, Z.; Margeta, K. *Seawater Steam Engine as a Prime Mover for Third Industrial Revolution*; LAP LAMBERT Academic Publishing: Saarbuecken, Germany, 2017.
23. Glasnovic, Z.; Margeta, K.; Omerbegovic, V. Artificial water inflow created by solar energy for continuous green energy production. *Water Resour Manag.* **2013**, *27*, 2303–2323. [CrossRef]
24. Glasnovic, Z.; Rogosic, M.; Margeta, J. A model for optimal sizing of solar thermal hydroelectric power plant. *Sol. Energy* **2011**, *85*, 794–807. [CrossRef]
25. Kummu, M.; Varis, O. The world by latitudes: A global analysis of human population, development level and environment across the north–south axis over the past half century. *Appl. Geogr.* **2011**, *31*, 495–507. [CrossRef]
26. Glasnovic, Z.; Margeta, J. Vision of Total Renewable Electricity Scenario. *Renew. Sust. Energy Rev.* **2011**, *15*, 1873–1884. [CrossRef]
27. Glasnovic, Z.; Urli, N.; Desnica, U.; Sesartic, M.; Miscevic, L.; Peric, N.; Firak, M.; Krcmar, S.; Etlinger, B.; Metikos, M.; et al. *Croatian Solar House: A Feasibility Study*; Rudjer Boskovic Institute: Zagreb, Croatia, 2002; ISBN 953-6690-24-1. (In Croatian)
28. Sesartic, M.; Glasnovic, Z.; Matijasevic, L.; Filipan, V.; Doerig, T.; Dejanovic, I.; Jukic, A. *Feasibility Study of the Application of Energy Efficiency Technologies in the Residential-Commercial Building Agria Osijek*; University of Zagreb: Zagreb, Croatia, 2007; ISBN 978-953-6470-29-7. (In Croatian)
29. Sesartic, M.; Glasnovic, Z.; Glasnovic, A.; Doerig, T.; Matijašević, L.; Filipan, V.; Jukić, A.; Margeta, K.; Dejanović, I.; Sesartić, A. *A New Standard for Energy Efficient Buildings in Croatia, Energy and the Environment 2008*; Croatian Solar Energy Association: Rijeka, Croatia, 2008.
30. Glasnovic, Z.; Sesartic, M.; Margeta, K. Straw as a building material. *Gradevinar* **2010**, *62*, 267–271.
31. Margeta, K.; Farkas, A.; Glasnovic, Z. Construction Materials Future. *Gradevinar* **2011**, *63*, 1009–1012.
32. City of Zagreb. GDP for the City of Zagreb and the Republic of Croatia, 2018, City Office for the Strategic Planning and Development of the City, Press Release, 4 March 2021. Available online: https://www.zagreb.hr/userdocsimages/arhiva/statistika/2020/bdp%202018/bdp%202018_web.pdf (accessed on 23 September 2021). (In Croatian).
33. Meteonorm 7.3.4. Worldwide Irradiation Data. Available online: https://www.meteonorm.com/ (accessed on 15 September 2021).
34. Premec, K. Climate data for Island Vis and Split in Croatia. *Meteorol. Hydrol. Serv.* **2014**, *57*, 1194–1209.
35. City of Zagreb. Statistical Yearbook of the City of Zagreb 2020, City Office for the Strategic Planning and Development of the City. 2020. Available online: https://www.zagreb.hr/UserDocsImages/1/SYCZ%202020_web.pdf (accessed on 23 September 2021).
36. City of Zagreb. Annual Energy Efficiency Plan of the City of Zagreb for 2017. Available online: https://knowledge-hub.circle-lab.com/article/4199?n=Annual-Energy-Efficiency-Plan-of-the-City-of-Zagreb (accessed on 23 September 2021). (In Croatian).

37. Dieckmann, S.; Dersch, J.; Giuliano, S.; Puppe, P.; Lüpfert, E.; Hennecke, K.; Pitz-Paal, R.; Taylor, M.; Ralon, P. LCOE reduction potential of parabolic trough and solar tower CSP technology until 2025. In *AIP Conference Proceedings*; American Institute of Physics: College Park, MD, USA, 2017; Volume 1850, p. 160004. [CrossRef]
38. Glasnovic, Z.; Sesartic, M. *Sustainable City Borongaj, Zagreb—Preliminary Design*; University of Zagreb: Zagreb, Croatia, 2008. (In Croatian)
39. COP26, Glasgow Climate Pact, Advance Version. Available online: https://unfccc.int/sites/default/files/resource/cma2021_L16_adv.pdf (accessed on 29 November 2021).
40. IEA. World Energy Statistics and Balances, Energy Technology Perspectives. Available online: https://globalabc.org/resources/publications/2020-global-status-report-buildings-and-construction (accessed on 5 October 2021).
41. Glasnovic, Z. Energy Efficient Building of the Faculty (FCET) as a Paradigm of Old Buildings of the City of Zagreb, Zagreb Energy Week, Faculty of Chemical Engineering and Technology, University of Zagreb, 10–16 May 2010. Available online: https://eko.zagreb.hr/UserDocsImages/zg_energetski_tjedan/2010/FKIT.pdf (accessed on 10 January 2022). (In Croatian).
42. Passive House Institute. "What Is a Passive House?", Passivhous Institute. Available online: https://passipedia.org/basics/what_is_a_passive_house (accessed on 5 October 2021).
43. Glasnovic, Z.; Margeta, K. *Revitalization of Hydro Energy: A New Approach for Storing Solar Energy, Solar Energy Storage*; Sørensen, B., Ed.; Elsevier: London, UK, 2015; Chapter 8; pp. 189–206, ISBN 978-0-12-409540-3.

Article

Examining the Relationship between Energy Consumption and Unfavorable CO$_2$ Emissions on Sustainable Development by Going through Various Violated Factors and Stochastic Disturbance–Based on a Three-Stage SBM-DEA Model

Wengchin Fong [1], Yao Sun [1] and Yujie Chen [2,*]

1. School of Business, Macau University of Science and Technology, Taipa, Macau 999078, China; anniefong.wsu@gmail.com (W.F.); crystal.yaosun@gmail.com (Y.S.)
2. Faculty of Hospitality and Tourism Management, Macau University of Science and Technology, Taipa, Macau 999078, China
* Correspondence: 2009853lbm20001@student.must.edu.mo

Abstract: The article applies a three-stage Slacks-Based Measure-Data Envelopment Analysis (SBM-DEA) pattern to examine the relationship between energy consumption and unfavorable CO$_2$ emissions on green sustainable development, for the 11 cities of the Guangdong-Hong Kong-Macao Greater Bay Area (GBA) during 2010–2016, by going through various violated factors and stochastic disturbance. Labor, capital and energy resource are chosen as input variables, while GDP and CO$_2$ emission as output variables. During the three phases consisting of the SBM-DEA model (first stage and third stage) and SFA analysis (second stage), CO$_2$ emission is considered as an unfavorable outcome, while stochastic statistical disturbances and external environmental influences are identified. The results show that the average efficiency of the GBA cities is 0.708, with only Shenzhen, Macao SAR and Hong Kong SAR having an efficiency of 1 during the whole study period. Based on the findings, suggestions are made for the GBA cities' sustainable development aspects.

Keywords: Guangdong-Hong Kong-Macao Greater Bay Area (GBA); energy efficiency; slack-based model (SBM); undesirable output; three-stage DEA; influencing factors

Citation: Fong, W.; Sun, Y.; Chen, Y. Examining the Relationship between Energy Consumption and Unfavorable CO$_2$ Emissions on Sustainable Development by Going through Various Violated Factors and Stochastic Disturbance–Based on a Three-Stage SBM-DEA Model. *Energies* **2022**, *15*, 569. https://doi.org/10.3390/en15020569

Academic Editors: Sergey Zhironkin and Michal Cehlar

Received: 10 December 2021
Accepted: 11 January 2022
Published: 13 January 2022

Publisher's Note: MDPI stays neutral with regard to jurisdictional claims in published maps and institutional affiliations.

1. Introduction

The World Commission on Environment and Development defined sustainable development as "development that meets current demands without jeopardizing future generations' ability to meet their needs" in 1987 [1]. In June 1992, in Rio de Janeiro, Brazil, at the United Nations Conference on Environment and Development, Agenda 21 first proposed sustainable development as the common development strategy of mankind toward the 21st century, transforming sustainable development strategy from a concept to a global scale of action and making sure that the most central part is economic sustainable development. Liu (1997) defines economic sustainable development as, "What we call sustainable development economy can be expressed as sustainable economic development, which should be the economy with the lowest ecological cost and social cost of economic development" [2]. Such economic sustainable development is economic development under the interests of future generations and the protection of natural resources. In addition to economic sustainability being at the core of sustainable development, energy consumption and environmental protection are also major focus of sustainable development strategies. Both show a one-way Granger causality and an inverted U-shaped relationship with sustainable development, respectively. In the early days, labor and energy-intensive industries were used for high economic development. But now, with sustainable development as a core value, cities are improving their energy consumption efficiency and increasing investment in environmental protection. At the same time, it is expected that they would attain sustainable development in the future. Here, the Greater Bay Area (GBA) in China is used as an example of sustainable development.

Figure 1 shows the total energy consumption, total CO_2 emission and total GDP of the GBA cities from 2010–2016. Although the total GDP kept on increasing since 2010, the total energy consumed has not changed tremendously. The amount of total CO_2 emission also demonstrated a decline to some extent.

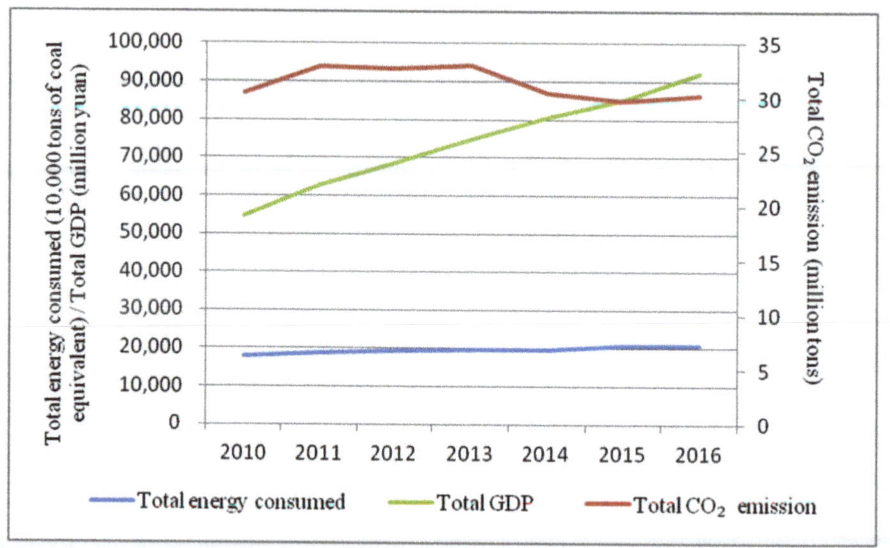

Figure 1. Total energy consumption, total CO_2 emission and total GDP of the GBA cities from 2010–2016. (Data Source: International Energy Agency, 2021).

Improving energy efficiency and considering undesirable output concurrently is one of the mainstreams in achieving the goal of sustainable development. However, most of the existing research literature is divided by provincial level or only includes 9 cities in Guangdong Province, China (Guangzhou, Shenzhen, Zhuhai, Foshan, Jiangmen, Zhaoqing, Huizhou, Dongguan, and Zhongshan), and few studies include Hong Kong SAR and Macau SAR, but GBA is a complete regional concept, so this study hopes to include data from Hong Kong SAR and Macau SAR by using the three-stage SBM-DEA model to study the complete DBA energy efficiency data for a total of 11 cities in GBA, the better usage of energy consumption toward green sustainable development can be investigated and use the data from these studies to provide better policy recommendations for those inefficient GBA cities, and eventually build an efficient and green GBA.

The followings are the research's main contributions: (1) This paper, with the best of my knowledge, is the first three-stage SBM-DEA model to be applied to discuss energy consumption and unfavorable CO_2 emissions on sustainable development from the various violated aspects, like industrial structure, energy structure etc. (2) When applying the model, previous researches focus on provincial or industrial analysis, while this study focuses on city-level evaluation, especially on GBA. (3) According to the UNEP Governing Council's definition of sustainable development, this research applies Hong Kong SAR and Macao SAR as decision-making units (DMUs) and then compare with other mainland cities. It is very important view of this research.

2. Literature Review

With the growing concern about adverse climate conditions and extreme weather due to global warming, a lot of efforts have been made to maintain the environmental sustainability. The GBA is considered one of the most economically dynamic regions in China since the reform and opening up. The development of GBA is of strategic importance in the development of China. In the early stages of development, most cities in the GBA achieved economic growth through labor- and energy-intensive industries, leading to over-exploitation of resources and environmental pollution. However, with the introduction of green and energy reforms in the Guangdong-Hong Kong-Macao Greater Bay Area in recent years, especially since the area consists of the cities of Guangdong as well as Hong Kong and Macao SARs, the study of the region has become representative. In February 2019, the state promulgated the Outline of the Development Plan of the Guangdong-Hong Kong-Macao Greater Bay Area, which states that priority is given to resource conservation, environmental protection, and natural restoration [3]. Therefore, how the government can improve the ecological environment and protect existing natural resources through various policies has become an important issue. Among them, controlling energy consumption is one of the keys to sustainable development, and this paper hopes to improve the energy efficiency of cities in the Guangdong-Hong Kong-Macao Greater Bay Area through policy recommendations, so as to achieve green and low-carbon development in the Guangdong-Hong Kong-Macao Greater Bay Area.

2.1. Sustainable Development

The IUCN initially established the notion of sustainable development in the World Conservation Strategy in 1980. This document states, "Sustainable development emphasizes the human use of the biosphere to manage it in order to meet the maximum sustainable interests of current generation while preserving its potential for the needs and desires of future generations" [4]. In 1989, the Governing Council of the United Nations Environment Program provided a more specific definition of sustainable development in its Statement on Sustainable Development: "It is defined as development that meets current demands without jeopardizing future generations' ability to meet their own needs, and it does not imply any violation of national sovereignty. According to the UNEP Governing Council, achieving sustainable development involves both domestic and international cooperation, include providing aid to developing nations in line with their national development programs' aims and development objectives. Furthermore, sustainable development entails a favorable international economic environment that promotes long-term economic growth and development in all countries, particularly in developing countries, which is critical for good environmental management. The maintenance, rational use, and enhancement of the natural resource base that promotes ecological stress tolerance and economic growth is also part of sustainable development. Furthermore, sustainable development refers to the integration of environmental concerns and considerations into development plans and policies, rather than a new kind of aid or development financing conditionality" [5].

In the above specific definition of sustainable development, it can be found that it encompasses various elements such as natural resources, economic growth, and the environment, which can be summarized as ecological development, economic development, and social development. Because of its central position in the sustainable development system, economic sustainable development has its unique definition. The British economist Barbier defines sustainable development from an economic perspective as "the maximization of net economic benefits while preserving natural resource quality and service provision" [6]. The British economist Pierce defines it as "economic development on the premise that natural capital remains unchanged, or that the use of resources today should not reduce real income in the future" [7]. The difference between the two is whether economic development is at the expense of resources and the environment. Barbier believes in maximizing economic benefits while preserving natural resource quality and using the money gained from environmental pollution and ecological damage as compensation for environmental and ecological construction. Pierce, on the other hand, believes that economic development should be done without destroying the world's natural resource base, and should not destroy before repairing. Yang (2002) defines sustainable economic development as "the continuous improvement of the economic welfare of the present generation on the basis of certain resources and environment, while being able to ensure that the economic welfare received by future generations is not less than the economic welfare enjoyed by the present generation" [8]. In addition to emphasizing the protection of natural resources, this definition also proposes another provision: emphasizing intergenerational equality.

The Greater Bay Area is an excellent object study for studying China's sustainable growth. The GBA includes the nine cities of Guangzhou, Shenzhen, Foshan, Zhuhai, Dongguan, Huizhou, Zhongshan, Jiangmen and Zhaoqing of the Guangdong Province, the Hong Kong Special Administrative Region (Hong Kong SAR) and the Macao Special Administrative Region (Macao SAR). GBA is considered as one of the most economically vibrant regions in China since the reformation. With a population of around 86 million, a total area of 56,000 km^2 and a GDP of USD 1669 billion in 2020 [9], the development of GBA has also its strategic importance in the development of China. In the early stage of development, most GBA cities achieved economic growth through labor and energy-intensive industries, resulting in over-exploitation of resources and pollution of the environment. However, with the introduction of green and energy reform in the GBA recently, especially with the GBA comprising of Guangdong municipalities as well as the Macao SAR and the Hong Kong SAR, the study on this area becomes representative. According to the February 2019 "Guangdong-Hong Kong-Macao Greater Bay Area Outline Development Plan" [3], priority is given to resource conservation, environmental protection and restoration of nature. Thus, efforts should be made to improve ecological and environmental quality, and then a greener and low-carbon development of GBA can be achieved.

In China, which is in the middle of industrialization, the most important factors hindering the implementation of sustainable development strategy are the significant share of high-energy-consumption industries, the unbalanced industrial structure, the backward technology and equipment, the large population base, the large number of poor people, and the generally low living standard of the people. Increasing industrial restructuring efforts, accelerating technology and equipment replacement, and actively advocating energy-saving production and lifestyle are the primary solutions to promote sustainable development in China.

Then, it is essential to do research on energy consumption efficiency for sustainable development, for the 11 cities of the Guangdong-Hong Kong-Macao Greater Bay Area (GBA) during 2010–2016, by going through various violated factors including investment, human capital, and environmental protection. Especially, compared with the high economic growth rate at the current stage, the low growth rate might lower the consumption of energy resources and other environmental aspects of nature, which play a catalytic role in the global process of sustainable development [10].

2.2. The Relationship between Energy Consumption, Environmental Protection, and Sustainable Development

Early Western mainstream economists fully recognized the role of capital, technological progress, labor, and institutions on economic growth, but did not pay attention to the impact of natural resources and environmental factors on economic growth. During the oil crisis in the 1970s, when economies suffered from economic growth difficulties of different degrees due to the scarcity of energy resources, natural resources and environmental factors began to be noticed by economists. The concept of sustainable development was introduced in the 1980s, and natural resources and environmental factors became important indicators for the analysis of sustainable development [11].

2.2.1. The Relationship between Sustainable Development and Energy Consumption

Energy is the material basis of social production and human survival, and is inextricably linked to the development of a country's economy. Sustainable development emphasizes the integrated and coordinated development of economy, resources, environment and population. Therefore, sustainable development of the energy economy means that the use and development of energy meet the needs of economic development without causing serious or irreversible damage to the ecosystem. It addresses the current generations' energy needs without posing a danger to the future generations' ability to satisfy their demands.

The original concept of a sustainable energy economy dates back to 1972, when the book Limits to Growth was published, which introduced the concept that social development and economic growth cannot be supported without energy, causing a heated controversy at that time [12]. Rashe and Tatom (1977) first introduced energy into the Douglas production function, hoping to uncover a fundamental pattern between economic growth and energy consumption [13]. Stem (1993) used a vector autoregressive (VAR) pattern with four variants (capital, energy consumption, labor, and GDP) and applied Granger multivariate causality tests to find a one-way Granger causality from energy consumption to GDP in the United States for 1947–1990 [14]. Willem et al., (2010) presented an explicit model of energy economic growth using a historical deductive approach in 2010, concluding that the current socio-economic model is unsustainable and that energy security is critical [15]. Zhang et al., (2011) examined the relationship between economic growth and energy consumption using the error correction model, cointegration test, and Granger causality test. They found that there is a significant one-way Granger causality of economic growth on energy consumption and that the two are in a long-term equilibrium relationship. [16]. Li et al., (2018) constructed a VAR model of the energy-economy-environment in Henan Province based on the statistical data of total energy consumption, GDP and industrial emissions from 2000–2014. Through impulse response function and variance decomposition, a long-term stable cointegration relationship between energy consumption, economic growth and environmental pollution in Henan Province is obtained, while energy consumption plays a certain positive impact on economic growth [17]. He et al., (2018) analyzed the relationship between energy consumption and economic growth in China since the 1950s by means of the elastic decoupling index and the generalized LMDI method and concluded that both total energy consumption and GDP growth exhibit an exponential growth curve [18]. Zhang and Wang (2018) examined the spatial variation of total factor energy efficiency in 30 provincial and municipal regions of China from 2006 to 2015 based on the haze constraint and concluded that economic development has significantly contributed to the improvement of energy efficiency [19]. Wang and Li (2019) studied Chinese coal cities and found a significant positive correlation between coal resources and economic growth in the sample cities [20]. Previous literature suggests a close relationship between energy consumption and sustainable economic development.

2.2.2. The Relationship between Sustainable Development and Environmental Protection

Environmental protection is the general term for human actions to ensure sustainable social and economic development and to solve real or potential environmental problems. Environmental sustainability is the goal of sustainable development that satisfies future generations' demands by protecting the environment and reducing ecological burdens through measures such as reducing pollutant emissions.

The most representative early study of the relationship between economic sustainability and environmental protection was the empirical verification of the environmental Kuznets curve by American economists Grossman and Kruger in 1990, which demonstrated the relationship between environmental pollution and economic growth is inverted U-shaped [21]. Subsequently, more studies demonstrated this curve. Shafik et al., (1992) found that the water quality of rivers in cities was deteriorating because of the fast growth of the economy; the concentration of suspended solids and sulfur dioxide in the atmosphere in cities showed an inverted U-shaped relationship with per capita income level; and sulfur dioxide emissions in cities increased with per capita income climbed [22]. Sherry and David (2008) studied the inverted U-shaped relationship between environmental pollution and economic growth from the perspective of theoretical analysis [23]. Salvador et al., (2010) used differential dynamics models to simulate the association between economic growth, CO_2 emissions and population, etc. [24]. Liang, et al., (2011) calculated the ecological environment index of 31 Chinese provinces and cities using entropy weight method and fuzzy comprehensive evaluation method, and analyzed the ecological environment status of different locations in the Yangtze River region, and carried out a study on the ecological environment sustainability with the Yangtze River basin as an example [25]. Feng et al., (2018) established a VAR model to analyze qualitatively and quantitatively the factors influencing carbon emissions in Beijing during 1996–2016, and the study found that carbon emissions per capita showed a positive relationship with GDP per capita [26]. Wang et al., (2018) studied the relationship between carbon emissions and GDP per capita in 27 provinces in China based on EKC theory. It was found that the relationship between carbon dioxide emissions and economic development of the 27 provinces as a whole, or geographically divided into eastern, central, and western provinces, met the inverted "U" curve of the EKC hypothesis [27]. According to Xu et al., (2019), green development is an expression of sustainable development, which is essentially a balanced relationship between the environment and the economy to promote economic development and ecological protection, so that the two complement each other [28]. Previous literature shows a close link between environmental protection and sustainable economic development, and it is mostly presented in an inverted U-shape.

2.2.3. Other Impacts on Sustainable Development

In addition to energy consumption and environmental protection factors, there are many external environmental factors that affect sustainable development, such as: industrial structure, energy structure, government financial expenditure, etc.

Tian (2007) proposed that the restructuring of industry forms a structural consumption of resources and thus leads to environmental changes, which largely affect the sustainable development of the economy [29]. Based on data from Shanxi Province, Wang and Gao (2018) used a double difference model to conclude that the increase in the degree of fiscal decentralization is conducive to the upgrading of industrial structure and thus has a positive impact on the ecological environment [30]. Yang (2000) analyzed the causal link between consumption of various energy sources and GDP (electricity, natural gas, oil, and coal) using the cointegration technique using sample data from 1954–1997 in Taiwan and found that there is a two-way Granger causality between GDP and coal and electricity and a one-way Granger causality from GDP to oil and natural gas to GDP [31]. Tian (2014) found that the fiscal expenditure policy has an indirect effect through influencing the economy and thus the ecological environment is significantly based on the analysis of fiscal revenue and expenditure data of Chinese provinces from 2008 to 2011 [32]. Jiang (2018) based on the analysis of data from 2007–2015 concluded that environmental spending has both economic and environmental nature and has a significant positive correlation with economic growth [33]. Hoff and Stiglitz (2001) argue that fiscal expenditure policies should focus on the updating and application of pollution treatment technologies to reduce pollution emission levels through technological innovation in order to enhance ecological protection and promote economic development [34]. Shi (2002) in "Recommendations on China's energy consumption decreased" proposed that opening up to the outside world has a significant role in improving the efficiency of energy use [35]. Xu (2004) selected Shandong Province as a sample to analyze the supply and demand of human capital in Shandong Province and made a qualitative and quantitative analysis study respectively calculated that the support capacity of human capital to sustainable development reached 66.65% [36].

2.3. SBM-DEA Model

Data Envelopment Analysis (DEA) is a non-parametric approach applied to obtain the efficiency of multiple Decision-Making Units (DMUs). It is like a peer group comparison using a frontier to determine efficient and inefficient units relatively. DEA has been widely used as a method in DMUs' efficiency analysis since the production function is not necessary for evaluation.

DEA was first introduced by Charnes, Cooper and Rhodes in 1978, and numerous researches have been done on methodological developments, practical applications, the status of variables and data variation, etc [37]. Since the first issued paper on using DEA in energy efficiency by Färe, Grosskopf and Logan (1983) on the relative efficiency of Illinois electric utilities, there are more and more researches on energy efficiency evaluation using DEA thereafter [38]. Bian and Yang (2010) used DEA models for estimating the aggregated efficiency of resources and the environment of the 30 provinces in China [39]. Guo et al., (2017) studied the dynamic DEA model to evaluate efficiencies based on fossil-fuel CO_2 emissions in OECD countries and China [40]. Iftikhar et al., (2018) applied the network DEA model under the free disability assumption for all undesirable outputs in 19 major economies to assess the energy and CO_2 emissions efficiency [41]. Mardani et al., (2017) also conducted an extensive review which indicated that DEA has shown to be a great evaluation tool for future analysis on energy efficiency issues, indicating that DEA is a common tool for energy and environmental efficiency researches [42].

Traditional DEA models include the CCR model and BBC model that deal with multiple inputs and multiple outputs to calculate efficiency. However, with the consideration of undesirable outputs such as atmospheric pollutants such as CO_2 emission, traditional DEA models do not perform well in increasing desirable outputs while decreasing undesirable outputs at the same time. Thus, Tone (2001) proposed a non-radial, non-angled slacks-based measure (SBM) model, which can improve radial models like the CCR model and BBC model that do not consider the slacks of inputs and outputs [43]. Du et al., (2016) constructed an SBM-DEA model to evaluate the total factor energy efficiency of 29 provincial-administrative regions of China during 1997–2011 and the influential factors [44]. Huang and Wang (2017) studied the total-factor energy efficiency of 276 cities in China during 2000–2012 by using a three-stage SBM model, with consideration of influential factors and undesirable output [45]. Shang et al., (2020) conducted a study using the SBM-DEA model to calculate the total factor energy efficiency including 30 provinces and municipalities of China from 2005 to 2016 [46]. Through previous literature, it is shown that SBM-DEA Model is a commonly used tool for assessing energy efficiency and can bring effective calculations for the energy efficiency assessment of DBA in this paper.

3. Methodological Framework

Various methods can be used to examine the energy efficiency regarding sustainability. One of the most direct ways is to evaluate the economic output in terms of energy input, i.e., GDP/energy consumption. In this study, however, a production possibility frontier theory is chosen as the method to examine energy efficiency. Instead of using only one input and one output to calculate efficiency, a nonparametric approach (DEA) is used, which has the advantage of measuring the relative energy efficiency of DMUs with multiple inputs and outputs. In addition, by application of a DEA method for efficiency analysis, the influences of external factors and stochastic disturbances can be removed.

Charles, Cooper and Rhodes are famous scholars in operation research who have developed a new systematic analysis approach known as Data Envelopment Analysis (DEA). During these years, different elaborations on DEA have been carried out by numerous researchers to tackle different situations and scenarios.

In a DEA model, each decision-making unit (DMU) is assumed to contain inputs $x \in R^n$, $y^g \in R^{S1}$ desirable inputs and $y^b \in R^{S2}$ undesirable outputs. The production possibility sets are thus defined as follows:

$$P = \left\{ \left(x, y^g, y^b \right) \ \middle| \ x \geq X\lambda, y^g \leq Y^g\lambda, y^b \geq Y^b\lambda, \lambda \geq 0 \right\} \tag{1}$$

which $\lambda \in R^n$ is the vector of intensity.

Based on the above production possibility set, a classic DEA model, known as a CCR model, can be obtained as follows:

$$\theta^* = \min\left[\theta - \varepsilon\left(\sum_{i=1}^{m} S_i^- + \sum_{r=1}^{s} S_r^+\right)\right]$$

$$\text{s.t. } \sum_{j=1}^{n} X_{ij}\lambda_j + S_i^- = \theta X_{ij}, \quad i = 1, 2, \cdots, m$$

$$\sum_{j=1}^{n} Y_{rj}\lambda_j - S_r^+ = Y_{rj}, \quad r = 1, 2, \cdots, s \tag{2}$$

$$\theta, \lambda, S_i^-, S_r^+ \geq 0; \quad j = 1, 2, \cdots, n$$

where S_i^- is the excess variable and S_r^+ is the insufficient variable respectively, making effective frontiers to expand horizontally or vertically to form the envelope. The variable θ represents the efficiency of the DMU.

Besides CCR, BCC model is another classic DEA model and is described as follow:

$$\min \theta_0$$

$$\text{s.t. } \sum_{j=1}^{n} \lambda_j x_{rj} \leq \theta_0 x_{i0}, \quad i = 1, 2, \cdots, m$$

$$\sum_{j=1}^{n} \lambda_j y_{rj} \leq y_{rj}, \quad r = 1, 2, \cdots, s \tag{3}$$

$$\sum_{j=1_j}^{n} = 1 \, j \geq 0, \quad j = 1, 2, \cdots, n$$

BBC model refers to pure technical efficiency, whereas CCR model refers to both scale efficiency and technical efficiency.

In this research, a three-stage DEA method is proposed to calculate energy consumption efficiency on sustainable development. The idea is to determine the comprehensive energy efficiency in phase 1, then eliminate the influence of stochastic disturbances and external environmental influences in the second stage, so as to enable the real efficiency to be evaluated in the third stage. The three-stage DEA methodology framework for the efficiency in the GBA cities is described in Figure 2.

3.1. The Initial Phase DEA: The Undesirable-SBM Pattern with Original Inputs

In comparison to CCR and BCC models, the undesirable-SBM model deals with input excess and output shortfall, according to Tone (2001) [43]. The formula is written as:

$$\rho^* = \min \frac{1 - \frac{1}{m}\sum_{i=1}^{m}\frac{S_i^-}{s_{i0}}}{1 + \frac{1}{S_1 + S_2}\left(\sum_{r=1}^{S_1}\frac{S_r^g}{y_{r0}^g} + \sum_{r=1}^{S_2}\frac{S_r^b}{y_{r0}^b}\right)} \tag{4}$$

$$\text{s.t. } \begin{cases} x_0 = X\lambda + S^- \\ y_0^g = Y^g\lambda - S^g \\ y_0^b = Y^b\lambda - S^b \\ \lambda \geq 0, \ S^- \geq 0, \ S^g \geq 0, \ S^b \geq 0 \end{cases} \tag{5}$$

From the above definitions, each DMU has m inputs, S_1 desirable outputs and S_2 undesirable outputs. The slacks of inputs, desirable outputs and undesirable outputs are S^-, S^g and S^b respectively. When a DMU is efficient, $S^- = 0$, $S^g = 0$ and $S^b = 0$. If either S^-, S^g and S^b is not equal to 0, then the objective function ρ^* is not 1, $0 \leq \rho^* < 1$, and this DMU is described as inefficient.

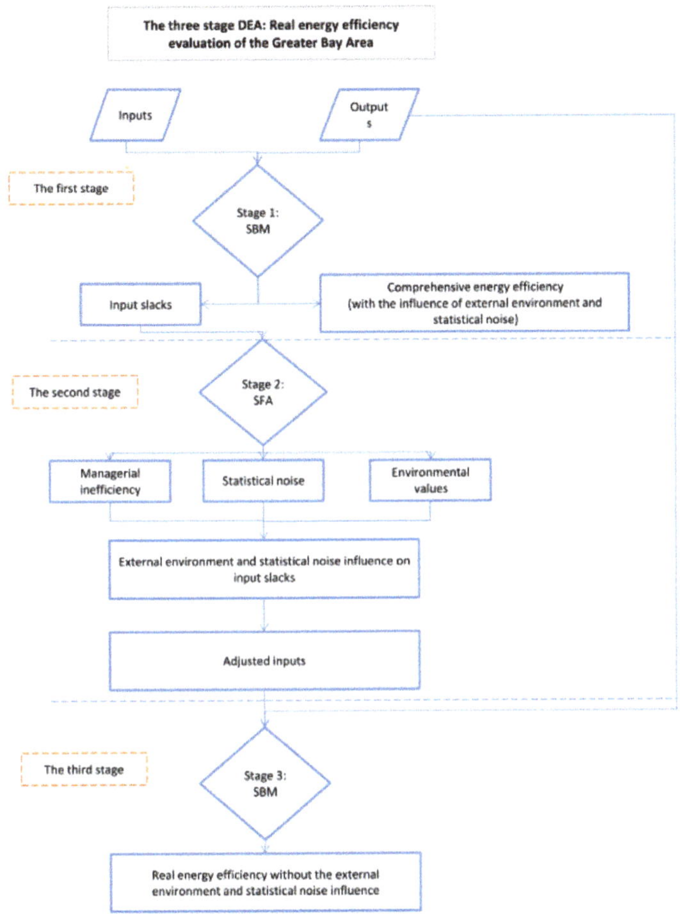

Figure 2. The three-stage DEA's methodological framework.

3.2. The Second Phase: Frontier Analysis with a Random Component

DEA is an ideal way to assess efficiency without having to know the production functions. However, as Fried et al., (2002) suggested, stochastic disturbances, environmental impacts, and managerial efficiency can all affect the efficiency calculated by DEA [47]. In order to deal with this disadvantage and to evaluate the real efficiency of each DMU, a SFA model is established to decompose the slacks in the first stage due to measurement errors in the input and output variables. To describe a SFA model, the environmental variables can be used as explanatory variables and the relaxation values in the first stage as interpreted variables:

$$S_{ni} = f^n(Z_i; \beta^n) + v_{ni} + \mu_{ni} \qquad n = 1, 2, \cdots N, \quad i = 1, 2 \cdots I \qquad (6)$$

In Equation (4), S_{ni} stands for the relaxation value of the n-th input of the i-th DMU and $f^n(Z_i; \beta^n)$ indicates the impact of the environmental variables on the relaxation value of the input and $Z_i = (z_{1i}, z_{2i}, \cdots z_{ki})$, $i = 1, 2, \cdots I$, and k are the external environmental variables. β^n is the coefficient of environmental variables. The term $v_{ni} + \mu_{ni}$ is the mixed error, in which v_{ni} is a random error term and $v_{ni} \sim N(0, \sigma_v^2)$; μ_{ni} is the management efficiency and $\mu_{ni} \sim N^+(0, \sigma_\mu^2)$. The term v_{ni} and μ_{ni} are independent and does not relate to each other. When $\gamma = \frac{\sigma_{\mu i}^2}{\sigma_{\mu i}^2 + \sigma_{vi}^2}$ is close to 1, it means that the difference in efficiency is affected by the managerial efficiency. However, when $\gamma = \frac{\sigma_{\mu i}^2}{\sigma_{\mu i}^2 + \sigma_{vi}^2}$ is near to 0, it suggests that efficiency remains primarily untouched by stochastic disruption.

The mixed error term can be further decomposed as follow [28]:

$$E[v_{ik}/v_{ik} + \mu_{ik}] = s_{ik} - f_i(z_k; \beta_i) - E[\mu_{ik ik} + \mu_{ik}]$$

Then, by applying the Dengyue's approach, $E[\mu_{ik}|v_{ik} + \mu_{ik}]$ can be solved as following [48]:

$$E[\mu_{ik}/v_{ik} + \mu_{ik}] = \frac{\sigma \lambda}{1 + \lambda^2} \left[\frac{\varphi\left(\frac{\varepsilon_k \lambda}{\sigma}\right)}{\varnothing\left(\frac{\varepsilon_k \lambda}{\sigma}\right)} + \frac{\varepsilon_k \lambda}{\sigma} \right]$$

with $\lambda = \frac{\sigma_u}{\sigma_v}$, $\varepsilon_k = v_{ik} + \mu_{ik}$, $\sigma^2 = \sigma_u^2 + \sigma_v^2$, φ, $\varnothing$ are the standard normal distribution's density and distribution functions respectively.

In order to adjust the input variables for real efficiency calculation under the same external environment, the SFA model is used to identify and exclude the associated environmental variables and stochastic disturbances. The formula for adjusted input variables X_{ni}^* is then described as:

$$X_{ni}^* = X_{ni} + \left[\max_i \{Z_i \beta^n\} - Z_i \beta^n \right] + \left[\max_i \{v_{ni}\} - v_{ni} \right] \tag{7}$$

where $n = 1, 2, \cdots N \, i = 1, 2, \cdots I$; X_{ni}^* denotes the adjusted input after the SFA model while X_{ni} indicates the initial input variable of the first phase; $\left| \max_i \{Z_i \beta^n\} - Z_i \beta^n \right|$ describes the external environmental variable adjustment and $\left| \max_i \{v_{ni}\} - v_{ni} \right|$ represents the adjusted stochastic disturbance value.

3.3. The Third Phase: The Undesirable-SBM Pattern with Adjusted Input Variables

During phase 3, the efficiency is once again evaluated using undesirable-SBM model but replacing the original inputs with the adjusted input variables X_{ni}^* instead. By comparing with the results of the first phase, the impact of stochastic disruption and external environmental variables have been removed as the outcomes of the third phase, indicating a real efficiency of the GBA cities.

4. Variables and Data Sources

The SBM-DEA Model's Input and Output Variables

During the examination of energy efficiency by using the SBM-DEA model, production assumptions are not necessary and multiple input and output variables can be handled at the same time. This is beneficial in cases when examining energy efficiency when details are uncertain. Table 1 shows a selection of input and output variables commonly used in DEA models.

Based on previous studies, the input variables can be justified by the total number of employees, the total investment in fixed assets and total energy consumption in this study.

Regarding output variables, both desirable output and undesirable output are included, especially with the global agenda of low carbon development. Therefore, both the environmental influence and the economic output should be considered. Hence, GDP is chosen as the desired output, while the undesirable output is CO_2 emissions based on Table 1. Table 2 presents the input and output variables that are chosen, while Table 3 demonstrates the relevant statistical descriptions.

Table 1. List of selected input and output variables commonly used in energy efficiency DEA models.

Author	Title	Input	Output
Khalili-Damghani, K., Tavana, M., & Haji-Saami, E. (2015)	A data envelopment analysis model with interval data and undesirable output for combined cycle power plant performance assessment	Fossil fuel	Energy, CO_2, SO_2, SO_3, NO_x
Zha, Y., Zhao, L., & Bian, Y. (2016)	Measuring regional efficiency of energy and carbon dioxide emissions in China: A chance constrained DEA approach	Labor, capital, coal, oil, natural gas	GDP, CO_2
Zhao, L., Zha, Y., Liang, N., & Liang, L. (2016)	Data envelopment analysis for unified efficiency evaluation: an assessment of regional industries in China	Capital input (Investment of fixed assets), labor, energy	Industrial production value, waste gas, wastewater
Chen, L., & Jia, G. (2017)	Environmental efficiency analysis of China's regional industry: a DEA based approach	Total number of employees of industry, total energy consumption, total investment in fixed assets of industry	GDP, total emissions of SO_2, industrial solid waste
Guo, X., Lu, C., Lee, J., & Chiu, Y. (2017)	Applying the dynamic DEA model to evaluate the energy efficiency of OECD countries and China	Land area, population, and energy use, carry-over variable, energy stock	CO_2 emission and GDP
Huang, J., Du, D., & Hao, Yu. (2017)	The driving forces of the change in China's energy intensity: An empirical research using DEA-Malmquist and spatial panel estimations	Energy, capital stock, labor, economic structure and GDP	GDP
Sueyoshi, T., & Yuan, Y. (2017)	Social sustainability measured by intermediate approach for DEA environmental assessment: Chinese regional planning for economic development and pollution prevention	Capital, labor and energy	Gross regional product, CO_2, SO_2, soot (dust), wastewater, COD, ammonia nitrogen
Sueyoshi, T., Yuan, Y., Li, A., & Wang, D. (2017)	Methodological comparison among radial, non-radial and intermediate approaches for DEA environmental assessment	Capital, labor, energy	Gross Regional Product, CO_2, SO_2, Smoke and dust, wastewater, COD, ammonia nitrogen
Zhao, L., Zha, Y., Wei, K., & Liang, L. (2017)	A target-based method for energy savings and carbon emission reduction in China based on environmental data envelopment analysis	Labor, capital stock, total energy consumption	GDP, CO_2
Cayir Ervural, B., Zaim, S., & Delen D. (2018)	A two-stage analytical approach to assess sustainable energy efficiency	Total renewable energy potential, network length, total installed power of renewable energy, transformer capacity	Gross energy generation from renewable sources, number of consumers, total exports, GDP per capita, human development index, total energy production, population, area
Iftikhar, Y., Wang, Z., Zhang, B., & Wang, B. (2018)	Energy and CO_2 emissions efficiency of major economies: A network DEA approach	Labor, capital, energy, population	CO_2 emission, middle income class, high income class, low income class
Khoshroo, A., Izadikhah, M., & Emrouznejad, A. (2018)	Improving energy efficiency considering reduction of CO_2 emission of turnip production: A novel data envelopment analysis model with undesirable output approach	Amount of energy for labor, machinery, diesel fuel, chemical fertilizers, seed and water	Turnip yield, GHG

Table 1. *Cont.*

Author	Title	Input	Output
Nadimi, R., & Tokimatsu, K. (2019)	Evaluation of the energy system through data envelopment analysis: Assessment tool for Paris Agreement	Heat, FEC, oil products, IMR, LEB, MYS, P_NRE, R_NRE, RE, IWA	CO_2, electricity, GDP, GNI per capita, QoL
Piao, S., Li, J., & Ting, C. (2019)	Assessing regional environmental efficiency in China with distinguishing weak and strong disposability of undesirable outputs	Labor (number of employees), water (total domestic and industrial water consumption), energy and capital	GDP, CO_2 (previous research), SO_2, wastewater and waste
Wu, J., Li, M., Zhu, Q., Zhou, Z., & Liang, L. (2019)	Energy and environmental efficiency measurement of China's industrial sectors: A DEA model with non-homogeneous inputs and outputs	Capital, labor, coal, oil, natural gas	Gross industrial output value, volatile hydroxyl-benzene, cyanide, COD, petroleum, ammonia-nitrogen, demand, petroleum, and ammonia-nitrogen
Yang, Z., & Wei, X. (2019)	The measurement and influences of China's urban total factor energy efficiency under environmental pollution: Based on the game cross-efficiency DEA	Productive capital stock	GDP, wastewater, SO_2, smoke and dust, pollutants (comprehensive index)
Zhao, H., Guo, S., & Zhao, H. (2019)	Provincial energy efficiency of China quantified by three-stage data envelopment analysis	Labor force (total amount of employees at year end), capital (total investment on fixed assets), energy consumption (total energy consumption)	GDP divided by amount of SO_2 emission
Zhou, Z., Xu, G., Wang, C., & Wu, J. (2019)	Modeling undesirable outputs with a DEA approach based on an exponential transformation: An application to measure the energy efficiency of Chinese industry	Energy, total asset	Total profit, industrial wastewater, industrial waste gas

Table 2. The choice of input and output variables.

Variable	Definition of Variables	Units
Inputs	Total number of employees	10,000 people
	Total investment in fixed assets	100 million yuan
	Total energy consumption	10,000 tons of coal equivalent
Desirable output	Gross domestic product (GDP)	100 million yuan
Undesirable output	Carbon dioxide (CO_2) emission	million tons

Table 3. The input and output variables, as well as the environmental elements, are statistically described.

Year	Variables	No. of Employees (10,000 ppl)	Investment on Fixed Assets (100 Million Yuan)	Energy Consumption (10,000 Tons of Coal Equivalent)	GDP (100 Million Yuan)	CO$_2$ Emission (Million Tons)	Secondary Industry GDP to Total GDP	Coal Consumption to Total Energy Consumption	R&D Investment (10,000 Yuan) to GDP	Local Fiscal Expenditure (10,000 Yuan) to GDP	Import and Export to GDP
2010	Average	359.17	1348.83	1641.30	4979.16	30.45	0.39	0.31	0.01	0.11	1.38
	Variance	60,080.56	1,152,070.33	2,032,778.57	23,002,949.16	715.36	0.05	0.06	0.00	0.00	1.13
	Maximum	758.14	3263.57	4775.60	14,928.92	99.32	0.63	0.76	0.03	0.17	3.60
	Minimum	31.48	230.10	98.35	965.12	2.13	0.05	0.00	0.00	0.06	0.23
2011	Average	373.25	1499.22	1716.27	5691.86	32.85	0.38	0.36	0.02	0.11	1.33
	Variance	68,132.36	1,390,138.24	2,205,935.34	28,291,300.48	774.88	0.05	0.07	0.00	0.00	1.13
	Maximum	828.86	3826.45	5013.40	16,257.63	103.90	0.63	0.94	0.04	0.15	3.67
	Minimum	32.76	297.09	98.00	1169.41	2.30	0.04	0.00	0.00	0.06	0.23
2012	Average	383.95	1700.31	1760.43	6223.67	32.67	0.37	0.35	0.02	0.11	1.26
	Variance	75,512.84	1,698,756.09	2,312,417.06	33,425,972.92	697.18	0.05	0.07	0.00	0.00	1.05
	Maximum	898.54	4348.50	5163.45	17,120.16	99.73	0.63	0.88	0.04	0.16	3.61
	Minimum	34.32	380.63	105.03	1279.64	2.34	0.04	0.00	0.00	0.07	0.23
2013	Average	400.43	1892.62	1799.00	6806.96	33.02	0.42	0.36	0.02	0.11	1.23
	Variance	88,584.02	1,900,441.68	2,425,523.24	38,804,566.21	700.95	0.04	0.09	0.00	0.00	1.04
	Maximum	967.14	4454.55	5333.57	17,971.06	100.18	0.62	0.98	0.04	0.15	3.56
	Minimum	36.10	448.20	98.11	1437.04	2.56	0.04	0.00	0.00	0.07	0.22
2014	Average	416.91	2062.03	1791.55	7317.69	30.41	0.42	0.35	0.02	0.12	1.17
	Variance	101,984.85	2,096,958.27	2,452,465.96	44,691,526.73	359.55	0.04	0.08	0.00	0.00	0.92
	Maximum	1034.58	4889.50	5496.46	18,993.87	66.02	0.63	0.83	0.04	0.16	3.49
	Minimum	38.81	678.07	107.66	1580.50	2.71	0.05	0.00	0.00	0.07	0.23
2015	Average	431.51	2299.13	1887.08	7758.39	29.70	0.41	0.31	0.02	0.15	1.11
	Variance	114,433.11	2,394,742.57	2,722,659.31	52,982,707.12	325.65	0.03	0.06	0.00	0.00	0.75
	Maximum	1100.80	5405.95	5688.89	20,155.98	67.33	0.62	0.74	0.04	0.22	3.19
	Minimum	39.65	726.85	117.57	1691.85	2.90	0.07	0.00	0.00	0.09	0.27
2016	Average	446.18	2496.35	1891.41	8378.46	30.22	0.40	0.31	0.02	0.14	0.97
	Variance	128,117.70	2,761,952.37	2,839,642.31	60,900,156.05	312.08	0.03	0.06	0.00	0.00	0.68
	Maximum	1165.73	5703.59	5852.60	20,930.51	65.55	0.61	0.68	0.04	0.23	3.05
	Minimum	38.97	640.43	121.83	1810.67	3.11	0.07	0.00	0.00	0.08	0.23

Note: No. of employees, investment in fixed assets and energy consumption are input variables; GDP and CO$_2$ emissions are output variables; the ratio of secondary industry GDP to total GDP, the ratio of coal consumption to total energy consumption, the ratio of R&D investment to GDP, the ratio of local fiscal expenditure to GDP and the ratio of import and export to GDP are environmental factors.

For differentiation, if the chosen input and output variables are appropriate, the Spearman's Rank correlation coefficients are calculated to demonstrate the strength of the link between inputs and outputs. The relevant outcomes in Table 4 show that the coefficients between input and output variables possess a positive correlation with significant values of 1% and 5%. Thus, the input and output variables have been chosen appropriately.

Table 4. Between input and output variables, the Spearman's rank correlation coefficients.

Inputs / Outputs	Labor	Capital	Energy
GDP	0.782 *** (0.004)	0.827 *** (0.0001)	0.627 ** (0.04)
CO_2 emission	0.873 *** (0.0004)	0.927 *** (0.00004)	0.709 *** (0.01)

Note: The significance level of *** and ** are 1% and 5% respectively. The relevant *p*-value is indicated by numbers in brackets.

External environmental variables do also exert influence on efficiency as mentioned by Liu et al. [49]. In terms of selecting these environmental variables, based on existing literatures, five major factors are analyzed in this research, which include industrial structure, energy structure, technology, government intervention and openness of economy [44,50,51]. In the SFA regression, Table 5 lists the factors chosen as explanatory variables.

Table 5. The selection of external environmental variables.

Factors	Explanatory Variables
Industrial structure IS	Ratio of secondary industry GDP to total GDP
Energy structure ES	Ratio of coal consumption to the total energy consumption
Technology T	Ratio of R&D investment to total GDP
Government intervention GI	Ratio of local fiscal expenditure to total GDP
Openness of economy OE	Ratio of import and export to total GDP

Secondary industries are known to be more energy intensive industries than tertiary industries. Thus, the industrial composition of a city exerts huge difference in energy demand and its energy efficiency. Besides, the types of energy resources used also affect the performance of outputs too. Coal is known to be an inexpensive but unclean source of energy. CO_2 emission is one of the environmental concerns when coal is being combusted tremendously during development. Thus, advancement in technology is necessary to clean up air pollution or even to utilize clean energy as a replacement. In this regard, the government can intervene with fiscal incentives to incur research and development in technology. Besides, the frequency of import and export often remarks the openness of an economy, hopefully with more technology transfer of goods and services taking place at the same time.

The data used in this research is mainly comprised of variables of 11 cities, including 9 Guangdong cities (Guangzhou, Shenzhen, Foshan, Zhuhai, Dongguan, Huizhou, Zhongshan, Jiangmen and Zhaoqing), and 2 special administrative regions, namely the Hong Kong Special Administrative Region (Hong Kong SAR) and Macao Special Administrative Region (Macao SAR) from 2010 to 2016. In most of the previous literatures on sustainable development of the cities of China, data of HK SAR and Macao SAR are often excluded due to statistical caliber. Therefore, in order to fulfill the consistency of input and output variables as well as for all cities in GBA, all data were derived from the official Statistical Yearbook of each Guangdong municipality, the Census and Statistics Department of the Hong Kong SAR and the Statistics and Census Service of the Macao SAR from 2010 to 2016. Since there is no official CO_2 emission data of the GBA cities available, relevant CO_2 data from 2010 to 2016 was directly obtained from research by Zhou et al. [52]. According to the study of Zhou et al., (2018), the total CO_2 emissions is calculated based on territorial emission accounting approach of the Intergovernmental Panel on Climate Change (IPCC), with inventories consisting of 17 kinds of fossil fuels, 47 socio-economic sectors and 7 industrial processes.

5. Results and Discussion

5.1. Results

5.1.1. The First Phase Undesirable-SBM Model: The Comprehensive Efficiency Calculation Results

During the first stage, with consideration of the undesirable output CO_2 emission, the undesirable-SBM pattern stated in article 3.1 was chosen to calculate the comprehensive efficiency of the 11 GBA cities from 2010–2016. The calculation was done with MaxDEA (Beijing Realword Software Company Ltd, Beijing, China) and the outcomes were displayed in Table 6.

Table 6. Comprehensive efficiencies of the GBA cities from 2010 to 2016.

Cities	2010	2011	2012	2013	2014	2015	2016	Average
Guangzhou	0.393	0.438	0.478	0.439	0.425	0.405	0.396	0.425
Shenzhen	1	1	1	1	1	1	1	1
Zhuhai	0.307	0.302	0.309	0.300	0.348	0.333	0.294	0.313
Foshan	0.374	0.371	0.378	0.348	0.359	0.373	0.354	0.365
Jiangmen	0.194	0.207	0.228	0.228	0.288	0.269	0.223	0.234
Zhaoqing	0.265	0.264	0.269	0.259	0.315	0.300	0.276	0.278
Huizhao	0.150	0.159	0.168	0.168	0.204	0.206	0.194	0.178
Dongguan	0.318	0.356	0.359	0.322	0.376	0.452	0.422	0.372
Zhongshan	0.203	0.215	0.231	0.243	0.344	0.328	0.315	0.269
Hong Kong SAR	1	1	1	1	1	1	1	1
Macao SAR	1	1	1	1	1	1	1	1
Average	0.473	0.483	0.493	0.482	0.514	0.515	0.498	0.494

5.1.2. The Second Phase: The Analysis of Influence of External Environmental Factors on Efficiency

During phase 2, the stochastic frontier analysis (SFA) approach is applied to analyze the influence of exterior environmental elements on the relaxation variables of inputs, namely the relaxations of the total number of employees, the slacks of total investment of fixed assets and the slacks of the total energy consumption. External environmental variables are elements that will affect energy efficiency, but not in a direct and controllable way. Five factors include the ratio of secondary industry GDP to total GDP, the ratio of coal consumption to the total energy consumption, the ratio of R&D investment to total GDP, the ratio of local fiscal expenditure to total GDP and the ratio of import and export to total GDP. The explanatory factors are these external environmental variables, while the slack inputs are considered as explained variables. Frontier 4.1 software is being used to create the SFA model. Table 7 shows the empirical findings.

After conducting the second stage SFA regression model, the input data in the first stage is adjusted with the coefficient values as shown in Table 7. With this adjustment, the effects from stochastic disturbances and external environmental elements can be eliminated. Adjusted input and original output variables can be run by the undesirable SBM-model once again to obtain the real energy efficiencies. The adjusted results of energy efficiency are compared with that of the original shown in Figure 3.

Table 7. SFA model parameters and estimation results in the second stage.

	Slacks		
Explanatory Variable	**Number of Employees**	**Investment in Fixed Assets**	**Energy Consumption**
Constant term	6.66 (42.56) ***	6.78 (3.72) ***	9.09 (24.11) ***
The ratio of secondary industry GDP to total GDP	−0.07 (−0.90)	1.97 (2.99) ***	0.23 (1.48)
The ratio of coal consumption to total energy consumption	−0.02 (−0.46)	−0.25 (−0.73)	0.05 (0.66)
The ratio of R&D investment to GDP	0.09 (2.24) **	0.25 (0.64)	0.15 (1.70) *
The ratio of local fiscal expenditure to GDP	−0.04 (−0.69)	−1.12 (−1.66) *	−0.07 (−0.54)
Import and export to GDP	−0.20 (−3.26) ***	−1.52 (−3.39) ***	−0.30 (−2.22) **
sigma-squared	12.08 (2.34) **	6.25 (2.12) **	18.32 (3.07) ***
gamma	0.99 (11414.37) ***	0.93 (26.08) ***	0.99 (4651.11) ***
LR test of the one-sided error	411.04 *	102.42 ***	359.24 ***

Note: The significance level of ***, ** and * are 1%, 5% and 10% respectively. The matching *t*-statistics of the computed parameters are indicated by numbers in brackets 5.1.3. The third phase: the real energy efficiency calculation results.

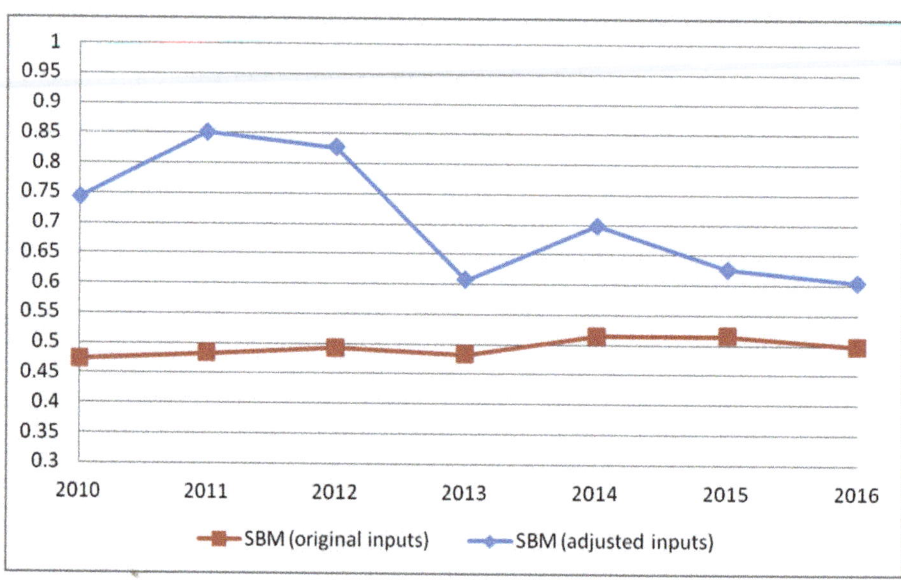

Figure 3. In the period 2010–2016, the average energy efficiency of the GBA cities' first stage (original inputs) and third stage (adjusted inputs) was compared.

5.2. Discussion

As seen in Table 5, Shenzhen, Macao SAR and Hong Kong SAR are operating at the efficient frontier from 2010–2016, with energy efficiencies equal to 1. This implies that Shenzhen, Macao SAR and Hong Kong SAR are efficient in terms of energy consumption. However, other cities such as Guangzhou, Zhuhai, Foshan, Jiangmen, Zhaoqing, Huizhao, Dongguan, and Zhongshan, their efficiencies are all below the average value of 0.494

from 2010–2016. Among inefficient cities, only Guangzhou and Dongguan are close to the average value, but still, the tendency of comprehensive efficiencies is not promising and have not been increasing obviously from 2010–2016. On the contrary, the comprehensive efficiencies of Shenzhen, Hong Kong SAR and Macao SAR demonstrate better performance in efficiencies and these cities have similar economic activities, industrial structure, technology and market openness.

According to previous literature demonstrating that there is a clear one-way Granger causality between sustainable economic development and energy consumption, the energy efficiency of the eight cities Guangzhou, Zhuhai, Foshan, Jiangmen, Zhaoqing, Huizhou, Dongguan and Zhongshan has been below the average from 2010–2016, and as the cities develop year by year, the low energy efficiency will have an impact on future sustainable economic development, and an efficient and green Energy consumption policy is an important issue at this stage to ensure sustainable development in the future.

From the SFA results, certain evaluated coefficients listed in Table 6 are significant at the 1%, 5%, and 10% significance levels respectively, indicating that certain external environmental conditions do have an impact on the GBA cities' energy efficiency. Besides, the correlation coefficients calculated in the SFA regression also represent the relationship between input relaxations and external environmental conditions.

For the GDP of secondary industries as a percentage of overall GDP, it shows a positive and 1% significant correlation with the slack of investment in fixed assets. An increase in the ratio of secondary industry GDP to the total GDP will increase the slack of the investment in fixed assets, which in turn bringing unfavorable impacts on energy efficiencies. Secondary industry is known to be an energy intensive industrial category and an adjustment policy in industrial structure to less energy intensive industry may have a positive effect on energy efficiencies of the GBA cities.

Coal combustion in energy generation is the known source of CO_2 emission. Thus, in Table 6, it shows that when the ratio of coal depletion to overall energy consumption rises, the slacks of energy consumption rise as well. However, the result is not significant.

R&D investment as a percentage of GDP shows positive and significant correlation coefficients with the slacks of number of employees and energy consumption. This finding is not similar to Zhao et al. [53], as investment of R&D usually encourages technology improvement and innovation that can contribute to efficiency in production process, causing a decrease in labor and investment demand, as well as more energy saving technologies. However, Liu (2020) mentioned that technological progress affects negatively on energy efficiency in her research [54]. One of the possible reasons maybe due to rebound effect, indicating that an increase in energy efficiency triggered by technology advancement is being offset by the increase in energy consumption demand.

The ratio of local fiscal expenditure to total GDP is negatively but not significantly correlated with the input slacks. An increase in government involvement in economic activities may be interpreted as incentives or drivers in energy efficiency improvement such as energy savings technology or implementation of relevant efficiency policies. The significant relationship between the ratio of local fiscal expenditure to GDP and investment in fixed assets signals the government's part in driving energy efficiency performance.

For the empirical study, the intensity of import and export activities shows a negative relationship with all the three input variables. Any increase in import and export as a percentage of GDP will decrease the slacks of inputs. Among all other environmental factors, import and export as a percentage of GDP appears to be the only factor that is significant to all the input variables of the GBA cities. It is worthwhile to note that the average import and export to GDP value of the GBA cities has been declining from 1.38 in 2010 to 0.97 in 2016 and the average energy efficiency has also dropped since 2010.

The average energy efficiency of the third stage during 2010–2016 has increased from 0.494 to 0.708, compared with the first phase. This explains that the overall energy efficiencies may have been underestimated in the first stage and are affected by the external environmental factors and stochastic disturbances. During phase 3 calculation, the average energy efficiencies of the GBA cities range between 0.604 and 0.851 throughout the study period from 2010–2016. From the first phase calculation shown in Figure 4, the values of energy efficiencies have increased but the average energy efficiency of the GBA cities does not improve during the period 2010–2016. In fact, the average energy efficiency had reached a peak of 0.85 in 2011 and since then, it started to drop to 0.605 in 2016. By eliminating the external environmental factors and stochastic disturbances after the third stage, a decrease in energy efficiency implies a real decline in managerial efficiency of the GBA cities since 2011.

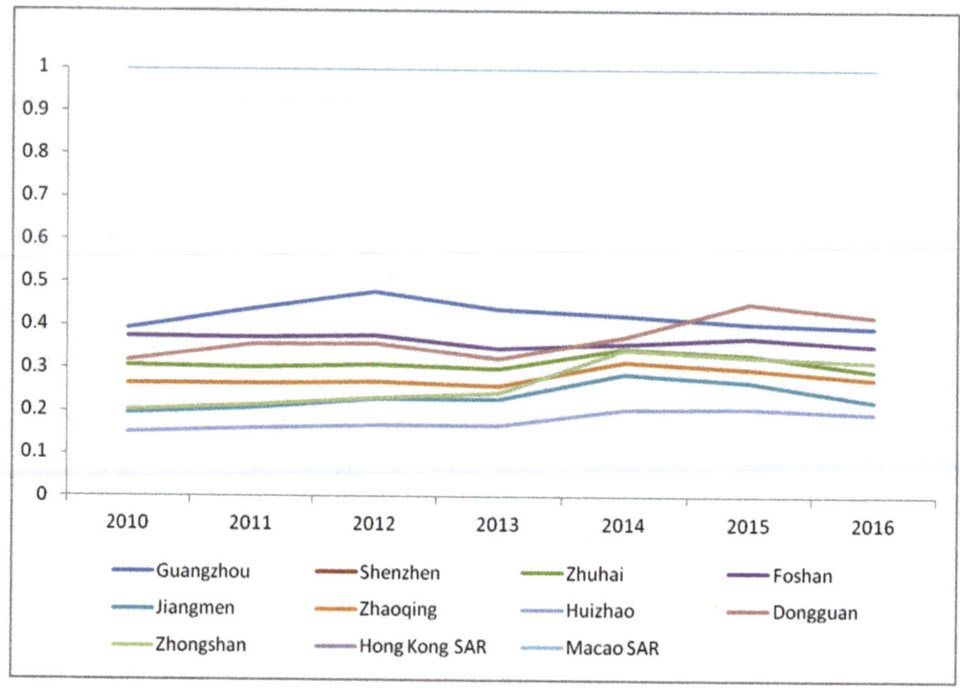

Figure 4. Energy efficiencies of the GBA cities in 2010–2016 (the first stage).

From Figure 5, it is shown that Shenzhen, Hong Kong SAR and Macao SAR are energy efficient cities, with an energy efficiency value of 1. When stochastic disturbances and external environmental elements are excluded after the SFA regression, the number of energy efficient cities varied from 4 to 7 in the studied years, including Shenzhen, Zhuhai, Zhaoqing, Zhongshan, Dongguan, Macao SAR and Hong Kong SAR. However, the number of cities with an average energy efficiency of 1 both in the first and the third stage of the SBM-DEA throughout 2010–2016 remain only 3, namely Shenzhen, Hong Kong SAR and Macao SAR.

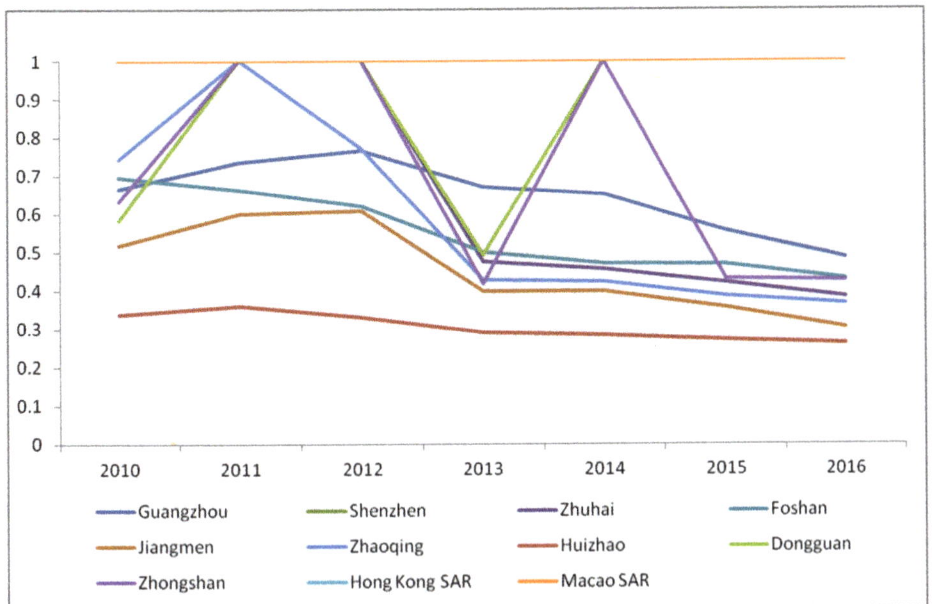

Figure 5. Energy efficiencies of the GBA cities in 2010–2016 (the third stage).

6. Conclusions and Policy Recommendations

6.1. Conclusions

The following conclusion can be obtained using a three-stage SBM-DEA model to determine energy efficiencies and investigate the influential elements of the GBA cities:

(1) Among the 11 GBA cities, only Shenzhen, Hong Kong SAR and Macao SAR have an energy efficiency of 1 from 2010–2016, both in the initial phase as well as in the third phase. This means that all these three cities operated at the efficient frontier during the study period. Other cities such as Guangzhou, Zhuhai, Foshan, Jiangmen, Zhaoqing, Huizhao, Dongguan, and Zhongshan, their energy efficiencies were all below the average value of 0.494 (the first stage) from 2010–2016;

(2) By eliminating the external environmental factors and stochastic disturbances by the stochastic frontier analysis during phase 2, the GBA cities' average energy efficiency during 2010–2016 has increased from 0.494 to 0.708 during phase 3 SBM-DEA model. This explains that energy efficiency has been underestimated at the first stage. Besides, Shenzhen, Hong Kong SAR and Macao SAR, Dongguan was the other GBA city that had an average energy efficiency of 0.87, which is above the average of all GBA cities. At the same time, a decreasing trend of energy efficiency after the third phase SBM-DEA pattern since 2010 had been observed. This implies a real decline in managerial energy efficiency of the GBA cities since 2010;

(3) Through the second stage stochastic frontier analysis, the influence of external environmental factors is investigated and among all, the ratio of import and export to GDP shows a negative and significant relationship to all three input variables, meaning an increase in the ratio will cause a decline in the input slacks, which favors energy efficiency.

6.2. Policy Recommendations

Based on the empirical analysis and conclusion, several policy recommendations are proposed for energy efficiency and sustainable growth:

(1) Acceleration of economic structure transformation. Instead of achieving the sole target of GDP growth, more emphasis should be made on environmental protection such as increasing the ratio of clean fuels and minimizing the emission of atmospheric pollutants such as CO_2 emission, especially under the agenda of sustainable development.

(2) Financial support from the government. Government intervention is always an important driver in encouraging energy-efficient and low-carbon production, especially during the initial stage in fixed assets investment. Government can also promote public awareness and enforce tougher environmental protection standards.

(3) Promotion of imports and exports. When products are less energy-intensive and environmentally friendly, an increase in imports and exports will not only facilitate a higher GDP, but also indirectly boost technological innovation and living standards.

This study, however, has a few limitations. First, this is the first research of this kind with the Guangdong-Hong Kong-Macao Bay Area cities as the DMUs and using the SBM-DEA model for energy efficiency evaluation, as far as we know. Thus, it is recommended that other DEA methods should be studied to examine the energy efficiency for comparison of results. Second, the structure of statistical data of mainland China and that of HKSAR and Macao SAR are quite different and may cause uncertainties in results. Further development of a homogeneous GBA statistical data collection system would be helpful to minimize the associated uncertainties. Last, different GBA cities exhibit different energy efficiency performances and are affected by different influential factors. Investigation on other social or economic influential factors for a wider aspect of policy formulation would be helpful.

Author Contributions: Conceptualization, W.F. and Y.S.; methodology, W.F. and Y.S.; software, W.F. and Y.S.; validation, W.F.; formal analysis, W.F. and Y.C.; investigation, W.F.; resources, W.F.; data curation, W.F. and Y.C.; writing—original draft preparation, W.F. and Y.S.; writing—review and editing, W.F. and Y.S.; visualization, W.F. and Y.S.; supervision, Y.C.; project administration, Y.C. All authors have read and agreed to the published version of the manuscript.

Funding: This research received no external funding.

Data Availability Statement: Not available.

References

1. World Commission on Environment and Development. *Our Common Future*; Jilin People's Publishing House: Jilin, China, 1997; p. 52.
2. Sihua, L. Some questions about sustainable development and sustainable economy. *Contemp. Financ.* **1997**, *6*.
3. Outline Development Plan for the Guangdong-Hong Kong-Macao Greater Bay Area. 2018. Available online: https://www.bayarea.gov.hk/en/outline/plan.html (accessed on 18 February 2021).
4. Guangyu, L.; Ziliang, S. *Economy, Environment and Law*; Science Publishing House: Karachi, Pakistan, 2000; p. 19.
5. Qigui, Z. *Sustainability Assessment*; Shanghai University of Finance and Economics Publishing House: Shanghai, China, 1999; pp. 19–20.
6. Barbier, E.B. *Economics. Natural Resource Scarcity and Development*; Earth Science: Hubei, China, 1989.
7. Pearce, D.; Wofford, J. *No End of the World—Economics—Environment and Sustainable Development*; China Financial and Economic Publishing House: Beijing, China, 1996; pp. 49–69.
8. Wenjin, Y. *Economic Sustainability Theory*; China Environmental Science Publishing House: Beijing, China, 2002; pp. 7–8.
9. Constitutional and Mainland Affairs Bureau, H.K.S. Guangdong-Hong Kong-Macao Greater Bay Area. 2018. Available online: https://www.bayarea.gov.hk/en/about/overview.html (accessed on 18 February 2021).
10. Lei, S.; Lituo, L.; Limao, W.; Fengnan, C.; Chao, Z.; Ming, S.; Shuai, Z. Scenario projection of China's energy consumption in 2050. *Nat. Resour. J.* **2015**, *30*, 361–373.
11. Lifeng, Z. Research on China's Energy Supply and Demand Forecasting Model and Development Countermeasures. Ph.D. Thesis, School of Economics, Capital University of Economics and Trade, Beijing, China, 2006.
12. Meadows, D.; Randers, J.; Meadows, D. *Limits to Growth*; Machinery Industry Publishing House: Shanghai, China, 2013.
13. Rashe, R.; Tatom, J. Energy resources and potential GNP. *Fed. Reserve Bank StLouis Rev.* **1977**, *59*, 68–76. [CrossRef]
14. Stern, D.I. Energy use and economic growth in the USA: A multivariate approach. *Energy Econ.* **1993**, *15*, 137–150. [CrossRef]
15. Nel, W.P.; van Zyl, G. Defining limits: Energy constrained economic growth. *Appl. Energy* **2010**, *87*, 168–177. [CrossRef]

16. Xinxin, Z.; Guangbin, L.; Lu, C. A study on the relationship between energy consumption and economic growth in China based on Granger test. *J. Shanxi Univ. Financ. Econ.* **2011**, *33*, 26–27.
17. Xiaofei, L.; Lichen, Z.; Kewen, L. Research on the dynamic relationship between energy—Economy—Environment in Henan Province based on VAR model. *J. Henan Univ. Technol.* **2018**, *1*, 48–54.
18. He, Z.; Yang, Y.; Song, Z.; Liu, Y. Mutual evolutionary trends and drivers of energy consumption and economic growth in China. *Geogr. Res.* **2018**, *8*, 1528–1540.
19. Zhang, Z.W.; Wang, Z.L. A study on spatially divergent measures of energy efficiency under technological heterogeneity and haze constraints. *East China Econ. Manag.* **2018**, *32*, 65–74.
20. Baoqian, W.; Jingya, L. An empirical study on the effect of "resource curse" in Chinese coal cities. *Stat. Decis. Mak.* **2019**, *35*, 121–125.
21. Grossman, G.M.; Krueger, A. Economic growth and the environment. *Q. J. Econ.* **1995**, *110*, 353–377. [CrossRef]
22. Shafik, N.; Bandayopadhyay, S. *Economic Growth and Environmental Quality: Time Series and Cross Country Evidence*; World Bank Publications: Washington, DC, USA, 1992.
23. Bartz, S.; Kelly, D.L. Economic growth and the environment: Theory and facts. *Resour. Energy Econ.* **2008**, *30*, 115–149. [CrossRef]
24. Puliafito, S.E.; Puliafito, J.; Grand, M. Modeling population dynamics and economic growth as competing species: An application to CO_2 global emissions. *Ecol. Econ.* **2008**, *65*, 602–615. [CrossRef]
25. Wei, L.; Konglai, Z. A study of ecological sustainability in the Yangtze river basin. *Explor. Econ. Issues* **2011**, *8*, 159–165.
26. Feng, M.; Yang, S.; Zifu, Z. VAR model analysis of carbon emission influencing factors—Based on Beijing city data. *Sci. Manag. Res.* **2018**, *36*, 78–81.
27. Wang, F.; Yang, X.; Yang, T. An empirical study on the relationship between carbon emissions and economic growth based on EKC hypothesis. *Ecol. Econ.* **2018**, *34*, 19–23.
28. Xianchun, X.; Xue, R.; Zihao, C. Big data and green development. *China Ind. Econ.* **2019**, *4*, 5–22.
29. Tao, T. A study on the relationship between industrial restructuring and the implementation of sustainable development strategy: The case of Hubei province. *Mod. Shopp. Mall* **2007**, *02*, 237.
30. Wang, L.Y.; Gao, Y.R. Fiscal decentralization and industrial structure upgrading: Empirical evidence from a quasi-natural experiment of "directly administered counties". *Financ. Trade Econ.* **2018**, *39*, 145–159.
31. Yang, H.Y. A note of the causal relationship between energy and GDP in Taiwan. *Energy Econ.* **2000**, *22*, 309–317. [CrossRef]
32. Dan, T. An Empirical Study of the Environmental Effects of Fiscal Balance Policy in China. Ph D Thesis, Wuhan University, Wuhan, China, 2014.
33. Nan, J. Can fiscal spending on environmental protection help achieve a win-win situation for both the economy and the environment? *J. Zhongnan Univ. Econ. Law* **2018**, *1*, 95–103.
34. Hoff, K.; Stiglitz, J.E. *Modern Economic Theory and Development*; World Bank: Washington, DC, USA, 2000.
35. Dan, S. Recommendations on China's energy consumption decreased. *China Econ. Trade Her.* **2002**, *18*, 26.
36. Yinliang, X. Analysis of Human Capital Supply and Demand in Shandong Province and its Ability to Support Sustainable Development. Master 'sThesis, Shandong Normal University, Jinan, China, April 2004.
37. Cook, W.D.; Seiford, L.M. Data envelopment analysis (DEA)—Thirty years on. *Eur. J. Oper. Res.* **2009**, *192*, 1–17. [CrossRef]
38. Färe, R.; Grosskopf, S.; Logan, J. The relative efficiency of Illinois electric utilities. *Resour. Energy* **1983**, *5*, 349–367. [CrossRef]
39. Bian, Y.; Yang, F. Resource and environment efficiency analysis of provinces in China: A DEA approach based on Shannon's entropy. *Energy Policy* **2010**, *38*, 1909–1917. [CrossRef]
40. Guo, X.; Lu, C.C.; Lee, J.H.; Chiu, Y.H. Applying the dynamic DEA model to evaluate the energy efficiency of OECD countries and China. *Energy* **2017**, *134*, 392–399. [CrossRef]
41. Iftikhar, Y.; Wang, Z.; Zhang, B.; Wang, B. Energy and CO_2 emissions efficiency of major economies: A network DEA approach. *Energy* **2018**, *147*, 197–207. [CrossRef]
42. Mardani, A.; Zavadskas, E.K.; Streimikiene, D.; Jusoh, A.; Khoshnoudi, M. A comprehensive review of data envelopment analysis (DEA) approach in energy efficiency. *Renew. Sustain. Energy Rev.* **2017**, *70*, 1298–1322. [CrossRef]
43. Tone, K. A slacks-based measure of efficiency in data envelopment analysis. *Eur. J. Oper. Res.* **2001**, *130*, 498–509. [CrossRef]
44. Du, H.; Matisoff, D.C.; Wang, Y.; Liu, X. Understanding drivers of energy efficiency changes in China. *Appl. Energy* **2016**, *184*, 1196–1206. [CrossRef]
45. Huang, H.; Wang, T. The total-factor energy efficiency of regions in China: Based on three-stage SBM model. *Sustainability* **2017**, *9*, 1664. [CrossRef]
46. Shang, Y.; Liu, H.; Lv, Y. Total factor energy efficiency in regions of China: An empirical analysis on SBM-DEA model with undesired generation. *J. King Saud Univ. Sci.* **2020**, *32*, 1925–1931. [CrossRef]
47. Fried, H.O.; Lovell, C.K.; Schmidt, S.S.; Yaisawarng, S. Accounting for environmental effects and statistical noise in data envelopment analysis. *J. Product. Anal.* **2002**, *17*, 157–174. [CrossRef]
48. Denyue, R. Three stage DEA model management inefficiency estimation note. *Stat. Res.* **2012**, *29*, 104–107.
49. Liu, H.; Zhang, Z.; Zhang, T.; Wang, L. Revisiting China's provincial energy efficiency and its influencing factors. *Energy* **2020**, *208*, 118361. [CrossRef]
50. Jie, H. Spatial network structure of energy and environmental efficiency in China and its influencing factors. *Resour. Sci.* **2018**, *40*, 14.

51. Li, H.; Fang, K.; Yang, W.; Wang, D.; Hong, X. Regional environmental efficiency evaluation in China: Analysis based on the Super-SBM model with undesirable outputs. *Math. Comput. Model.* **2013**, *58*, 1018–1031. [CrossRef]
52. Zhou, Y.; Shan, Y.; Liu, G.; Guan, D. Emissions and low-carbon development in Guangdong-Hong Kong-Macao Greater Bay Area cities and their surroundings. *Appl. Energy* **2018**, *228*, 1683–1692. [CrossRef]
53. Zhao, H.; Guo, S.; Zhao, H. Provincial energy efficiency of China quantified by three-stage data envelopment analysis. *Energy* **2019**, *166*, 96–107. [CrossRef]
54. Qingqing, L. *Green Total Factor Energy Efficiency Measurements and Impact Factors*; Jilin University: Jilin, China, 2020.

Article

A Strategy for Planned Product Aging in View of Sustainable Development Challenges

Małgorzata Niklewicz-Pijaczyńska [1,*], Elżbieta Stańczyk [1,*], Anna Gardocka-Jałowiec [2,*], Zofia Gródek-Szostak [3,*], Agata Niemczyk [4,*], Katarzyna Szalonka [1,*] and Magdalena Homa [1,*]

1. Faculty of Law, Administration and Economics Institute of Economic Sciences, University of Wroclaw, Uniwersytecka 22/26 Street, 50-145 Wrocław, Poland
2. Faculty of Economics and Finance, University of Bialystok, Warszawska 63, 15-062 Bialystok, Poland
3. Department of Economics and Enterprise Organization, Cracow University of Economics, 27 Rakowicka Street, 31-510 Krakow, Poland
4. Department of Tourism, Institute of Management at the Cracow University of Economics, 27 Rakowicka Street, 31-510 Krakow, Poland
* Correspondence: malgorzata.niklewicz-pijaczynska@uwr.edu.pl (M.N.-P.); elzbieta.stanczyk@uwr.edu.pl (E.S.); a.gardocka@uwb.edu.pl (A.G.-J.); grodekz@uek.krakow.pl (Z.G.-S.); niemczya@uek.krakow.pl (A.N.); katarzyna.szalonka@uwr.edu.pl (K.S.); magdalena.homa@uwr.edu.pl (M.H.)

Abstract: In this paper, the issue of the deliberate aging of products by manufacturers is discussed. Deliberate aging consists in intentionally planning or designing a product with an artificially limited lifetime in order to force consumers to replace it faster. The resulting rapid acceleration of the cycle of obtaining and utilizing consumer goods has serious consequences in the form of negative externalities. For this reason, the conscious aging of products is now recognized as the cause of unjustified consumption, generating huge economic and social costs and leading to the devastation of the natural environment and excessive exploitation of natural resources. Thus, it is in clear contradiction to the model of sustainable development. The aim of this paper was to identify the purchasing attitudes of buyers in the durable goods market. For its implementation, a pilot questionnaire study, which covered a representative group of 354 respondents, was carried out. The results indicate that the factors that influence the purchase of restitution goods depend on the type of product and the consumer's income. At the same time, about two thirds of the respondents recognized the problem of the deliberate aging of products. In their opinion, the goods produced in the autarkic economy were more durable and their life cycle was much longer. The results obtained require further empirical verification carried out in comparative studies.

Keywords: circular economy; sustainable development; planned obsolescence; environment; classification trees

Citation: Niklewicz-Pijaczyńska, M.; Stańczyk, E.; Gardocka-Jałowiec, A.; Gródek-Szostak, Z.; Niemczyk, A.; Szalonka, K.; Homa, M. A Strategy for Planned Product Aging in View of Sustainable Development Challenges. *Energies* **2021**, *14*, 7793. https://doi.org/10.3390/en14227793

Academic Editors: Sergey Zhironkin and Michal Cehlar

Received: 4 November 2021
Accepted: 17 November 2021
Published: 21 November 2021

1. Introduction

In the third decade of the 21st century, the circular economy has gained importance as an extension of the sustainable development concept by taking into account environmental protection and providing useful business solutions [1–3]. In this respect, product and process innovation has been identified as a key factor for growth and environmental sustainability [4]. An important role in the implementation of the sustainable and balanced growth assumptions is played by the value for the consumer and self-identity, which have gained a leading role in theory and practice [5,6]. Identity can be defined as the way an individual perceives himself or herself and how they decide to follow the values and behavior of the group of people with whom they want to be or feel connected [7]. In this context, identity can be defined as a key component of assessing the preferences and value-building process of an individual [8,9]. In the area of environmental or "green" issues, due to their growing importance and specificity, there is a general agreement among researchers that self-identity takes the form and name of "green identity" [7,10–12], referring to a "consumer's overall assessment of the product or service net benefit, between what is received and what is provided, based on the consumer's environmental wishes, balanced expectations and green needs" [11]. These research findings suggest that the perceived value of green products may be influenced by the consumer's value system—in particular, the degree to which the consumer identifies as an environmentally responsible person. At the same time, producers frequently use controversial practices that undermine the idea of sustainable development and reduce quality, and thus also consumer welfare. One of such activities is the phenomenon of planned obsolescence [13].

The strategy of the planned aging of products is an element of well-thought-out business activities [14,15], which leads to a decrease in their usefulness against economic [16], social and environmental premises [17]. The resulting rapid acceleration of the cycle of obtaining and utilizing consumer goods has serious consequences in the form of negative externalities [18]. Whether it is the accumulation of waste and pollution, greenhouse gas emissions, water eutrophication, destruction of the ozone layer or depletion of natural resources, all of these lead to ecological imbalance [19,20]. Undoubtedly, the planned aging of products contributes to the waste of natural resources and the degradation of the natural environment as a whole. This is the result of: (1) the use of new resources and energy to produce new goods; (2) sending an increasing amount of products to landfill; and (3) the use of new energy in the case of recycling.

In the case of planned aging, the usefulness of products is shortened against economic, social and environmental rationale. The resulting rapid acceleration of the cycle of obtaining and utilizing consumer goods has serious consequences in the form of negative externalities. The accumulation of waste and pollution, emission of greenhouse gases, water eutrophication, destruction of the ozone layer and depletion of natural resources lead to disturbances in the ecological balance. On the other hand, the continuous, artificially provoked exchange of products as an alternative to their repair increases household expenses, promotes consumerism by creating a dissonance between the real needs and the possibilities of the buyer [21] and eliminates jobs in the repair industry [21,22]. Thus, even if the argument that planned obsolescence helps save limited resources per unit produced (through the use of cheaper components or the use of technological progress) is accepted [23], it is impossible to ignore the fact that, on the whole, it leads to an increase in total production by reducing costs and lowering prices, stimulating a further demand growth [24]. It is worth noting that the issue of responsibility for the negative effects of planned product aging does not rest only with the producers. Perceiving the buyer only as a victim of a well-thought-out business strategy is definitely short sighted due to the fact that the decision to exchange a good is complex [24] and depends on the subjective assessment of the consumer [20,25–29]. For this reason, the problem of planned obsolescence has attracted the attention of many stakeholders, including the media, community and consumer protection organizations, academics, businesses and institutions, and has been the subject of wide public debate at local and international levels [30]. One of its effects is the gradual introduction of institutional solutions aimed at combating the abuses arising in this area and improving the effectiveness of their prosecution by law [31].

The aim of this paper was to identify the purchasing attitudes of buyers in the durable goods market. The authors wanted to assess whether the experiences of using household goods are positive or negative from the point of view of durability, functionality and usability. It was assumed that, due to the costs of restitution purchases and environmental protection, customers were guided not so much by the price of the product, but by the functionality and durability of the goods.

2. Literature Review

Planned aging of products, combined with the reluctance of individuals to postpone consumption in time, is now recognized as the cause of the development of consumerism which consists in the purchase of goods unjustified by real human needs, without taking into account social, environmental and individual costs [32]. It is the result of taking advantage of the emotional character of buyers' behavior by producers [32,33]. The emergence of consumerism coincided with the concept of "consumer society" promoted by advertisers in the 1920s and stimulated by the development of consumer credit. Consumers consistently had a new lifestyle imposed on them which focused on material values and a rapid increase in the standard of living [34]. Since then, the phenomenon of product aging developed in two directions. First, producers started to produce lower-quality products [35]. Secondly, companies tried to stimulate consumption using psychological mechanisms, encouraging buyers to get rid of products against functional and economic reasons for their further use [36]. With the passage of time, the growing awareness of the negative consequences of consumerism [37] and the progressive degradation of the natural environment meant that the problem of the planned aging of products began to be raised in a new context—excessive, unjustified exploitation of resources. Thus, it became one of the fundamental issues in the concept of sustainable development, whose important elements are sustainable consumption and production [38]. The change in consumer attitudes is confirmed by Eurobarometer surveys of June 2014. Their results show that 77% of EU consumers declared that they would rather repair damaged goods than buy new ones. They would also like to have easy access to unambiguous information on the durability of the products purchased and the possibility of their repair [23]. Meanwhile, according to a report published by the European Parliament, the average lifetime of products is progressively shortening. In the case of smartphones, toys and shoes, it is 1–2 years, computers and outerwear 3–4 years, vacuum cleaners and dishwashers 5–6 years and cars and refrigerators 7–10 years [23,25]. Today's household items are durable from 2 to 12 years, although they are made of materials that should remain usable for at least half a century—meaning 99% of the products will become obsolete over time, which will cost people from EUR 40,000 to 50,000 over the course of their lifetime [25]. For this reason, although the strategy of the planned aging of products is generally not prohibited by law, and manufacturers are free to determine the level of durability of their products [38], it is increasingly postulated that it should be regulated as a separate area of anti-consumer practices [39,40]. However, the situation is complicated by the fact that unambiguous identification of planned aging cases is extremely difficult and is based more on suspicions than on real examples [39,41]. An additional complication in the process of proving planned obsolescence is that corporate strategies are protected by industrial secrecy. Moreover, by 2015, no legal tools useful for tracking fraud had been developed, and attempts to use the method of comparing the viability of objects over time were not sufficiently reliable [23,42].

The year 2015 turned out to be a breakthrough. It was then that the French Parliament took a decisive action to combat the planned aging of products, which, as part of the larger movement against the planned aging of products in the European Union, introduced an obligation to declare the expected life of products and inform consumers about how long spare parts for them will be manufactured. In the same year, the French Assemblée Nationale imposed a fine of EUR 300,000 and a prison sentence of up to two years on manufacturers planning to fail their products in advance [43]. On the other hand, in 2016, a law was introduced according to which manufacturers are obliged to repair or replace a defective product free of charge under warranty within two years from the date of original purchase [44]. Thus, France not only became the leader in the application of sanctions against abuses related to the deliberate aging of products, but also the first country where the concept of planned obsolescence was defined for the purposes of law, assuming that it consists in the deliberate introduction of a defect, a planned stoppage, technical limitations, incompatibilities or other obstacles to repair [45]. Similar solutions of a national nature are also currently being introduced by other countries. For instance, the Spanish Foundation for Energy and Sustainable Innovation Without Planned Obsolescence grants companies a special ISSOP label confirming that they produce environmentally friendly goods and services without planned obsolescence, thus contributing to the reduction in emissions and the proper management of waste [46]. In Sweden, on the other hand, special tax breaks for the repair of used goods, including electronic equipment and clothes, have been introduced; e.g., VAT has been reduced from 25% to 12% when repairing bicycles, while the cost of repairing audio and video equipment and household appliances is tax deductible [47]. Even before the solutions adopted by France, in 2013, the European Economic and Social Committee announced that it was considering a total ban on the planned aging of products. The crowning argument was that replacing products that are supposed to stop working within two or three years of their purchase is an obvious waste of energy and resources and a way of generating unnecessary pollution [48–50]. The "Best Practices for Built-in Aging and Shared Consumption", convened in Madrid one year later, called for sustainable consumption to be recognized as a consumer right in EU legislation [50]. The postulate to introduce a labeling system indicating the durability of the product would give consumers the right to decide whether they prefer to buy a cheap product or a more expensive but more durable one [49]. On 4 July 2017, the "Resolution of the European Parliament on a longer life cycle of products: benefits for consumers and businesses" was adopted. It emphasizes that at all stages of the life cycle of products, a balance must be found between extending the life cycle of products, turning waste into secondary raw materials, industrial symbiosis, innovation [51], consumer demand, environmental protection and growth policy. At the same time, the development of increasingly resource-efficient products must not encourage shortening the life cycle of products or their premature disposal. On the other hand, elements such as product durability, extended warranty, availability of spare parts, ease of repair and exchangeability of components should be included in the manufacturer's commercial offer and meet the diverse needs, expectations and preferences of consumers.

As a result of the legislative measures taken, the EU member states, under Directive 1999/44/EC of the European Parliament and of the Council of 25 May 1999 on certain aspects of the sale of consumer goods and related guarantees, introduced provisions granting consumers a longer warranty protection [52]. On the other hand, from April 2021, regulations governing the lifetime of consumer products intended for households, such as computers, televisions, refrigerators, washing machines and dishwashers, came into force. In the future, it is planned to create a system to detect planned aging of products and to reorganize production to make it more sustainable. New products are to be durable, designed for energy efficiency, maintained and updated, and made of recycled materials. According to the Vice President of the European Commission for Jobs, Growth, Investments and Competitiveness J. Katainen, the following can be saved by European households: an average of EUR 150 per year, a reduction of over 46 million tons of CO_2 equivalent and energy savings equal to the annual energy consumption in Denmark by 2030 [53]. Although the problem of the planned aging of products has been present in economic considerations for a long time, the research undertaken in this area is fragmentary and focuses primarily on its supply side. It relates to a small material scope [54], a selected, narrow group of respondents [55] and a specific market structure [56] and is of the nature of regional analyses [57] without the possibility of universal scaling. Rare attempts to model the process mainly concern the high technology industry [58]. In the Polish literature on the subject, this problem is actually only taken up by Ryś from the perspective of consumer identification [59]. Bearing in mind the literature review and the measures cited in the fight against the planned aging of products, it is important to monitor the level of consumer awareness of environmental responsibility and consumer sensitivity to environmental issues, and the subject of sustainable development in the context of the strategy and promotion of an economy based on the reuse and repair of products.

3. Materials and Methods

3.1. Research Material

In the period from November 2020 to January 2021, a pilot study, whose purpose was to check the correctness of the assumed research procedure, the selection of respondents, the adopted variable indicators or the research tools used, was carried out. This was particularly important because the planned main study will be extensive (costly and time consuming). The preliminary examination conducted was to be a source of introductory information on the correctness of the adopted research procedure. In connection with the above, the source of empirical data in the analysis was a random sample of 384 respondents—Polish residents aged 19 and older, who were owners or users of durable products that were in their possession for at least 2 years. The main purpose of the questionnaire study was to determine:

[1] To what extent respondents noticed signs of planned aging of products by producers so that customers would buy durable goods more often;

[2] Whether the perception of this phenomenon depended on the identified attitudes of respondents (buyers of durable goods), especially towards the life expectancy of selected durable products.

In particular, an attempt was made to identify the main determinants of the perception of product aging by producers and distinguish respondent profiles in terms of attitudes towards preferences/criteria when purchasing durable goods, and noticing signs of planned aging of products. Therefore, the pilot survey questionnaire consisted of 2 parts: the first was devoted to customer purchasing decisions in the context of product life expectancy; the second was devoted to purchasing determinants of durable goods in a household. It included 20 questions, appropriately grouped thematically in terms of the period of functioning/use of selected products by respondents, self-assessment of the products used (in terms of quality of workmanship, esthetics, ease of use, energy efficiency), self-assessment of product brands in terms of durability, the importance of the lifespan of durable products, the main determinants of the purchase of selected products and opinions on the deliberate aging of products by producers. In addition to general questions, the questionnaire contained specific questions for selected durable products, such as: a washing machine, a refrigerator, a cooker, an electric kettle, a dishwasher, a vacuum cleaner, a computer, a mobile phone, a car and a bicycle. In the questionnaire, both closed and semi-open, single- or multiple-choice questions were used.

The survey was conducted through an electronic form, on an internet web platform. Participation in the study was anonymous.

3.2. Methodology Research

The analysis assumed that the explained (dependent) variable was the perception of the phenomenon of the deliberate aging of products by producers so that customers would buy durable goods more often (Y). This variable was assigned a Likert value scale depending on the respondent's answer to whether they agreed with the above statement:

1. Definitely not;
2. Probably not;
3. It's hard to say
4. Rather yes;
5. Definitely yes.

On the other hand, the group of independent variables was selected based on the questions included in the questionnaire concerning the following:

- Purchase of any durable product within the last two years (variable X_1 of the values 1—yes, 0—no);
- The main determinant of purchases of durable goods (X_2 variable concerning price, appearance, functionality, durability, innovation, quality, with values 1—yes, 0—no, for 10 individual products such as: a washing machine, a refrigerator, a cooker, an electric kettle, a dishwasher, a vacuum cleaner, a computer, a mobile phone, a car, a bicycle);
- Self-assessment of the importance of the life expectancy/functioning of the possessed durable products (X_3 variable with a scale of values: 1—very small; 2—small; 3—neither large nor small; 4—large; 5—very large);

- Self-assessment of the duration of operation of the equipment used in comparison to the assumed time on the day of purchase (X_4 variable with a scale of values: 1—definitely shorter; 2—neither longer nor shorter; 3—rather longer; 4—definitely longer);
- Self-assessment of used products in the household in terms of: quality of workmanship, esthetics, ease of use, energy efficiency (the X_5 variable relating to quality, esthetics, ease of use, energy efficiency, with values 1—very low level, 2—poor, 3—satisfactory, 4—good, 5—very good, for each product, such as a washing machine, a fridge, a cooker, an electric kettle, a dishwasher, a vacuum cleaner, a computer, a mobile phone, a car, a bicycle);
- Dealing with the breakdown of a product after the warranty period: buying a new one and throwing away the broken product, buying a new one and storing the broken product at home, repairing broken hardware and carrying product's use on (X_6 variable for these three situations: a new one–broken out, new one–broken at home, no new one–repair broken one, each with values 1– yes, 0—no);
- Belief in the correctness of the phrase: "many products available in the market are of lower quality than the products manufactured before the political transformation in Poland, i.e., before 1990" (variable X_7 with values 1—definitely not; 2—probably not; 3—hard to say; 4—rather yes; 5—definitely yes);
- Belief in the correctness of the phrase: "products made in Asia are less durable than those made in Poland" (variable X_8 with values 1—definitely not; 2—rather not; 3—it's hard to say; 4—rather yes; 5—definitely yes).

On the other hand, the control variables were as follows:

- Respondent's gender (X_9; W—female, M—male);
- Age group (X_{10}; 19–29; 30–49; 50–69; 70 years and more);
- Education (X_{11}; 1—elementary completed, no education; 2—lower secondary; 3—basic vocational; 4—post-secondary, secondary vocational and general education; 5—higher);
- Net income per person in the household (X_{12}; up to PLN 2000; PLN 2001—4000; PLN 4001—6000; PLN 6001 and more);
- Place of residence (X_{13}; 4—cities with more than 100,000 inhabitants; 3—cities with up to 100,000 inhabitants).

The first stage included the analysis of the independence of the dependent variable and the independent variables adopted.

To verify whether the two selected classification criteria of the group of respondents (pairs of variables: Y and one of the variables X_1 or X_{12}, which are qualitative in nature) are independent at the adopted significance level of $\alpha = 0.05$, appropriate multi-way tables were constructed, and the Pearson chi-square test of independence was used. In a situation where the size of the multi-way table was below 10, the chi-square and NW (maximum likelihood) tests were used [60]. The decision to reject/accept the null hypothesis was based on the probability p (p-value) determined for the established value of the empirical test statistic χ^2. If the p-value was less than or equal to the adopted significance level $\alpha = 0.05$, then the null hypothesis concerning the independence of X and Y was rejected in favor of the alternative hypothesis concerning the dependence of X and Y.

For this purpose, an algorithm for building classification trees, one of the data mining methods, was applied. The second stage of the analysis included an attempt to identify the main causes related to the phenomenon of the deliberate aging of products and, consequently, to distinguish relatively homogeneous classes of respondents taking into account the criteria resulting from the study (attitudes and behavior of users of durable goods). Its use in discriminant and regression analysis was initiated by Breiman et al. [60]. Currently, this method is increasingly used in many different research fields, not only in the field of statistics and econometrics but also in medicine, e.g., in determining survival predictions [61,62]. The use of classification trees in strategies of segmentation of individuals, e.g., recipients of health services, for which appropriate segmentation into homogeneous subgroups may constitute a basis for targeted interventions by the health service [63,64], or to identify significant food choices that affect health and define respondent profiles in terms of food choices [65], is of great interest. This wide use of classification trees is primarily related to the simplicity of the interpretation of results. This simplicity is valuable mainly because of the simple "model" that explains why observations are classified or predicted the way they are. In this type of analysis, classification tree methods can disclose relationships between several variables that would not be detected by other analytical techniques. Therefore, in the study presented, the classification tree algorithm was used to analyze the relationship of each explanatory variable (X_i, $i = 1, \ldots , 12$) with the dependent variable (Y) and to determine whether the respondents belonged to disjoint segments, or qualitative classes of the dependent variable (Y, perceiving the deliberate aging of products by producers) on the basis of measurements of explanatory variables (X_i, $i = 1, \ldots , 12$, i.e., selected attitudes and behaviors of buyers of durable goods).

Due to the qualitative and categorical nature of the dependent variable Y and the goal of classifying, the analysis was based on classification and regression trees. Unlike other statistical methods, classification trees do not require any preliminary assumption as to the nature of the relationship between the predictors and the dependent variable. Moreover, these methods are well suited to data mining tasks, where there is often little a priori knowledge, and there are no sound theories or assessments of what variables are related and how. This was a premise for the use of classification trees, and the variables that were selected for the model during the construction of the classification tree and that determined the subsequent divisions of the surveyed population of respondents (in the appropriate nodes of the tree) and profiled the relevant subsets of the surveyed population were, at the same time, the variables that determined the perception of signs of planned aging of products by manufacturers. The final division of the population studied is illustrated by the end nodes—the lists containing information on: the number of respondents assigned to individual classes and histograms of the frequency distribution of the dependent variable. In order to obtain a relatively simple classification tree, it was necessary to stop the procedure of recursive group division before achieving full homogeneity of segments and classes. For this purpose, the FACT (Fast Algorithm for Classification Trees) direct stop rule was used for a given fraction of objects—5% of the studied population.

The pilot nature of the study naturally resulted in a smaller sample size than in the target study; moreover, the limited duration of the study resulted in no response, which, ultimately, was the reason for the use of the simplified dependent variable. It turned out to be necessary to recode the five-valued original variable Y into a binary form.

The value of 1 that defines agreement with the opinion that producers deliberately age products so that customers buy durable goods more often was assigned by combining answers to one category ("rather yes" and "definitely yes"). On the other hand, possible responses to the primary variable in two variants: "probably not" and "hard to say' (no answer was recorded as "definitely no" in the pilot study), were assigned the value of 0.

Of course, such a simplification results in a loss of information, but due to the pilot nature of the study, the use of the simplified model is sufficient since it allows indicating the variables that, during the tree construction, had the greatest classification power, dividing significantly different respondents into classes. On this basis, despite the adopted simplification, it can be assumed that these are the variables that determine the respondents' attitude towards the opinion on aging products. In this sense, the simplified model can be used to test the analytical method, which in the target study will be based on the original dependent variable.

In addition, in order to avoid excessive detail in the analysis of all 10 products included in the study (such as: washing machine, refrigerator, stove, electric kettle, dishwasher, vacuum cleaner and computer, mobile phone, also a car, a bicycle) for each of the main determinants of the purchase of durable goods an auxiliary synthetic variable was constructed. The synthetic variable was the sum of the values of the variables recorded for individual

10 products (frequency of indicating a particular determinant).Thus, the synthetic variable X_2 for each main purchasing determinant has the form

$$X_2^{synthetic} = \sum_{i=1}^{10} X_{2(D)}^i,$$ (1)

where:

i—a product (1—a washing machine, 2—a fridge, 3—a cooker, 4—an electric kettle, 5—a dishwasher, 6—a vacuum cleaner, 7—a computer, 8—a mobile phone, 9—a car, 10—a bicycle);

D—Determinant that stands for price, appearance, functionality, durability, brand, innovativeness and quality. For each purchasing determinant, the X_2 variable may take values from 0 (when purchasing any product, the respondent did not use a given determinant) to 10 (when purchasing each of the aforementioned products, the respondent used the same determinant).

Similarly, in the case of the self-assessment of individual products used in the household in terms of: quality of workmanship, esthetics, ease of use and energy efficiency, an auxiliary synthetic variable was constructed as the average of the values recorded for the 10 individual products. Thus, the variable X_5 for each criterion of the evaluation has the form

$$X_5^{synthetic} = \sum_{i=1}^{10} X_{5(K)}^i$$ (2)

where:

K—Assessment criterion that stands for quality, esthetics, ease of use and energy efficiency. For each criterion to assess the products used, the X_5 variable may take values from 1 (each of the 10 analyzed products received the lowest score, i.e., 1—poor level) to 5 (each product received the highest score, i.e., 5—very good).

The calculations were conducted in the Statistica software purchased by universities under the Site License academic license. For the classification tree, the C&RT (Classification & Regression Trees) module was used, which is a complete implementation of the CART method introduced by Breiman and his colleagues from Berkeley [60].

4. Results and Discussion

Of the total number of respondents (owners/users of durable products for at least 2 years) who answered the questionnaire, 72% were women and 28% were men. This group was dominated by people aged 30–49, who constituted 41% of the sample. The lowest percentage of respondents was recorded in the case of the oldest age groups: for people aged 50–69, it was 9%, and for people aged 70 and older, it was 3%. More than 2/3 of the respondents declared the net income per person in their household with a value of, at most, PLN 4000 (including the net income for about half of the people, which ranged from PLN 2001 to PLN 4000). When considering the place of residence of the respondents, 75% were urban residents, including 39% who were inhabitants of cities with more than 100,000 inhabitants.

According to the information collected, 32% of respondents declared that they had purchased at least one durable product in the last two years. This subset was dominated by consumers who noticed the problem of the deliberate aging of products by producers so that customers would buy more often (they accounted for 64.5%). Over 10 times fewer respondents did not notice this problem, including not a single respondent who definitely did not agree with the occurrence of such a phenomenon (see Table 1).

Table 1. Opinions of respondents on the deliberate aging of products and the differences in product quality in Poland compared to the period before 1990 or compared to products from Asia.

Specification	Manufacturers Deliberately Age Products So That Customers Buy Durable Goods More Often (Y)	Many Products Available on the Market in Poland Are of Lower Quality Than the Products Manufactured before the Political Transformation, i.e., before 1990. (X_7)	Products Manufactured in Asia Are Less Durable Than Those Made in Poland (X_8)
		[%]	
Definitely no	-	2.5	0.8
Rather no	5.0	7.4	14.9
Hard to say	13.2	9.1	23.1
Rather yes	33.9	31.4	25.6
Definitely yes	30.6	31.4	17.4
No answer	17.4	18.2	18.2

Source: own study.

According to nearly 63% of the consumers in the survey, many products available in the market in Poland are of a lower quality than the products manufactured before the political transformation, i.e., before 1990 (compared to 10% of consumers who did not agree with this statement) (see Table 1). It is worth noting that the percentage of consumers who noticed the problem of the deliberate aging of products by producers was different among buyers of individual products in the last two years. This problem was most noticed by respondents who, in the last two years, purchased products such as a washing machine, a mobile phone and, the least, a TV set, a cooker or a car (see Figure 1).

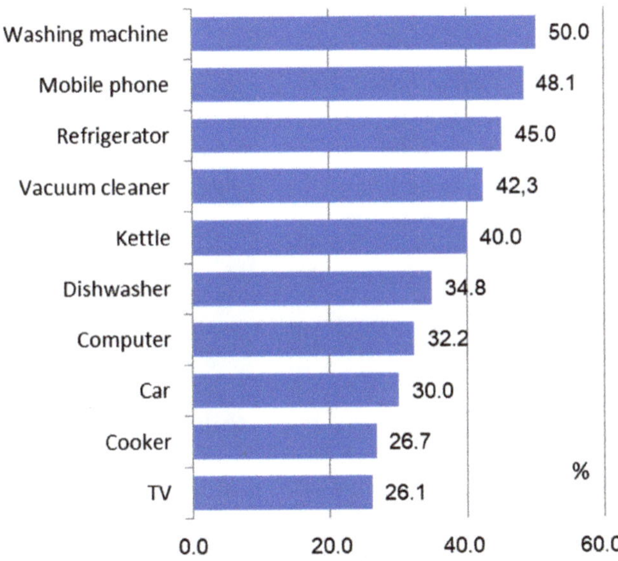

Figure 1. Percentage of buyers of selected durable goods who strongly agreed with the statement that producers deliberately age products so that customers buy more often.

For most products, functionality and durability were the main determinants of purchases: for instance, in the case of purchasing a washing machine or a cooker (see Table 2). However, in the case of a kettle or a mobile phone, the look was the main determinant, and in the case of a car purchase, the price was the main determinant. The results confirm that the distribution of determinants varies depending on the respondents' income (see Figure 2). In particular, for the vast majority of items, for a lower income level, the most important determinant among those analyzed when buying items was the price (especially in the case of a washing machine, a car or a computer). The respondents had different preferences when buying a cooker, a mobile phone and a bicycle. Even with lower incomes, innovation (when buying a cooker or a mobile phone) and durability (when buying bicycle) played a big role in these cases. On the other hand, people with the highest incomes were guided by the brand (except for the purchase of a car, in the case of which, irrespective of income, the price is of great importance).

Table 2. Main determinants of purchases of selected durable products.

Product	Percentage of Respondents for Whom the Main Purchasing Determinant Was:						
	Price	Appearance	Functionality	Durability	Brand	Innovativeness	Quality
				%			
Washing machine	28.9	13.2	43.8	34.7	14.9	12.4	31.4
Mobile phone	21.5	29.8	47.1	24.0	24.0	24.0	24.8
Refrigerator	23.1	27.3	38.8	36.4	12.4	10.7	27.3
Vacuum cleaner	23.1	9.9	47.9	24.0	11.6	14.9	24.8
Kettle	22.3	28.9	22.3	20.7	5.8	6.6	12.4
Dishwasher	13.2	9.9	30.6	28.9	6.6	6.6	21.5
Computer	21.5	19.8	37.2	23.1	15.7	17.4	24.0
Car	28.9	28.1	26.4	28.1	22.3	14.9	24.8
Cooker	16.5	21.5	35.5	28.1	6.6	13.2	24.0
TV	16.5	21.5	28.9	21.5	16.5	21.5	23.1
Bicycle	24.8	24.8	22.3	18.2	7.4	5.8	18.2

Source: own study.

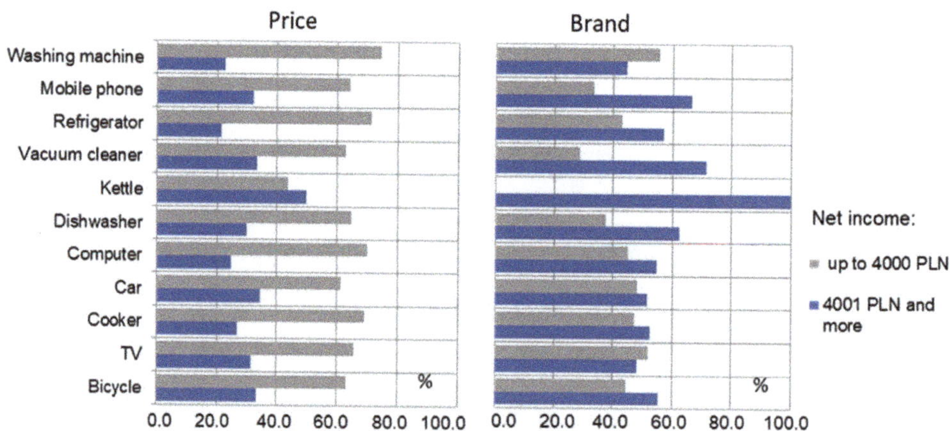

Figure 2. The main determinants of purchases (price and brand) depending on the amount of income: own study.

In the next stage of the analysis, for the subset of respondents who declared that they had purchased durable products, the following multi-way table (Table 3) presents the frequency of responses in terms of the expected length of operation of the used equipment compared to the assumed time on the day of purchase, and in terms of opinions on the deliberate aging of products by manufacturers.

In the analyzed subset of respondents, 59% expected that the purchased product would function longer than the assumed time on the day of purchase, while the second group, representing 17% of the respondents, expected that the purchased product would function for a shorter period of time. The distinguished groups of respondents differed in the frequency of perceiving the problem of product aging by producers. A total of 61% of the respondents from the first group agreed with the statement that producers deliberately aged their products so that customers would buy more often (25% strongly agreed). This percentage in the group of people expecting a shorter period of functioning of the purchased products was slightly higher—67% (including 38% definitely agreeing with the above statement).

The 324 users of the durables covered by the survey were also asked about what they would do if the product broke after the warranty period. The following results were obtained: purchase a new one and dispose of the defective product—15.1%; purchase a new one and store the broken product at home—5.7%; and repair broken equipment and continue its use—18.8%.

The distributions of respondents' answers presented in Table 4 indicate structural differences between the distinguished groups. The problem of product aging was most noticed in the group of people who declared that they had repaired broken equipment and that it would be continued to be used if the product broke after the warranty period—75% (answers "yes" and "definitely yes"). Among people who declared buying a new product and storing a broken product at home, this percentage was 50%, while among people who bought a new product and threw away the broken product, it was 60%.

Based on the responses of the respondents presented in Table 5, in terms of their self-assessment of owned and used selected products in a household, it was found that most of the products' highest rating related to the quality of workmanship, while the worst rating was in terms of energy efficiency. The opinions obtained from users of durable goods differed according to the type of product. On average, in terms of the quality of workmanship and esthetics, mobile phones and TV sets had the highest scores, and vacuum cleaners had the lowest; in terms of energy efficiency, the lowest scores were recorded for cars and vacuum cleaners, and the highest for dishwashers and refrigerators.

In order to verify the relationship between the dependent variable, namely, the perception of the problem of the deliberate aging of products (Y), and the explanatory variables selected for analysis ($X_1, \ldots X_{12}$), e.g., selected respondents' attitudes or control variables, the Pearson chi-square test of independence was applied. The results of the calculated values of the test statistics (Pearson chi-square or NW chi-square) and p-value (p value) for individual variables are presented in Tables 6 and 7.

The applied χ^2 test showed the existence of a correlation between the dependent variable, namely, the perception of the problem of the deliberate aging of products, and the variables: (1) belief in the correctness of the phrase: "many products available in the market are of a lower quality than products manufactured before the political transformation, i.e., before 1990"; and (2) in a situation where the purchased product broke after the warranty period, one would repair the broken equipment and use it further. Moreover, a significant relationship was found between the dependent variable and the control variable of age.

Considering the main determinants of purchasing the selected durable products declared by the respondents, on the basis of the value of the relevant test statistics, a relationship between variables—noticing the problem of the deliberate aging of products (Y) and deciding to purchase mainly based on price in relation to a washing machine, a refrigerator, an electric kettle and a dishwasher—was found. In the case of the variable relating to durability as the main determinant of purchases, a significant relationship occurred for the purchase of a washing machine, a refrigerator and a cooker.

Table 3. Opinions of respondents on the deliberate aging of products depending on the expected life of the equipment used.

Estimated Length of the Functioning of the Used Equipment Compared to the Assumed Time on the Day of Purchase (X_4)	The Degree of Acceptance of the Phrase "Manufacturers Deliberately Age Products So That Customers Make Purchases More Often" (Y)					Total
	Rather No	Rather Yes	Hard To Say	Definitely Yes	No Answer	
Definitely shorter than T *	-	-	-	1	-	1
Rather shorter than T	2	1	6	7	4	20
Neither longer nor shorter than T	-	5	10	11	3	29
Rather longer than T	3	8	21	14	11	57
Definitely longer than T	1	2	4	4	3	14
Total	6	16	41	37	21	121

* T—assumed time on the day of purchase. Source: own study.

Table 4. Structure of attitudes of respondents' opinions in terms of perceiving the problem of deliberate product aging by producers in subgroups determined by the type of declared behavior when the product broke after the warranty period.

Opinion on the Correctness of the Phrase "Manufacturers Deliberately Age Products So That Customers Make Purchases More Often" (Y)	Procedure When the Product Breaks Down after the Warranty Period (X_6):		
	Purchase a New One and Throw Away the Broken Product	Purchase a New One and Store the Broken Product at Home	Repair Broken Equipment and Use It Further
Rather no	2	1	5
Hard to say	9	6	8
Rather yes	20	6	29
Definitely yes	15	5	25
No answer	12	4	5
Total	58	22	72

Source: own study.

Table 5. Self-assessment of owned and used products in the household in terms of: quality of workmanship, esthetics and energy efficiency.

Product	Average Score *			Percentage of Respondents for Whom the Product Was Rated at Least Good		
	Quality	Esthetics	Energy Saving	Quality	Esthetics	Energy Saving
Washing machine	4.36	4.29	4.06	87.0	84.4	73.1
Refrigerator	4.43	4.46	4.07	89.6	89.0	76.5
Vacuum cleaner	4.05	3.96	3.85	76.6	71.5	66.0
Dishwasher	4.35	4.37	4.04	86.2	83.7	77.7
Cooker	4.38	4.30	4.05	86.1	83.3	74.3
TV	4.53	4.56	4.09	90.8	91.4	74.2
Car	4.31	4.21	3.70	88.5	84.0	64.3
Computer	4.50	4.53	4.07	89.9	88.0	75.5
Mobile phone	4.57	4.66	4.12	91.8	91.8	76.1

* A 5-point grading scale was used: 1—very poor level, 2—poor, 3—satisfactory, 4—good, 5—very good. Source: own study.

Table 6. Results of the chi-square test of independence for the dependent variable Y and the selected explanatory variables X_i.

Variable X_i	Value χ^2	p-Value
Gender (X_9)	0.234	$p = 0.628$
Age (X_{10})	8.756 *	$p = 0.032$ *
Education (X_{11})	0.476	$p = 0.788$
Net income per person in the household (X_{12})	0.307	$p = 0.958$
Place of residence (X_{13})	0.778	$p = 0.854$
Purchase of any durable product in the last two years (X_1)	3.781 *	$p = 0.050$ *
Self-assessment of the duration of operation of the equipment used compared to the assumed time on the day of purchase (X_4)	3351	$p = 0.500$
Belief in the correctness of the phrase: "many products available in the market are of a lower quality than the products manufactured before the political transformation, i.e., before 1990" (X_7)	10.352 *	$p = 0.0348$ *
Belief in the correctness of the phrase: "products manufactured in Asia are less durable than those made in Poland" (X_8)	4.746	$p = 0.314$
Procedure when the purchased product breaks down after the warranty period:		
Purchase a new one and throw out the broken one (X_6, a new one–broken out)	0.049	$p = 0.825$
Purchase a new one and store the broken product at home (X_6, a new one–broken at home)	2.647	$p = 0.104$
Repair broken equipment and continue its use (X_6, no new one–repair broken one)	7.200 *	$p = 0.007$ *

Source: own study. * The test result is statistically significant (significant relationship between X_i and Y).

Table 7. The results of the chi-square independence test for the dependent variable, namely, perceiving the problem, and the variable regarding the choice of durability or price as a purchasing determinant.

Product	The Main Determinant of Purchasing a Product—Durability X_2 (Durability)		The Main Determinant of Purchasing a Product—Price X_2 (Price)	
	Value χ^2	p-Value	Value χ^2	p-Value
Washing machine	3.393 *	$p = 0.065$ *	6.237 *	$p = 0.0125$ *
Refrigerator	3.963 *	$p = 0.046$ *	5.486 *	$p = 0.019$ *
Vacuum cleaner	2.609	$p = 0.106$	2.688	$p = 0.101$
Kettle	0.101	$p = 0.751$	5.080 *	$p = 0.0242$ *
Dishwasher	0.205	$p = 0.650$	4.415 *	$p = 0.035$ *
Cooker	6.021 *	$p = 0.014$ *	2.646	$p = 0.103$
TV	2.850	$p = 0.091$	2.646	$p = 0.103$
Mobile phone	0.007	$p = 0.932$	1.499	$p = 0.220$
Computer	1.502	$p = 0.220$	3.484	$p = 0.0619$

Source: own study. * The test result is statistically significant (significant relationship between X_i and Y).

In this paper, the development of a classification tree was carried out in order to verify the significance of the influence of the relation of each explanatory variable (X_i, $i = 1, \ldots 12$) on the dependent variable (Y) and to define disjoint segments and qualitative classes of the dependent variable (Y, perceiving the deliberate aging of products by producers) on the basis of measurements of explanatory variables (X_i, $i = 1, \ldots 12$, i.e., selected attitudes and behaviors of buyers of durable goods). During the construction of the model, cases of non-responses were omitted. As a result of the development of the classification tree, a ranking of the importance of variables (Table 8) based on one-dimensional divisions (where 0 means low importance, and 100 means high importance) was obtained.

According to the model obtained (Figure 3) and the ranking of explanatory variables (Table 8), the following factors had the greatest impact on the perception of the problem of the deliberate aging of durable products: (1) belief in the correctness of the phrase: "many products available in the market are of a lower quality than the products manufactured before the transformation system, i.e., before 1990, and (2) being guided by price and durability when making purchases. The gender of the respondent turned out to be the least significant, along with the manner and procedure in case the product broke after the warranty period: buying a new one and throwing away the broken product. Finally, after applying the adopted stop rule, six significant variables were taken into account in the structure of the classification tree, which had the highest classification power when creating the model (which is equivalent to the fact that they were crucial in the division of the entire population into classes of respondents, which differed significantly in terms of perceiving the problem of the deliberate aging of products). As a result of the procedure carried out, using appropriate measures of the quality of the division of the analyzed group of respondents according to the assessment of products' condition, 10 classes of respondents were distinguished ($C_1, \ldots C_{10}$)—see end nodes (lists) in Figure 3 and the set of classes of respondents in Table 9.

Table 8. Ranking of importance of variables relating to durability.

Variable	Importance
Belief in the correctness of the phrase: "many products available in the market are of a lower quality than the products manufactured before the political transformation, i.e., before 1990" (X_7)	100
Price is the main determinant of purchasing durable products X_2 (Price)	57
Durability is the main determinant of purchasing durable products X_2 (Durability)	54
Self-assessment of owned and used products in the household in terms of energy efficiency X_5 (Energy efficiency)	53
Belief in the correctness of the phrase: "products manufactured in Asia are less durable than those made in Poland" (X_8)	51
Self-assessment of owned and used products in the household in terms of the quality of workmanship X_5 (Quality)	49
Self-assessment of the duration of operation of the equipment used compared to the assumed time on the day of purchase (X_4)	44
Self-assessment of owned and used products in the household in terms of esthetics X_5 (Look)	44
Procedure in case the product breaks down after the warranty period: repairing the broken equipment and using it further, described by the variable X_6 (no new one–repair broken one)	39
Age (X_{10})	37
Net income per person in the household (X_{12})	37
Procedure in case the product breaks down after the warranty period: buying a new one and storing the broken product at home, X_6 (a new one–broken at home)	37
Place of residence (X_{13})	23
Education (X_{11})	11
Purchase of any durable product in the last two years (X_1)	8
Gender (X_9)	6
Procedure in case the product broke after the warranty period: buy a new one and throw away the broken product, X_6 (a new one–broken out)	6

Source: own study.

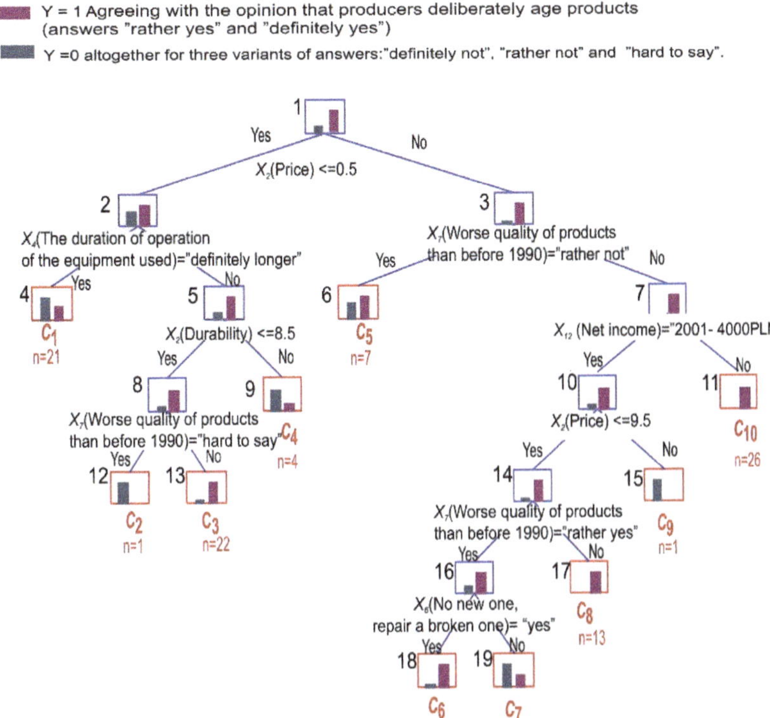

Figure 3. Classification tree for identifying the deliberate aging of products by manufacturers. Source: own study.

Table 9. Classes of respondents agreeing with the statement about the deliberate aging of products by producers so that customers buy durable goods more often.

Class	Characteristics	People Share
C_1	Number = 21 units (18% of the studied subset). The price was not taken into account when purchasing any durable product. It was expected that the possessed durable products would function longer than the assumed time on the day of purchase.	38%
C_2	Number = 1 unit (approximately 1% of the studied subset). The price was not taken into account when purchasing any durable product. It was not expected that the possessed durable products would function longer than the assumed time on the day of purchase. For 8 or fewer products out of 10 analyzed, durability was the main determinant of purchases. No opinion on the lower quality of products available in Poland than before 1990.	0%
C_3	Number = 22 units (19% of the studied subset). The price was not taken into account when purchasing any durable product. It was not expected that the possessed durable products would function longer than the assumed time on the day of purchase. For 8 or fewer products out of 10 analyzed, durability was the main determinant of purchases. They agreed with the statement that many products available on the market in Poland are of a lower quality than products manufactured before 1990.	90%

Table 9. *Cont.*

Class	Characteristics	People Share
C_4	Number = 4 units (3% of the total number of respondents). The price was not taken into account when purchasing any durable product. It was not expected that the possessed durable products would function longer than the assumed time on the day of purchase. For over 8 products out of 10 analyzed, durability was the main determinant of purchases.	25%
C_5	Number = 7 units (6% of the total number of respondents). When buying at least one durable product, the price was the main determinant. They disagreed with the statement that many products available on the market in Poland are of a lower quality than products manufactured before 1990.	57%
C_6	Number = 9 units (8% of the total number of respondents). When buying at least one durable product, the price was the main determinant. They agreed with the statement that many products available on the market in Poland are of a lower quality than the products manufactured before 1990 (or there was no opinion on this subject). In a situation where the product broke after the warranty period, the broken equipment was repaired and used further.	89%
C_7	Number = 3 units (3% of the total number of respondents). When buying at least one durable product, the price was the main determinant. They agreed with the statement that many products available on the market in Poland are of a lower quality than the products manufactured before 1990 (or there was no opinion on this subject). In a situation where the product broke after the warranty period, new equipment was purchased.	33%
C_8	Number = 13 units (11% of the total number of respondents). When buying at least one durable product, the price was the main determinant. They strongly agreed with the statement that many products available on the market in Poland are of a lower quality than products manufactured before 1990. Net income per person in the household ranged from PLN 2001 to PLN 4000.	100%
C_9	Number = 1 unit (about 1% of the total number of respondents). When buying at least one durable product, the price was the main determinant. They agreed with the statement that many products available on the market in Poland are of a lower quality than the products manufactured before 1990 (or there was no opinion on this subject). Net income per person in the household ranged from PLN 2001 to PLN 4000.For all products, price was the main determinant of purchases.	0%
C_{10}	Number = 26 units (22% of the total number of respondents). When buying at least one durable product, the price was the main determinant. They agreed with the statement that many products available on the market in Poland are of a lower quality than the products manufactured before 1990 (or there was no opinion on this subject). Net income per person 4001 PLN and more.	100%

Source: own study.

5. Conclusions

Durable goods have become common and widely available products for households. According to this survey, every third respondent purchased a DTU in the last two years. Restitution purchases of products generate the need to pay attention to the price, durability, functionality, brand and look. These determinants depend on the type of product. Therefore, for example, when buying a kettle or a mobile phone, the look is important, and when buying a car, the price and the look are important. The results confirm that the ranking of purchasing determinants changed depending on the respondents' income. Important/significant determinants of purchases also turned out to be innovation (when buying a cooker or a mobile phone) and durability (when buying a bicycle), and in the case of a subset of people, respondents with the highest income were guided by the brand (apart from buying a car, in which, regardless of income, the price is important). A total of 65% of the respondents noticed the problem of the deliberate aging of products. Over 60% of the respondents took the position that the products produced in the autarkic economy (before 1990) were more durable and their life cycle was much longer.

This research indicates the features of durable products that are desired by consumers (durability, functionality, price, look, brand). This results in managerial implications: taking up measures to reduce the strategies of shortening the life of products (i.e., focusing on durability and long-term functionality) will contribute to the improvement in the company's position in the market as an enterprise responding to consumer expectations and that, at the same time, is environmentally friendly. A special role in terms of practical implications is played by building eco-energy awareness of purchased products among young people, as in the future, they will be responsible for implementing energy-saving solutions in their households and the entire economy [66]. These activities are in line with the sustainable and circular economy system. Generating sales by shortening the life of products is unethical and also unprofitable. In the era of transparency of information and opinions issued on the Internet by users, customers will buy valuable products. Increased awareness of buyers who evaluate the value of products with a short lifetime and are aware of their short lifetime affects the rational management of energy resources in the household. The production of subsequent devices with a short lifetime translates into additional energy consumption because the production of each new product means a certain energy cost and additional emissions of pollutants into the atmosphere, resulting from the production process of all the necessary elements.

The results of this research indicate the growing awareness of consumers in the field of responsibility for the environment and sensitivity to the issues of ecological, sustainable development. Therefore, re-evaluating the existing strategies and promoting an economy based on the reuse and repair of products seems to be an increasingly urgent challenge. However, for consumers to trust producers, legal solutions should be introduced. They should ensure transparency of solutions and implement the possibility of institutional quality control of durable goods.

The proposed method of classification trees showed, inter alia, that among the distinguished classes of respondents in terms of attitudes towards preferences when purchasing durable goods, there were classes consisting mainly of people who noticed signs of planned aging of products. Their common feature was that when shopping, they were guided mainly by price and the belief that many products currently available in the Polish market are of a lower quality than products manufactured before 1990.

This research did not identify the purchasing attitudes of durable goods buyers. The conducted questionnaire study was exploratory. This pilot study also allowed testing the possibility of applying the proposed analytical method (classification trees), which could be used in the actual study, with a much larger size, taking into account the categorical dependent variable in the model without converting it into a binary variable.

In subsequent studies, it is recommended to conduct in-depth research on an extended sample and to extend the territorial scope to the international dimension. Moreover, taking into account the importance of the problem under consideration, it is justified to monitor consumers' attitudes towards the aging of products on an ongoing basis. In this aspect, it is important to study changes in environmental awareness and the level of responsibility for the consequences of the choices made (including paying attention to consumers' tendency to perceive consumption in a qualitative manner).

Author Contributions: Conceptualization, M.N.-P., E.S., M.H., K.S., A.G.-J., Z.G.-S. and A.N.; methodology, M.N.-P., E.S., M.H., K.S., A.G.-J., Z.G.-S. and A.N.; validation, M.N.-P., E.S., M.H., K.S., A.G.-J., Z.G.-S. and A.N.; formal analysis, M.N.-P., E.S., M.H., K.S., A.G.-J., Z.G.-S. and A.N.; investigation, M.N.-P., E.S., M.H., K.S., A.G.-J., Z.G.-S. and A.N.; resources, M.N.-P., E.S., M.H., K.S., A.G.-J., Z.G.-S. and A.N.; writing—original draft preparation, M.N.-P., E.S., M.H., K.S., A.G.-J., Z.G.-S. and A.N.; writing—review and editing, M.N.-P., E.S., M.H., K.S., A.G.-J., Z.G.-S. and A.N.; visualization, M.N.-P., E.S., M.H., K.S., A.G.-J., Z.G.-S. and A.N.; supervision, M.N.-P., E.S., M.H., K.S., A.G.-J., Z.G.-S. and A.N.; project administration, Z.G.-S. All authors have read and agreed to the published version of the manuscript.

Funding: The publication was financed by a subsidy for the University of Wroclaw, Funds for the "Excellence Initiative - Research University" program. and by a subsidy granted to the University of

Bialystok. This publication was financed by a subsidy granted to the Cracow University of Economics (6/ZZE/2021/POT).

Institutional Review Board Statement: Not applicable.

Informed Consent Statement: Not applicable.

Data Availability Statement: Data sharing not applicable.

Conflicts of Interest: The authors declare no conflict of interest.

References

1. Ghisellini, P.; Cialani, C.; Ulgiati, S. A review on circular economy: The expected transition to a balanced interplay of enviromental and economic systems. *J. Clean. Prod.* **2016**, *114*, 11–32. [CrossRef]
2. Singh, P.; Giacosa, E. Cognitive biases of consumers as barriers in transition towards circular economy. *Manag. Decis.* **2019**, *57*, 921–936. [CrossRef]
3. Taušová, M.; Čulkova, K.; Tauš, P.; Domaracka, L.; Seňová, A. Ocena efektywnego wykorzystania materiałów z punktu widzenia celów polityki środowiskowej UE. *Energies* **2021**, *14*, 4759. [CrossRef]
4. Xie, X.; Huo, J.; Zou, H. Green process innovation, green product innovation, and corporate financial performance: A content analysis method. *J. Bus. Res.* **2019**, *101*, 697–706. [CrossRef]
5. Johe, M.H.; Bhullar, N. To buy or not to buy: The roles of self-identity, attitudes, perceived behavioral control and norms in organic consumerism. *Ecol. Econ.* **2016**, *128*, 99–105. [CrossRef]
6. Zou, L.W.; Chan, R.Y.K. Why and when do consumers perform green behaviors? An examination of regulatory focus and ethical ideology. *J. Bus. Res.* **2019**, *94*, 113–127. [CrossRef]
7. Whitmarsh, L.; O'Neill, S. Green identity, green living? The role of pro-environmental self-identity in determining consistency across diverse pro-environmental behaviours. *J. Environ. Psychol.* **2010**, *30*, 305–314. [CrossRef]
8. Dermody, J.; Koenig-Lewis, N.; Zhao, A.L.; Hanmer-Lloyd, S. Appraising the influence of pro-environmental self-identity on sustainable consumption buying and curtailment in emerging markets: Evidence from China and Poland. *J. Bus. Res.* **2018**, *86*, 333–343. [CrossRef]
9. Cheng, Y.; Awan, U.; Ahmad, S.; Tan, Z. How do technological innovation and fiscal decentralization affect the environment? A story of the fourth industrial revolution and sustainable growth. *Technol. Forecast Soc. Chang.* **2021**, *162*, 120398. [CrossRef]
10. Barbarossa, C.; de Pelsmacker, P. Positive and negative antecedents of purchasing eco-friendly products: A comparison between green and non-green consumers. *J. Bus. Ethics* **2016**, *134*, 229–247. [CrossRef]
11. Chen, Y.-S.; Chang, C.-H. Enhance green purchase intentions: The roles of green perceived value, green per-ceived risk, and green trust. *Manag. Decis.* **2012**, *50*, 502–520. [CrossRef]
12. Khare, A. Antecedents to green buying behaviour: A study on consumers in an emerging economy. *Mark. Intell. Plan.* **2015**, *33*, 309–329. [CrossRef]
13. Guiltinan, J. Creative Destruction and Destructive Creation: Environmental Ethics and Planned Obsolescence. *J. Bus. Ethics* **2009**, *89*, 19–28. [CrossRef]
14. London, B. Ending the Depression through Planned Obsolescence. *Rev. MAUSS* **2014**, *44*, 47–50. [CrossRef]
15. Waldman, M. Durable Goods Pricing When Quality Matters. *J. Bus.* **1996**, *69*, 489. [CrossRef]
16. Bulow, J. An Economic Theory of Planned Obsolescence. *Q. J. Econ.* **1986**, *101*, 729–749. [CrossRef]
17. Fels, A.; Falk, B.; Schmitt, R. Social Media Analysis of Perceived Product Obsolescence. *Procedia CIRP* **2016**, *50*, 571–576. [CrossRef]
18. Sherif, Y.S.; Rice, E.L. The search for quality: The case of planned obsolescence. *Microelectron. Reliab.* **1986**, *26*, 75–85. [CrossRef]
19. Boland, M. Water and the Environment. *Forbes* **2001**, *168*, 60–62. Available online: https://www.forbes.com/asap/2001/0910/060.html (accessed on 29 October 2021).
20. Box, J.M.F. Extending Product Lifetime: Prospects and Opportunities. *Eur. J. Mark.* **1983**, *17*, 34–49. [CrossRef]
21. Boone, D.; Lemon, K.; Staelin, R. The Impact of Firm Introductory Strategies on Consumers' Perceptions of Future Product Introductions and Purchase Decisions. *J. Prod. Innov. Manag.* **2001**, *18*, 96–109. [CrossRef]
22. Cardoso Satyro, W.; Sacomante, J.B.; Contador, J.C.; Telles, R. Planned obsolescence or planned resource depletion? A sustainable approach. *J. Clean. Prod.* **2018**, *195*, 744–752. [CrossRef]
23. European Parliament Resolution of 4 July 2017 on a Longer Lifetime for Products: Benefits for Consumers and Companies. Available online: https://www.europarl.europa.eu/doceo/document/TA-8-2017-0287_EN.html (accessed on 29 October 2021).
24. Kuppelwiesera, V.G.; Klaus, P.; Manthiou, A.; Boujena, O. Consumer responses to planned obsolescence. *J. Retail. Consum. Serv.* **2019**, *47*, 157–165. [CrossRef]
25. Płoska, R. Ekologiczne i społeczne skutki planowego postarzania produktów. *Zarządzanie Finans.* **2018**, *16*, 210.
26. Su, X.; Zhang, F. On the Value of Commitment and Availability Guarantees When Selling to Strategic Consumers. *Manag. Sci.* **2009**, *55*, 713–726. [CrossRef]
27. Roels, G.; Su, X. Optimal Design of Social Comparison Effects: Setting Reference Groups and Reference Points. *Manag. Sci.* **2014**, *60*, 606–627. [CrossRef]

28. Plambeck, E.; Wang, Q. Implications of Hyperbolic Discounting for Optimal Pricing and Scheduling of Unpleasant Services that Generate Future Benefits. *Manag. Sci.* **2013**, *59*, 1927–1946. [CrossRef]
29. Erat, S.; Bhaskaran, S. Consumer Mental Accounts and Implications to Selling Base Products and Add-ons. *Mark. Sci.* **2012**, *31*, 801–818. [CrossRef]
30. ManagementStudyGuide.com an Educational Portal. Available online: https://www.managementstudyguide.com/argument-against-planned-obsolescence.htm (accessed on 29 October 2021).
31. Malinauskaite, J.; Erdem, F.B. Planned Obsolescence in the Context of a Holistic Legal Sphere and the Circular Economy. *Oxf. J. Leg. Stud.* **2021**, *41*, 719–749. [CrossRef] [PubMed]
32. Gardocka-Jałowiec, A. Czas w konsumpcji-rozważania teoretyczne. Zeszyty Naukowe Uniwersytetu Ekonomicznego w Katowicach. *Studia Ekon.* **2017**, *326*, 41–54.
33. Damasio, A. *Descartes' Error: Emotion, Reason and the Human Brain*; Penguin Books Ltd: London, UK, 2005; ISBN 014303622X.
34. Hassin, R.; Haviv., M. *To Queue or Not to Queue: Equilibrium Behavior in Queueing Systems*; International Series in Operations Research & Management Science; Springer: New York, NY, USA, 2003. [CrossRef]
35. Waldman, M. Durable Goods Theory for Real World Markets. *J. Econ. Perspect.* **2003**, *17*, 131–154. [CrossRef]
36. Amaldoss, W.; Jain, S. Conspicuous Consumption and Sophisticated Thinking. *Manag. Sci.* **2005**, *51*, 1449–1466. [CrossRef]
37. Agrawal, V.V.; Kavadias, S.; Toktay, L.B. The Limits of Planned Obsolescence for Conspicuous Durable Goods. *Manuf. Serv. Oper. Manag.* **2016**, *18*, 216–226. [CrossRef]
38. Orbach, B. The Durapolist Puzzle: Monopoly Power in Durable-Goods Market. *Yale J. Regul.* **2004**, *21*, 67–118.
39. Wells, W.C. *Antitrust and the Formation of the Postwar World*; Columbia University Press: New York, NY, USA, 2003; ISBN 9780231123990.
40. Borenstein, D.; MacKieMason, J.; Netz, J. Antitrust Policy in Aftermarkets. *Antitrust Law J.* **1995**, *63*, 455–482.
41. Calcott, P.; Walls, M. Waste, Recycling, and Design for Environment: Roles for Markets and Policy Instruments. *Resour. Energy Econ.* **2005**, *27*, 287–305. [CrossRef]
42. Bates-Prince, J. Interview: The True Story of France's Fight against Planned Obsolescence. Available online: https://buymeonce.com/blogs/articles-tips/interview-france-fight-planned-obsolescence (accessed on 29 October 2021).
43. Cissé, S.; Metcalf, C.; de Meersman, G.; Nance, R. Marly (March 31, 2020). In the Crosshairs: Planned Obsolescence. Available online: https://www.linklaters.com/pl-pl/insights/blogs/digilinks/2020/march/in-the-crosshairs---planned-obsolescence (accessed on 29 October 2021).
44. French Law on Energy Transition for Green Growth. Available online: https://www-legifrance-gouv-fr.translate.goog/jorf/article_jo/JORFARTI000031044819?_x_tr_sl=en&_x_tr_tl=pl&_x_tr_hl=pl&_x_tr_pto=nui,sc (accessed on 29 October 2021).
45. The Battle Against Planned Obsolence. Available online: www.activesustainability.com/sustainable-development/battle-against-planned-obsolescence/ (accessed on 29 October 2021).
46. The EESC Calls for a Total Ban on Planned Obsolescence. 2013. Available online: https://ec.europa.eu/commission/presscorner/detail/en/CES_13_61 (accessed on 19 November 2021).
47. Vidales, R. Lavadoras con Muerte Anunciada. El Pais. 2 November 2014. Available online: https://elpais.com/economia/2014/10/31/actualidad/1414761553_335774.html (accessed on 29 October 2021).
48. Study. New Attitudes Towards Consumption: Best Practices in the Domain of Built-in Obsolescence and Collaborative Consumption. Available online: https://www.eesc.europa.eu/en/our-work/publications-other-work/publications/new-attitudes-towards-consumption-best-practices-domain-built-obsolescence-and-collaborative-consumption (accessed on 29 October 2021).
49. Consultative Commission on Industrial Change—Presentation. Available online: https://web.archive.org/web/20160610025704/http://www.eesc.europa.eu/?i=portal.en.ccmi (accessed on 29 October 2021).
50. Niklewicz-Pijaczyńska, M. Aktywność innowacyjna i patentowa w ujęciu regionalnym. *Przedsiębiorczość Zarządzanie* **2014**, *15*, 333–343.
51. Directive 1999/44/EC of the European Parliament and of the Council of 25 May 1999 on Certain Aspects of the Sale of Consumer Goods and Associated Guarantees. Available online: https://eur-lex.europa.eu/legal-content/EN/TXT/?uri=CELEX%3A01999L0044-20111212 (accessed on 19 November 2021).
52. New Rules Make Household Appliances More Sustainable. Press Release. Available online: https://ec.europa.eu/commission/presscorner/detail/en/ip_19_5895 (accessed on 29 October 2021).
53. Aczel, A.D.; Sounderpandian, J. *Complete Business Statistics*, 6th ed.; McGraw-Hill: Boston, MA, USA, 2006; ISBN 9780071244169.
54. Iizuka, T. An Empirical Analysis of Planned Obsolescence. *J. Econ. Manag. Strategy* **2007**, *16*, 191–226. [CrossRef]
55. Adrion, M.; Woidasky, J. Planned Obsolescence in Portable Computers—Empirical Research Results. In *Cascade Use in Technologies*; Pehlken, A., Kalverkamp, M., Wittstock, R., Eds.; Springer: Berlin/Heidelberg, Germany, 2018. [CrossRef]
56. Utaka, A. Planned obsolescence and marketing strategy. *Manag. Decis. Econ.* **2000**, *21*, 339–344. [CrossRef]
57. Aladeojebi, T.K. Planned Obsolescence. *Int. J. Sci. Eng. Res.* **2013**, *4*, 1504–1508. Available online: https://www.ijser.org/researchpaper/Planned-Obsolescence.pdf (accessed on 19 November 2021).
58. Zallio, M.; Berry, D. Design and Planned Obsolescence. Theories and Approaches for Designing Enabling Technologies. *Des. J.* **2017**, *20*, S3749–S3761. [CrossRef]
59. Ryś, A. Planowane postarzanie produktu—Analiza zjawiska. *ZN WSH Zarządzanie* **2015**, *1*, 121–128.

60. Breiman, L.; Friedman, J.H.; Olshen, R.A.; Stone, C.J. *Classification and Regression Trees*; Brooks/Cole Publishing: London, UK, 1984; ISBN 0412048418.
61. Lamborn, K.R.; Chang, S.M.; Prados, M.D. Prognostic factors for survival of patients with glioblastoma: Recursive partitioning analysis. *Neurol. Oncol.* **2004**, *6*, 227–235. [CrossRef]
62. Hess, K.R.; Abbruzzese, M.C.; Lenzi, R.; Raber, M.N.; Abbruzzese, J.L. Classification and regression tree analysis of 1000 consecutive patients with unknown primary carcinoma. *Clin. Cancer Res.* **1999**, *5*, 3403–3410. [PubMed]
63. Lemon, C.S.; Roy, J.; Clark, M.A.; Friedmann, P.D.; Rakowski, W. Classification and regression tree analysis in public health: Methodological review and comparison with logistic regression. *Ann. Behav. Med.* **2006**, *26*, 172–181. [CrossRef]
64. Kuhn, L.; Page, K.; Ward, J.; Worrall-Carter, L. The process and utility of classification and regression tree methodology in nursing research. *J. Adv. Nurs.* **2014**, *70*, 1276–1286. [CrossRef] [PubMed]
65. Szalonka, K.; Stańczyk, E.; Gardocka-Jałowiec, A.; Waniowski, P.; Niemczyk, A.; Gródek-Szostak, Z. Food Choices and Their Impact on Health and Environment. *Energies* **2021**, *14*, 5460. [CrossRef]
66. Gródek-Szostak, Z.; Malinowski, M.; Suder, M.; Kwiecień, K.; Bodziacki, S.; Vaverková, M.D.; Maxianová, A.; Krakowiak-Bal, A.; Ziemiańczyk, U.; Uskij, H.; et al. Energy Conservation Behaviors and Awareness of Polish, Czech and Ukrainian Students: A Case Study. *Energies* **2021**, *14*, 5599. [CrossRef]

energies

Article

Analysis of the EU-27 Countries Energy Markets Integration in Terms of the Sustainable Development SDG7 Implementation

Aurelia Rybak [1,*], **Aleksandra Rybak** [2] and **Spas D. Kolev** [3]

[1] Department of Electrical Engineering and Industrial Automation, Faculty of Mining, Safety Engineering and Industrial Automation, Silesian University of Technology, 44-100 Gliwice, Poland

[2] Department of Physical Chemistry and Technology of Polymers, Faculty of Chemistry, Silesian University of Technology, 44-100 Gliwice, Poland; aleksandra.rybak@polsl.pl

[3] School of Chemistry, The University of Melbourne, Parkville, VIC 3010, Australia; s.kolev@unimelb.edu.au

* Correspondence: aurelia.rybak@polsl.pl

Abstract: The article presents the results of research related to the SDG7 sustainable development implementation analysis. The goal is to provide affordable and clean energy. Its implementation will allow for development that will simultaneously provide the possibility of economic growth and the achievement of an optimal level of citizens' health and life. The research was conducted for the countries of the European Union EU-27. During the analysis, the indicators proposed by Eurostat were used. The research aimed to examine the progress in EU member states' energy markets integration. In order to carry out the indispensable research, it was necessary to use a spatial information system. Cluster analysis, as well as TSA analysis, were applied. The conducted research made it possible to verify the posed hypotheses and showed that the energy transformation process of the EU-27 countries is so complicated and heterogeneous that it has given rise to new independent and unique clusters. The authors also verified the adopted set of SDG7 achievement indicators using multiple regression. Additional indicators were also proposed that could complement the set and clarify its analyses.

Keywords: energy cluster; sustainable development; GIS; affordable and clean energy; multiple regression

check for **updates**

Citation: Rybak, A.; Rybak, A.; Kolev, S.D. Analysis of the EU-27 Countries Energy Markets Integration in Terms of the Sustainable Development SDG7 Implementation. *Energies* **2021**, *14*, 7079. https://doi.org/10.3390/en14217079

Academic Editors: Sergey Zhironkin and Michal Cehlar

Received: 1 October 2021
Accepted: 22 October 2021
Published: 29 October 2021

Publisher's Note: MDPI stays neutral with regard to jurisdictional claims in published maps and institutional affiliations.

1. Introduction

The idea of sustainable development arose due to concerns about the condition of the Earth's ecosystem, which was threatened by human activity. Economic growth had a negative impact on the natural environment, so it became necessary to strike a balance between economic development and care for the natural environment.

The article focuses on the seventh goal (SDG7), i.e., affordable and clean energy. The degree of SDG7 implementation was presented in the Sustainable Development Report 2021 [1]. The progress has been recognized as on track or maintaining SDG achievement, for example in Iceland, New Zealand, Switzerland, and Norway. In the case of the European Union, only a few member states have succeeded in this area, such as Finland, Denmark, Bulgaria, Portugal, and Sweden.

For each of the goals, Eurostat has assigned indicators that make it possible to assess the degree of their implementation. Eurostat has adopted a set of indicators to monitor and verify the progress of individual EU countries towards the SDG7. These indicators are:

- Primary energy consumption
- Final energy consumption
- Final energy consumption in households per capita
- Energy productivity
- Share of renewable energy in gross final energy consumption by sector
- Energy import dependency by products

- Population unable to keep home adequately warm by poverty status
- Greenhouse gas emissions intensity of energy consumption [2,3].

In addition to the eight fundamental indicators, the set also includes additional subgroups, such as energy import dependency by products (coal, crude oil, and natural gas), primary energy consumption per capita, or final energy consumption in households per capita.

The European Union places the emphasis on actions aimed at achieving all the goals of sustainable development. Therefore, they were included in the policy and vision of the European Union until 2030 [4]. The research aimed to examine the progress in EU member states' integration of energy markets. It was assumed that the degree of the SDG7 achievement depended on the country for which the analyses were conducted. Therefore, in the article, a cluster analysis was used. This enabled an assignment of the analyzed countries to groups similar to each other in terms of the level of the analyzed indicators. The analysis was carried out for the EU-27 in two years: 2000 as the base year and 2019. This allowed for dynamic analysis. Cluster analysis was conducted using the hierarchical method in two variants, without and with weights of the considered factors. To determine the importance of indicators, an expert survey was carried out. In the next step the set of indicators adopted by Eurostat was verified.

2. Literature Review

The concept of sustainable development was first used in the 1980s. Its authors are considered to be Pearce D.W., Barbier E.W., Markandya A. [5]. It meant at the time the activity undertaken in order to meet the needs of the society [6]. The United Nations then disseminated the idea after the Rio De Janeiro Conference in 1992. The result of this summit was the creation of the principles presented in the Action Program—Agenda 21 [7]. Since then, the term has changed constantly [8–10]. In the face of the enormous challenges that the world is facing in the 21st century, solutions should be sought that would make it possible to ensure a dignified and safe life for everyone now and in the future [11]. The global problems that must be dealt with are the humanitarian crisis and environmental degradation. Today, the concept of sustainable development is understood as 17 goals that enable actions to be taken to benefit humanity, its well-being, and the protection of the natural environment. These goals are described in the 2030 Agenda for Sustainable Development from 2015 [12].

- No poverty
- No hunger
- Good health and well-being
- Quality education
- Gender equality
- Clean water and sanitation
- Affordable and clean energy
- Decent work and economic growth
- Industry, innovation, and infrastructure
- Reduced inequality
- Sustainable cities and communities
- Responsible consumption and production
- Climate action
- Life below water
- Life on land
- Peace, justice, and strong institutions
- Partnership for the goals [13].

Their implementation will allow for the development of such an economy that will give the possibility of economic development, and at the same time, achieve the optimal level of health and life of citizens. Such an economy also guarantees no negative impact of human activities on the natural environment [14]. The implementation of the set goal is primarily intended to reduce the amount of fuel and consumed energy, increase the share of renewable energy sources in the energy mixes of countries, and reduce greenhouse gas emissions [15–17]. On the other hand, sustainable energy management consists of two complementary elements: energy efficiency and the use of renewable energy sources [18].

Access to energy carriers has a huge impact on the possibility of eradicating poverty. Access to energy enables economic development and is synonymous with citizens' access to health care and education. This is evidenced, inter alia, by the correlation between the Human Development Index HDI [19] and energy consumption per capita in a given country [20,21]. The development of the world and the achievement of SDG goals will require access to energy. Therefore, one of the most important goals included in the SDG is the seventh goal: clean and available energy [22], the implementation of which has been analyzed and presented in the article.

Very serious and more and more real threats such as degradation of the natural environment, dying species of plants and animals, climate change, and increasingly dangerous weather phenomena and natural disasters have drawn attention to the concept of sustainable development, which has become very important and is being implemented all over the world [23–25].

The Sustainable Development Goals (SDG) have been set for the whole world, both for developed and developing countries. They are based on three pillars: economic growth, social inclusion, and environmental protection [22]. Due to the limited amount of fossil energy resources, sustainable development is becoming increasingly important. Optimal consumption of energy resources is significant due to the growing demand for energy [26,27]. The level of achievement of individual goals varies depending on the country in question. The presented research focuses on analyzing the implementation of the Sustainable Development Goals in the EU-27 European Union countries. According to the treaties of the European Union, sustainable development means synergistic activities in the field of economic development, as well as environmental protection and policy. These activities are aimed at meeting the needs of EU citizens today, but also in the future [28].

3. Materials and Methods

The research was preceded by a literature review in sustainable development, cluster analysis, and multiple regression. Subsequently spatial and statistical data were collected. Then, a database was built that was used to conduct spatial analyses in the spatial information system. Clusters were built, and the results were analyzed. Subsequently, multiple regression was used to verify the set of data accepted for analysis. The methodology used in the research is presented in the diagram below (Figure 1).

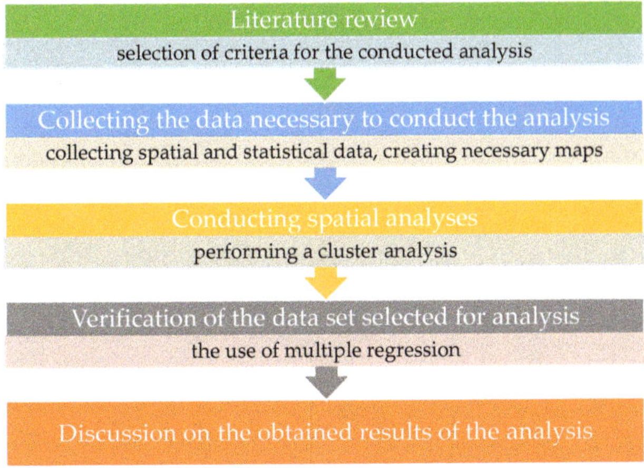

Figure 1. Diagram showing the methodology used during the research.

In conclusion, the authors of the article put forward and verified during the research the following hypotheses:

Hypothesis 1 (H1). *Activities related to implementing SDG7 will unify the analyzed indicators within the EU.*

Hypothesis 2 (H2). *The countries of the "Old" and the "New" Union will implement sustainable development differently.*

The presented research required using methods in cluster analysis, Trend Surface Analysis TSA, and multiple regression. The tools and used methods are described below.

3.1. Trend Surface Analysis

Trend Surface Analysis TSA is one of the best-known and frequently used global surface fitting procedures. It was used for the first time in 1927 by O. Loverson. In the TSA model, the independent variables are the geographical coordinates of the object, and the z_{ij} is any attribute characterizing the studied phenomenon [29–32]. The general form of the model takes the form [33]:

$$z_{ij} = f(x_i, y_i) + \varepsilon_{ij} \tag{1}$$

where:

z_{ij}—dependent variable,
x_i, y_i—geographic coordinates,
ε_{ij}—random component.

TSA was used to characterize the examined attributes for the EU-27 countries. It made it possible to present the local diversity of the analyzed feature in the individual member states.

3.2. Cluster Analysis

Cluster analysis is one of the methods of multivariate comparative analysis. It allows for comparison between the objects in terms of selected attributes. Moreover, it enables the determination of objects groups with similar characteristics [34]. Cluster analysis can be carried out concerning both point and area features. In the presented research the algorithm used grouped area units corresponding to individual countries of the European Union. Cluster analysis allows dividing a set of objects into clusters and groups of features. This set is defined as [35]:

$$O = \{w_i = (w_{i1}, \ldots, w_{ij}) | i = 1, \ldots N \} \tag{2}$$

where:

w_i—is the j-dimensional feature vector of an object that belongs to the set O.

Groups are determined using following principle: objects included in one group are homogeneous, while objects in two different groups are characterized by heterogeneity. The methods for grouping objects fall into two categories of hierarchical and non-hierarchical methods. In the presented studies, hierarchical methods were used [36–39]. In the first phase, the algorithms assume that each analyzed object is a separate cluster. In the next step, similar clusters are combined. Successive iterations are repeated until similar clusters are combined into one. The main issue of multivariate comparative analysis is the normalization of the input variables. The normalization task is to transform the values of variables expressed in different units to bring them to mutual comparability [40]. Subsequently, the Chebyshev distance was used to calculate the distance between the objects, which made it possible to compare the data subjected to normalization [41]:

$$d(o_i, o_j) = max_i \left| z_{ik} - z_{jk} \right| \tag{3}$$

where:

i, k—number of objects,

j—number of variables,

z_{ik}, z_{jk}—normalized values of j-th features of objects i and k.

The determined distances between individual objects form the distance matrix D_o [42]:

$$D_o = \begin{bmatrix} d_{11} \ldots d_{12} \ldots d_{1m} \\ d_{21} \ldots d_{22} \ldots d_{2m} \\ \ldots \ldots \ldots \ldots \\ d_{n1} \ldots d_{n2} \ldots d_{nm} \end{bmatrix} \tag{4}$$

This matrix is used to compute clusters. In order to define clusters, the grouping method should be used. In this case, e.g., the nearest neighbor method, the median method, the center of gravity method, and the Ward method can be selected. The single linkage method was used in the presented research. The method enables the determination of the distance between two clusters by determining the distance between two closest neighbors—objects that have been assigned to different clusters. Using the single linkage method also identifies outliers. The algorithm of the hierarchical unweighted method allows for the automatic selection of the optimal number of clusters. However, to determine the number of clusters in the case of the weighted hierarchical method, the elbow method was used [43–45]. This method is expressed in terms of Sum of Squared Error:

$$SEE = \sum_{k=1}^{k} \sum_{x_i \in S_k} \parallel x_i - C_k \parallel_2^2 \tag{5}$$

where:

k—number of clusters,

x_i—object assigned to a given cluster S_k,

C_k—cluster centroid.

The first step in the elbow method is to determine the centroids for the available objects randomly. After calculating the Euclidean distance, the objects are assigned to the nearest centroid, thus creating groups k. Next, further centroids are determined for the resulting groups. Points are reassigned to the nearest centroid. The procedure is repeated until the position of the centroids stabilizes [43].

3.3. Expert Survey

The weighted cluster analysis requires the determination of individual SDG7 indicators weights. For this purpose, an expert survey was carried out—in accordance with the principles of the Delphi method. The respondents participating in the study were experts in the discipline of environmental engineering, mining, and energy, as well as the discipline of chemical science. The examination of experts' opinions made it possible to obtain results for an undiagnosed problem for which no empirical data was available. The survey was anonymous. Experts also had the opportunity to supplement the presented set of indicators with those which, according to them, should be included in the survey. Experts were asked to weight each indicator on a 0–100% scale. The sum of the weights for all indicators assessed by one expert should be 100%. The weight of the indicators was determined according to the formula [46]:

$$w_{iw} = \frac{w_i}{\sum_{i=1}^{N} w_i} \tag{6}$$

where:

w_{iw}—weight of the i-th parameter,

w_i—the importance of the i-th parameter.

3.4. Verification of the SDG7 Indicators Set

After the survey is completed, the compliance of the results obtained from individual experts should be verified. For this purpose, the variation coefficient was determined, which made it possible to indicate the level of the obtained opinions unanimity [47]:

$$V = \frac{s}{\bar{x}} \tag{7}$$

where:

s—standard deviation,
$\bar{x}$—mean value.

In the next step, it was necessary to determine which indicators are statistically significant variables explaining the analyzed phenomenon. Therefore, the multiple regression presented in the next section was used.

3.5. Trend Multiple Regression

TSA models can also constitute multiple regression when several explanatory variables are introduced into the model. Multiple regression was performed for the analyzed spatial and statistical data. The explained and explanatory variables constituted a set of indicators for the achievement of SDG7. The following equation can describe the multiple regression model [48–50]:

$$z = \alpha_0 + \alpha_1 x_1 + \ldots + \alpha_i x_i + \varepsilon \tag{8}$$

where:

z—dependent variable,
α_i—model parameters,
x_i—explanatory variables,
ε—random component.

Multiple regression allows to detect the relationship between explanatory variables(x_i) and explained variables (z). Therefore, it enables investigating which explanatory variables best describe the dependent variable. It also allows determining the trend and the rest of the model. A trend is a long-term development tendency, and it is a consequence of a constant set of factors influencing the variable z. The rest of the model, on the other hand, is white noise, random fluctuations that cannot be explained and controlled. The result of the analysis is a three-dimensional, smoothed surface approximation.

When deciding whether a variable is statistically significant, the value of Student's t-test and the p-level (probability) should be considered:

- If $p \leq \alpha$, the null hypothesis that the variable is insignificant should be rejected and an alternative hypothesis of significance of the variable adopted.
- If $p > \alpha$, there is no reason to reject the null hypothesis of irrelevance.

The backward selection method was used. It consisted in introducing all explanatory variables into the model and then removing statistically insignificant variables from the model [43].

3.6. Analysis of SDG7 Achievement Indicators

The research was divided into two stages. In the first stage, the indicators of SDG7 were analyzed in terms of their changes in the years 2000–2019. The level of homogeneity of these indicators in the individual member states was also examined. The most important conclusions are described below.

The indicators of the achievement of SDG7 are presented in the Table 1 and Figure 2. Figure a presents the values of indicators in 2000, while b is the year 2019.

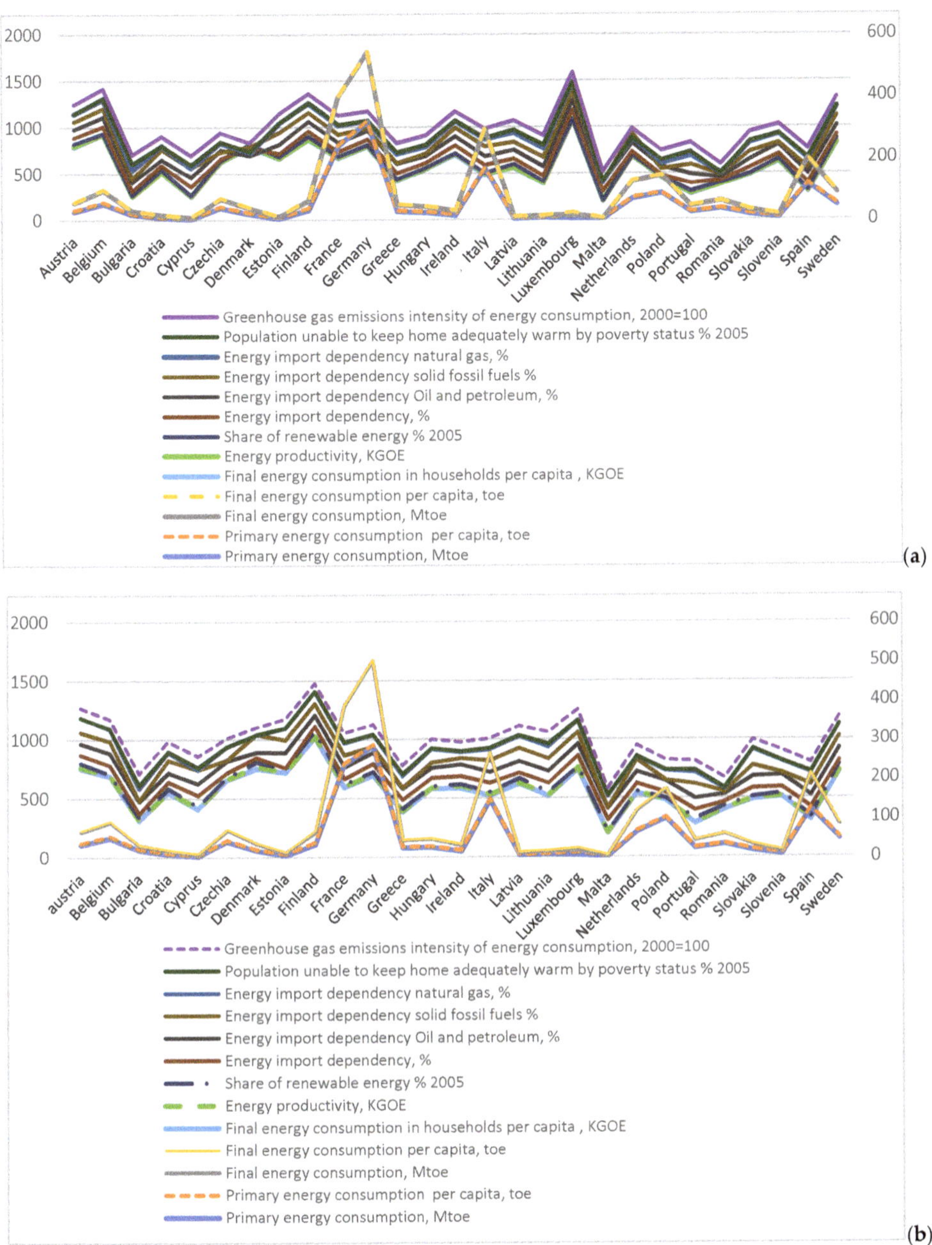

Figure 2. Value of target 7 achievement indicators in 2000 (**a**) and 2019 (**b**).

Table 1. Goal 7 achievement indicators.

No.	Indicator
1	Primary energy consumption, Mtoe
2	Primary energy consumption per capita, toe
3	Final energy consumption, Mtoe
4	Final energy consumption per capita, toe
5	Final energy consumption in households per capita, KGOE
6	Energy productivity, KGOE
7	Share of renewable energy % 2005
8	Energy import dependency, %
9	Energy import dependency Oil and petroleum, %
10	Energy import dependency solid fossil fuels %
11	Energy import dependency natural gas, %
12	Population unable to keep home adequately warm by poverty status % 2005
13	Greenhouse gas emissions intensity of energy consumption, 2000 = 100

The energy market of the European Union member states is significantly diversified. This can be seen when analyzing the indicators of achieving SDG7.

To enable a visual analysis of this differentiation, TSA models were used. TSA models allowed the analysis of the homogeneity and variability of the explanatory variables. Additionally, the coefficient of variation of the indicators used during the analysis was determined. It has been noticed that the level of primary energy consumption is the most diverse in the individual member states (Figure 3). On the other hand, greenhouse gas emissions are the most similar in the EU-27 (Figure 4).

Figure 3. Primary energy consumption in the EU-27 in 2019.

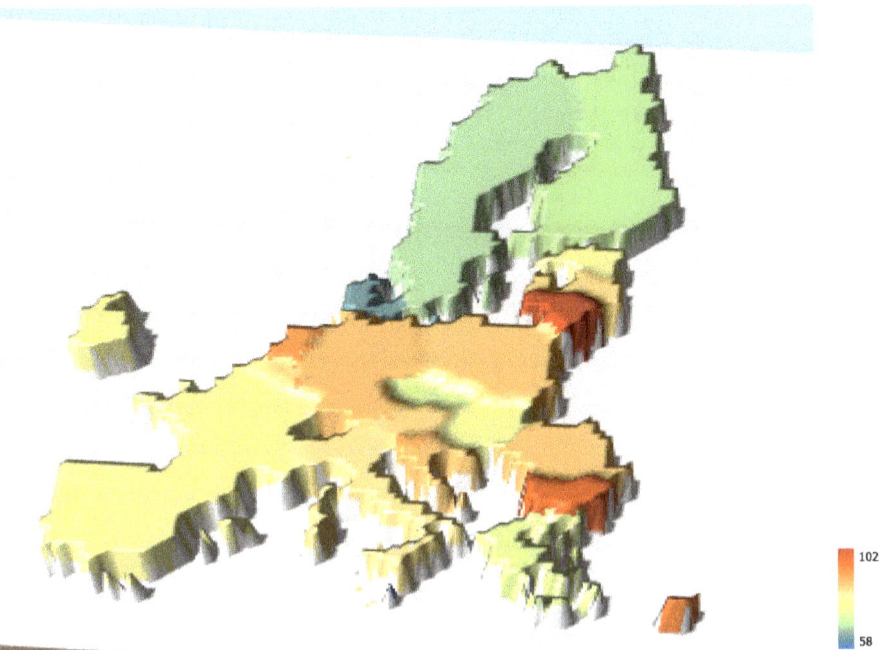

Figure 4. Greenhouse gas emissions in the EU-27 in 2019.

The average consumption of primary and final energy in the EU countries slightly decreased in 2000–2019. The activities of the European Union are focused on the use of energy-saving technologies and thermal modernization of buildings. However, this feature's differentiation level in the EU-27 is the highest and amounts to 138%.

Greenhouse gas emissions in the EU in the years 2000–2019 decreased by less than 20%. It was possible mainly due to the reduction of the share of coal in the energy mixes of the member states and the increase in the share of renewable energy sources and natural gas. The coefficient of variation in greenhouse gas emissions for the EU-27 was 12% in 2019 and was the lowest among the analyzed indicators (Figure 3).

The energy mixes of individual countries have different compositions. Therefore, the level of energy import dependency is also very heterogeneous. Energy import dependency means that the EU cannot self-satisfy its energy needs. The lowest level of dependency on imports occurs in the case of coal. In 2019, it amounted to an average of 75% for the EU-27. On the other hand, the most significant import dependency occurs in crude oil, where the mean value is 95%. This level is very stable and has been maintained since 2000 [51,52]. The European Union obtains fossil fuels mainly from Russia, and the low level of diversification of import sources poses a threat to the EU's energy security.

The analysis of the energy productivity index showed that in the analyzed period, the average productivity level increased by 45%. The level of productivity diversification in individual countries, determined using the variation coefficient, was 50% in 2019. It is one of the most important indicators included in the analyzed set. The increase in productivity will reduce the consumption of primary and final energy and reduce greenhouse gas emissions.

In the case of the renewable energy share in the energy mixes of the EU-27 countries, an average 75% increase can be noticed in 2000–2019. In most EU countries, the emphasis has been on the development of renewable energy sources (RES) for many years [53]. On the other hand, in some countries, such as the countries of the Visegrad Group, this development was intensified after they acceded to the EU. RES will continue to be developed in connection with the growing energy demand and the need to implement Sustainable Development Goals in environmental protection. The coefficient of variation for the share of renewable energy is 52%.

The economic situation of households significantly influences the adequate heating of households. The analysis of the population unable to keep home adequately warm by poverty status indicator shows that the percentage of households affected by this problem in Europe is declining. However, there are countries where it is still high. In the analyzed period, the value of the indicator decreased on average by about 40%. The coefficient of variation, in this case, is 94% for 2019.

The situation in countries where coal occupies an essential position in energy mixes may deteriorate in terms of problems with sufficient heating of households due to phasing out this fuel. The energy obtained by burning coal is one of the cheapest. This applies to both heat and electricity. Therefore, the authors believe that new technological solutions should be sought that will allow the further use of coal in the mix and enable efficient and clean coal combustion. These methods include membrane techniques. They are an excellent solution due to their economy, efficiency, versatility, and ease of use [54–59].

The discovered regularities allowed for the correct interpretation of the results of spatial analyses carried out in the second stage.

4. Results and Discussion

4.1. Cluster Analysis

The conducted research aimed to examine the progress of the EU-27 countries on the way to transforming their economies and energy systems following the concept and goals of sustainable development. They concerned the availability of energy as well as the achievement of goals related to climate protection. It was assumed that actions taken in connection with the implementation of SDG7 should unify and integrate the EU-27 countries and that differences between the "Old" and the "New" Union countries may arise. A spatial information system was used to verify the hypotheses and perform the necessary analyses. Quantum GIS 3.20.1 Odense was used [60]. In the first step, it was necessary to obtain both spatial and statistical data that would enable the planned research to be carried out. Statistical data were obtained from the Eurostat database [2], while vector maps of individual EU-27 member states were obtained from the Free vector world & country maps website [61]. Individual maps of countries were combined, creating a vector map of the entire European Union. Then, statistical data on eight indicators of the achievement of SDG7 were added to the database related to map (Figure 5).

Cluster analysis was performed using the hierarchical method in two variants:

1. Not weighted—where each of the considered attributes has a weight of 1. In this case, the algorithm enables the automatic determination of the number of clusters.
2. Weighted—in this case, the attributes considered during the analysis are given weights from 0–1. For this purpose, an expert survey was carried out. Ten experts participated in the study, and their competencies made it possible to answer the question: which of the indicators are considered the most important and which are least important. The respondents were asked to rate a set of all 13 indicators. The results of the study are presented in the Table 2.

Experts considered the energy import dependency as the most important of the indicators taken into account. It was given a weight of 14%. Energy productivity, import dependency, and greenhouse gas emissions were also considered the most critical factors that will affect SDG7. For the least important indicators, experts considered dependence on oil and natural gas imports and population unable to keep home adequately warm by poverty status.

Figure 5. Vector map of the European Union.

Expert assessments have shown that the most important factor in the survey is the import of energy resources. This is because only some European countries have sufficient energy resources or sufficiently developed RES to become independent from energy sources outside the EU. This, in turn, poses a severe threat to the energy security of EU countries. Additionally, it was indicated that the import of coal was the most important in import dependency indicators. This is related to the nationality of the experts participating in the study. Since the power industry in Poland is mainly based on coal, the inability to obtain its own fuel and the need to rely only on raw material from Russia would pose a severe threat to energy security. It is not only characteristic for Poland but also other countries such as the Czech Republic, Greece, or Germany, where coal is an essential component of the energy mix [62]. For the same reasons, dependence on natural gas and oil was considered the least important. Additionally, this result was influenced by the growing share of renewable energy sources in the mixes of the EU-27 countries. It is assumed that they are to form the basis of the energy mixes of EU countries in the future. The share of the population unable to adequately heat their homes was also considered insignificant. In this case, in most EU countries, such a phenomenon occurs in the amount lower than 10%. The highest level of the indicator in EU is specific for Bulgaria and amounts (in 2019) to 30%. Due to the above, the indicator was considered insignificant.

In order to conduct the cluster analysis, the attribute-based clustering tool was used. This made it possible to consider the distance between individual objects and additional features of each country during the construction of clusters. A set of 13 indicators constituted these attributes. The not weighted cluster analysis was carried out for both 2000 and 2019. The analysis results are presented in the Figures 6 and 7. For both years, the same assumptions were made as to the grouping method and the method of determining the distance between objects. As a result of the analysis, considering selected attributes (presented in the Table 2), 13 clusters were created for 2000 and 15 clusters for 2019. In 2000, the most numerous cluster was cluster no. 3: France, Spain, Italy, Poland, the Czech Republic, Hungary, and Greece. In 2019, Poland, the Czech Republic, and Hungary were classified into separate group. There was also a differentiation in the case of Lithuania, Latvia, and Estonia.

Table 2. Expert survey results.

Indicator	Indicator Unit	Weight (%)	Coefficient of Variation
Primary energy consumption	Mtoe	6.2	0.32
Primary energy consumption per capita	toe	6.0	0.31
Final energy consumption	Mtoe	5.6	0.31
Final energy consumption per capita	toe	6.4	0.34
Final energy consumption in households per capita	KGOE	7.7	0.40
Energy productivity	KGOE	1.1	0.43
Share of renewable energy in gross final energy consumption	% (base year 2005)	8.8	0.27
Energy import dependency	%	14.4	0.31
Oil and petroleum products import dependency	%	4.2	0.33
Solid fossil fuel import dependency	%	9.8	0.23
Natural gas import dependency	%	5.2	0.50
Population unable to keep home adequately warm by poverty status	% (base year 2005)	5.1	0.43
Greenhouse gas emissions intensity of energy consumption	2000 = 100	10.0	0.50

The analysis showed that despite the EU countries' efforts to implement SDG7, a serious discrepancy in their actions appeared. This is evidenced by the growing number of designated clusters from 13 to 15 in 2019.

Clusters formed by countries that were part of the EU before 2000, and those that systematically joined the EU since 2004, are visible. Due to the vast diversity of specific indicators, some countries have been classified as separate objects. An example, in this case, is Denmark (cluster no. 14) due to its low level of dependence on oil imports, the highest level of dependence on coal imports, and a level of dependence on natural gas imports lower than 0. Luxembourg is also classified as a separate cluster (no. 15) due to its almost 100% level of dependence on imported energy resources. The cluster no. 3, which includes Poland, the Czech Republic, Hungary, and Slovenia, is distinguished from other countries mainly by a significant increase in dependence on the import of raw materials, a dynamic increase in energy productivity, and a substantial decrease in population unable to keep home adequately warm by poverty status indicator. In the countries of the Visegrad Group, energy productivity grew faster and energy intensity slower than changes in these parameters for the entire EU, mainly as a result of the economic transformation. The countries of the "Old" Union were separated from cluster no. 7 in 2019 because they systematically implemented the policy consistent with SDG7 long before the countries of the "New" Union. Therefore, the dynamics of changes in their case are slower. Moreover, cluster no. 2 from 2000 (Sweden and Finland) was split in 2019. This was mainly due to a 36% increase in population unable to keep home adequately warm by poverty status in Sweden. In the case of cluster no. 1 (2000), Lithuania, Latvia, and Estonia, in 2019, were not classified as one cluster mainly due to the seven-fold decrease in Estonia's dependence on imports, its dynamic development in the field of renewable energy sources, and a reduction in the level of the population unable to keep home adequately warm by poverty status.

In the next step, a hierarchical cluster analysis was performed, considering the weights of individual attributes. In this case the elbow method was used. For the data from 2000, 7 clusters were created, and for the year 2019, 11 clusters.

The results of the analysis are presented in the Figures 8 and 9.

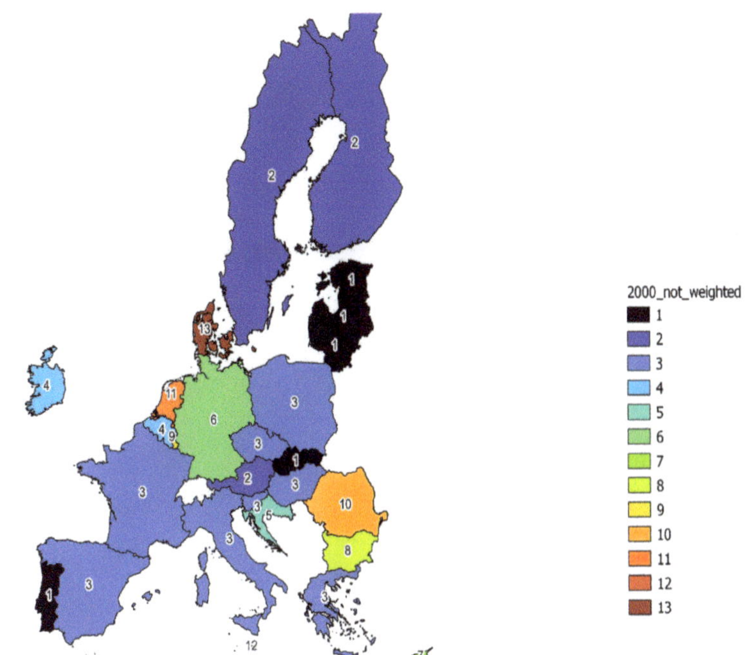

Figure 6. Unweighted EU-27 cluster analysis for the year 2000.

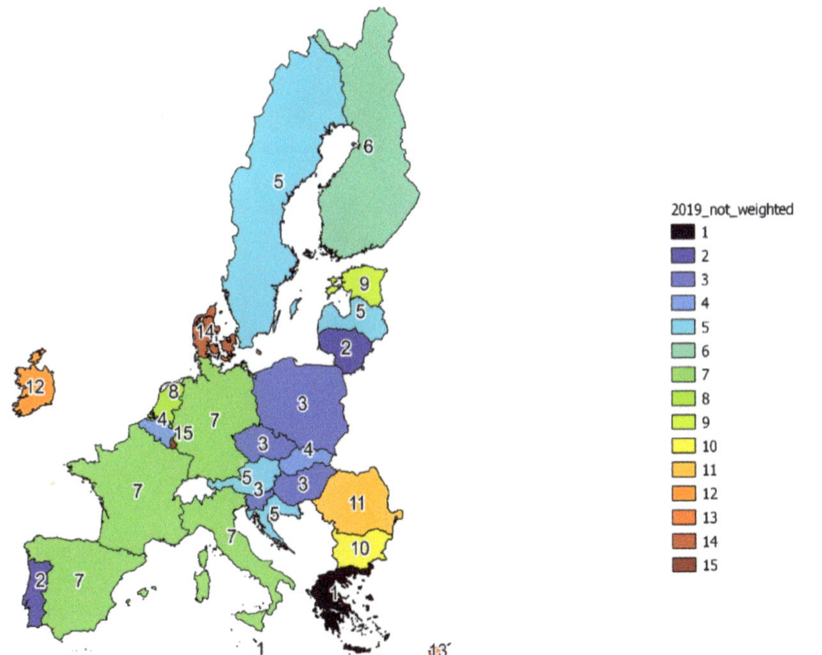

Figure 7. Unweighted EU-27 cluster analysis for the year 2019.

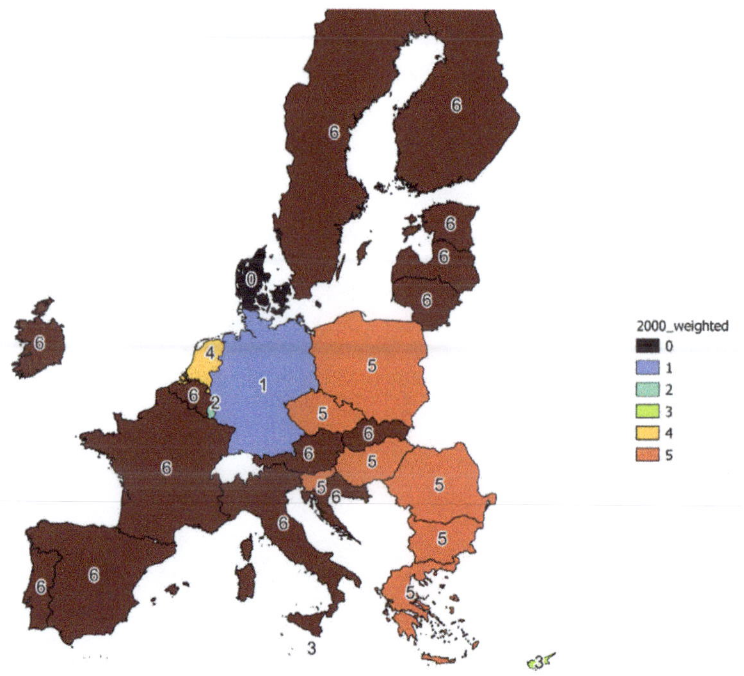

Figure 8. Weighted cluster analysis for the year 2000.

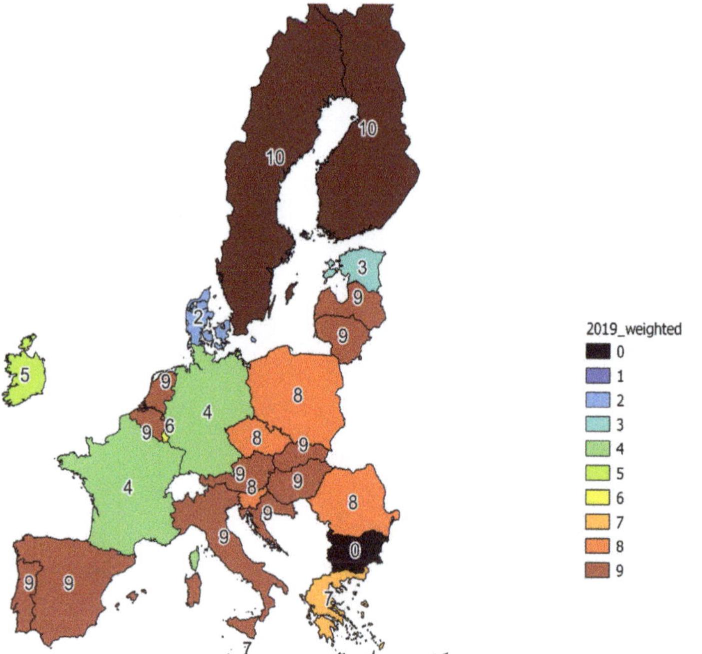

Figure 9. Weighted cluster analysis for the year 2019.

Considering the importance of the attributes used in the analysis, an increase in the number of clusters was again observed. The number of clusters increased from 7 to 11. There are also clusters composed of only one object, such as Denmark or Luxembourg. In 2000, clusters of countries that were part of the Union before 2000 and after 2004 were again visible. However, after 15 years, i.e., in 2019, this division disappeared, and the following clusters have been designated:

Cluster no. 0—Bulgaria

Cluster no. 1—Cyprus

Cluster no. 2—Denmark

Cluster no. 3—Estonia

Cluster no. 4—Germany, France

Cluster no. 5—Ireland

Cluster no. 6—Luxembourg

Cluster no. 7—Malta, Greece

Cluster no. 8—Poland, Czech Republic, Slovenia, Romania

Cluster no. 9—Lithuania, Latvia, Portugal, Spain, Italy, Belgium, the Netherlands, Croatia, Hungary, Austria, Slovakia

Cluster no. 10—Sweden, Finland

Bulgaria has been classified as a separate cluster due to the lowest level of energy productivity and the highest level of the population unable to keep home adequately warm by poverty status. Moreover, Cyprus and Estonia are separate clusters due to one of the highest levels of dependence on imported energy resources, a factor that has received the highest weight. Ireland, on the other hand, has the highest level of energy productivity. Cluster no. 8 constitutes the majority of the Visegrad group countries in which the factors of the highest importance are at a similar level, i.e., energy productivity, import dependence and greenhouse gas emissions. Cluster no. 9 is characterized by a similar level of dependence on the import of energy resources (approx. 70%). On the other hand, for cluster no. 10, the emission of greenhouse gases at the level of 70% is characteristic and the highest share of renewable energy sources.

4.2. Verification of the SDG7 Indicators Set

During the expert's survey, it was noticed that experts disagreed on the assessment of individual indicators. Additionally, experts found the set of indicators to be incomplete. They also stated that the indicators should be differentiated for energy availability and its impact on the natural environment. Taking into account the opinion of experts, the strengths and weaknesses of the SDG7 indicator set have been identified. They are presented in the Table 3.

First of all, the experts pointed out that the set of indicators should be supplemented with indicators that will inform about the availability of energy in economic terms. It is also difficult to deduce from the current set of indicators whether sufficient energy is available in a given country. The set covers the level of consumption extensively but does not take supply into account. It was also noticed that a very important factor in the aspect of clean energy is the analysis of the number of technologies developed and implemented in the EU member states. Such technologies include, inter alia, technologies related to the production and acquisition of clean energy, as well as clean coal technologies (CCT). CCT should be used along the entire coal and energy distribution chain. They can be used at the stage of coal extraction and mechanical processing, hard coal combustion and waste management.

One of the methods of producing clean coal is its enrichment, that is, separating the coal from waste rock and excesses. These are the most cost-effective ways to eliminate pollution to date. Greenhouse gases generated during the combustion of coal pose a serious threat to the natural environment. At this stage, CCTs focus on the removal of carbon dioxide. This can take place before or after the combustion of coal, as well as during combustion. CCT technologies used at the coal processing stage are developing the most dynamically, and they can be used primarily in the process of:

- Combustion—oxy-combustion, power plants with fluidized bed boilers, supercritical power plants with steam boilers [63],
- Coal gasification—this category includes ground and underground gasification [64,65],
- Flue gas cleaning—for example carbon dioxide sequestration, desulphurization, dedusting [66,67].

Table 3. Strengths and weaknesses of the SDG7 indicators set.

Strengths	Weaknesses
The set is very comprehensive in terms of energy consumption. It takes into account the level of primary and final energy consumption in the scale of the entire country, but also taking into account energy consumption per capita	None of the indicators provide information on the price of energy for households and industry, which is of key importance in terms of energy availability
One of the indicators is energy productivity, which is perceived as one of the most important sources of additional energy	There is no information on the certainty of energy supply, the level of energy production in a given country and excess production over energy demand
A very important indicator is also share of renewable energy and greenhouse gas emissions intensity of energy consumption in terms of the ecological aspect of energy acquisition and sustainable energy production	The indicators also do not include information on clean energy generation methods, e.g., CCT or RES technology patents and the number of technologies implemented in the member states in a given year
Energy import dependency allows to visualize the energy security of the EU-27 countries	

The coefficient of variation was determined for the results of the expert's survey (Table 2). For all indicators, the coefficient of variation exceeds 10%, which determines the threshold beyond which the respondents' opinions should be considered inconsistent. For 85% of the indicators, it exceeds 30%. This means that the experts disagreed, and their assessments were not alike. Therefore, the authors of the article decided to verify the adopted set of indicators. For this purpose, multiple regression was used.

SDG7 combines the two most essential postulates concerning the energy sector of the European Union. This is, firstly, access to energy in an amount appropriate to the recipients. Secondly, this energy must be clean, i.e., energy production must not have a negative impact on the natural environment. Because the set of indicators proposed by Eurostat has not been differentiated in this respect, the authors decided to define which indicators affect availability and which affect the ecological dimension related to energy production. For this purpose, multiple regression was carried out. The SAGA multiple regression analysis—raster and predictor raster's algorithm was used to make the application of multiple regression possible with simultaneous consideration of the spatial dimension [68]. Both the dependent variable and the explanatory variables were introduced to the model in the form of raster maps. As a result of the algorithm, a new raster layer was created, which presents the values based on the regression. As the constructed maps were created in the form of vector maps, they were converted first. It made it possible to transform vector layers into raster layers. For this purpose, the rasterize-vector to raster algorithm was used. When executing individual maps for explanatory variables, the option burn-in value was selected. This made it possible to create raster maps, where individual pixels were assigned the values of selected attributes. In this way, 13 raster layers were created as input data to the multiple regression analysis algorithm. The authors considered the set of indicators in terms of energy availability first. In this case, primary energy consumption per capita was used as the explained variable (Model 2). In terms of clean energy, the emission of greenhouse gases was assumed as the dependent variable (Model 1). All the remaining 12 indicators were entered into the model each time. Thanks to this, it was possible to determine which of the explanatory variables are statistically significant in terms of purity and energy availability.

Table 4 presents the results of multiple regression. The variables that turned out to be significant for explaining the explained variable are marked with asterisks. In the case of energy availability, it was six indicators, and in the case of clean energy 4 out of 12 proposed indicators.

Table 4. Multiple regression analysis results.

Indicator	Model 1 p-Value	Model 2 p-Value
Primary energy consumption		*
Primary energy consumption per capita (model 2 explained variable)	*	
Final energy consumption		*
Final energy consumption per capita		***
Final energy consumption in households per capita	**	**
Energy productivity		
Share of renewable energy in gross final energy consumption by sector		
Energy import dependency		***
Oil and petroleum products import dependency	***	
Solid fossil fuel import dependency		
Natural gas import dependency		
Population unable to keep home adequately warm by poverty status	**	
Greenhouse gas emissions intensity of energy consumption (model 1 explained variable)		*

The number of asterisks means: * that $p < 0.1$, ** that $p < 0.05$, *** $p < 0.01$.

The constructed regression models are characterized by high accuracy and reliability. The MAPE (mean absolute percentage error) is less than 10% [69]. From among the created models, those with the smallest information criterion were selected. The Akaike, Hannan-Quinn and Schwarz information criteria were used [70]. Model errors are presented in the Table 5.

Table 5. Multiple regression models 1 and 2 error.

Indicator	Model 1	Model 2
MAPE (mean absolute percentage error), %	10%	9%
RMSE (root mean square error)	10	0.3
Akeike information criterion	213	23
Hannan-Quinn information criterion	214	26
Schwarz information criterion	218	31

Multiple regression models were used to create raster maps of the explained variables presented in the Figures 10 and 11. They present the value of the dependent variable predicted based on the values of the explanatory variables that were used to carry out the analysis. Multiple regression also makes it possible to build forecasts of dependent variables in a selected time horizon (period for which the forecast is determined). After introducing only statistically significant variables into the model, the obtained values of explanatory variables were calculated on the basis of multiple regression parameters. It was noticed that the results of the forecast correspond to the empirical values. In the case of greenhouse gas emission, the highest emission level was recorded in Lithuania, and the lowest in Finland. In the case of primary energy consumption per capita, the highest energy consumption was recorded in Finland, and the lowest in Lithuania. The results of the empirical and forecast values are presented in Table 6. The low values of the model errors and the clear compliance of the forecasted values with the empirical values indicate the need to verify the validity of the constructed set of indicators.

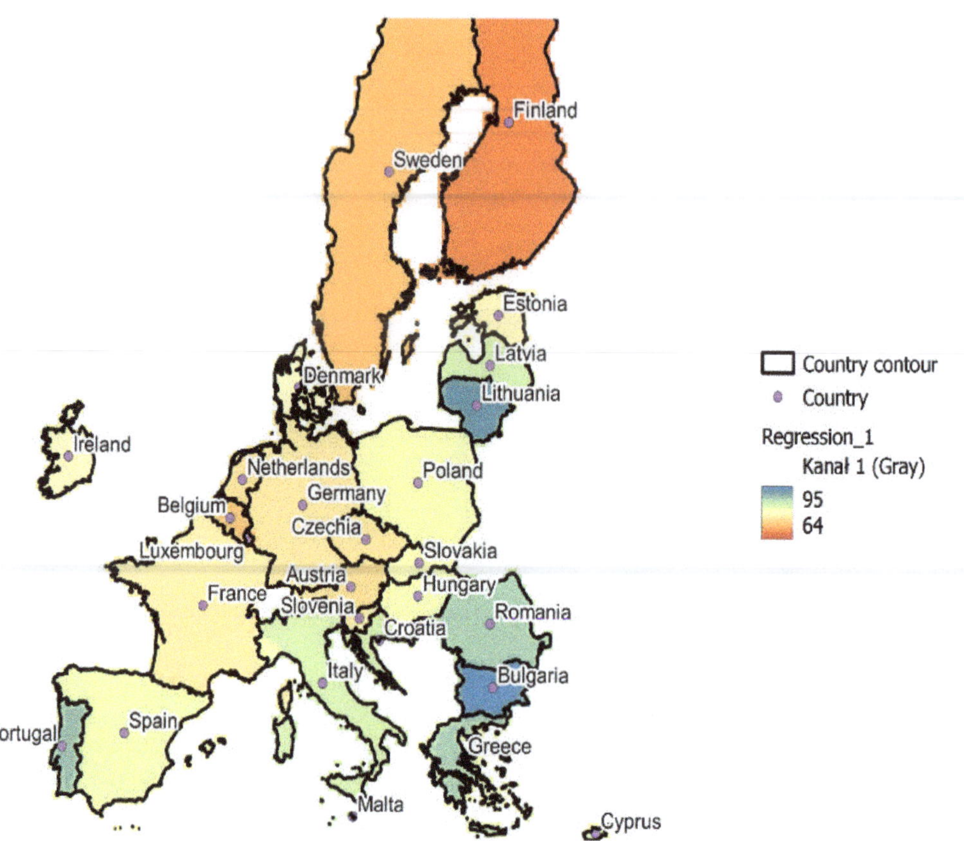

Figure 10. Results of multiple regression of greenhouse gas emissions for the EU-27 (2019).

Table 6. Empirical and forecast values of selected indicators.

Indicator, Country	Unit	Forecast	Empirical Value
greenhouse gas emission		2000 = 100	
Lithuania		92	102
Finland		69	69
primary energy consumption per capita		toe	
Lithuania		0.85	2.25
Finland		5.67	5.81

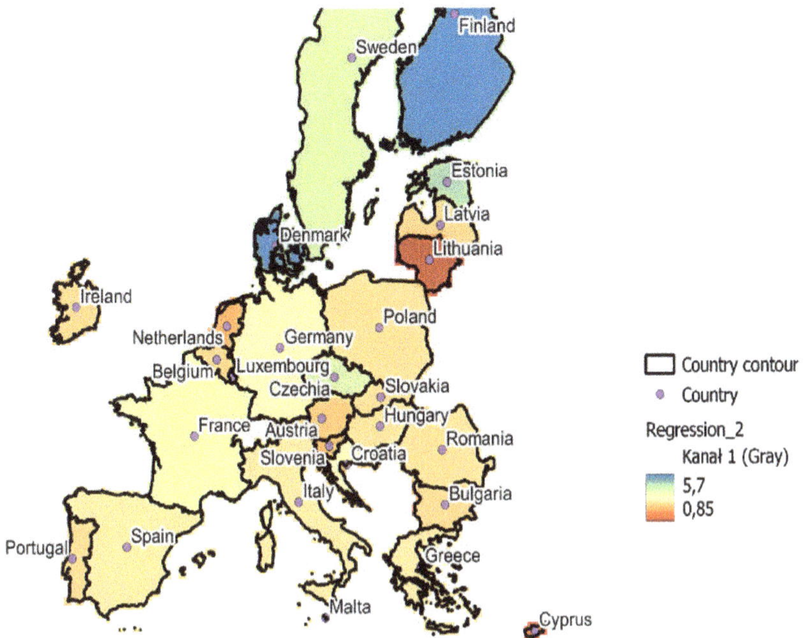

Figure 11. Results of multiple regression of primary energy consumption per capita for the EU-27 (2019).

5. Conclusions

In the presented research, cluster analysis was used to examine the progress of the EU-27 countries towards achieving SDG7. The conducted research showed that the member states differ significantly in terms of the applied attributes. Moreover, in the period 2000–2019, this differentiation increased. The process of energy transformation of the EU-27 countries is so complicated and heterogeneous that it has created new independent and unique clusters. This situation was mainly influenced by the composition of the member states' energy mix, energy productivity, dependence on imports, and a population unable to keep home adequately warm by poverty status indicator. The main differences were observed in the case of the "Old" and "New" EU countries. Therefore, actions should be taken that would make it possible to standardize the changes taking place in individual EU countries.

The discovered growing diversity may pose a threat to the timely achievement of the SDG7 goal. The main factor that delays the implementation of this goal is the growing tendency of dependence on imported energy carriers. Therefore, it is imperative to look for alternative energy sources that could take on the role of fossil fuels. It is also necessary to take further actions to increase the level of diversification of obtaining energy carriers and increase the share of renewable energy sources in the energy mix. The share of renewable energy sources is already systematically increasing, which positively impacts the pace of implementation of the set goal of sustainable development. Moreover, the increase in energy productivity affects the implementation of SDG7, reducing the energy sector's negative impact on the natural environment while reducing the demand for energy resources.

The level of SDG7 achievement set out in the Sustainable Development Report 2021 is consistent with the market integration demonstrated with the use of cluster analysis conducted by the authors. Sweden and Finland stand out positively compared to the EU-27, where this goal has already been recognized as achieved. Most of the "Old" EU countries were classified as a group "challenges remain". This is the case, for example of Germany, France, Italy, and Spain. The countries of the Visegrad Group have been classified as "significant challenges remain". On the other hand, the country that formed a separate cluster is Luxembourg as the only EU country classified as "major challenges remain".

The cluster analysis showed that in 2000–2019, the European union member states and their activities did not lead to the unification of clusters in the EU. On the contrary, the pace and direction of changes in individual countries have led to the creation of additional clusters. Therefore, the authors decided to verify the set of indicators and the validity of their selection. The conducted multiple regression analysis showed that the set of indicators proposed as a measure of the achievement of SDG7 should be verified. It has been proven that this set should be separated and supplemented with separate indicators for the aspect of available and clean energy. The conducted experts survey showed that some indicators might be irrelevant when assessing the achievement of SDG7. The authors believe that the set of indicators should be enriched with new components. In the case of energy availability, these should undoubtedly be energy prices, which will determine whether consumers can purchase raw materials and electricity available in a given country. In terms of clean energy, an indicator that should complement the set of indicators should be the number of implemented technologies that enable clean energy production. This is in line with the goals to be achieved in the European Union in 2030. First of all, it concerns establishing and expanding the existing international cooperation to conduct research on clean technologies, including clean coal technologies, ensuring access to modern energy services and infrastructure and universal access to affordable, reliable, and modern energy services. In conclusion the research confirmed hypothesis 2, while hypothesis 1 on the unification of energy markets in the EU-27 countries was rejected.

Author Contributions: Conceptualization, A.R. (Aleksandra Rybak), A.R. (Aurelia Rybak) and S.D.K.; methodology, A.R. (Aleksandra Rybak) and A.R. (Aurelia Rybak); software, A.R. (Aleksandra Rybak) and A.R. (Aurelia Rybak); writing—original draft preparation, A.R. (Aleksandra Rybak), A.R. (Aurelia Rybak) and S.D.K. All authors have read and agreed to the published version of the manuscript.

Funding: The work was elaborated in the frames of the statutory research 06/010/BK_21/0046.

Institutional Review Board Statement: Not applicable.

Informed Consent Statement: Not applicable.

Data Availability Statement: The data presented in this study are available on request from the corresponding author. The data are not publicly available due to the extremely large size.

Conflicts of Interest: The authors declare no conflict of interest.

References

1. Ki-moon, B. *Sustainable Energy for All: A Vision Statement by Ban Ki-Moon Secretary-General of the United Nations*; United Nations: New York, NY, USA, 2011.
2. Eurostat Database. Available online: https://ec.europa.eu/eurostat/web/main/data/database (accessed on 1 September 2021).
3. Johnston, R.B. Arsenic and the 2030 Agenda for sustainable development. In Arsenic Research and Global Sustainability—Proceedings of the 6th International Congres on Arsenic in Environment AS, Stockholm, Sweden, 19–23 June 2016; pp. 12–14.
4. *Communication from the Commission to the European Parliament, the Council, the European Economic and Social Committee and the Committee of the Regions Next Steps for a Sustainable European Future: European Union Action for Sustainability*; SWD/2016/0390 final; European Commission: Strasbourg, France, 2016.
5. Pearce, D.W.; Barbier, E.W.; Markandya, A. *Sustainable Development*; Earthscan: London, UK, 1990.
6. World Commission on Environment and Development. *Report of the World Commission on Environment and Development*; World Commission on Environment and Development: New York, NY, USA, 1987.
7. Sustainable Development Goals. Available online: https://sustainabledevelopment.un.org/content/documents/Agenda21.pdf (accessed on 10 May 2021).
8. Rosen, M.A. Energy sustainability: A pragmatic approach and illustrations. *Sustainability* **2009**, *1*, 55–80. [CrossRef]
9. UNDP. *World Energy Assessment: Energy and the Challenge of Sustainability, United Nations Development Programme*; United Nations: New York, NY, USA, 2000.
10. Turkson, C.; Acquaye, A.; Liu, W.; Papadopoulos, T. Sustainability assessment of energy production: A critical review of methods, measures, and issues. *J. Environ. Manag.* **2020**, *264*, 110464. [CrossRef] [PubMed]

11. Hillerbrand, R. Why Affordable Clean Energy Is Not Enough. A Capability Perspective on the Sustainable Development Goals. *Sustainability* **2018**, *10*, 2485. [CrossRef]
12. United Nations. Available online: https://sdgs.un.org/goals (accessed on 10 June 2021).
13. Sustainable Development in the European Union—Monitoring Report on Progress towards the SDGs in an EU Context—2020 Edition. Available online: https://ec.europa.eu/eurostat/web/products-statistical-books/-/ks-02-20-202 (accessed on 17 October 2021).
14. United Nations. *A/RES/70/1, United Nations General Assembly*; United Nations: New York, NY, USA, 2015.
15. Wang, Z.; Zhu, Y.; Zhu, Y.; Shi, Y. Energy structure change and carbon emission trends in China. *Energy* **2016**, *115*, 369–377. [CrossRef]
16. McCollum, D.L.; Krey, V.; Riahi, K. An integrated approach to energy sustainability. *Nat. Clim. Chang.* **2011**, *1*, 428–429. [CrossRef]
17. von Hippel, D.; Suzuki, T.; Williams, J.H.; Savage, T.; Hayes, P. Energy security and sustainability in Northeast Asia. *Energy Policy* **2011**, *39*, 6719–6730. [CrossRef]
18. Şalvarlı, M.; Salvarli, H. For Sustainable Development: Future Trends in Renewable Energy and Enabling Technologies. In *Renewable Energy—Resources, Challenges and Applications*; IntechOpen: London, UK, 2020.
19. Human Development Report. Available online: hdr.undp.org/en/content/human-development-index-hdi (accessed on 17 October 2021).
20. Osborne, D. *The Coal Handbook: Towards Cleaner Production*; 2 Coal Utilization; Woodhead Publishing: Cambridge, MA, USA, 2013.
21. Marcondes dos Santos, H.T.; Perrella Balestieri, J.A. Spatial analysis of sustainable development goals: A correlation between socioeconomic variables and electricity use. *Renew. Sustain. Energy Rev.* **2018**, *97*, 367–376. [CrossRef]
22. Nerini, F.F.; Tomei, J.; To, L.S.; Bisaga, I.; Parikh, P.; Black, M.; Borrion, A.; Spataru, C.; Broto, V.C.; Anandarajah, G.; et al. Mapping synergies and trade-offs between energy and the sustainable development goals. *Nat. Energy* **2018**, *3*, 10–15. [CrossRef]
23. Bebbington, J.; Unerman, J. Achieving the United Nations Sustainable Development Goals: An enabling role for accounting research. *Account. Audit. Account. J.* **2018**, *31*, 2–24. [CrossRef]
24. Osborn, D.; Cutter, A.; Ullah, F. Universal Sustainable Development Goals. In *Understanding the Transformational Challenge for Developed Countries*; Report of a Study by Stakeholder Forum; Stakeholder Forum: Herne Bay, UK, 2015.
25. Indicators for Sustainable Development: Guidelines and Methodologies. Available online: https://www.un.org/esa/sustdev/natlinfo/indicators/guidelines.pdf (accessed on 17 October 2021).
26. Skytt, T.; Nielsen, S.N.; Froling, M. Energy flows and efficiencies as indicators of regional sustainability—A case study of Jamtland, Sweden. *Ecol. Indic.* **2019**, *100*, 74–98. [CrossRef]
27. Kavouridis, K.; Koukouzas, N. Coal and sustainable energy supply challenges and barriers. *Energy Policy* **2008**, *36*, 693–703. [CrossRef]
28. International Atomic Energy Agency; United Nations Department of Economic and Social Affairs; International Energy Agency; Eurostat, European Environment Agency. *Energy Indicators for Sustainable Development: Guidelines and Methodologies*; International Atomic Energy Agency: Vienna, Austria, 2005.
29. Li, J.; Heap, A.D. Spatial interpolation methods applied in the environmental sciences: A review. *Environ. Model. Softw.* **2014**, *53*, 173–189. [CrossRef]
30. Sen, Z. *Spatial Modeling Principles in Earth Sciences*; Springer: Berlin/Heidelberg, Germany, 2009.
31. Hota, R.N. Trend surface analysis of spatial data. *Gondwana Geol. Mag.* **2014**, *29*, 39–44.
32. Dessaint, F.; Caussenal, J.P. Trend surface analysis: A simple tool for modelling spatial patterns of weeds. *Crop. Prot.* **1994**, *13*, 433–438. [CrossRef]
33. Watson, G.S. Trend-surface analysis. *J. Int. Assoc. Math. Geol.* **1971**, *3*, 215–226. [CrossRef]
34. Bolin, J.H.; Edwards, J.M.; Finch, W.H.; Cassady, J.C. Applications of cluster analysis to the creation of perfectionism profiles: A comparison of two clustering approaches. *Front. Psychol.* **2014**, *5*, 343. [CrossRef]
35. Suchecka, J. Statystyka przestrzenna. In *Metody Analizy Struktur Przestrzennych*; C.H. Beck: Warszawa, Poland, 2014.
36. Nielsen, F. Hierarchical Clustering. In *Introduction to HPC with MPI for Data Science*; Springer: Cham, Switzerland, 2016. [CrossRef]
37. Köhn, H.F.; Hubert, L.J. *Hierarchical Cluster Analysis*; Statistics Reference Online; Wiley StatsRef: Hoboken, NJ, USA, 2015.
38. Arbolino, R.; Carlucci, F.; Cirà, A.; Ioppolo, G.; Yigitcanlar, T. Efficiency of the EU regulation on greenhouse gas emissions in Italy: The hierarchical cluster analysis approach. *Ecol. Indic.* **2017**, *1*, 115–123. [CrossRef]
39. Bluszcz, A.; Manowska, A. The Use of Hierarchical Agglomeration Methods in Assessing the Polish Energy Market. *Energies* **2021**, *14*, 3958. [CrossRef]
40. Kotu, V.; Deshpande, B. *Classification. Data Science*, 2nd ed.; Morgan Kaufmann: Cambridge, MA, USA, 2019; pp. 65–163.
41. Xueyan, L.; Nan, L.; Sheng, L.; Jun, W.; Ning, Z.; Xubin, Z.; Kwong-Sak, L.; Lixin, C. Normalization Methods for the Analysis of Unbalanced Transcriptome Data: A Review. *Front. Bioeng. Biotechnol.* **2019**, *7*, 358.
42. Suchecki, B. Ekonometria przestrzenna. In *Metody i Modele Analizy Danych Przestrzennych*; C.H. Beck: Warszawa, Poland, 2010.
43. Syakur, M.A.; Khotimah, B.K.; Rochman, E.M.S.; Satoto, B.D. Integration K-means clustering method and elbow method for identification of the best customer profile cluster. *IOP Conf. Ser. Mater. Sci. Eng.* **2018**, *336*, 012017. [CrossRef]
44. Thorndike, R.L. Who belongs in the family? *Psychometrika* **1953**, *18*, 267–276. [CrossRef]
45. Nainggolan, R.; Perangin-angin, R.; Simarmata, E.; Tarigan, A. Improved the performance of the K-means cluster using the sum of squared error (SSE) optimized by using the Elbow method. *J. Phys Conf. Ser.* **1361**, *2019*, 012015. [CrossRef]

46. Zimon, D.; Kruk, U. The use of the CSI method to examine the logistic customer service on the example of a selected organization. *Logistics* **2015**, *3*, 5094–5101.

47. Everitt, B.S.; Skrondal, A. *The Cambridge Dictionary of Statistics*; Cambridge University Press: Cambridge, UK; New York, NY, USA, 1998; ISBN 978-0521593465.

48. Moore, A.W.; Anderson, B.; Das, K.; Wong, W.K. *Chapter 15—Combining Multiple Signals for Biosurveillance, Handbook of Biosurveillance*; Academic Press: Cambridge, MA, USA, 2006; pp. 235–242.

49. Nugus, S. *Chapter 5—Regression Analysis, CIMA Professional Handbook, Financial Planning Using Excel*, 2nd ed.; CIMA Publishing: Burlington, MA, USA, 2009; pp. 59–74.

50. Águila, M.M.; Benítez-Parejo, N. Simple linear and multivariate regression models. *Allergol. Immunopathol.* **2011**, *39*, 159–173. [CrossRef]

51. Rybak, A.; Manowska, A. Przyszłość gazu ziemnego jako substytutu węgla w aspekcie bezpieczeństwa energetycznego Polski. *Wiadomości Górnicze* **2017**, *68*, 144–152.

52. Rybak, A.; Manowska, A. The future of crude oil and hard coal in the aspect of Polands energy security. *Polityka Energetyczna* **2018**, *21*, 141–154. [CrossRef]

53. *The Promotion of Electricity Produced from Renewable Energy Sources in the Internal Market, Directive 2001/77/EC of the European Parliament and the Council*; European Parliament and the Council: Brussels, Belgium, 2001.

54. Rybak, A.; Rybak, A.; Sysel, P. Modeling of Gas Permeation through Mixed-Matrix Membranes Using Novel Computer Application MOT. *Appl. Sci.* **2018**, *8*, 1166. [CrossRef]

55. Kusworo, T.D.; Budiyono, A.F.; Ismail, A.; Mustafa, A. Fabrication and Characterization of Polyimide-CNTs hybrid membrane to enhance high performance CO_2 separation. *Int. J. Sci. Eng.* **2015**, *8*, 115–119.

56. Hasebe, S.; Aoyama, S.; Tanaka, M.; Kawakami, H. CO_2 separation of polymer membranes containing silica nanoparticles with gas permeable nano-space. *J. Membr. Sci.* **2017**, *536*, 148–155. [CrossRef]

57. Sarfraz, M.; Ba-Shammakh, M. A novel zeolitic imidazolate framework based mixed-matrix membrane for efficient CO_2 Separation under wet conditions. *J. Taiwan Inst. Chem. Eng.* **2016**, *65*, 427–436. [CrossRef]

58. Sridhar, S.; Smith, B.; Ramakrishna, M.; Aminabhavi, T.M. Modified poly (phenylene oxide) membranes for the separation of carbon dioxide from methane. *J. Membr. Sci.* **2006**, *280*, 202–209. [CrossRef]

59. Xiang, L.; Pan, Y.; Jiang, J.; Chen, Y.; Chen, J.; Zhang, L.; Wang, C. Thin poly(ether-block-amide)/attapulgite composite membranes with improved CO_2 permeance and selectivity for CO_2/N_2 and CO_2/CH_4. *Chem. Eng. Sci.* **2017**, *160*, 236–244. [CrossRef]

60. QGIS. Available online: https://qgis.org/pl/site/index.html (accessed on 10 May 2021).

61. Free Vector Maps. Available online: https://freevectormaps.com/world-maps/europe/WRLD-EU-010004?ref=more_region_map (accessed on 10 May 2021).

62. Eurostat. Available online: https://ec.europa.eu/eurostat/statistics-explained/index.php?title=Energy_production_and_imports#Production_of_primary_energy_decreased_between_2009_and_2019 (accessed on 15 May 2021).

63. Uchida, T.; Goto, T.; Yamada, T.; Kiga, T.; Spero, C. Oxyfuel combustion as CO_2 capture technology advancing for practical use—Callide Oxyfuel Project. *Energy Procedia* **2013**, *37*, 1471–1479. [CrossRef]

64. Porada, S.; Grzywacz, P.; Czerski, G.; Kogut, K.; Makowska, D. Ocena przydatności polskich węgli do procesu zgazowania. *Energy Policy J.* **2014**, *17*, 89–102.

65. Sobczyk, E.J.; Wot, A.; Kopacz, M.; Frączek, J. Clean Coal Technologies—A chance for Poland's energy security. Decision-making using AHP with Benefits, Opportunities, Costs and Risk Analysis. *Miner. Resour. Manag.* **2017**, *33*, 27–48. [CrossRef]

66. Roddy, D.J.; Younger, P.L. Underground coal gasification with CCS: A pathway to decarbonising industry. *Energy Environ. Sci.* **2010**, *3*, 400–407. [CrossRef]

67. Viebahn, P.; Vallentin, D.; Höllerm, S. Prospects of carbon capture and storage (CCS) in India's power sector—An integrated assessment. *Appl. Energy* **2014**, *117*, 62–75. [CrossRef]

68. Saga. Available online: http://www.saga-gis.org/saga_tool_doc/2.2.0/statistics_regression_8.html (accessed on 15 May 2021).

69. Myttenaere, A.; Golden, B.; Grand, B.; Rossi, F. Mean Absolute Percentage Error for regression models. *Neurocomputing* **2016**, *192*, 38–48. [CrossRef]

70. Profillidis, V.A.; Botzoris, G.N. *Chapter 6—Trend Projection and Time Series Methods Modeling of Transport Demand*; Elsevier: Amsterdam, The Netherlands, 2019; pp. 225–270.

Article

Experimental Evaluation of an Innovative Non-Metallic Flat Plate Solar Collector

Radim Rybár [1], Martin Beer [1,*][ID], Tawfik Mudarri [1], Sergey Zhironkin [2,3,4], Kamila Bačová [1] and Jaroslav Dugas [1]

[1] Institute of Earth Sources, Faculty of Mining, Ecology, Process Control and Geotechnology, Technical University of Košice, Letná 9, 042 00 Košice, Slovakia; radim.rybar@tuke.sk (R.R.); tawfik.mudarri@tuke.sk (T.M.); kamila.bacova@student.tuke.sk (K.B.); jaroslav.dzugas@student.tuke.sk (J.D.)

[2] Institute of Trade and Economy, Siberian Federal University, Svobodny Av. 79, 660041 Krasnoyarsk, Russia; szhironkin@sfu-kras.ru

[3] School of Core Engineering Education, National Research Tomsk Polytechnic University, Lenina St. 30, 634050 Tomsk, Russia; zhironkin@tpu.ru

[4] Mezhdurechensk Branch, T.F. Gorbachev Kuzbass State Technical University, 36 Stroiteley St., 652881 Mezhdurechensk, Russia; zhironkinsa@kuzstu.ru

* Correspondence: martin.beer@tuke.sk

Abstract: The present article deals with the concept of the non-metallic flat plate liquid solar collector and its evaluation. The innovative concept lies in the elimination of metal parts of the solar collector and their replacement by the foam glass block, which significantly reduces the energy and material demands of the production process. The evaluation of the collector took place in two phases, the first was focused on the numerical evaluation, which resulted in the compilation of a theoretical curve of the efficiency of the solar collector. The second phase was focused on verifying the basic functionality of the concept based on the results obtained from experimental tests of the collector, which confirmed the functionality of the concept and revealed several areas that will need to be addressed in the further development of the prototype.

Keywords: renewable energy; solar flat plate liquid collector; non-metallic

Citation: Rybár, R.; Beer, M.; Mudarri, T.; Zhironkin, S.; Bačová, K.; Dugas, J. Experimental Evaluation of an Innovative Non-Metallic Flat Plate Solar Collector. *Energies* **2021**, *14*, 6240. https://doi.org/10.3390/en14196240

Academic Editor: George Kosmadakis

Received: 8 September 2021
Accepted: 28 September 2021
Published: 30 September 2021

1. Introduction

One of the essential conditions for the sustainable development of modern society is the necessity to ensure the energy supply in the required quantity and quality while minimizing possible negative impacts on the environment [1–6]. From an environmental point of view, the continued use of fossil fuels is one of the biggest risks [7–14]. The biggest environmental, but also industrial, challenge facing modern society is the gradual replacement of fossil fuels and, respectively, drastic reduction in fossil fuel consumption. One of the ways to achieve this goal is the wider use of renewable energy sources in the energy mix of individual countries [15–21]. In terms of consumption of primary energy sources, RES (renewable energy sources) provided more than 7000 TWh in 2019, which represents more than 11% of total energy production [22]. When looking at electricity production, RES provided 26.2% of the total electricity production [23]. From the whole range of renewable energy sources, one of the greatest theoretical, but also real, technical potentials is solar energy [24]. This source is possible to use in many ways, mostly to produce electricity using PV systems or heat using solar collector systems [25–31]. In the European Union, the installed capacity of photovoltaics is 117 GWe [32] and 36.1 GWth in thermal solar systems [33], the vast majority of which are installed in smaller scales at the family houses, etc. The present article deals with the use of solar energy to produce heat in the form of hot water, which is produced using an innovative non-metallic solar collector that was designed and manufactured by the authors. The concept of a non-metallic solar collector is based on the experience the authors gained in the process of assessing the technical and operational parameters of commercially produced and available flat plate solar collectors. From this experience, it can be stated that the currently produced flat plate liquid solar collectors are built exclusively on a metal basis, apart from air collectors and low-efficiency swimming pool solar heaters that are made of polypropylene or other plastics. In the construction of the absorber, metals with high thermal conductivity, i.e., copper and aluminum, are used [34]. The use of metallic materials is associated with an increase in energy [35] and technological complexity of production [36], high prices [37] of manufacturing equipment, and last, but not least, the high thermal conductivity of metallic parts, which can negatively affect the increased rate of heat losses from the collector body. Due to these facts, the authors needed to find a different conceptual strategy for capturing and obtaining heat by the body of the solar collector with the exclusion of metal elements from the energy-conversion part of the collector.

It can be assumed that the replacement of metal construction materials with cheaper non-metallic materials will lead to a reduction in the total cost of production of the solar collector, which should be reflected in the final price of the product. However, this can occur only while maintaining a similar operational efficiency in the solar collector. Another positive aspect is the reduction of manufacturing energy consumption by excluding the most energy-intensive materials from the construction of the solar collector, which is directly reflected in the reduction of the energy payback period of the solar system. The reduction of energy required for production has a positive environmental effect in an area that is often a subject of criticism in relation to renewable energy sources, namely, the considerable energy and, partly, the raw material intensity of the manufacturing process [38]. The mentioned energy intensity is often the result of the use of highly sophisticated technological processes and design solutions. The reduction of manufacturing energy consumption by the conversion of construction materials towards less energy-intensive represents only a partial, but time-immediate, elimination of the above-mentioned phenomenon.

The concept of the non-metallic flat plate liquid solar collector proposed by the authors is presented at the level of structural design, prototype construction, numerical evaluation, and experimental operation, which was carried out to verify the basic functionality of the concept, in particular the structural integrity of the individual parts.

2. Materials and Methods

2.1. Basic Concept of Non-Metallic Solar Collector

The basic philosophy of the proposed design is based on the change in the material and construction layout of the proposed solar collector. The presented concept consists of four main construction parts—foam glass insulating block with a top surface coating, transparent glazing, pillars, and connecting fittings. In the presented concept, the black non-selective surface coating of the foam glass insulating block acts as an absorber and, therefore, creates a photothermal converter. On this surface, shortwave solar radiation is converted into heat, which is directly transferred to the heat transfer medium. Figure 1 (left) shows, in a partial cross-section view, the basic arrangement of parts and the energy balance of the solar collector. The figure depicts heat fluxes representing heat gains and losses, where q_s is solar radiation, $q_{h,loss}$ is the heat loss from the transparent cover and the back of the collector, $q_{opt,loss}$ are optical losses caused by reflection and absorption of radiation through the transparent cover, q_{abs} is the heat flux dissipated by the absorber (i.e., useful heat), t_m is the temperature of the heat transfer medium, t_{abs} is the temperature of the absorber, τ is the transmittance of the solar collector glazing.

The heat transfer from the surface of the absorber to the body of the foam glass block is prevented by the high thermal resistance of the foam glass. In comparison to a conventional absorber, which is characterized by significant heat loss from the front of the absorber in the presented design, this loss is eliminated by the structural layout of parts. The back of the absorber is thermally insulated comparable to a conventional solar collector. A heat transfer medium flows in front of the top surface of the absorber, dissipating heat from the absorber body. Above the heat transfer medium is the lower glass of the transparent cover, which contributes to the elimination of heat loss due to its thermal resistance. Thus, the upper surface of the lower glass is equivalent to the surface of the metal absorber of a conventional solar collector, however, the surface temperature of the glass is lower and, at the same time, the heat flux through the glass is lower.

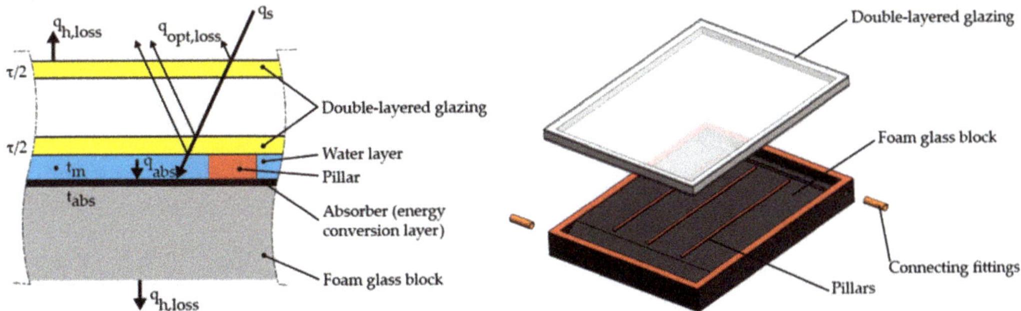

Figure 1. Cross-section of a proposed non-metallic solar collector with depicted basic parts and energy balance (**left**) and exploded view of the 3D model of the proposed solar collector (**right**).

In the above-mentioned arrangement of the absorber, energy gain is achieved by direct absorption of solar radiation during its passage through the layer of heat transfer liquid, when a fraction of UV radiation is used. The heating of the heat transfer media also occurs under the diffuse component of the solar radiation, while a higher degree of extinction is manifested in the part of the spectrum with a longer wavelength (red light). Part of the light that penetrates the water is scattered; a part is absorbed and converted into heat. However, this method of heating cannot be overestimated due to the very thin layer of liquid. From the point of view of the passage of solar radiation through a thin layer of liquid, it is possible to speak of a negligible influence, concerning the values of light transmittance at wavelengths that are characteristic of solar radiation [39].

2.2. Prototype of a Non-Metallic Solar Collector

The designed and constructed prototype of a non-metallic solar collector consists of four main parts, as is depicted in Figure 1 (right):

(a)   conversion-insulating foam glass block fulfilling the role of a frame with a sealing-delimiting surface,
(b)   transparent cover,
(c)   pillars,
(d)   connecting fittings.

Due to the limited offer of foam glass, the prototype has dimensions of 600×450 mm, but the concept can be used analogously in the construction of a collector of classic dimensions (i.e., 2000×1000 mm). In the proposed concept of the solar collector, we considered the use of a block construction of the solar collector body, which integrates the thermal insulation and the back enclosure frame, respectively, into one block. This block is also part of the heat transfer fluid distribution system, which functions as an absorber and, respectively, a photothermal converter. The key part of the proposed collector, the conversion-insulating block, is made of foam glass. This material has good thermal insulation properties and, in comparison with other insulating materials, has the advantage of higher compressive strength, chemical inertness, and water resistance at low weight ($\rho = 120$ to 165 kg.m^{-3}) [40]. Foam glass is produced by foaming glass crumbs together with pulverized coal. Since this insulating material has closed pores, it is completely watertight in its entire volume and, at the same time, non-absorbent for all liquids. For this reason, neither its thermal resistance nor its volume changes over time. Foam glass has the highest compressive strength among thermal insulation materials (compressive strength 0.7 to 1.2 MPa depending on the type).

At the same time, it has high rigidity, is practically incompressible, and is resistant to extreme temperatures. In the case of the proposed collector, the block has a plate shape, and its thickness is determined by ensuring sufficient thermal resistance of the block and sufficient rigidity of the structure (frame function). Due to the value of the coefficient of thermal conductivity λ, which in the case of foam glass is 0.038 to 0.048 W.m^{-1}.K^{-1} [41], it was possible to consider the total thickness of the insulating part of the block of 60 mm. The foam glass block also contains openings for connection fittings and tear-shaped recesses serving as a distribution or the collecting section of the flow channels as seen in Figure 2. Foam glass properties are summarized in Table 1.

Table 1. Basic foam glass properties [40].

Type of Foam Glass (Commercial Name)	T4	S3	F	W&F
Bulk density [kg.m^{-3}]	120	135	165	100
Thermal conductivity coefficient [W.m^{-1}.K^{-1}]	0.04	0.044	0.048	0.038
Compressive strength [MPa]	0.7	0.9	1.2	0.35
Flexural strength [MPa]	0.4	0.5	0.6	0.4
Flexural modulus [MPa]	800	1200	1500	600
Coefficient of longitudinal expansion (K^{-1})	9×10^{-6}	9×10^{-6}	9×10^{-6}	9×10^{-6}
Specific heat capacity [kJ.kg^{-1}.K^{-1}]	0.84	0.84	0.84	0.84
Diffusion resistance factor [-]		70,000		

In the production process of the prototype, foam glass with the trade name T4 was used. The upper side of the block was made of foam glass, respectively, the side in contact with the heat transfer medium was treated with a 1 mm thick black, hygroscopic, heat-resistant polyurethane coating. The black matte coating on the foam glass acts as a solar absorber and increases the durability of the foam glass structure. The collector area was divided into four flow channels, which are delimited by pillars made of silicone sealant. The width of the pillar is 7 mm; its length is 415 mm, and its height is 3 mm, i.e., the thickness of the layer of heat transfer medium is 3 mm. The width of the four flow channels is 97.25 mm. The thickness of the heat transfer medium was chosen to keep the liquid volume of the solar collector at an acceptable level, in this case, 0.81 L.

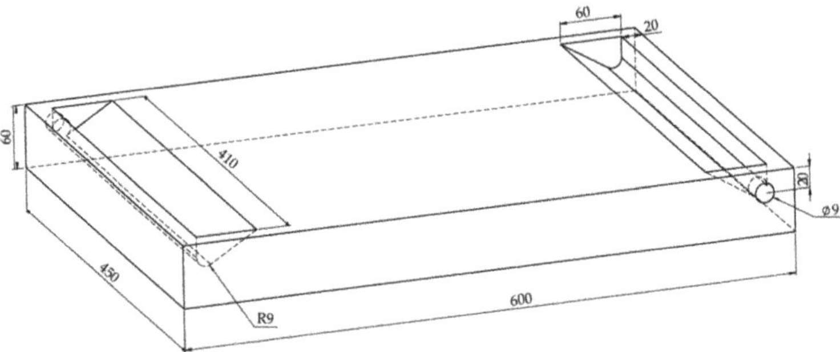

Figure 2. Foam glass block with dimensions of the constructed prototype.

The upper part of the collector is formed by insulating double-layered glazing, which consists of a pair of 4 mm thick Solarglass glasses with a transmissivity value $\tau = 0.78$ and a heat transfer coefficient k = 0.8 W.m^{-2}.K^{-1}. The use of a transparent cover, which is characterized by the use of two glasses with evacuated space between them (as is depicted in Figure 3 right), was conditioned by the effort to minimize heat transfer from the top of the solar collector and, thus, by the effort to reduce heat loss of the proposed device.

Since the inner glazing of the transparent cover is in direct contact with the liquid, it must withstand thermal and pressure stress; for that reason, it was a specifically chosen glass, designed for solar applications. Although in conventional collectors, this glass is not in contact with the heat transfer medium, its chemical composition and production process makes it possible to assume its resistance in the proposed use, which was verified by a test operation in which no problems with the transparent cover appeared. The pressure of water in the hydraulic circle was set to 650 kPa, following the design of a typical flat plate solar collector. The gap between the pair of glasses is 20 mm. The dimensions of the glazing correspond to the size of the collector, i.e., 600 × 450 mm. The transparent cover was attached to the foam glass block with a silicone sealant, which transmits the hydrostatic pressure of the heat transfer fluid. The proposed solar collector was connected to the hydraulic circuit using connecting fittings made of brass pipes with a diameter of 18 mm, which form the only metal part of the proposed solar collector. The structural design and the dimensions of the parts are shown in Figure 3, where the proposed solar collector is presented in front, cross-section view, and in the detailed view of the individual layers of the solar collector.

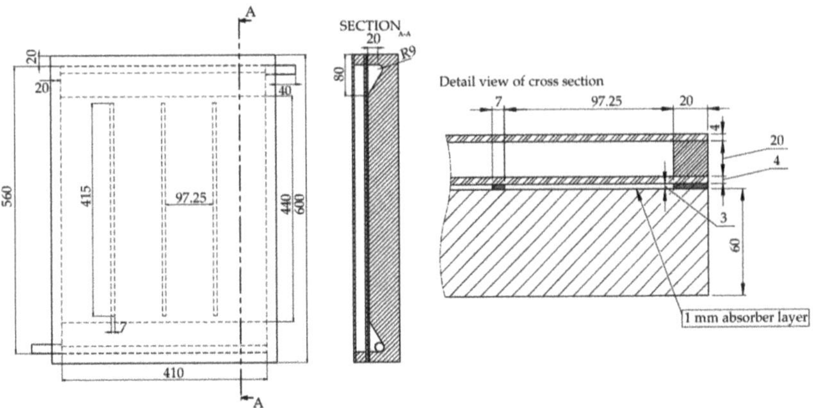

Figure 3. The structural design and the dimensions of the parts.

The heat transfer medium is introduced into the solar collector by a copper connection fitting, located in the lower-left corner. The significant teardrop-shaped widening, visible in Figure 2, allows the developing flow from the circular cross-section into the rectangular cross-section flow channels of the solar collector without the formation of significant turbulent areas. In the main part of the solar collector, the flow of the heat transfer medium is divided into four streams; this segmentation allows the designers to achieve a uniform volumetric flow rate of the heat transfer medium. This ensures homogeneous heating of the medium by the incident solar radiation. The basic building element of the proposed solar collector is a block made from the prepared raw foam glass plate. The raw foam glass block had to be adjusted to the required dimensions, discussed above. The production also included a surface treatment, which consisted of a surface roughness treatment and chamfering of the edges. The next step was to make the desired shape of the distribution, respectively, the collection channel of teardrop shape, and to drill holes enabling the introduction of copper pipe sections serving as a connection with the hydraulic circuit of the measuring apparatus. After these treatments, the whole block was painted with a hydrophobic coating. After the coating had hardened, copper fittings were glued with a hydrophobic epoxy. The final step was the creation of dividing pillars and the installation of a transparent glass cover. In this step, water-soluble auxiliary blocks were used, which ensured the distance of the transparent cover from the foam glass block to provide the required gap, since the pillars would not be able to maintain the weight of the transparent cover in the unhardened state. The advantage of the procedure we used is that after the pillars have hardened, the auxiliary blocks are dissolved in water and flushed from the internal volume of the collector. The manufactured foam glass block with connection fittings and raw foam glass material are depicted in Figure 4.

Figure 4. The prototype of the foam glass block with surface treatment (**left**) and raw foam glass plates (**right**).

2.3. Methodology

The evaluation of the collector took place in two phases: the first was focused on the numerical evaluation, which resulted in the compilation of a theoretical curve of the efficiency of the solar; the second phase was focused on verifying the basic functionality of the concept, based on the results obtained from experimental tests of the collector. The evaluation of the collector was carried out in the months May and June, at the Renewable Energy Center, Faculty BERG. This center is located in Košice, Slovakia (latitude 48°43′ N and longitude 21°15′ E); the area is characterized as a temperate climate zone. A prototype of the proposed solar collector was installed on a measuring apparatus and connected to the hydraulic circuit shown in Figure 5. The inclination of the collector was 45°, which represents the optimal value for a given geographical location and selected time of year [42].

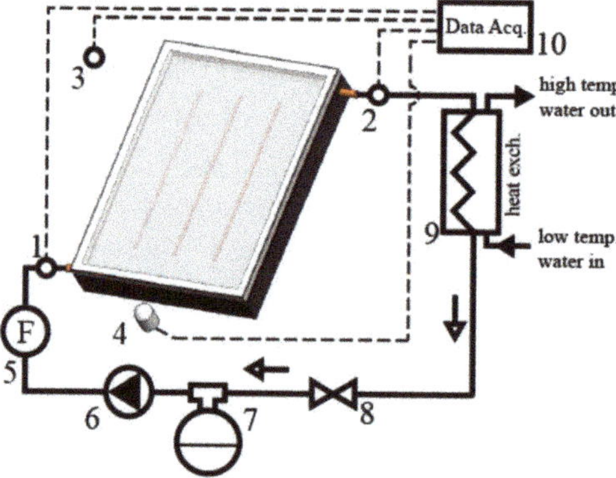

Figure 5. The schematic diagram of the hydraulic circuit and measuring apparatus with depicted positions of measuring devices (1, 2—inlet and outlet thermocouples, 3—ambient air thermocouple, 4—solarimeter, 5—flowmeter, 6—circulation pump, 7—expansion vessel, 8—flow control valve, 9—heat exchanger, 10—data acquisition system).

The assembly of the hydraulic circuit was based on the principle that the circulation pump must operate in the cold branch of the hydraulic circuit, and that its heat output must not affect the heat transfer medium behind the temperature probes. The assembly of the solar collector connected to the hydraulic circuit was placed on a semi-mobile platform, which represented certain limitations in terms of available space, which necessitated some compromises such as the relative position of the valve and the circulation pump. In the presented arrangement, the service life of the pump could be reduced during long-term operation due to cavitation, but in the short-term operation of a given experimental system, this is not a major problem.

The basic characteristic that defines the solar collector in terms of its use is the solar collector thermal efficiency curve. Based on the shape and parameters of the curve, solar collectors can be evaluated and compared with each other. The thermal efficiency of a solar collector is generally defined as the ratio of the energy transferred by a heat transfer medium over a period of time, to the product of a solar collector absorber area and the solar radiation incident on the solar collector under steady-state conditions. This ratio can be termed as an energy balance equation and has the form [43,44]:

$$\eta = \frac{q_A}{G \cdot A} = \frac{\dot{m} \cdot c \cdot \Delta T}{G \cdot A} = \tau \cdot a - k \cdot \frac{t_m - t_E}{G} \tag{1}$$

where η is solar collector thermal efficiency, q_A is the heat flux dissipated by the absorber (i.e., the useful heat flux dissipated by the heat transfer medium), G is a solar radiation intensity, A is an absorber area, product $\tau \cdot a$ is referred to as solar collector optical efficiency, k is the coefficient of heat loss, t_m is a mean heat transfer medium temperature, t_e is an ambient temperature. The ratio $(t_m - t_E)/G$ [$m^2.K.W^{-1}$] is called the reduced temperature difference determined for the solar collector and is one of the basic parameters for evaluating the thermal efficiency of solar collectors under different operating conditions [45]. Based on these theoretical findings, the basic linear curve of collector efficiency is numerically defined results section.

In the tests process, water was used as the heat transfer medium. The measuring apparatus consisted of a series of KIMO TTKE-363 thermocouples (type K, range $-40\,°C$ to $+400\,°C$) placed according to Figure 5 in the hydraulic circuit, which recorded the temperature of the heat transfer medium at the inlet and outlet of the collector and the ambient temperature, and in the body of the proposed solar collector according to Figure 6, which records the temperature of the heat transfer medium in the various locations in the proposed solar collector.

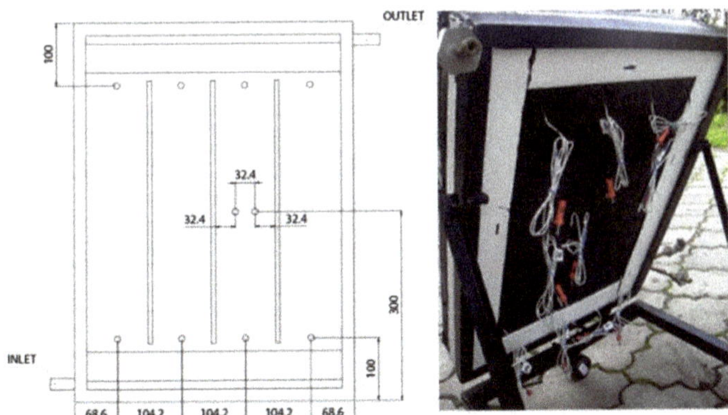

Figure 6. Position of thermocouple probes and real image of the backside of the prototype.

Thermocouple data were recorded by the KIMO AMI 300 data acquisition system with a thermocouple module with an uncertainty of $\pm0.8\,°C$. The volumetric flow rate of the heat transfer medium was set using a TACOSETTER 100 rotameter with a measurement uncertainty of $\pm5\%$ (F.S.). The intensity of solar radiation was recorded with a KIMO SL200 solarimeter, which operates with 5% accuracy and a recording frequency of 2/s. The solar cell operates in the range from 400 to 1300 nm. When measuring the intensity of solar radiation, it was necessary to compromise by using a solarimeter at the expense of a pyranometer. This compromise was forced by the availability of a solarimeter and, respectively, the unavailability of a pyranometer in the workplace. The used solarimeter operates in the wavelength range from 400 to 1300 nm, i.e., it covers the visible and most of the IR part of the spectrum from the solar spectral irradiance point of view. Although wavelengths above 1300 nm affect the energy balance of the solar collector, it is possible in our opinion, to accept the use of this device. This is mainly because the values of solar radiation intensity were no longer used in the evaluation of the thermal technical properties of the solar collector, such as the calculation of thermal efficiency. The intensity of solar radiation was mainly data that characterizes the insulating conditions of a specific measuring day.

3. Results

The numerical evaluation of the proposed solar collector was performed by numerical methods, which resulted in a theoretical curve of the thermal efficiency of the solar collector and the determination of the theoretical value of the stagnation temperature of the solar collector. The calculation itself used a widely accepted methodology based on the values of the overall transparency and absorptivity of the transparent cover and the heat transfer coefficient of the foam glass block.

The optical losses of the proposed solar collector are given by the product of the solar transmittance coefficient of the collector glazing $\tau = 0.78$ and the solar radiation absorption coefficient according to [46] $a = 0.94$. The total optical loss then has a coefficient value $(\tau \cdot a) = 0.73$. The maximum value of the reduced temperature difference $(t_m - t_E)/G$ corresponds to the stagnation state, which is the state of operation of the solar collector when the heat transfer medium does not circulate through the solar collector, but the absorber is exposed to solar radiation and, thus, captured heat is not removed. The value of the maximum stagnation state temperature $t_{A,max}$ is given by Equation (2), which is based on the above-mentioned Equation (1).

$$t_{A,max} = \frac{\tau \cdot a \cdot G}{k} + t_E, \tag{2}$$

In the conditions of Central Europe, the maximum temperature $t_{A,max}$ of the solar collector in equilibrium state is determined either experimentally in tests of the solar collector or by calculation for the solar radiation intensity of $G = 1000$ W.m^{-2} [47] and the ambient temperature of $t_e = 30\ °C$ [48]. The presented calculation considers the heat transfer coefficient $k = 3$ Wm^{-2}.K^{-1}, which corresponds to the lowest quality level of the double-glazing structure of the transparent cover and the value of the heat transfer coefficient on the back of the collector $k_2 = 0.4$ Wm^{-2}.K^{-1} (normally for conventional solar collectors, k_2 is at the level of 0.3 to 0.5 Wm^{-2}.K^{-1} [48]).

Subsequently, the total heat transfer coefficient of the solar collector is $k = 3.4$ W.m^{-2}.K^{-1}. However, to calculate the maximum temperature of the absorber, due to the higher surface temperatures according to [48], it is necessary to correct the value of k using a correction factor c, which has the value $c = 1.12$ for k in the presented range. Then, the corrected total heat transfer coefficient of the collector is $k' = 3.8$ W.m^{-2}.K^{-1}. Then, $t_{A,max}$ is 222 °C, and the value of the parameter x, i.e., the reduced temperature difference $(t_m - t_E)/G$ is 0.222.

Based on these data, it is possible to compile a theoretical linear curve of the thermal efficiency of the presented solar collector, which is shown in Figure 7. A tabular representation of the thermal efficiency calculated with the use of the energy balance equation and input data according to [48] is given in Table 2.

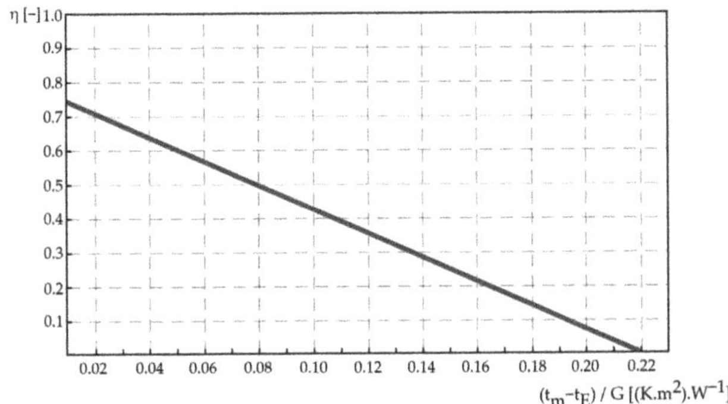

Figure 7. The theoretical linear curve of the proposed solar collector thermal efficiency depending on x parameter.

Table 2. The thermal efficiency of the proposed solar collector for selected temperature difference and intensity of solar radiation.

G [W.m^{-2}]	η [-]								
	$\Delta t = t_m - t_E$ [°C]								
	0	10	20	30	40	50	60	70	80
200	0.73	0.54	0.35	0.16	0.00	0.00	0.00	0.00	0.00
400	0.73	0.64	0.54	0.45	0.35	0.26	0.16	0.06	0.00
600	0.73	0.67	0.60	0.54	0.48	0.41	0.35	0.29	0.22
800	0.73	0.68	0.64	0.59	0.54	0.49	0.45	0.40	0.35
1000	0.73	0.69	0.65	0.62	0.58	0.54	0.50	0.46	0.43

4. Discussion

4.1. Basic Evaluation of the Functionality of the Solar Collector

The functionality of the prototype of the proposed solar collector was evaluated based on the results of performed tests, which were specifically designed to reveal possible deficiencies in the concept before the actual evaluation of the thermal efficiency of the solar collector, which quantifies the operation of the proposed solar collector in terms of energy efficiency. The basic functionality was evaluated by a test that can be characterized by the stagnation state of the solar collector, and tests the monitored changes in the temperature of the heat transfer medium in the longitudinal profile of the solar collector, as well as changes in the temperature of the transparent cover and the rear wall of the insulation block.

The purpose of the tests marked as A, B, C was to verify the structural integrity of the foam glass block, which is affected by the fluid pressure and to verify the functionality and tightness of the connection between the foam glass block and the transparent cover. Tests also verifed the increase in the temperature of the heat transfer medium in the proposed solar collector, as well as the temperature of the outer surfaces of the solar collector, which would indicate critical points when occurring increased heat losses. The stagnant and quasi-stagnant state exposes the solar collector to increased thermal and pressure-induced stresses, which allow the manifestation of various design and manufacturing errors.

4.1.1. Test A

This test aimed to confirm the assumption that the change in solar radiation intensity has a characteristic effect on the development of temperature parameters of the solar temperature system, and to determine the maximum temperature values that the solar collector can reach, under-insulated conditions, with the limited quasi-stagnation flow. The input parameters of the test are characterized by a clear to semi-clear sky, with an ambient temperature of 24 °C. The initial temperature of the heat transfer medium was 22 °C, and its flow rate reached an almost negligible value in terms of the transferred heat output. In numerical terms, the flow rate of the heat transfer medium was set to 0.001389 L.s^{-1}. The test was performed between 11:47 and 13:30 when the solar collector was exposed to sunlight and the surrounding atmospheric influence. However, the recording of values itself took place in three intervals as shown in Table 3.

Table 3. Temperatures of each thermocouple probes (T_n) and values of solar radiation intensity (I).

Time of the Day	Temperature [°C]										I [W.m^{-2}]
	T_1	T_2	T_3	T_4	T_5	T_6	T_7	T_8	T_9	T_{10}	
11:47–11:57	23.1	23.1	23.2	23.4	25.1	25.1	27.9	32.0	32.0	31.5	253
12:52–13:02	57.0	58.7	60.0	60.2	65.7	65.7	71.3	73.3	73.9	72.2	824
13:20–13:30	50.7	53.9	55.3	56.9	73.7	84.8	90.7	93.0	93.5	93.0	806

The measurement confirmed the good functionality of the device, from the point of view of achieving high temperature values of the heat transfer medium in all profiles and, respectively, measuring points. An increase in temperature with time and increasing intensity of sunlight is also visible.

4.1.2. Test B

The purpose of this measurement was to obtain outputs whose values describe the relationship between the temperature development of the transparent cover and the rear insulating part of the collector, in connection to increasing the temperature of the working substance, in a certain time interval, under specific ambient conditions. The input parameters of the test were again characterized by a clear to semi-clear sky with low clouds and an ambient temperature of 21.0 °C, the initial temperature of the working substance was 21.3 °C, and a quasi-stagnant flow state (0.001389 L.s^{-1}) was also considered again. The test was performed from 9:02 to 9:52 and the data were evaluated in ten-minute intervals. During the test, the value of solar radiation reached 500 W.m^{-2} ($\pm$5 W.m^{-2} fluctuations). The evaluated prototype was equipped with thermocouple probes at the middle of the transparent cover area and, secondly, in the middle of the back wall of the solar collector area. The values of the measured temperatures are summarized in Table 4.

Table 4. The temperature of thermocouple probes located in the inlet (T_1), in the middle of the transparent cover ($T_{\text{trans\_cover}}$), in the middle of the back wall of the insulation layer ($T_{\text{back\_wall}}$), and in the middle of the flow channel (T_5).

Time of the Day	Temperature [°C]			
	T_1	$T_{\text{trans\_cover}}$	$T_{\text{back\_wall}}$	T_5
9:02	30.2	19.4	26.6	39.3
9:12	42.5	29.1	28.1	50.5
The ambient temperature rises to 22 °C.				
9:22	56.9	29.7	27.7	62.2
9:32	70.7	29.7	27.6	75.1
The ambient temperature rises to 22.5 °C.				
9:42	78.6	34	29.6	87.3
9:52	84.5	34.3	30.3	89.1

During the measurement, the ambient temperature increased; this condition is also recorded in Table 4. By analyzing the measurements, it was found that there is a slight increase in temperature both on the side of the transparent cover and on the side of the insulating layer, which can be seen in Figures 8 and 9. However, their increment is minimal and is mainly related to the balance of the temperature of the monitored areas with the ambient temperature, as well as to the influence of diffuse radiation emitted from various reflecting surfaces of the surrounding environment.

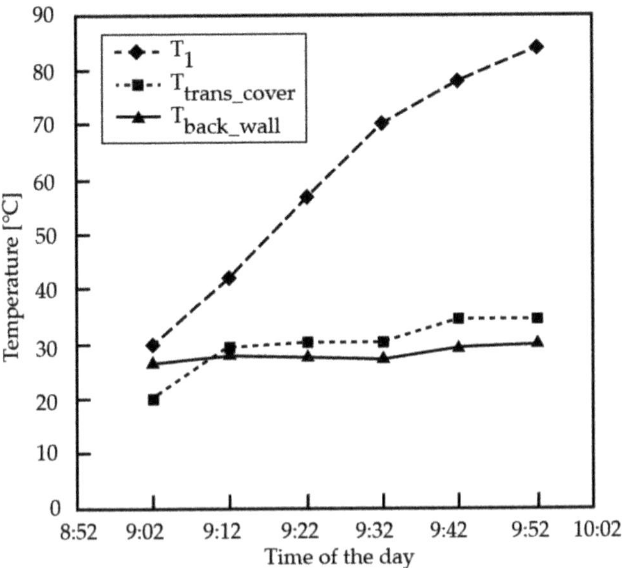

Figure 8. Temperature development of thermocouple probe in inlet position, transparent cover, and back wall of insulation block.

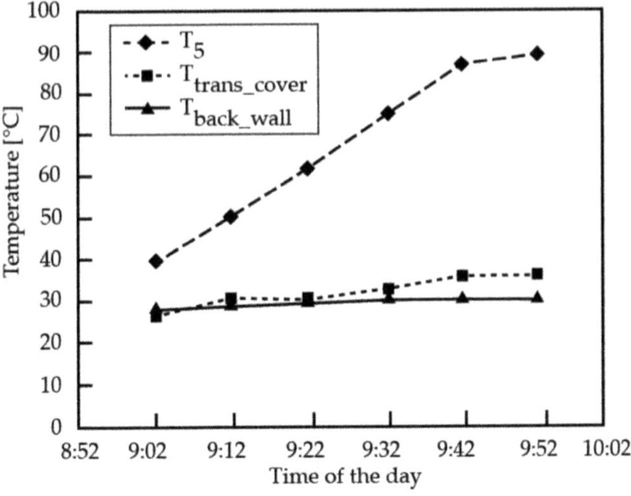

Figure 9. Temperature development of thermocouple probe in inlet position, transparent cover, and back wall of insulation block.

4.1.3. Test C

The measurement was performed to characterize the temperature increase in individual levels of the collector field, due to the change in the amount of accumulated energy, due to the predominance of the diffuse component of solar radiation. In this test, semi-cloudy weather with more pronounced diffusion radiation prevailed; the ambient temperature was 23 °C and the initial temperature of the working substance was 16.4 °C. Again, the solar collector was tested under a quasi-stagnation state, with a minimum flow rate of 0.001389 L.s^{-1}. The measurement results summarized in Table 5 show that the individual layers of the collector array are also affected by the change in the intensity of solar radiation. For evaluation, we chose individual thermocouples from each level (T_1, T_5, T_7), which in the monitored interval reached the temperatures listed in the table of values.

Table 5. Temperatures of each thermocouple probes (T_n) and values of solar radiation intensity (I).

No. of Measurement	Time of the Day	Temperature [°C]										I [W.m^{-2}]
		T_1	T_2	T_3	T_4	T_5	T_6	T_7	T_8	T_9	T_{10}	
1	11:43–11:52	21.7	23.4	25.1	27.1	35	35.6	40.5	43.5	44	43.5	439
2	12:03–12:11	21.6	22.7	23.3	23.0	28.5	28.4	34.1	35.5	35.6	35.1	341
3	12:16–12:26	25.7	23.9	24.5	25.6	30.5	30.1	34	35.1	36.4	37.5	240
4	12:43–12:53	28.8	29.3	29.8	30.6	35.5	36.0	38.3	39.1	39.5	38.6	404

From the graphical representation in Figure 10, it is evident that there is a comparable influence of individual outputs at all levels, in terms of the amount of incident solar energy. The characteristic change occurs only at the initial value of the T_1 probe, which is located in channel One. It is most subject to the influence of the inflow of heat transfer medium at a given time. After repeated measurements, when the temperature has stabilized, this change no longer manifests itself, and the effect of the intensity of solar radiation is, in this case, comparable to the others.

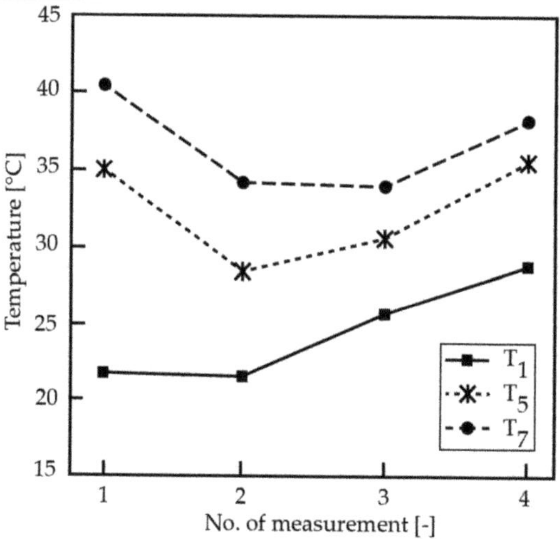

Figure 10. Temperature increases at individual solar collector levels.

4.2. Identification of Possible Design Risks and Their Elimination

During the production of the prototype, and also in the experiments, several problem areas were identified, in terms of the concept or manufacturing. These problems will need to be addressed before the next phase of research or commercial deployment.

The anticipated risks are:

1. Ensuring the transparency of the heat transfer medium distribution system in the solar collector. The transparency of this element has a significant effect on the thermal

efficiency of the heat exchange processes in the absorber. Possible sources of pollution are:

- Incrustation on the walls of the absorber channels. Incrustation on the glass surface has a direct effect on the level of penetration of solar radiation into the absorber area (reduction of radiation intensity, an increase in optical losses of the collector). Incrustation on the conversion surface of the foam glass block reduces the rate of conversion of solar radiation into heat. The elimination measure consists of the use of a high-purity heat transfer medium without incrustation additives. The proposed solar collector cannot be operated in an open hydraulic circuit in which directly heated water flows through the collectors.

- Development of algae or other organisms causing gradual deterioration of the heat transfer fluid or the solar collector itself. This risk is particularly acute, as it can be assumed that the conditions in the absorber channels will be suitable for the life and reproduction of simple living organisms, especially green algae and cyanobacteria. It is not possible to rely on reaching and exceeding the pasteurization temperature during the operation of the solar system, because, in times of prevailing optimal conditions (low solar radiation intensity and, therefore, the low temperature of heat transfer medium), organisms can multiply in a short time and cause deterioration of the entire system. Probably the only effective measure to ensure permanently abiotic conditions is to add a suitable inhibitor to the heat transfer medium. However, the inhibitor must not affect the required properties of the heat transfer fluid and must also be suitable for use in water heating systems. From the above, it can be assumed that the collectors will not be able to operate in Drain-back systems.

2. Ensuring the stability and coherence of the solar collector as a result of the pressure action of the heat transfer fluid. The very nature of the proposed concept implies the greater sensitivity of the construction to the increased fluid pressure in comparison with the metal pipes of conventional solar collectors. The elimination of this risk has its limits, given by the concept itself. Increasing the area of the pillars naturally entails a reduction in the net area of the absorber and, thus, a reduction in the energy production of the solar collector. The strength parameters of the pillars are determined by the type of used material and its bonds with the surface. Here, it is possible to search for more durable materials and a method of surface preparation for the best possible bonding of the joint. Another way to eliminate this problem is to operate the collector in such conditions that it is not exposed to an enormous increase in pressure due to heating of the heat transfer medium or hydrostatic pressure action of the fluid column at a high vertical configuration of the gravitational hydraulic circuit. A possible way to eliminate this risk is to use the collector in a thermosiphon configuration with an integrated tank located just above the collector, where the advantage of low hydraulic resistance of the collector could also be used.

3. The high weight of the solar collector. If we consider a solar collector with an area 2×1 m and a foam glass block with a thickness of 6 cm with a bulk density of 140 kg.m^{-3} (interval is 125–150 kg.m^{-3}), the weight of the foam glass block is approximately 17 kg. The weight of transparent double-layered glazing with a glass thickness of 4 mm with a total area of 2×2 m^2 is 40 kg. The total gross weight of the solar collector is then about 60 kg. Compared to conventional flat plate solar collectors, it is about 15 kg more. The possibility of weight reduction is limited mainly by the use of two glasses; however, it is possible to consider lightweight glass or thinner borosilicate glass. In this calculation was considered the bulk density of the foam glass at the upper interval. When using a lighter type, together with lower thermal conductivity or thickness, (however, it is necessary to take into account the strength limits), it is possible to partially reduce the weight of the foam glass block.

5. Conclusions

The present paper deals with the design and evaluation of an innovative concept of a non-metallic flat plate liquid solar collector. The proposed concept considers the exclusion of metal parts from the energy-conversion part of the solar collector and their replacement by a foam glass block, and presents a change in philosophy regarding conducting the heat transfer medium. It can be assumed that the replacement of metal construction materials with cheaper non-metallic materials will lead to a reduction in the total cost of production of the solar collector, which should be reflected in the final price of the product. However, this must occur while maintaining a similar operational efficiency of the solar collector. Another positive aspect is the reduction of manufacturing energy consumption by excluding the most energy-intensive materials from the construction of the solar collector, which is directly reflected in the reduction of the energy payback period of the solar system. The key part of the proposed solar collector, the conversion-insulating block, is made of foam glass. This material has good thermal insulation properties and, in comparison with other insulating materials, has the advantage of higher compressive strength, chemical inertness, and water resistance at low weight The prototype designed and built by the authors was evaluated theoretically and by performed tests. The numerical evaluation resulted in the construction of a theoretical linear curve of the thermal efficiency of the solar collector and the calculation of the stagnation temperature. The corrected total heat transfer coefficient of the collector is $k' = 3.8$ W.m^{-2}.K^{-1} and $t_{A,max}$ is 222 °C. Subsequently, a functional prototype of the proposed solar collector was installed in an experimental measuring apparatus, enabling the tested operation of the solar collector and recording of its basic operating characteristics. The evaluation of the collector was carried out in the months of May and June, at the Renewable Energy Center, Faculty BERG. This center is located in Košice, Slovakia (latitude 48°43′ N and longitude 21°15′ E); the area is characterized as a temperate climate zone. The data obtained from the simulated operation verified the basic functionality of the proposed concept, but it is necessary to perform further measurements during the simulated operation, as well as measurements that will result in the compilation of the thermal efficiency curve of the proposed solar collector. As part of the evaluation, the risk points of the design were identified and theoretically discussed, which could reduce the overall efficiency of the solar collector or limit its application potential.

Author Contributions: Conceptualization, R.R. and M.B.; methodology, M.B.; validation, M.B.; data curation, M.B., T.M., S.Z., K.B. and J.D.; writing—original draft preparation, R.R. and M.B.; writing—review and editing, R.R. and M.B.; visualization, M.B. All authors have read and agreed to the published version of the manuscript.

Funding: This paper was created in connection with the project VEGA 1/0290/21 Study of the behavior of heterogeneous structures based on PCM and metal foams as heat accumulators with application potential in technologies for obtaining and processing of the earth resources. This research was funded by the Scientific Grant Agency of the Ministry of Education, Science, Research and Sport of the Slovak Republic (VEGA). This paper was created in connection with the project 048TUKE-4/2021 Universal educational—competitive platform.

Institutional Review Board Statement: Not applicable.

Informed Consent Statement: Not applicable.

Conflicts of Interest: The authors declare no conflict of interest.

References

1. Armeanu, D.Ş.; Vintilă, G.; Gherghina, Ş.C. Does Renewable Energy Drive Sustainable Economic Growth? Multivariate Panel Data Evidence for EU-28 Countries. *Energies* **2017**, *10*, 381. [CrossRef]
2. Garshasbi, S.; Haddad, S.; Paolini, R.; Santamouris, M.; Papangelis, G.; Dandou, A.; Methymaki, G.; Portalakis, P.; Tombrou, M. Urban mitigation and building adaptation to minimize the future cooling energy needs. *Sol. Energy* **2020**, *204*, 708–719. [CrossRef]
3. Abd Alla, S.; Simoes, S.G.; Bianco, V. Addressing rising energy needs of megacities—Case study of Greater Cairo. *Energy Build.* **2021**, *236*, 110789. [CrossRef]

4. Nadimi, R.; Tokimatsu, K. Fundamental energy needs quantification across poor households through simulation model. *Energy Procedia* **2019**, *158*, 3795–3801. [CrossRef]

5. Vogt Gwerder, Y.; Marques, P.; Dias, L.C.; Freire, F. Life beyond the grid: A Life-Cycle Sustainability Assessment of household energy needs. *Appl. Energy* **2019**, *255*, 113881. [CrossRef]

6. Wei, C.; Li, C.; Löschel, A.; Managi, S.; Lundgren, T. Digital technology and energy sustainability: Impacts and policy needs. *Resour. Conserv. Recycl.* **2021**, *170*, 105559. [CrossRef]

7. Oryani, B.; Koo, Y.; Rezania, S. Structural Vector Autoregressive Approach to Evaluate the Impact of Electricity Generation Mix on Economic Growth and CO2 Emissions in Iran. *Energies* **2020**, *13*, 4268. [CrossRef]

8. Nie, X.; Zhao, T.; Su, Y. Fossil fuel carbon contamination impacts soil organic carbon estimation in cropland. *Catena* **2021**, *196*, 104889. [CrossRef]

9. Solarin, S.A. An environmental impact assessment of fossil fuel subsidies in emerging and developing economies. *Environ. Impact Assess. Rev.* **2020**, *85*, 106443. [CrossRef]

10. Silva, L.F.O.; Santosh, M.; Schindler, M.; Gasparotto, J.; Dotto, G.L.; Oliveira, M.L.S.; Hochella, M.F., Jr. Nanoparticles in fossil and mineral fuel sectors and their impact on environment and human health: A review and perspective. *Gondwana Res.* **2021**, *92*, 184–201. [CrossRef]

11. Gaete-Morales, C.; Gallego-Schmid, A.; Stamford, L.; Azapagic, A. Life cycle environmental impacts of electricity from fossil fuels in Chile over a ten-year period. *J. Clean. Prod.* **2019**, *232*, 1499–1512. [CrossRef]

12. Schroth, A.W. Impacts of Anthropocene Fossil Fuel Combustion on Atmospheric Iron Supply to the Ocean. *Encycl. Anthr.* **2018**, *1*, 103–113.

13. Hanif, I. Impact of fossil fuels energy consumption, energy policies, and urban sprawl on carbon emissions in East Asia and the Pacific: A panel investigation. *Energy Strategy Rev.* **2018**, *21*, 16–24. [CrossRef]

14. Machol, B.; Rizk, S. Economic value of U.S. fossil fuel electricity health impacts. *Environ. Int.* **2013**, *52*, 75–80. [CrossRef]

15. Hori, K.; Kim, J.; Kawase, R.; Kimura, M.; Matsui, T.; Machimura, T. Local energy system design support using a renewable energy mix multi-objective optimization model and a co-creative optimization process. *Renew. Energy* **2021**, *96*, 105174. [CrossRef]

16. Ahn, K.; Chu, Z.; Lee, D. Effects of renewable energy use in the energy mix on social welfare. *Energy Econ.* **2021**, *96*, 105174. [CrossRef]

17. Cany, C.; Mansilla, C.; Mathonnière, G.; da Costa, P. Nuclear contribution to the penetration of variable renewable energy sources in a French decarbonised power mix. *Energy* **2018**, *150*, 544–555. [CrossRef]

18. Faúndez, P. Renewable energy in the equilibrium mix of electricity supply sources. *Energy Econ.* **2017**, *67*, 28–34. [CrossRef]

19. Hong, S.; Bradshaw, C.J.A.; Brook, B.W. Global zero-carbon energy pathways using viable mixes of nuclear and renewables. *Appl. Energy* **2015**, *143*, 451–459. [CrossRef]

20. Milstein, I.; Tishler, A. Intermittently renewable energy, optimal capacity mix and prices in a deregulated electricity market. *Energy Policy* **2011**, *39*, 3922–3927. [CrossRef]

21. Tafarte, P.; Das, S.; Eichhorn, M.; Thrän, D. Small adaptations, big impacts: Options for an optimized mix of variable renewable energy sources. *Energy* **2014**, *72*, 80–92. [CrossRef]

22. *BP Statistical Review of World Energy 2019*; BP plc: London, UK, 2019; pp. 5–10.

23. *Renewable Energy Statistics 2019*; The International Renewable Energy Agency: Abu Dhabi, UAE, 2019; pp. 380–398.

24. Perez, R.; Perez, M. A fundamental look at energy reserves for the planet. *Int. Energy Agency SHC Programme Sol. Update* **2009**, *50*, 2–3.

25. Wu, Y.; Li, C.; Tian, Z.; Sun, J. Solar-driven integrated energy systems: State of the art and challenges. *J. Power Sources* **2020**, *478*, 228762. [CrossRef]

26. Leonzio, G. Solar systems integrated with absorption heat pumps and thermal energy storages: State of art. *Renew. Sustain. Energy Rev.* **2017**, *70*, 492–505. [CrossRef]

27. Nema, P.; Nema, R.K.; Rangnekar, S. A current and future state of art development of hybrid energy system using wind and PV-solar: A review. *Renew. Sustain. Energy Rev.* **2009**, *13*, 2096–2103. [CrossRef]

28. Alahmer, A.; Ajib, S. Solar cooling technologies: State of art and perspectives. *Energy Convers. Manag.* **2020**, *214*, 112896. [CrossRef]

29. Karimirad, M.; Rosa-Clot, M.; Armstrong, A.; Whittaker, T. Floating solar: Beyond the state of the art technology. *Sol. Energy* **2021**, *219*, 1–2. [CrossRef]

30. Al-Shahri, O.A.; Ismail, F.B.; Hannan, M.A.; Hossain Lipu, M.S.; Al-Shetwi, A.Q.; Begum, R.A.; Al-Muhsen, N.F.O.; Soujeri, E. Solar photovoltaic energy optimization methods, challenges and issues: A comprehensive review. *J. Clean. Prod.* **2021**, *284*, 125465. [CrossRef]

31. Ravi Kumar, K.; Krishna Chaitanya, N.V.V.; Sendhil Kumar, N. Solar thermal energy technologies and its applications for process heating and power generation—A review. *J. Clean. Prod.* **2021**, *282*, 125296. [CrossRef]

32. Jäger-Waldau, A. *PV Status Report 2019, EUR 29938 EN*; Publications Office of the European Union: Luxembourg, 2019; pp. 10–12.

33. *Solar Heat Markets in Europe (Summary 2019), Solar Heat Europe*; ESTIF: Brussels, Belgium, 2019; pp. 1–6.

34. Evangelisti, L.; De Lieto, R.; Asdrubali, V.F. Latest advances on solar thermal collectors: A comprehensive review. *Renew. Sustain. Energy Rev.* **2019**, *114*, 109318. [CrossRef]

35. Horodníková, J.; Khouri, S.; Rybár, R.; Kudelas, D. TESES rules as a tool of analysis for chosen OZE projects. *Acta Montan. Slovaca* **2008**, *13*, 350–356.
36. Kosai, S.; Yamasue, E. Global warming potential and total material requirement in metal production: Identification of changes in environmental impact through metal substitution. *Sci. Total Environ.* **2019**, *651*, 1764–1775. [CrossRef] [PubMed]
37. Taušová, M.; Horodníková, J.; Khouri, S. Financial analysis as a marketing tool in the process of awareness. *Acta Montan. Slovca* **2007**, *12*, 258–263.
38. Kommalapati, R.; Kadiyala, A.; Shahriar, M.T.; Huque, Z. Review of the Life Cycle Greenhouse Gas Emissions from Different Photovoltaic and Concentrating Solar Power Electricity Generation Systems. *Energies* **2017**, *10*, 350. [CrossRef]
39. Said, Z.; Sajid Hossain, M.; Abd Rahim, N. Evaluating the Optical Properties of TiO2 Nanofluids for a Direct Absorption Solar Collector. *Numer. Heat Transf. Appl.* **2015**, *67*, 1–18.
40. König, J.; Lopez-Gil, A.; Cimavilla-Roman, P.; Rodriguez-Perez, M.A.; Petersen, R.; Østergaard, M.B.; Iversen, N.; Yue, Y.; Spreitzer, M. Synthesis and properties of open- and closed-porous foamed glass with a low density. *Constr. Build. Mater.* **2020**, *247*, 118574. [CrossRef]
41. Arriagada, C.; Navarrete, I.; Lopez, M. Understanding the effect of porosity on the mechanical and thermal performance of glass foam lightweight aggregates and the influence of production factors. *Constr. Build. Mater.* **2019**, *228*, 116746. [CrossRef]
42. Jacobson, M.Z.; Jadhav, V. World estimates of PV optimal tilt angles and ratios of sunlight incident upon tilted and tracked PV panels relative to horizontal panels. *Sol. Energy* **2018**, *169*, 55–66. [CrossRef]
43. Vengadesan, E.; Senthi, R. A review on recent developments in thermal performance enhancement methods of flat plate solar air collector. *Renew. Sustain. Energy Rev.* **2020**, *134*, 110315. [CrossRef]
44. Khedher, N.B. Experimental Evaluation of a Flat Plate Solar Collector Under Hail City Climate. *Eng. Technol. Appl. Sci. Res.* **2018**, *8*, 2750–2754. [CrossRef]
45. Chaji, H.; Ajabshirchi, Y.; Esmaeilzadeh, E.; Zeinali Heris, S.; Hedayatizadeh, M.; Kahani, M. Experimental Study on Thermal Efficiency of Flat Plate Solar Collector Using TiO2/Water Nanofluid. *Mod. Appl. Sci.* **2013**, *10*, 60–69. [CrossRef]
46. Tiwari, G.N.; Tiwari Shyam, A. *Handbook of Solar Energy, Theory, Analysis and Applications*; Springer: Singapore, 2016; pp. 171–246.
47. Cihelka, J. *Solar Thermal Technology*; T. Malina Publishing: Praha, Czech Republic, 1994; pp. 64–71.
48. Rybár, R. Use of Alternative Energy Sources in Urban Areas. Doctoral Dissertation, Technical University of Košice, Košice, Slovakia, 2001.

Article

Evaluation of Communal Waste in Slovakia from the View of Chosen Economic Indicators

Beáta Stehlíková [1], Katarína Čulková [2,*], Marcela Taušová [2], Ľubomír Štrba [3] and Eva Mihaliková [4]

1   Department of Information, Technical University Košice, FBERG, Letná 9, 042 00 Košice, Slovakia; beata.stehlikova@tuke.sk
2   Department of Earth Resources, Technical University Košice, FBERG, Letná 9, 042 00 Košice, Slovakia; marcela.tausova@tuke.sk
3   Department of Geotourism, Technical University of Košice, FBERG, Letná 9, 042 00 Košice, Slovakia; lubomir.strba@tuke.sk
4   Department of Economy, Faculty of Public Administration, Pavol Jozef Šafarik University in Košice, Popradská 66, 040 01 Košice, Slovakia; eva.mihalikova1@upjs.sk
*   Correspondence: katarina.culkova@tuke.sk; Tel.: +421-556023116

Abstract: Waste treatment consists of activities required to make sure that waste has the least practical impact on the environment. In Slovakia, more than 50% of waste is in storage. Waste development depends on the economic situation of the state. In Slovakia, there is economic intolerance of waste treatment due to the weaker economic situation of the inhabitants. The goal of this contribution is to study the development of waste production in Slovakia in regard to economic indexes of households with the aim of improving waste management. The goal is achieved by searching for a relation between economic indexes and households by a correlation matrix and by verification of polynomial dependence. According to the results of the statistical importance, we found similarity of the regions in chosen indexes by using of cluster analysis. By this method a sustainable economy and healthy environment is guaranteed and waste is used to produce energy.

Keywords: waste economy; productivity of energy sources; circular economy; waste treatment; Slovakia

Citation: Stehlíková, B.; Čulková, K.; Taušová, M.; Štrba, Ľ.; Mihaliková, E. Evaluation of Communal Waste in Slovakia from the View of Chosen Economic Indicators. *Energies* **2021**, *14*, 5052. https://doi.org/10.3390/en14165052

Academic Editors: Michal Cehlar and Sergey Zhironkin

Received: 16 June 2021
Accepted: 11 August 2021
Published: 17 August 2021

Publisher's Note: MDPI stays neutral with regard to jurisdictional claims in published maps and institutional affiliations.

1. Introduction

Everyone produces waste; some less, some more. The richer the country, the more waste produced by its inhabitants. The more developed the country, the higher the rate of recycled waste [1]. In northern and western European countries waste stocks almost do not exist any more. However, in countries of eastern and southern Europe more than half of the waste is in storage [2]. Slovakia is no exception. The amount of municipal waste produced increases every year [3]. The most recent data of the Statistical Office of the Slovak Republic for 2019 show up to 434.63 kg of municipal waste per capita and a high percentage of this waste in landfill (50.45%). The annual reduction of land filling has been positive; for example, in 2017 up to 61.43% of municipal waste was in landfill in Slovakia.

The European Union wants to increase waste recycling by one half of the total waste. This means we need to deal with the waste, since waste treatment allows it to be further processed into reusable materials and energy, thereby maximizing the savings of primary resources and energy, as well as reducing environmental burdens [4]. This process is referred to as recycling, a key part in the modern waste reduction hierarchy. Waste treatment must be evaluated from the view of environmental, technical, as well as economic points of view to be useful for decision-making in the case of pollution impairing living environment [5].

The basis of recycling is a properly set up waste sorting system. Thorough sorting of waste at the source provides the "cleanest" waste for recycling [6]. Activities, related to municipal waste management, including sorting, in Slovakia are provided by the local government. Its interest should be to increase communities' discipline concerning generation and sorting of municipal waste, as the level of sorting also depends on the amount of the fee paid by the municipality for depositing the waste in a landfill. This, in turn, has an impact on setting the level of the municipal waste fee for the population. Sorting can thus be considered an important step towards waste recovery. The classification process can be influenced by several factors, from building the environmental awareness of the population to socio-economic factors related to the size of the household, the level of education and also the household income [7].

The EU goal is to live in 2050 in balance with the ecologic limits of the planet and to have waste-to-energy plants that burn waste to produce energy (steam in the boiler, etc.) that can be used to generate electricity [8–10].

The present contribution is orientated to communal waste production in individual regions of Slovakia with the goal to find out the development of waste production in Slovakia in regard to economic indexes of households following a consideration of obstacles and possibilities for improving waste management. The goal of the contribution results from the actual situation: In the field of waste economy, Slovakia is significantly behind many EU countries due to a low level of recycling and a high level of municipal waste storage. According to the European Commission, the problem is in the system of collection and waste sorting, as well as worse waste treatment in socially weaker regions. The legislation on waste in Slovakia should be stricter to move from landfill to recycling and waste reduction across the whole country.

2. Present State of Problem Solving

Various authors research waste treatment and waste economy. Waste recycling can be done by composting when producing fertilizing products for the market of EU [11]. By this method, composting could mitigate the environmental risk of waste creation. Paul et al. [12] studied waste attributes, such as pollution index, ecological risk index and geo-accumulation index and also revealed adverse impacts of the deeper landfill layers on the surrounding environment and found the industrial value of waste through economic evaluation of waste treatment technologies.

Deterioration of soil quality presents consequences of open waste dumping which have resulted in growing public concern [13]. Warguła et al. [14] conducted research and development works on reducing the environmental impact of machines used to process branches in urbanized areas. They show that it is possible to reduce fuel consumption [14,15] and the emission of harmful exhaust gases [14,16,17] generated during these processes by introducing innovative fuel supply systems [14,15] or by using alternative fuels to drive machines [16–18]. The process of processing branches on the premises of households is often a requirement for collecting green waste or supporting the process of transport, storage and processing, e.g., by burning or composting thanks to reducing the volume [19]. Dangi et al. [20] studied waste creation from restaurants, hotels, schools and 72 streets, and found a greater potential for recovery of organic wastes via composting and recycling.

Roychoudhury et al. [21] incorporated a symbiotic waste reduction system with the efficient functionalism of micro and macro-conversion agents which is attributed to the synthesis of waste-derived co-products from the same platform. Their study presents a novel approach, which is capable of addressing a wide fraction of solid waste and domestic sewage, and is capable of strengthening the sustainable waste management strategy [22,23].

Worldwide, vast development of rural areas has taken place, but due to the limited economic development level, some villages lack facilities for waste treatment [24], while it is easy to find practical and mature technology to treat one kind of waste in rural area, but there is a lack of overall consideration for recycling of produced and organic waste in the village.

Moreover, electrical and electronic equipment waste can be an important source of recycling [25]. However, according to D'Adamo et al. [26], the management of electronic components still presents many challenges. In order to raise awareness about the situation in electronic and electrical waste, Pérez-Beliz et al. [27] analyzed and found an extensive number of articles have been published around the world.

To make sure of the success of any policies aiming to decrease living environment pollution, it is also necessary to resolve the growing crisis of water pollution necessary. Water treatment surveys show vast differences in different regions and cities, as water is treated more thoroughly in larger cities [28]. High and fast economic increase and urbanization worldwide have influenced water consumption; the trend has increased in recent years. In this connection, Bian et al. [29] performed an analysis of regional water use and treatment in China with the aim of finding efficient waste water systems, as well as finding inefficiencies. According to Dong et al. [30] the issue could be solved by considering plant size, capacity, climate type and environmental impact. Warguła et al. [31] indicate that the increase in nutrients for aquatic plants resulting from inefficient wastewater treatment systems or too intense fertilization of fields is a serious problem for the tourism industry, due to contamination of beaches and water reservoirs, e.g., with seaweed. Wastewater and liquid waste present an important service problem in municipalities. In this area, Vialkova et al. [32] evaluated and confirmed an energy-efficient economy for microwave irradiation, using municipal wastewater sediments [32,33].

In the area of waste tire dumps also threaten human health and living environments [34]. Worldwide, waste from tires amounts to 1.3–1.5 billion tons/year and by the end of 2025 is expected to amount to more than 2.5 billion tones. In the area of waste, tire dumps also threaten human health and living environments. In 2013, the EU countries reached 3.6 million tons of used tires. The cheapest treatment method at present is landfill and is considered to be a major threat for the environment and public health.

Therefore, the problems of waste tire recycling must be solved as well. In this connection Karagoz et al. [35] found that tire pyrolytic oil production from waste tires is important from the viewpoint of both waste management and protection of fossil fuel resource depletion. Moreover, Symeonides et al. [36] evaluated the tire waste management system in Cyprus to strengthen the circular economy [36–38].

It is profitable to study waste production in individual countries. For example, with the extensive application of decentralized sewage treatment systems in China, a study must show the environmental economic value of resources and sustainability of systems to improve the efficiency of environmental resource use and decrease or avoid waste [39]. In this area, Yang et al. [40] evaluated the environmental economic value and the sustainability of a decentralized sewage treatment plant and found a remarkable economic benefit. The reuse of treated water can bring environmental benefits that are consistent with the requirements of sustainable development; the method has a very wide application prospect in developing countries [41,42].

The presented literature review shows there are no detailed studies addressing economic indexes of households influencing waste treatment development, such as average disposable equivalent incomes of households, waste costs, net money expenditure of households, people below the poverty line per one household, measure of poverty risk, etc. This creates the space for the presented contribution research, offering an evaluation of the waste treatment from the view of productivity in communal and separated waste production and productivity of waste sorting, as well as an evaluation of the waste cost development in individual regions, which could bring attention to possible regional developments, contributing to the sustainability of the regions.

3. Materials and Methods

The goal of the contribution is to evaluate the development of communal waste creation and to evaluate the waste sorting of individual elements of communal waste. The evaluation was done in accord with households and indexes, connected incomes and expenses of households, available from the Statistical Office [43] and other economic indexes related to inhabitants' standards of living [44,45]. The evaluation is realized both for Slovakia as a whole, as well as according to the regional structure. Regions are presented as the eight following counties: Bratislavský, Trnavský, Trenčiansky, Nitriansky, Žilinský, Banskobystrický, Prešovský and Košický. The indexes are used per one household and summary indexes are used for the development of waste creation as well. The contribution also verifies the statistical importance of the factors (region and year) to the value of waste creation per household and economic indexes relating to the inhabitants' standard of living concerning the household [46,47]. We searched for the existence of a relation between indexes with a correlation matrix and verified for chosen indexes the existence of polynomial dependence. According to the results of the statistical importance we searched for the similarity of the regions in chosen indexes using a cluster analysis. The evaluation for Slovakia was carried out by region and according to development over time (years).

3.1. Indexes of the Analysis

The values of the indexes were obtained from the Statistical Office of Slovakia. Some rate indexes, used in the analysis, were calculated from the data available from the Statistical Office.

Number of inhabitants—value obtained from the website of the Statistical Office of Slovakia. Due to the fact that the number of inhabitants, as registered in the data, is the same in several different time periods, for the evaluation and comparison we chose three periods: 2006, 2012 and 2018. The difference between individual periods is the same, 6 years. In the case of other time periods, the results may be slightly different. Discussion of the exactness and reliability of the results will be mentioned in the conclusion of the paper.

Amount of communal waste in tons per year—index obtained from the website of the Statistical Office of Slovakia. From the index the volume of communal waste in tons per one household is determined; this value is also used as the productivity of communal waste.

Volume of separated waste in tons per year—index obtained from website of the Statistical Office of Slovakia. From the index the volume of separated waste in tons per one household is determined; the value is further used as the productivity of separated waste.

Productivity of waste sorting presents the index, calculated as the volume of separated waste in tons per one household/volume of communal waste in tons per one household.

Net incomes of household in euros—the index was obtained from the website of the Statistical Office of Slovakia.

Net money expenditure of households in euros—calculated from the expenses of households in euros per person per month times the average number of persons in the household times 12.

Waste cost in euros per kilogram and year—calculated from the waste cost index in euros per year/waste in kilogram: Waste costs = net expenses in euros/communal waste in kg per 1 household.

Average disposable equivalent income of households (euros/month) was obtained from the website of Statistical Office of Slovakia.

People below the poverty line per 1 household—calculated from the index of people under level of poverty/number of households, obtained from the website of the Statistical Office of Slovakia.

Measure of poverty risk: 60% median (%)—measure obtained from the Statistical Office of Slovakia [48].

The development of waste productivity was evaluated with the summary index of Laspeyers, Paasche and Fisher by comparing three periods: 2006, 2012 and 2018. In a graphical illustration of the indexes for each region there is value of the index for Slovakia due to the possibility to compare the situation in each region.

Analysis of development of waste creation was carried out by summary index. The index of communal waste volume in tons per one household is recorded as production of communal waste. Volume of produced waste in a time period and territorial unit will be marked with the sign:

$$w_{u,p}$$

where u is territorial unit and p is analyzed period.

For types of waste shortage, we will use the abbreviations CW (communal waste) and SW (separated waste); shortages and types of waste will be explained with reference to their place of use.

Number of households in a time period in the frame of a territorial unit will be marked as:

$$nh_{u,p} \tag{1}$$

where u is territorial unit and p is analyzed period.

Waste production in a time period in the frame of a territorial unit will be:

$$wp_{u,p} = \frac{w_{u,p}}{nh_{u,p}} \tag{2}$$

Summary index of waste productivity with weights from the basis period is:

$$I_V^{(L)} = \frac{\sum_{u=1}^{n} \frac{w_{u,0}}{wp_{u,0}}}{\sum_{u=1}^{n} \frac{w_{u,0}}{wp_{u,1}}} \tag{3}$$

Summary index of waste productivity with weights from the common period is:

$$I_V^{(P)} = \frac{\sum_{u=1}^{n} \frac{w_{u,1}}{wp_{u,0}}}{\sum_{u=1}^{n} \frac{w_{u,1}}{wp_{u,1}}} \tag{4}$$

3.2. Analysis of Economic and Waste Indexes from the View of Verification of Statistical Importance of Regions and Year to the Development of Index Values

The economic and waste indexes used in the analysis are as follows:

- Communal waste in t per 1 household—komnadom
- Separated waste in t per 1 household—sepnadom
- Separated waste in t per 1 household/communal waste in t per 1 household—prodsep
- Waste cost in €—waste cost
- Average household disposable income in €/month—pdepdom
- Net money expenditure of households in €—cpvd
- People below the poverty line per 1 household—ospodchnadom
- Measure of poverty risk: 60% median (%)—mieraregch
- Level of significance 0.05 was used for all statistical analysis.

The analysis consisted of two steps. The first step was the ANOVA analysis, testing whether factors of the territorial unit (region) and periods are statistically important. The second step was the choice of indexes that would be used in a cluster analysis. We chose the indexes that would be statistically important for a region. According to the results for the factor ´year´ we chose to cluster analyze two periods for which the difference between various indexes would be statistically important. ANOVA test of the statistical hypothesis of the importance of the difference between the medium values during the process of choosing according to the region and period was as follows:

$$H0 : \mu_1 = \mu_2 = \ldots = \mu_n \tag{5}$$

$$HA : \mu_i \neq \mu_j \tag{6}$$

ANOVA test can be used only in the case of homogeneity of variances and normality of residuals.

4. Results

4.1. Evaluation of Communal Waste Productivity

As for the evaluation of communal waste productivity, Figure 1 illustrates calculated values of communal waste productivity in regions in the three analyzed periods. Regions are ranked by comparison with the country.

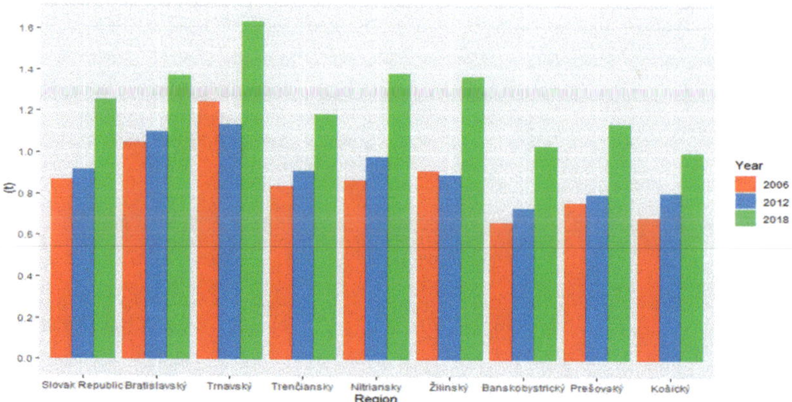

Figure 1. Productivity of communal waste in Slovakian regions.

In 2012 there is obvious smooth growth and in 2018 rapid growth of the index against the earlier period. The greatest volume of communal waste per household in 2018 was registered in the Trnavský region, the lowest in the Košický region. To evaluate the development of waste productivity, we used volume extensive indexes. The results of the indexes are given in Table 1.

Table 1. Indexes of communal waste production according to region.

Basic Period	Common Period	Laspeyres Extensive Index	Paasche Extensive Index	Fisher Extensive Index
2006	2012	1.062	1.067	1.064
2006	2018	1.455	1.460	1.458
2012	2018	1.228	1.371	1.298

Waste production between periods developed as follows:

- In 2012 it increased against basis period 2006 by 6.4%.
- In 2018 it increased against basis period 2006 by 45.75%.
- In 2018 it increased against 2012 by 29.7%.

When using summary index of waste productivity with weights from the basis period (Laspeyres extensive index), in comparing 2006 and 2018 we registered a vivid difference against the average Fisher index. In the common period 2018 the index grew against the basis period 2012 by 22.81%. Moreover, using the Paasche index, we registered a difference for the same basic and common period. In the common period 2018 the index grew against the basis period 2012 by 37.11%. The growth of waste productivity between 2012 and 2018 was caused by an increase in awareness of waste treatment in the society. Communal waste of production in time development is given in Figure 2.

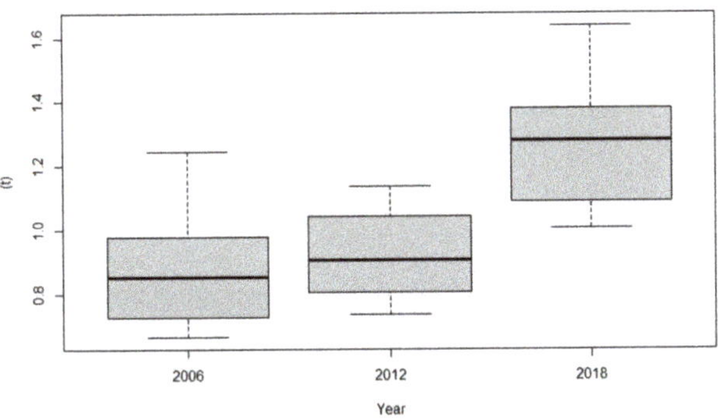

Figure 2. Development of communal waste per household.

4.2. Evaluation of Separated Waste Productivity

Figure 3 illustrates the calculated value of separated waste productivity for regions in three analyzed periods. In 2012 there is obvious smooth growth and in 2018 rapid growth of the index against the earlier period.

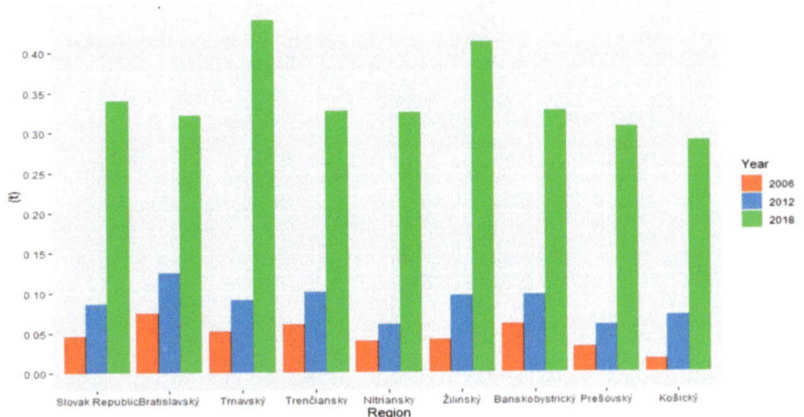

Figure 3. Productivity of separated waste.

To evaluate the development of waste production we uses extensive indexes (see Table 2).

Table 2. Indexes of communal waste production.

Basic Period	Common Period	Laspeyres Extensive Index	Paasche Extensive Index	Fisher Extensive Index
2006	2012	1.914	2.162	2.034
2006	2018	7.359	8.912	8.098
2012	2018	0.335	4.106	1.172

Production of waste sorting between periods developed as follows:

- In 2012 it increased against basis period 2006 by 103.4%
- In 2018 it increased against basis period 2006 by 709.8%.
- In 2018 it increased against 2012 by 17.2%.

Production of waste sorting in time development is given in Figure 4.

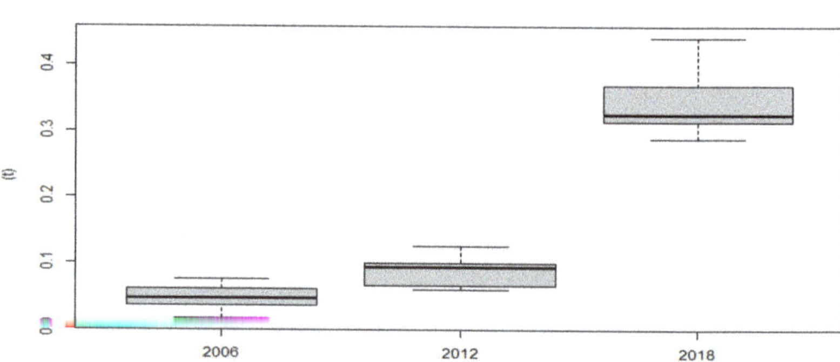

Figure 4. Development of separated waste per household.

4.3. Evaluation of Waste Sorting Productivity

Productivity of waste sorting was registered as sep/kom per household and year as percentage expression.

Figure 5 shows data of waste sorting productivity in time development according to the regions. Productivity of waste sorting between analyzed periods developed as shown in Figure 6 as follows: Support of waste sorting is visible from the increase of sorted waste by communal waste volume. According to data in Figures 5 and 6 we can state the following: In 2006 the rate of separated waste by communal waste changed by around 5%; the highest rate was in the region Banskobystrický—9%; the lowest in the region Košický—2%. In 2012 the value of the median changed over 10%; the highest separation was in the region Banskobystrický—13%; waste was sorted the least in the region Nitriansky—6%. In 2018 the medium value changed by over 27%; maximum 31% of the sorting belonged to region the Banskobystrický; waste was sorted the least in the regions Bratislavský and Nitriansky—23%.

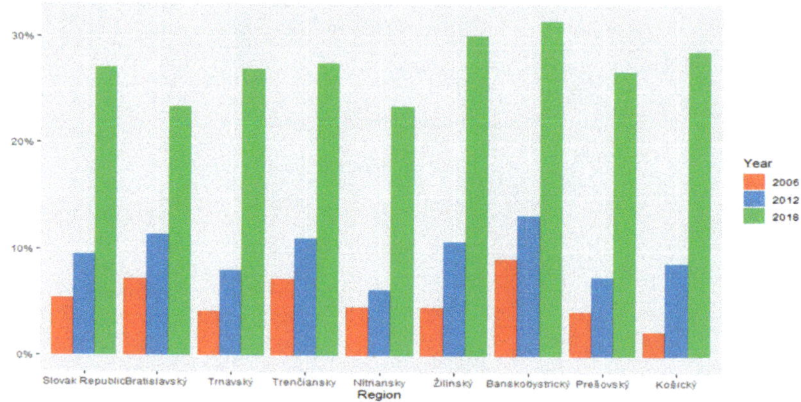

Figure 5. Development of waste sorting productivity according to the regions.

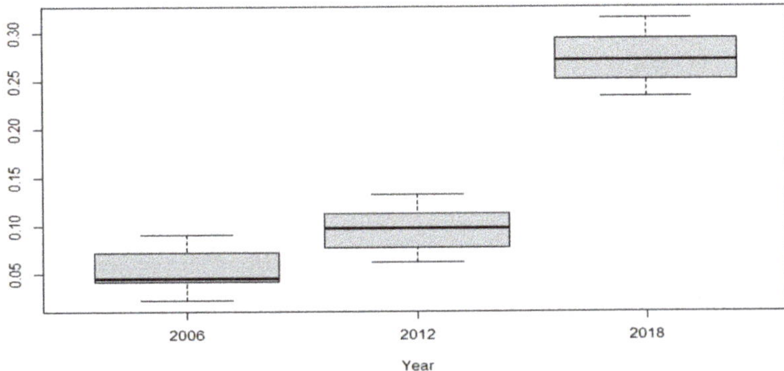

Figure 6. Development of waste sorting productivity.

4.4. Evaluation of Waste Cost

The index was created with aim of comparing the volume of communal waste in monetary terms. It presents how many euros of net income from households is produced per one kilogram of waste.

Figures 7 and 8 address the following: While in 2006 and 2012 the value of the waste cost moved in regions by between 11 and 14 €, in 2018 it was around 8–11 €. Due to the volume of communal waste per household net incomes also increased; the index development addresses the gradually decreasing production of waste over time.

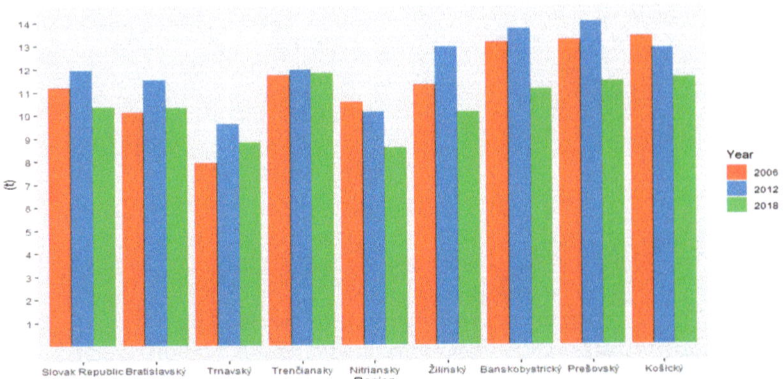

Figure 7. Development of waste cost.

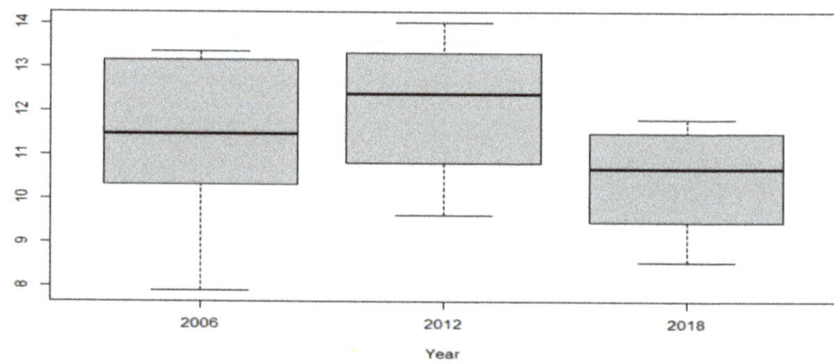

Figure 8. Development of waste cost in time (in €/kg).

The total value of waste cost changed due to the change of costs by weight in relation to the basis period—the Laspeyers index—due to the influence of waste volume change. The total value of waste cost changed also due to the influence of cost change with weight in relation to the common period—the Paasche index—due to the influence of price change.

Table 3 mentions evaluated fictitious sales of the waste, the value of the waste in prices and the volume according to the selected periods for the Paasche and Laspeyers indexes. The indexes speak of the increased waste value due to the changes in price (on decreasing the kilograms of the waste). Due to the influence of the change in waste volume, the value of the waste increased by 44% between 2006 and 2018, and by 37% between 2012 and 2018.

Table 3. Indexes of waste cost development in analyzed periods.

Basic Period	Common Period	Laspeyres Extensive Index	Paasche Extensive Index	Fisher Extensive Index
2006	2012	1.060	1.053	1.056
2006	2018	1.450	1.443	1.448
2012	2018	1.374	1.374	1.371

4.5. Analysis of Economic and Waste Indexes from the View of Verification of Statistical Importance of Regions and Year to the Development of Indexes Value

Table 4 mentions the result of the ANOVA test for chosen economic and waste indexes. We mention the value p, the probability of zero hypothesis rejection, in case it is correct. The periods and regions are the factors of the choice. Except as mentioned, the table mentions results of the p value of the Bartlett test of homogeneity of dispersion and the Shapiro Wilk test of residue normality, which means verification of whether assumptions are achieved and ANOVA test can be used.

Table 4. Results, p values of ANOVA test and verification of assumptions tests.

Test	Komnadom	Prodsep	Waste Cost	Pdedom	Cpvd	Ospodchnadom	Mieraregch
ANOVA/region	0.000	0.021	0.000	0.000	0.000	0.000	0.005
ANOVA/period	0.000	0.000	0.001	0.000	0.000	0.412	0.846
Bartlett test of variances homogeneity/period	0.994	1.000	0.780	1.000	0.952	0.922	0.958
Bartlett test of variances homogeneity/region	0.602	0.781	0.692	0.713	0.456	0.364	0.254
Shapiro–Wilk normality test of residuals	0.455	0.295	0.420	0.591	0.420	0.368	0.325

According to the results, as shown in the table, for the region and indexes—komnadom, prodsep, waste cost, pdedom, cpvd, ospodch, mieraregch—a valid alternative hypothesis confirmed that there was at least one pair of regions for which the medium values were statistically different. According to the results, as shown in the table, for the region and indexes—komnadom, prodsep, waste cost, pdedom, cpvd, ospodch—a valid alternative hypothesis confirmed that there was at least one pair of regions for which the average values were statistically different. The results of the p value Bartlett test of dispersion homogeneity and the Shapiro–Wilk test of residual normality are as follows: All p values were higher than the determined level of importance, which means the hypothesis of dispersion homogeneity cannot be rejected and the hypothesis of residual normality cannot be rejected either. The assumptions for ANOVA test were realized.

For the index sepnadom not all the assumptions of dispersion homogeneity were realized, nor for residual normality; therefore, the ANOVA test cannot be used. The results of the index evaluation are outlined in the following. To find out the differences between indexes, we carried out a pair comparison for period and region. In the following tables, we outline the results of the comparison of p-values, the probability of rejection of zero hypotheses—concerning whether the consistency between the medium values of the two populations is correct. Evaluation and rejection were realized at the level of importance 0.05. The results of pair comparison for the period and indexes, when at least one pair of periods is statistically important, are as follows (see Table 5):

Table 5. Pair comparison of indexes and period.

Period	Komnadom	Sepnadom	Podsep	Waste Cost	Pedom	Cpvd
Y_2012-Y_2006	0.339	0.157	0.001	0.089	0.000	0.000
Y_2018-Y_2006	0.000	0.000	0.000	0.072	0.000	0.000
Y_2018-Y_2012	0.000	0.036	0.000	0.001	0.060	0.000

According to the results, presented in the table, it is obvious that three indexes have statistically important differences between 2006 and 2012. Between 2006 and 2018 and between 2012 and 2018 five indexes have statistically important differences. In accordance with the results, we carried out a cluster analysis for 2018 and 2012. Results of the pair comparison of the indexes for the region were obtained when for at least one pair of regions there was a statistically important difference. The results are given by Table 6.

According to the results, presented in the table, we can say that for the region pairs Žilinský–Trenčiansky, Žilinský–Nitriansky, Trenčiansky–Nitriansky and Košický–Banskobystrický there is no significant difference in the values of any index. Index prodsep registered significant difference only for one pair of regions Nitriansky–Banskobystrický; therefore, it was not used in the cluster analysis. The results, presented in the table, offer an assumption for cluster analysis use; there are regions that could be more similar from the view of the chosen indexes.

Table 6. Results of pair comparison of indexes for regions.

Pair	Komnadom	Prodsep	Waste Cost	Pedom	Cpvd	Ospodchnadom	Mieraregch
Bratislavský–Banskobystrický	0.000	0.223	0.260	0.000	0.000	0.001	0.000
Košický–Banskobystrický	1.000	0.099	0.414	0.989	0.623	1.000	0.980
Nitriansky–Banskobystrický	0.002	0.011	0.004	0.962	1.000	0.755	0.668
Prešovský–Banskobystrický	0.600	0.059	0.035	0.670	0.416	0.012	0.730
Trenčiansky–Banskobystrický	0.058	0.628	0.351	0.027	0.376	0.021	0.006
Trnavský–Banskobystrický	0.000	0.078	0.000	0.187	0.308	0.036	0.011
Žilinský–Banskobystrický	0.003	0.588	0.017	0.672	0.969	0.960	0.430
Košický–Bratislavský	0.000	0.999	1.000	0.000	0.000	0.000	0.001
Nitriansky–Bratislavský	0.539	0.666	0.300	0.000	0.000	0.009	0.007
Prešovský–Bratislavský	0.001	0.990	0.919	0.000	0.000	0.000	0.000
Trenčiansky–Bratislavský	0.022	0.988	1.000	0.000	0.006	0.474	0.706
Trnavský–Bratislavský	0.059	0.997	0.007	0.000	0.007	0.331	0.540
Žilinský–Bratislavský	0.335	0.992	0.738	0.000	0.001	0.003	0.016
Nitriansky–Košický	0.004	0.909	0.179	1.000	0.453	0.654	0.989
Prešovský–Košický	0.856	1.000	0.773	0.254	1.000	0.017	0.249
Trenčiansky–Košický	0.132	0.872	1.000	0.113	0.019	0.015	0.037
Trnavský–Košický	0.000	1.000	0.004	0.553	0.015	0.026	0.063
Žilinský–Košický	0.007	0.898	0.540	0.979	0.171	0.914	0.909
Prešovský–Nitriansky	0.044	0.977	0.910	0.183	0.279	0.001	0.060
Trenčiansky–Nitriansky	0.505	0.246	0.218	0.162	0.536	0.301	0.164
Trnavský–Nitriansky	0.002	0.948	0.400	0.680	0.454	0.437	0.257
Žilinský–Nitriansky	1.000	0.273	0.989	0.996	0.996	1.000	1.000
Trenčiansky–Prešovský	0.749	0.728	0.836	0.001	0.010	0.000	0.000293
Trnavský–Prešovský	0.000	1.000	0.058	0.009	0.007	0.000	0.000
Žilinský–Prešovský	0.086	0.766	1.000	0.056	0.093	0.002	0.028
Trnavský–Trenčiansky	0.000	0.810	0.005	0.941	1.000	1.000	1.000
Žilinský–Trenčiansky	0.729	1.000	0.614	0.438	0.895	0.131	0.312
Žilinský–Trnavský	0.001	0.842	0.118	0.964	0.835	0.206	0.453

Note: For sepnadom the assumption was not met; therefore, the ANOVA test was not used to verify the factors' importance for the region and the year nonparametric Friedman rank test was used with the result p value = 0.04052. For pair comparison for the region Nemenyi's test of multiple comparisons for independent samples (tukey) was used successfully when any pair of regions confirmed a statistically important difference. Pair comparison of Nemenyi's test of multiple comparisons for independent samples (tukey) for the factor 'period' in the table is for the factor 'years'.

The following indexes were used in the cluster analysis: Komnadom, nakladovost, pedom, cpvd, ospodch and mierareg chudoby. These factors, which were originally considered, were not included: Sepnadom and prodsep. A cluster analysis was carried out for 2018 and 2012 with a subsequent comparison of results.

Analysis of development in years and of which region is the weakest and which is strongest can be done according to the mentioned graphs and the following tables. As can be seen in Figure 9: The first four graphs present indicators of waste creation, sorting and costing. All indicators were described in detail in an earlier analysis, from which results the following:

In 2018 compared to 2012 there was increase of communal waste per household, which means increased production of waste. Waste sorting, volume of separated waste per one household and the rate of separated waste by communal waste increased. Costs of waste decreased, which means kilogram production of waste in 2018 cost less to a household.

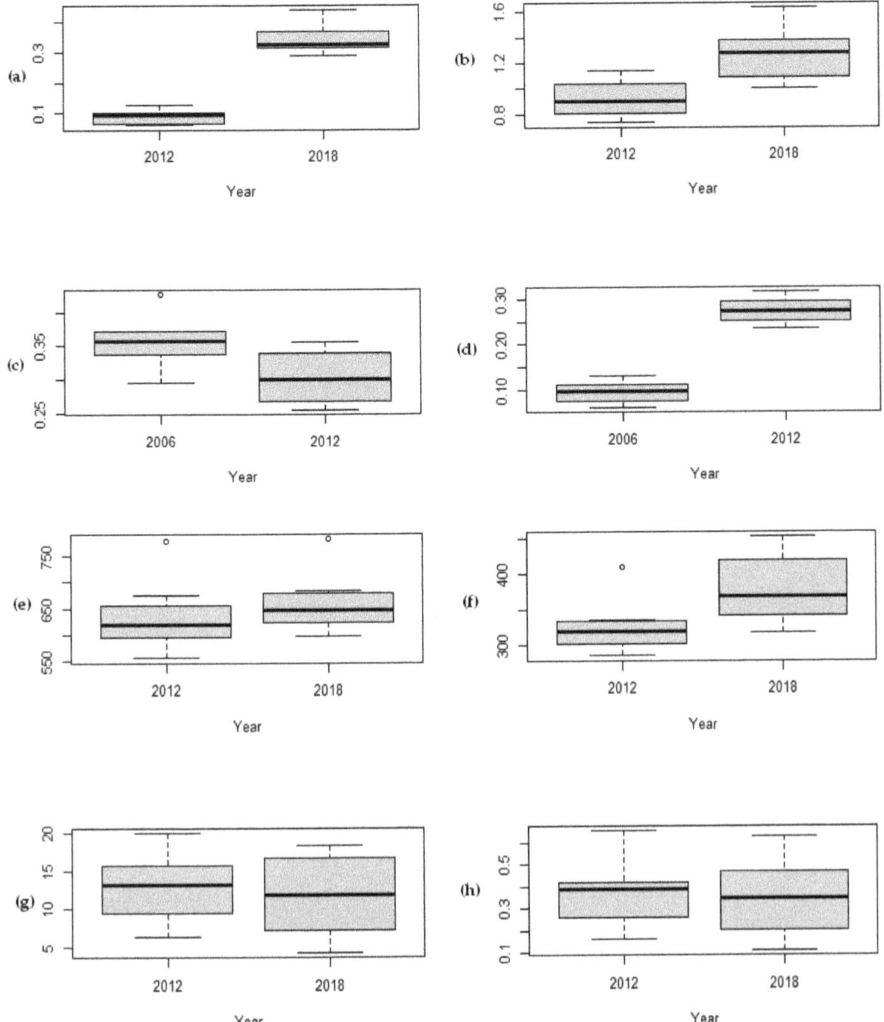

Figure 9. Comparison of the indicators for 2012 and 2018. Note: (**a**) Separated waste in t per 1 household; (**b**) communal waste in t per 1 household; (**c**) waste cost in €; (**d**) ratio of separated/communal waste; (**e**) average household disposable income in €; (**f**) net money expenditure of households in €; (**g**) at risk of poverty percentage; (**h**) people below the poverty line per 1 household.

Indicators concerning the financial situation of the households in regions and at risk of poverty are presented in the second group of four graphs.

Average disposal incomes in 2012 changed in each region by between 555 and 778 €. In 2018 the lower level changed by over 596 € and the upper level by over 782 €. In both cases, the minimal value was recorded in the region Prešovský and the maximal value was recorded in the region Bratislavský—registered by outliers. The second highest value was in 2012—674 € for the region Trenčiansky; and in 2018 for the region Trnavský—682 €.

Net money expenditure in 2012 was between 9919 and 12,663 €; the highest value is the outlier, belonging to the region Bratislavský; distribution is asymmetric (skew = 0.55).

In 2018 the levels moved upwards from 11,476 € to 14,432 €. Distribution is more symmetric (skew −0.22).

The next two indicators, presenting the state of poverty—at risk of poverty rate and people below the poverty line per one household—developed similarly.

At risk of poverty rate: In 2018 the minimum decreased from 6.30 to 4.30; the maximum decreased as well from 19.90 to 18.40. The box width in the graph shows that the differences between the most and the least threatened regions decreased. In spite of there being a registered decrease in the average value and the levels of the indicator, the value increased in the regions Banskobystrický and Košický.

The indicator "people below the poverty line per one household" in 2018 did not decrease compared to 2012 in the regions Žilinský, Banskobystrický and Košický. The results are provided in Tables 7–10.

Table 7. Input data in 2012.

Region	Komnadom	Waste Cost	Pdedom	Cpvd	Ospodchnadom	Mierarchud
Bratislavský	1.099	11.52	778.17	12,663.38	0.161	6.3
Trnavský	1.138	9.61	637.69	10,935.85	0.289	10.6
Trenčiansky	0.915	11.93	674.01	10,909.04	0.231	8.3
Nitriansky	0.983	10.10	600.71	9919.25	0.419	15.9
Žilinský	0.895	12.94	636.79	11,577.80	0.367	12.7
Banskobystrický	0.735	13.70	592.2	10,069.38	0.417	15.6
Prešovský	0.802	14.01	555.88	11,231.08	0.652	19.9
Košický	0.810	12.85	603.05	10,418.23	0.405	13.5

Table 8. Description characteristics in 2012.

Indicator	Mean	SD	Median	Min	Max	Range	Skew	Kurtosis	SE
komnadom	0.92	0.14	0.90	0.74	1.14	0.40	0.26	−1.60	0.05
waste cost	12.80	1.61	12.39	9.61	14.10	4.40	−0.32	−1.61	0.57
pdedom	634.81	67.95	619.92	555.88	778.17	222.29	0.94	−0.25	24.20
cpvd	10,965.50	888.78	10,922.44	9919.25	12,663.38	2744.13	0.55	−0.92	314.23
ospodchnadom	0.37	0.15	0.39	0.16	0.65	0.49	0.44	−0.76	0.05
mierarchud	12.85	4.40	13.10	6.30	19.90	13.60	0.02	−1.37	1.56

Table 9. Input data in 2018.

Region	Komnadom	Waste Cost	Pdedom	Cpvd	Ospodchnadom	Mieraregch
Bratislavský	1.375	10.31	782.18	14,167.32	0.112	4.3
Trnavský	1.638	8.81	682.31	14,432.99	0.229	7.9
Trenčiansky	1.187	11.80	675.23	14,007.71	0.183	6.6
Nitriansky	1.383	8.56	658.4	11,834.70	0.292	10.8
Žilinský	1.369	10.09	633.33	13,817.62	0.399	12.9
Banskobystrický	1.036	11.08	615.11	11,476.04	0.465	17.6
Prešovský	1.144	11.40	596.63	13,045.08	0.629	18.4
Košický	1.005	11.57	630.98	11,629.90	0.475	15.8

Table 10. Description characteristics in 2018.

Indicator	Mean	SD	Median	Min	Max	Range	Skew	Kurtosis	SE
komnadom	1.27	0.21	1.28	1.1	1.64	0.63	0.29	−1.34	0.08
nakladovost	10.45	1.24	10.69	8.56	11.80	3.24	−0.41	−1.63	0.441
pdedom	659.27	57.62	645.86	596.63	782.18	185.55	0.98	−0.18	20.37
cpvd	13,051.42	1232.80	13,431.35	11,476.04	14,432.96	134.50	−0.22	−1.98	435.86
ospodchnadom	0.35	0.17	0.35	0.11	0.63	0.52	0.15	−1.53	0.06
mierarchud	11.79	5.26	11.85	4.30	18.40	14.10	−0.06	−1.79	1.86

According to the calculated description characteristics we can find out how the regions changed between 2012 and 2018.

Figure 10 shows that some regions are, from the perspective of pair indexes (productivity of waste and cpvd—net incomes of households), closer than others. We calculated the Euclidean distance of all region pairs from the values of the indexes, chosen for the cluster analysis. Values of mutual distances are presented in Table 11. Euclidean distance is calculated from the standardized variables, which have medium value 0 and standard deviation 1. Individual regions present points of multivariate space, created by chosen indexes. In 2012 the closest regions were Trenčiansky and Žilinský. The next closest pair was regions Košický and Banskobystrický. Region Prešovský was close to region Košický and Banskobytrický. Region Bratislavský was not close to any of the regions.

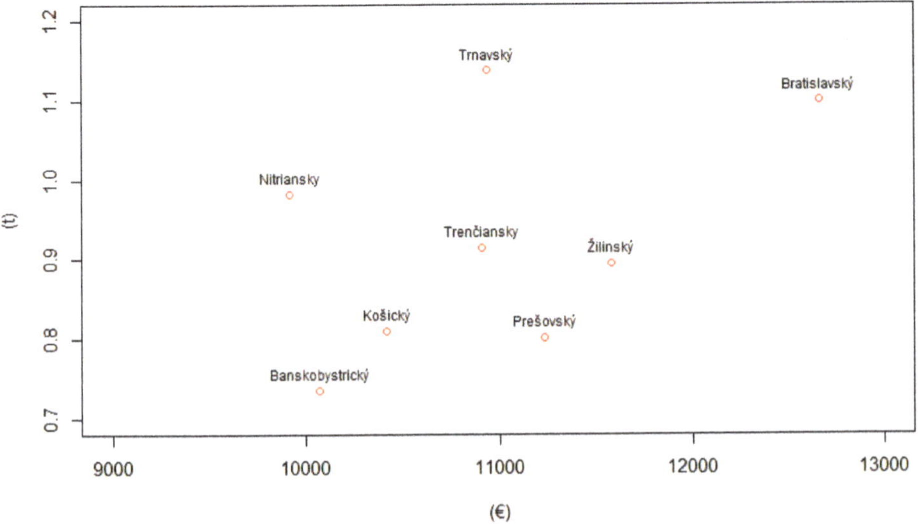

Figure 10. Communal waste development according to the net expenses of household in 2012.

Table 11. Values of mutual distances 2012.

Region	Bratislavský	Trnavský	Trenčiansky	Nitriansky	Žilinský	Banskobystrický	Prešovský
Trnavský	3.35						
Trenčiansky	2.90	2.28					
Nitriansky	5.06	2.26	2.92				
Žilinský	3.56	2.86	1.76	2.82			
Banskobystrický	5.63	4.21	3.07	2.83	2.31		
Prešovský	6.36	5.00	4.52	3.66	2.97	2.39	
Košický	4.81	3.30	2.24	2.24	1.55	0.98	2.59

As for the Euclidean distance of regions, in the cluster analysis for 2018 we see that some regions are close; their Euclidean distance is around 1; other regions are remote (see Table 12).

Table 12. Euclidean distance of regions in cluster analysis 2018.

Region	Bratislavský	Trnavský	Trenčiansky	Nitriansky	Žilinský	Banskobystrický	Prešovský
Trnavský	2.63						
Trenčiansky	2.459	3.247					
Nitriansky	3.574	2.555	3.447				
Žilinský	3.492	2.343	2.481	2.02			
Banskobystrický	5.161	4.871	3.626	3.182	2.784		
Prešovský	5.400	4.741	3.768	3.802	2.421	1.719	
Košický	4.935	4.899	3.309	3.363	2.872	0.621	1.772

According to the results we constructed a dendogram. Clusters show a graphical presentation of a double graph: Waste productivity and net incomes of households. Cluster dendogram in 2018 (see Figure 11) and Cluster dendogram in 2012 (see Figure 12) provided information for Figures 13 and 14. Figure 13 is for the cluster dendogram Figure 11, and Figure 14 is for the cluster dendogram Figure 12.

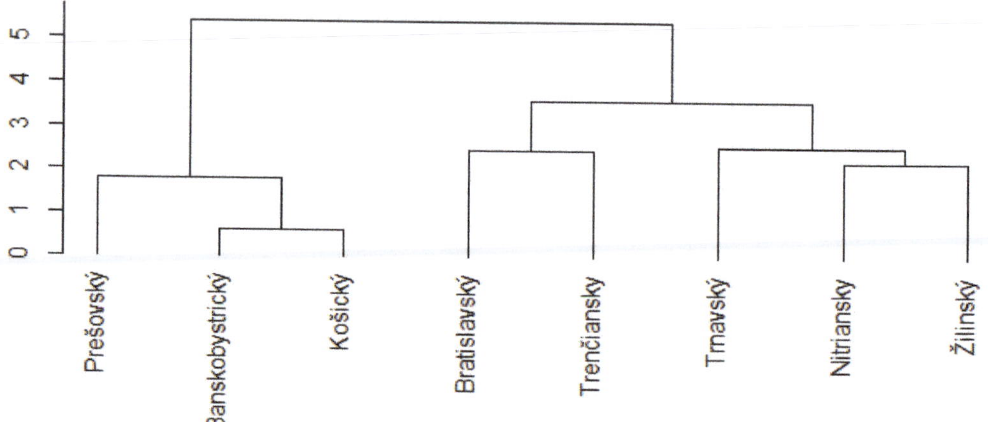

Figure 11. Cluster dendogram in 2018.

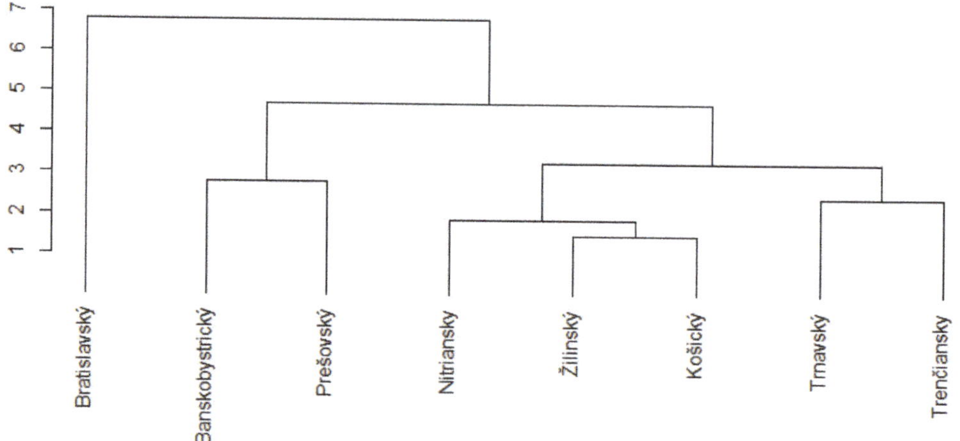

Figure 12. Cluster dendogram in 2012.

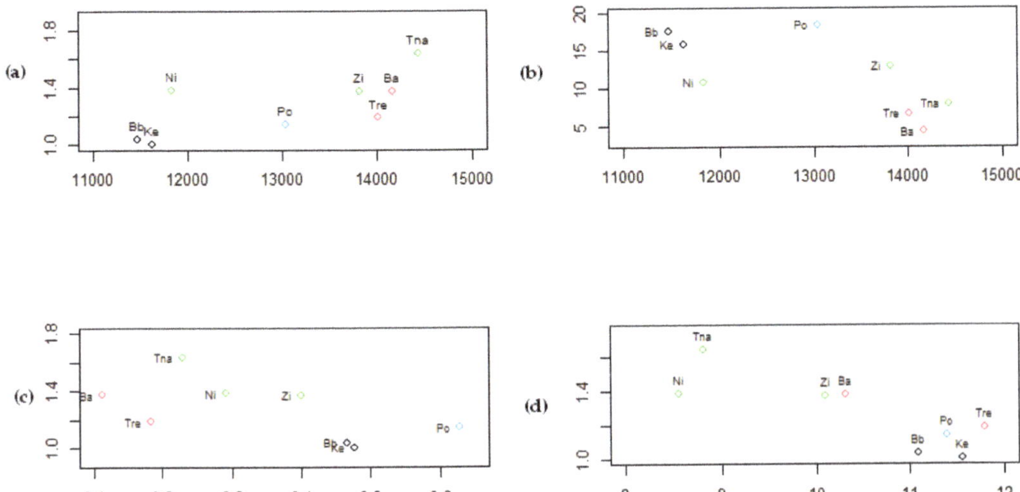

Figure 13. Comparison of indexes after clustering for 2018: (**a**) Communal waste per household in t per year according to net expenses of household in (€); (**b**) measure of poverty risk—60% of median at percentage (%) according to net expenses of household in €; (**c**) communal waste per household in t per year according people below the poverty line per 1 household; (**d**) communal waste per household in t per year according to waste costs in € with relation to kg.

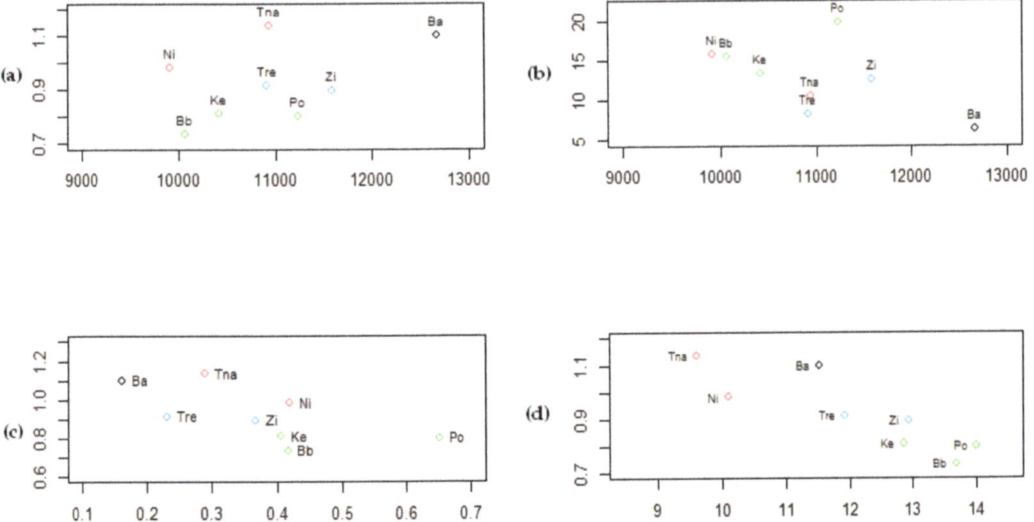

Figure 14. Comparison of indexes after clustering 2012: (**a**) Communal waste per household in t per year according net expenses of household in €; (**b**) measure of poverty risk—60% of median at percentage (%) according to net expenses of household in €; (**c**) communal waste per household in t per year according to people below the poverty line per 1 household; (**d**) communal waste per household in t per year according to waste costs in € with relation to kg.

Figures 13 and 14 illustrate pairs of indexes for individual regions. The same colors of points at Figures 13 and 14 mean individual clusters according to cluster dendrograms. As for the regions' names, for convenience, the following abbreviations are used: Bratislavský as Ba, Trnavský as Tna, Trenčiansky as Trc, Nitriansky as Nl, Žilınský as Zi, Banskobystrický as Bb, Prešovský as Po, Košický as Ke. Similar colors of points mean regions with the same cluster in the dendogram.

In 2012 the closest regions were Banskobystrický and Prešovský, Trnavský and Trenčiansky, Košický and Žilinský. Every region was close to Bratislavský.

Figure 13 shows a comparison of indexes after clustering.

When comparing dendograms in 2012 and 2018 we see a slight regrouping of region groups. While in 2012 the region Bratislavský was in any group, in 2018 it presents a pair with region Trenčiansky.

Figure 14 illustrates how the groups, created by clustering, are structured for the following pairs of indexes:

5. Discussion

The most dominant fraction in communal waste is biological decomposted waste [49]. The organic fraction in the waste amounts to around 48–69% [50]. Therefore, great attention is given to the sorting of biological decomposted waste and its evaluation. A pilot project, carries out in Ghane, showed that sorting into proper baskets would be effective in many areas, since it achieved on average more than 80% in the case of "biologically removable elements, except paper waste" and more than 75% in "other waste". High efficiency of waste sorting meant that using a one-way sorting system was proper for the households that participated. Treatment of biological waste separated from the source is still an important task due to the high rate of contamination [51].

The system of waste sorting in Slovakia is not unified. In Slovakia there are over 2800 autonomous regions and it is not possible to create a unified document for waste treatment. In some autonomies, every waste element must be separated, in other autonomies the costs of sorting and waste treatment are saved and waste is collected in common. It depends on the capacity of the autonomy to deal with waste sorting companies. On the other hand all autonomies have a goal to achieve in the area of waste economy—to be sorting 65% of waste before 2025. Presently waste sorting is at only 36%.

The Slovak Republic is aware of the importance and significance of sorting and is constantly improving the relevant systems. From the beginning of 2021, a ban was issued on the disposal of waste by landfill, unless it has undergone treatment. At the same time, waste sorting at the source is also considered waste treatment.

Awareness of the population about the need and methods of sorting plays a big role in this. For example it is necessary to teach households to sort and to recognize the recycling labels on the packaging in order to know what can be separated and how. If households are not used to separating waste, they must gradually learn new habits such as sorting biodegradable waste. This follows from the obligation of municipalities from 2021 to make sure of separate collection of biodegradable municipal waste.

Another role of municipalities in increasing the rate of waste sorting is to have a proper set up of the system [52]. It also turns out, also on the basis of experience in developed countries, that the quality of waste sorting is increased by the application of a "door to door" system, which is based on bag collection in which anonymity is eliminated. Another motivating element in the sorting process is waste charges, which should be set fairly according to the "pay for what you throw away" pattern. Residents would thus be motivated to decrease unsorted waste.

A good "mirror" for the public is also information about the production of municipal waste, sorting and management. These should show as realistically as possible that this is possible only by precise waste records, which should be provided in Slovakia by a fully functional waste management information system from 2022.

Living standard in Slovakia in 2018 increased to 78% of the EU average. Incomes and consumption of households is increasing. Most expenses are orientated towards the household and food. Consumption of food is also increasing, causing communal waste.

There is a dependence between living standard and production and sorting of communal waste, which is confirmed by the presented results of individual regions of Slovakia.

On the other hand, the more economically developed the region, the more production of communal waste; this is illustrated for example in the regions Bratislavský, Trnavský and Trenčiansky. Those regions belong to the western part of Slovakia, in which there is higher living standard for inhabitants, developed industry, better infrastructure and considerably higher inflow of foreign investment [53,54]. This approach can also be linked to the possibilities of effective use of GIS data [55] that are commonly used in such research. It has been shown that the amount of waste is growing with increasing household incomes, from which the increase in their expenditures also stems. In addition to the inhabitants and household motivation to sort waste, employees of enterprises must also be motivated to properly treat waste [56].

This paper defined new possibilities for waste evaluation through economic variables. The fact that the presented contribution deals only with economic indexes of households creates the space for the next research project: To find out the behavior of households when throwing away and separating communal waste. However, problems of household waste should also be considered, such as, wood-based waste, which means, for example, furniture [57].

6. Conclusions

The use and consumption of goods are natural and necessary parts of human societies. The more a society and a country are developed, the higher the consumption of goods, and consequently the market produces a higher volume of products. In this connection, the waste economy belongs to the one of the most important areas: Sustainable development. The hierarchy of the waste economy results from the need to minimize negative impacts to the living environment. This brings a decrease in storing and an increase in waste recycling. Various researchers are dealing with the possibilities of recycling. Proper sorting is the foundation of good recycling. Metals, paper, glass and plastics are the most common materials that are separated and have high recycling potential.

The presented paper dealt not only with the production and sorting of municipal waste in Slovakia but also with economic factors that may affect it.

In Slovakia there are rather considerable differences between individual regions, which is reflected in the research indicators, including measure of poverty risk. Results of the analysis show there is long-term poverty risk measure in the regions Prešovský, Košický and Banskobystrický. This could be caused by a lower living standard, higher unemployment, higher marginalized groups of inhabitants or a higher number of families with many children and not full families. The regions are typical due to the lower incomes, as well as lower volume of waste per household.

The results of the contribution can be used not only for evaluation of waste development over time in Slovakia, but also for evaluation of the position of Slovakia in the frame of the EU, when the situation shows the waste treatment in Slovakia is weak in comparison with the EU. This demands a change in legislative decrees and an increase in waste treatment awareness. This should result in verified systems that run effectively in the EU countries, when the systems are effective from the view of financial investment in relation to the achieved results of waste recycling.

Since the volume of mixed communal waste is not changing considerably over time, but the volume of communal waste is growing, the volume of other elements in communal waste must also be growing. Such a situation demands future research from the perspective of trends of communal waste growth, as well as from the perspective of assumed development of GDP, according to which there is the aim to find a growing trend of communal waste creation in a future period.

Author Contributions: Conceptualization, B.S., E.M.; methodology, B.S., K.Č.; software, B.S., M.T.; validation, E.M.; formal analysis, B.S., K.Č.; resources, B.S.; data curation, E.M.; writing—original draft preparation, M.T., and K.Č.; writing—review and editing K.Č.; visualization, M.T.; supervision, Ľ.Š.; project administration, K.Č.; funding acquisition, M.T. All authors have read and agreed to the published version of the manuscript.

Funding: This research was funded by grant project VEGA, grant number 1/0264/21 and grant project VEGA, grant number 1/0797/20.

Institutional Review Board Statement: Not applicable.

Informed Consent Statement: Not applicable.

Data Availability Statement: Not applicable.

Acknowledgments: Contribution presents partial result of grant project VEGA 1/0515/18, VEGA 1/0797/20 and VEGA 1/0264/21.

Conflicts of Interest: The authors declare no conflict of interest.

References

1. Dong, J.; Chi, Y.; Zou, D.; Fu, C.; Huang, Q.; Ni, M. Comparison of municipal solid waste treatment Technologies form a life cycle perspective in China. *Waste Manag. Res.* **2014**, *32*, 13–23. [CrossRef]
2. Alan, Y.; Gao, G.P.; Gaur, V. Does inventory productivity predict future stock returns? A retailing industry perspective. *Manag. Sci.* **2014**, *60*, 2416–2434. [CrossRef]
3. Badran, M.F.; El-Haggar, S.M. Optimization of municipal solid waste management in Port Said—Egypt. *Waste Manag.* **2006**, *26*, 534–545. [CrossRef] [PubMed]
4. Adamisin, P.; Kotulic, R.; Mura, L.; Kravcakova Vozarova, I.; Vavrek, R. Managerial approaches of environmental projects: An empirical study. *Pol. J. Manag. Stud.* **2018**, *17*, 27–38. [CrossRef]
5. Gomes, L.A.; Santos, A.F.; Pinheiro, C.T.; Góis, J.C.; Quina, M.J. Screening of waste materials as adjuvants for drying sewage sludge based on environmental, technical and economic criteria. *J. Clean. Prod.* **2020**, *259*, 120927. [CrossRef]
6. Pearce, D. Green Economics. *Environ. Values* **1992**, *1*, 3–13. Available online: http://www.environmentandsociety.org/node/5454/ (accessed on 1 September 2018). [CrossRef]
7. Čech, J. Economic Growth and Evaluation of Living Environment. *Acta Montan. Slovaca* **2007**, *12*, 194–204. Available online: https://actamont.tuke.sk/pdf/2007/n3/4cech.pdf/ (accessed on 20 October 2018). (In Slovak).
8. Ardolino, F.; Lodato, C.; Astrup, T.F.; Arena, U. Energy recovery from plastic and biomass waste by means of fluidized bed gasification: A life cycle inventory model. *Energy* **2018**, *165*, 299–314. [CrossRef]
9. Winans, K.; Kendall, A.; Deng, H. The History and Current Applications of the Circular Economy concept. In *Renewable and Sustainable Energy Reviews*. Available online: http://linkinghub.elsevier.com/retrieve/pii/S1364032116306323/ (accessed on 29 June 2017).
10. Zhou, Z.; Tang, Y.; Dong, J.; Chi, Y.; Ni, M.; Li, N.; Zhang, Y. Environmental performance evolution of municipal solid waste management by life cycle assessment in Hangzhou, China. *J. Environ. Manag.* **2018**, *227*, 23–33. [CrossRef]
11. Pecorini, I.; Peruzzi, E.; Albini, E.; Doni, S.; Macci, C.; Masciandaro, G.; Iannelli, R. Evaluation of MSW compost and digestate mixture for a circular economy application. *Sustainability* **2020**, *12*, 3042. [CrossRef]
12. Paul, S.; Choudhury, M.; Deb, U.; Pegu, R.; Das, S.; Bhattacharya, S.S. Assessing the ecological impacts of ageing on hazard potential of solid waste landfills: A green approach through vermitechnology. *J. Clean. Prod.* **2019**, *236*, 117643. [CrossRef]
13. Ali, S.M.; Pervaiz, A.; Afzal, B.; Hamid, N.; Yasmin, A. Open dumping of municipal solid waste and its hazardous impacts on soil and vegetation diversity at waste dumping sites in Islamabad city. *J. King Saud Univ. Sci.* **2014**, *26*, 59–65. [CrossRef]
14. Warguła, Ł.; Kukla, M.; Lijewski, P.; Dobrzyński, M.; Markiewicz, F. Influence of Innovative Woodchipper Speed Control Systems on Exhaust Gas Emissions and Fuel Consumption in Urban Areas. *Energies* **2020**, *13*, 3330. [CrossRef]
15. Warguła, Ł.; Krawiec, P.; Waluś, K.J.; Kukla, M. Fuel Consumption Test Results for a Self-Adaptive, Maintenance-Free Wood Chipper Drive Control System. *Appl. Sci.* **2020**, *10*, 2727. [CrossRef]
16. Warguła, Ł.; Kukla, M.; Lijewski, P.; Dobrzyński, M.; Markiewicz, F. Influence of the Use of Liquefied Petroleum Gas (LPG) Systems in Woodchippers Powered by Small Engines on Exhaust Emissions and Operating Costs. *Energies* **2020**, *13*, 5773. [CrossRef]
17. Warguła, Ł.; Kukla, M.; Lijewski, P.; Dobrzyński, M.; Markiewicz, F. Impact of Compressed Natural Gas (CNG) Fuel Systems in Small Engine Wood Chippers on Exhaust Emissions and Fuel Consumption. *Energies* **2020**, *13*, 6709. [CrossRef]
18. Warguła, Ł.; Kukla, M.; Krawiec, P.; Wieczorek, B. Reduction in Operating Costs and Environmental Impact Consisting in the Modernization of the Low-Power Cylindrical Wood Chipper Power Unit by Using Alternative Fuel. *Energies* **2020**, *13*, 2995. [CrossRef]

19. Rogov, A.A.; Kublitckii, A.O.; Tokareva, M.V. The forest park areas maintenance activities improvement in metropolis, based on information and production technologies. In Proceedings of the 2019 IEEE Conference of Russian Young Researchers in Electrical and Electronic Engineering (EIConRus), Saint Petersburg and Moscow, Russia, 28–31 January 2019; pp. 1453–1457.

20. Dangi, M.B.; Pretz, C.R.; Urynowicz, M.A.; Gerow, K.G.; Reddy, J.M. Municipal solid waste generation in Kathmandu, Nepal. *J. Environ. Manag.* **2011**, *92*, 240–249. [CrossRef]

21. Roychoudhury, A.; Kumar, N.A.; Arutchelvan, V. Integrated waste management through symbiotic culture: A holistic approach. *IOP Conf. Ser. Earth Environ. Sci.* **2019**, *344*, 0120455. [CrossRef]

22. Brzeszczak, A.; Imiołczyk, J. Ratio analysis of Poland's sustainable development compared to the countries of the European Union. *Acta Oecon. Univ. Selye* **2016**, *5*, 31–41.

23. Hazra, T.; Goel, S. Solid waste management in Kolkata, India: Practices and challenges. *Waste Manag.* **2009**, *29*, 470–478. [CrossRef]

24. Zhou, H.; Shen, Y.; Meng, H.; Zhao, L.; Cheng, H.; Huo, L.; Dong, S. Research on rural waste recycling mode and its evaluation in village. *Trans. Chin. Soc. Agric. Eng.* **2018**, *34*, 206–212.

25. Velenturf, A.P.M.; Jopson, J.S. Making the business case for resource recovery. *Sci. Total Environ.* **2018**, *648*, 1031–1041. [CrossRef] [PubMed]

26. D´Adamo, I.; Rosa, P.; Terzi, S. Challenges in waste electrical and electronic equipment management: A profitability assessment in three European countries. *Sustainability* **2016**, *8*, 633. [CrossRef]

27. Pérez-Belis, V.; Bovea, M.D.; Ibáñez-Forés, V. An in-depth literature review of the waste electrical and electronic equipment context: Trends and evolution. *Waste Manag. Res.* **2015**, *33*, 3–29. [CrossRef]

28. Pan, D.; Hong, W.; Kong, F. Efficiency evaluation of urban wastewater treatment: Evidence from 113 cities in the Yangtze River economic belt of China. *J. Environ. Manag.* **2020**, *270*, 110940. [CrossRef] [PubMed]

29. Bian, Y.; Yan, S.; Xu, H. Efficiency evaluation for regional urban water use and wastewater decontamination systems in China: A DEA approach. *Resour. Conserv. Recycl.* **2014**, *83*, 2014, 15–23. [CrossRef]

30. Dong, X.; Zhang, X.; Zeng, S. Measuring and explaining eco-efficiencies of wastewater treatment plants in China: An uncertainty analysis perspective. *Water Res.* **2017**, *112*, 195. [CrossRef] [PubMed]

31. Warguła, Ł.; Wieczorek, B.; Kukla, M.; Krawiec, P.; Szewczyk, J.W. The Problem of Removing Seaweed from the Beaches: Review of Methods and Machines. *Water* **2021**, *13*, 736. [CrossRef]

32. Vialkova, E.; Zemlyanova, M.; Fugaeva, A. Treatment and utilization of liquid communal waste in the cities. In Proceedings of the MATEC Web of Conferences, České Budějovice, Czech Republic, 6–7 November 2018; Volume 212, p. 030052018.

33. Nayal, F.S.; Mammadov, A.; Ciliz, N. Environmental assessment of energy generation from agricultural and farm waste through anaerobic digestion. *J. Environ. Manag.* **2016**, *184*, 389–399. [CrossRef]

34. Šourková, M.; Adamcová, D.; Winkler, J.; Vaverková, M.D. Phytotoxicity of Tires Evaluated in Simulated Conditions. *Environments* **2021**, *8*, 49. [CrossRef]

35. Karagoz, M.; Uysal, C.; Agbulut, U.; Saridemir, S. Energy, exergy, economic and sustainability assessments of a compression ignition diesel engiene fueled with tire pyrolytic oil-diesel blends. *J. Clean. Prod.* **2020**, *264*, 121724. [CrossRef]

36. Symeonides, D.; Loizia, P.; Zorpas, A.A. Tire waste management system in Cyprus in the framework of circular economy strategy. *Environ. Sci. Pollut. Res.* **2019**, *26*, 35445–35460. [CrossRef]

37. Abreu, M.C.S.; Ceglia, D. On the implementation of a circular economy: The role of institutional capacity-building through industrial symbiosis. *Resour. Conserv. Recycl.* **2018**, *138*, 99–109. [CrossRef]

38. Ellen MacArthur Foundation. Towards A Circular Economy: Business Rationale For An Accelerated Transition. Available online: https://www.ellenmacarthurfoundation.org/assets/downloads/TCE_Ellen-MacArthur-Foundation-9-Dec-2015.pdf/ (accessed on 29 June 2017).

39. Martinico-Perez, M.F.G.; Schandl, H.; Tanikawa, H. Sustainability indicators from resource flow trends in the Philippines. *Resour. Conserv. Recycl.* **2018**, *138*, 74–86. [CrossRef]

40. Yang, L.; Kong, F.L.; Xi, M.; Li, Y.; Wang, S. Environmental economic value calculation and sustainability assessment for constructed rapid infiltration system based on energy analysis. *J. Clean. Prod.* **2017**, *167*, 582–588. [CrossRef]

41. Troschinetz, A.M.; Mihelcic, J.R. Sustainable recycling of municipal solid waste in developing countries. *Waste Manag.* **2009**, *29*, 915–923. [CrossRef]

42. Zhuonan, S. Research on guarantee mechanism of waste concrete recycling logistics mode in Beijing city. International Conference on Logistics, Informatics and Service Science. Available online: http://toc.proceedings.com/28971webtoc.pdf/ (accessed on 14 December 2017).

43. Eurostat. Available online: http://ec.europa.eu/eurostat/en/data/browse-statistics-by-theme (accessed on 1 March 2021).

44. Zeller, T.L.; Kostolansky, J.; Bozoudis, M. A changed taxonomy of retail financial ratios. *J. Appl. Bus. Res.* **2017**, *33*, 1069–1080. [CrossRef]

45. Samson, A.A.; Mary, J.; Yemisi, B.F.; Erekpitan, I.O. The impact of working capital management on the profitability of small and 45medium scale enterprises in Nigeria. *Res. J. Bus. Manag.* **2012**, *6*, 61–69. [CrossRef]

46. Hebák, P.; Hustopecký, J.; Pecáková, I.; Plašil, M.; Prúša, M.; Řezankova, H.; Vlach, P.; Svobodová, A. *Multi-Dimensional Statistic Methods 3*; Praha, Informatorium: Prague, Czech Republic, 2007. (In Czech)

47. Feng, M.; Li, C.; McVay, S.E. Does ineffective internal control over financial reporting affect a firm´s operations? Evidence from firm´s inventory management. *Account. Rev.* **2015**, *90*, 529–557. [CrossRef]

48. Senova, A.; Pavlik, T.; Pavolova, H.; Domaracka, L.; Kuzevic, S. The evaluation process of production risks in the manufacturing plant. In Proceedings of the International Multidisciplinary Scientific GeoConference Surveying Geology and Mining Ecology Management (SGEM), Vienna, Austria, 27–29 November 2017; Volume 17, pp. 919–926.

49. Kulisz, M.; Kujawska, J. Prediction of Municipal Waste Generation in Poland Using Neural Network Modeling. *Sustainability* **2020**, *12*, 10088. [CrossRef]

50. Miezah, K.; Obiri-Danso, K.; Kádár, Z.; Fei-Baffoe, B.; Mensah, M.Y. Municipal solid waste characterization and quantification as a measure towards effective waste management in Ghana. *Waste Manag.* **2015**, 15–27. [CrossRef] [PubMed]

51. Lopes, A.C.P.; Robra, S.; Müller, W.; Meier, M.; Thumser, F.; Alessi, A.; Bockreis, A. Comparison of two mechanical pre-treatment systems for impurities reduction of source-separated biowaste. *Waste Manag.* **2019**, *100*, 66–74. [CrossRef] [PubMed]

52. Straka, M.; Khouri, S.; Rosová, A.; Čulková, K. Utilization of computer simulation for waste separation design as a logistics system. *Int. J. Simul. Model.* **2018**. [CrossRef]

53. Feranecová, A.; Manová, E.; Meheš, M.; Simonidesová, J.; Stašková, S.; Blaščák, P. Possibilities of harmonization of direct taxes in the EU. *Invest. Manag. Financ. Innov.* **2017**, *14*, 191–199.

54. Cucchiella, F.; D´Adamo, I.; Lenny Koh, S.C.; Rosa, P. A profitability assessment of European recycling processes treating printed circuit boards from waste electrical and electronic equipments. *Renew. Sustain. Energy Rev.* **2016**, *64*, 749–760. [CrossRef]

55. Bindzárová Gergeľová, M.; Kuzevičová, Ž.; Kovanič, Ľ.; Kuzevič, Š. Automation of Spatial Model Creation in GIS Environment. In *Inżynieria Mineralna*; Polskie Towarzystwo Przeróbki Kopalin: Krakow, Poland, 2014; Volume 33, pp. 15–22, ISSN 1640-4920.

56. Jurkasová, Z.; Cehlár, M.; Khouri, S. Tools for organizational changes managing in companies with high qualified employees. *Prod. Manag. Eng. Sci.* **2016**, 409–412.

57. Warguła, Ł.; Kukla, M. Determination of maximum torque during carpentry waste comminution. *Wood Res.* **2020**, *65*, 771–784. [CrossRef]

 *energies*

Article

Evaluation of the Effective Material Use from the View of EU Environmental Policy Goals

Marcela Taušová , **Katarína Čulková** * , **Peter Tauš** , **Lucia Domaracká** and **Andrea Seňová**

Institute of Earth Resources, Faculty of Mining, Ecology, Process Control and Geotechnologies,
Technical University of Košice, Park Komenského 19, 042 00 Košice, Slovakia; marcela.tausova@tuke.sk (M.T.);
peter.taus@tuke.sk (P.T.); lucia.domaracka@tuke.sk (L.D.); andrea.senova@tuke.sk (A.S.)
* Correspondence: katarina.culkova@tuke.sk; Tel.: +421-556-023-116

Abstract: Humanity is dependent on natural resources. Use and productivity of these resources plays an important role in energy savings and circular economy. The goal of this contribution is to evaluate productivity of resources in the frame of EU countries. Single analysis deals with data from the publicly available portal database and collected data were processed in the statistical software JMP. The trend of development and analysis of variability and linear dependence helped to create cluster analysis and comparison of the EU countries. The results from the view of average value of the indicator registered the growth, and from the view of variability the statistically important differences were verified for EU member states. Some pairs of indicators recorded positive, while some pairs recorded negative linear dependence. Cluster analysis shows two groups of countries—the first one with positive results, having the lowest tax burden in the case of energy taxes and environment, and the second one with negative results, having the highest tax burden of environmental and energy policy. The results are useful for a proper setting of energy and environmental goals that can increase the effectiveness of resource productivity in the countries studied.

Keywords: productivity of sources; environmental goals; energy efficiency; circular economy

Citation: Taušová, M.; Čulková, K.; Tauš, P.; Domaracká, L.; Seňová, A. Evaluation of the Effective Material Use from the View of EU Environmental Policy Goals. *Energies* **2021**, *14*, 4759. https://doi.org/10.3390/en14164759

Academic Editors: Wadim Strielkowski and Dalia Štreimikienė

Received: 17 June 2021
Accepted: 2 August 2021
Published: 5 August 2021

1. Introduction

Effective use of raw materials means the sustainable use of the limited resources of the Earth. Humanity is dependent on natural resources—metals, minerals, fuels, water, soil, wood, fertile soil, clean air and biodiversity. All mentioned resources present significant input that makes possible the operation of our economy. Raw materials are the prerequisite for the production of enterprises. Therefore, the science of materials management relates to economic benefit of an enterprise from the view of reducing costs and promoting the efficiency of materials management, which further reflects the important role of the materials classification in materials procurement [1]. Proper choice of the materials also plays a significant role in energy savings and high-quality suppliers [2,3].

Increasing the effectiveness of materials use is a key approach to providing economic growth and new workplaces in Europe [4]. It brings many economic possibilities, including decreasing costs and increasing competition [5]. Therefore, we must find new tactics toward this goal in all steps of the value chain: to improve management of resource stocks; to decrease inputs; to optimize production processes, management and business methods; to improv logistics; to change the calculation of consumption and decrease waste; and to develop new products and services.

Effective use of raw materials can help to stimulate technological innovation, increase employment in the rapidly developing sector of ecological technologies, open new export markets and bring benefits for consumers through sustainable products [6]. Effective use of raw materials also means the sustainable use of limited resources while minimizing impacts to the living environment [7,8]. This enables the creation of more products with higher value, while using less materials. This approach is a result of the main initiative of Europe to use a resource effectively as a part of Strategy Europe 2020, a strategy of EU growth for an intelligent, inclusive, and sustainable economy [9,10]. It supports the transition to sustainable growth through a low-carbon economy, effectively using the available resources. The plan for Europe to use resources effectively is one of the basic elements of the initiative of effective resources use. A road map was created that determined the framework for performing future activities. The plan also outlined structural and technological changes to be made by 2050, including milestones that should be achieved by 2020.

Fast and reliable evaluation of the materials' efficiency is necessary [11]. Moreover, the announcement, "towards circular economy" (also referred to as circularity, which means an economic system that tackles global challenges like climate changes, biodiversity loss, waste and pollution) supports further basic transition in EU, compared to a linear economy, where the resources are not only extracted, but not reused and returned to production. In a circular economy, the resources must be returned to the circle to be used for the longest possible time. Measurements leading to the effective use of resources and the minimization of waste are determined in this study. The circular economy is presently studied by a number of authors, throughout the world and in Europe [12,13].

The goal of this contribution is to evaluate a level of the indicator, expressing source productivity in the frame of individual EU countries during 10 analyzed years, and to define a trend of the development and discover connections with other indicators, evaluated at the EU level. The structure of the contribution consists of an investigation of the present state of problem solving followed by main research, consisting of three steps: basic analysis of resource productivity in EU, analysis of linear dependence of resource productivity indicators, and cluster analysis according to the similar behavior clusters that have been created.

2. Present State of Problem Solving

In the last decades the EU established a broad scale of legal decrees in the area of the living environment. This required focus on the Triple Bottom Line approach, measuring the impacts of business on 3P (people, planet, profit) criteria [14–16].

Due to these decrees, the aforementioned pollution of air, water and soil considerably decreased. The legal decrees were modernized, and concerned mainly chemical elements, limiting the use of various toxic or dangerous substances. Presently, inhabitants of EU have the best quality of water worldwide and more than 18% of the EU area is registered as protected areas of nature. However, there is still a number of problems needing structural solutions.

The seventh Environmental Action Program (EAP) includes European policy in the area of the living environment to the year 2020, with a goal to give a long-term vision of EU to 2050 [17]. The goal is to live in 2050 properly, in the frame of the ecological limits of the planet. The prosperity of humanity and a healthy environment will be the result of an economy focused on the sustainable use of natural resources, and, protecting, evaluating, and renewing biodiversity. This growth, paired with a decrease of CO_2 emissions, and responsible use of resources, will create a secure and sustainable global society.

Three key goals are determined in the frame of the program:

- To protect, preserve and increase the natural capital of the EU;
- To change the EU to an ecological and competitive low-carbon economy, effectively using available resources;
- To protect EU inhabitants from the problems connected with the living environment and risks of health and comfort.

At the same time "four activators" can help the EU to meet these goals:

- Better implementation and holding of legislation;
- More frequent and wise investments into policy regarding the living environment and climate protection;
- Full integration of environmental demand and considerations to other policies [18].

From the view of an aforementioned number of authors, evaluating materials use and their impact on the living environment from different considerations is of utmost importance. Joensuu et al. (2020) determined sectors that are most responsible for materials extractions and use [19]. This must be evaluated due to the impact it has on the protection of the living environment, as well as in reusing and recycling of materials. According to Chen et al. (2020), materials use requires safe and effective evaluation with the goal of helping to develop innovative products [20].

Rajca et al. (2020) studied materials use for refuse-derived fuel in order to use it for energy purposes [21,22]. Materials can also be used in waste. In this area, Girondi et al. (2020) studied use of biomass in ceramic materials, observing that it was possible to save costs for the ceramic industry [23]. Tian et al. (2020) studied materials use from the view of temperature [24]. According to their results, economic evaluations are significant criteria in the marketing area of any industrial product produced from raw materials.

During the economic evaluation of materials use it is necessary to also consider technical criteria and material properties [25]. Factor analysis quality and assessments of the innovations using can be used during the economic evaluation of materials from the view of innovative materials [26].

Presently, there is a drive to advance technology, such as the use of better and cheaper material [27]; increasing material efficiencies and pairing that with maintaining a clean environment for people, as well as the life cycle cost of the alternative energy proposals are compared.

Wang et al. (2019) recommend using eco-friendly materials in different industries; however, data for each eco-friendly material are managed individually, causing inefficiency, increased costs, and potentially greater environmental impacts associated with material and resource choice [28].

Long et al. (2018) studied the workability, static and dynamic mechanical properties, and environmental impact of materials from the view of thermoplastic behavior, finding through an economic and ecological evaluation a benefit to the environment [29]. Guo et al. (2018) summarized and discussed materials use based on the analysis of economic indicators, including initial cost, operating cost, revenue, subsidy, and energy cost [30]. Haider and Bhat (2020) studied the linkage between material and energy efficiency and total factor productivity, finding that not all states are equally energy intensive [31]. The increase of total factor productivity is associated with a lower level of energy per unit of output.

Material and energy consumption do not have a direct causal relationship to gross domestic product [32], which is necessary for follow-up of sustainable development. A study by Belke [33] found a long-running relationship between material and energy consumption and GDP. Productivity in the area of material recycling must also be researched [34], bringing contributions to the improvement of municipal solid waste management. Moreover, material use and recycling must be regarded in the construction industry especially [35,36], affecting entrainment factors such as energy consumption, carbon footprints, and overall construction operation productivity. The productivity of material use in the rest of the production steps can be increased. For example, Dini et al. (2018) studied how wood can be used throughout its entire production so that its production can be more economical [37].

During this research, we obtained results from previous research aiming to analyze the effectiveness of resource use at the level of individual EU states in context of the goals determined in the frame of environmental policy in 2020. Consequently, the goal of this contribution is also to define the intensity of the economic transformation necessary to influence the circular economy toward meeting the EU environmental policy goals.

3. Materials and Methods

In the frame of single analysis, we obtained results from the continuously published values of chosen indicators from the portal https://ec.europa.eu/eurostat/data/database (20 June 2020) during all available years and for all available member states. Collected data had been registered, selected, and adopted in the database, and created in MS Excel editor according to the demands of the statistical software JMP, where adapted data was transmitted and then analyzed [38].

The collected data presents the results of the eight chosen indicators during 1995–2018 (Table 1). The constructed database consists of 3765 data points and any indicator is defined for concrete EU member state and concrete year. As we can see from Table 1, the extent of published data for each indicator is considerably different, while the volume of data connects with an incompleteness of countries assigned, or the publishing of data for some indicators for any second year, as in the case of measure of recycling, waste production on GDP, and measure of waste stocking. Extend and structure of obtained data had been adapted by analysis choice and results formulation.

Table 1. Structure of collected data.

Indicator	Goal	Number of Measurements	Analyzed Period
Productivity of sources	increase	703	2000–2018
Measure of material use in cycle	increase	507	2010–2016
Index eco-innovation	increase	252	2010–2018
Measure of recycling	increase	120	2010, 2012 2014, 2016
Waste production on GDP	decrease	287	2004, 2006, 2008, 2010, 2012, 2014, 2016
Measure of waste stocking	decrease	120	2010, 2012, 2014, 2016
Environmental taxes	–	888	1995–2018
Energy taxes on GDP	–	888	1995–2018

(Source: own processing in MS Excel editor).

The process of the analysis is as follows:

1. Analysis of the main indicator—productivity of sources:
   - Trend of development;
   - Graphical analysis—cartographer;
   - Analysis of the variability—nonparametric Kruskal-Wallis test.
2. Mutual analysis of a group of indicators:
   - Analysis of linear dependence—pair correlation analysis of indicators;
   - Cluster analysis;
   - Comparison of the countries.

Resource and material productivity represent the main indicator of the evaluation table in effectiveness of resources. It is used for the monitoring of the steps taken toward effective use of resources in the EU. The indicator is defined as gross domestic product (GDP) divided by domestic material consumption (DMC). DMC measures total volume of materials directly used in the economy, defining the annual volume of raw materials as that extracted from a domestic territory into the local economy, plus all physical import, minus all physical export. It is necessary to underline that "consumption" in DMC means net, not last consumption. DMC does not include inflows connected with import and export of raw materials and products with origin outside the local economy.

Since productivity of sources is calculated as GDP divided by DMC, measuring units are GDP units over DMC units.

$$\text{Productivity of sources} = \frac{\text{GDP}}{\text{DMC}}[\text{PPS·kg}^{-1}] \tag{1}$$

$$\text{DMC} = \text{Mining} + \text{Import} - \text{Export [kg]} \tag{2}$$

$$\text{Productivity of sources} = \frac{\text{GDP}}{\text{Mining} + \text{Import} - \text{Export}}[\text{PPS·kg}^{-1}] \tag{3}$$

PPS, standards of purchase power, present fictitious "currency" units that remove differences in purchase power, and by this way eliminate differences in price levels in individual countries used during comparison between countries.

4. Results

4.1. Basic Analysis of Sources Productivity in EU

The indicator of resource productivity is defined as the GDP and DMC rate in a concrete year and country. Due to the achievement of environmental policy goals in the EU, the effort of any country is to increase the value of the indicator.

Analysis of the published results in the database Eurostat of the indicator during 2000–2018 in EU member states had been evaluated from the view of success of each country orientation in an effort to use the resources effectively. With the use of a cartographer it is possible to make visual comparisons of the countries. In 2006, Switzerland and the Netherlands belonged to the countries with the highest values of the indicator. Ten years later there is considerable change in the color scale of the countries, but still the best results are recorded in Switzerland and Netherlands. However, some additional countries are entering to this group, such as Italy, England, Spain and France (Figure 1).

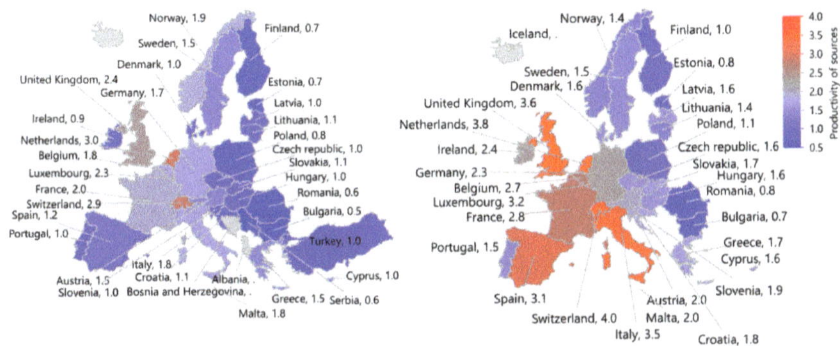

Figure 1. Cartographer—resources productivity in 2006 and 2016.

At the same time, the average value of the indicator moved from 1.17 (in 2000) to 2.35 (in 2018), which presents 200% growth during 18 years (see Figure 2). EU countries are then on average able to produce double the volume of production compared to in 2000, with the same inputs.

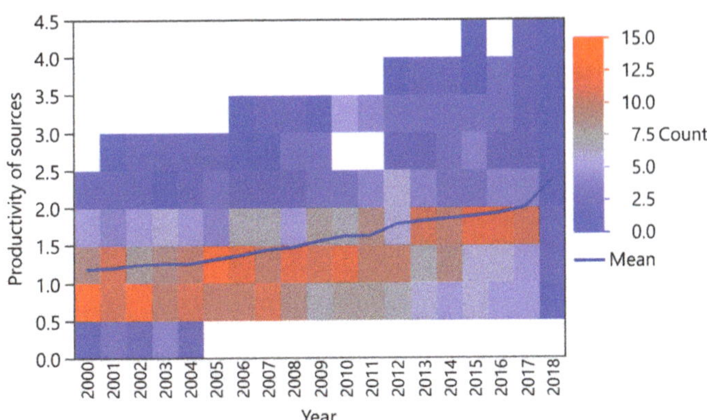

Figure 2. Average resources policy in EU in 2000–2018.

According to the analysis of variability through the Kruskal-Wallis test, the statistically important differences were verified for EU member states according to achieved results of the resources productivity (see Figure 3).

1-Way Test, ChiSquare Approximation		
ChiSquare	**DF**	**Prob>ChiSq**
486.3820	35	<0.0001

Figure 3. Results of Kruskal-Wallis test.

It is also illustrated by Figure 4, that the results of individual EU states are considerably different. The EU average in 2017 is at the level 2.2; significantly over-average values were registered in the Netherlands, Switzerland, United Kingdom, Italy, Luxembourg. Luxembourg at the same time belongs among countries where during the last years there is a registered undesirable trend of the indicator decrease. Moreover, Hungary, Albania, Norway and Cyprus belong here (red ellipse). We see here countries, where results of the indicator are under EU average, but the trend is significantly growing. Here belongs, for example, Slovakia, Greece, Czech Republic, Croatia and North Macedonia.

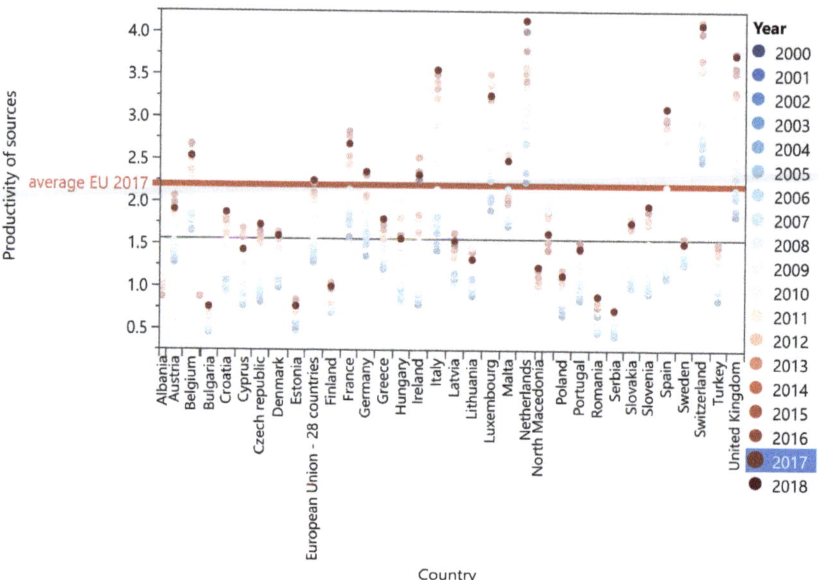

Figure 4. Graphical comparison of resources productivity in EU in 2000–2018.

Overall, we can conclude that productivity of resources has, in most countries, an increasing trend. However, the level of the indicator proportionally different in comparing the countries. We tried to find out what the reason is for such differences by analysis of other indicators and the relations between them, which could significantly influence productivity of resources in these countries. It means choosing an indicator that defines the transformation of the economy toward a circular economy, such as:

- Production of the waste on GDP [kg·1000 €$^{-1}$];
- Measure of material use in circulation [%];
- Environmental taxes [%];
- Index eco-innovation [%];
- Measure of waste stocking [%];

- Measure of recycling [%];
- Taxes from energy per GDP.

The following descriptive statistics give basic statistical characteristics of the indicators (average, standard deviation, summary, minimum, maximum), included in the following analysis (Table 2).

Table 2. Descriptive statistics (Univariate simple statistics) of the analyzed indicators.

Column	Mean	Std Dev	Sum	Minimum	Maximum
Waste production without main mineral waste on GDP (kg/1000€)	107.317	134.378	12,878.0	20.0000	779.000
Measure of material using in circulation (%)	8.5232	6.4263	954.600	0.7000	29.0000
Taxes of energy/GDP	0.0193	0.0052	2.3186	0.0095	0.0329
Environmental taxes % (rate of incomes from total taxes)	7.3443	1.7936	910.690	2.4100	11.6600
Index eco-innovation (%)	89.2545	30.2192	9818.00	29.0000	149.000
Measure of waste stocking %	32.9712	23.4796	3429.00	1.0000	85.0000
Measure of recycling %	49.5922	16.5259	5108.00	10.0000	87.0000
Productivity of sources	1.8827	0.8331	225.927	0.6607	3.9902

(Source: own processing in software JMP).

4.2. Analysis of Linear Dependence of Indicators

With the use of correlation analysis, we searched the existence of the relationships between indicators and looked for possible reasons for the achieved values in the EU and at the level of individual states. The aim is to limit indicators that could participate at the Europe orientation to achieve environmental policy goals in area of increasing effective resource use. Analysis of linear dependence of indicators was performed in the software JMP by using pair correlation analysis. We analyzed all indicators mutually, and obtained statistically important correlations. Results of the analysis went through further searching, while finding important correlations with the coefficient of correlations $r > 0.5$ (positive linear dependence = red area) and $r < -0.5$ (negative linear dependence = blue area), which is illustrated by Table 3.

Results of the analysis showed in some pairs strong positive linear dependence between indicators with the coefficient of correlation at the level max 0.64. It means the following pairs of indicators:

- Productivity of sources and measure of recycling with the coefficient correlation of $r = 0.56$, and productivity of sources and measure of material use in a cycle with the coefficient correlation of $r = 0.64$. In the case that we would see productivity of sources as a variable indicator, it means that the productivity of sources is growing when in the country the measurement of recycling is growing. The same applies in the case of material use in cycle. States that have high measure of recycling and use a higher volume of material in a cycle, record at the same time a higher value of productivity of sources.
- Measure of recycling and measure of material use in a cycle with the coefficient correlation of $r = 0.56$. Countries with higher measure of recycling use have at the same time more recycled material in production processes.
- Waste productivity and measure of stocking with the coefficient correlation of $r = 0.54$. A positive rate is confirmed at the same time in the case of measure of recycling and waste production. States that record high volume of waste production on GDP also have a high level of stocking.

Table 3. Results of pair correlation of indicators.

Variable	by Variable	Correlation	Signif Prob
Productivity of sources	Measure of material use in cycle (%)	0.6408	<0.0001 *
Productivity of sources	Measure of recycling %	0.5643	<0.0001 *
Measure of recycling %	Measure of material use in cycle (%)	0.5560	<0.0001 *
Measure of waste stocking %	Waste production without main mineral waste on GDP (kg/1000€)	0.5437	<0.0001 *
Measure of recycling %	Index eco-innovation (%)	0.4885	<0.0001 *
Energy taxes/GDP	Measure of material use in cycle (%)	0.4862	<0.0001 *
Productivity of sources	Energy taxes/GDP	0.4649	<0.0001 *
Productivity of sources	Index eco-innovation (%)	0.4518	<0.0001 *
Index eco-innovation (%)	Energy taxes/GDP	0.4168	<0.0001 *
Measure of waste stocking %	Environmental taxes % (rate of incomes from total taxes)	0.3991	<0.0001 *
Index eco-innovation (%)	Measure of material use in cycle (%)	0.3967	<0.0001 *
Environmental taxes % (rate of incomes from total taxes)	Waste production without main mineral waste on GDP (kg/1000€)	0.3256	0.0003 *
Measure of recycling %	Energy taxes/GDP	0.2782	0.0044 *
Measure of material use in cycle (%)	Waste production without main mineral waste on GDP (kg/1000€)	−0.0186	0.8456
Energy taxes/GDP	Waste production without main mineral waste on GDP (kg/1000€)	−0.2042	0.0279 *
Environmental taxes % (rate of incomes from total taxes)	Measure of material use in cycle (%)	−0.2761	0.0032 *
Measure of recycling %	Environmental taxes % (rate of incomes from total taxes)	−0.2938	0.0026 *
Productivity of sources	Environmental taxes % (rate of incomes from total taxes)	−0.3086	0.0006 *
Measure of waste stocking %	Energy taxes/GDP	−0.3150	0.0011 *
Environmental taxes % (rate of incomes from total taxes)	Energy taxes/GDP	−0.3237	0.0003 *
Productivity of sources	Measure of waste stocking %	−0.4253	<0.0001 *
Index eco-innovation (%)	Waste production without main mineral waste on GDP (kg/1000€)	−0.4593	<0.0001 *
Productivity of sources	Waste production without main mineral waste on GDP (kg/1000€)	−0.4672	<0.0001 *
Index eco-innovation (%)	Environmental taxes % (rate of incomes from total taxes)	−0.4932	<0.0001 *
Measure of waste stocking %	Measure of material use in cycle (%)	−0.5160	<0.0001 *
Measure of recycling %	Waste production without main mineral waste on GDP (kg/1000€)	−0.5184	<0.0001 *
Measure of waste stocking %	Index eco-innovation (%)	−0.7499	<0.0001 *
Measure of recycling %	Measure of waste stocking %	−0.8621	<0.0001 *

(Own processing according to the ES data from software JMP. Signif Prob: Significance probabilities correlation; * draws attention to a statistically significant correlation).

Results of the analysis also pointed in some pairs to the strong negative linear dependence between indicators with the coefficient correlation at the level max−0.86. It means the following pairs of indicators:

- Measure of recycling and measure of stocking with the coefficient correlation of r = −0.86. Countries that achieve a high measure of stocking, record at the same time low measure of recycling and logically vice versa.
- Measure of stocking and index of eco-innovation with the coefficient correlation of r = −0.75. In this case, measure of stocking decreases with the growing index of eco-innovation in the state.
- Measure of recycling and waste production on GDP with the coefficient correlation of r = 0.52. States that produce more waste, record at the same time lower measure of recycling, and on the other hand countries that have a high measure of waste recycling are trying at the same time to produce less waste on GDP.
- Measure of waste stocking and measure of material use in a cycle with the coefficient of correlation r = −0.52. States that meet a high measure of stocking use at the same time have less material in a cycle, or vice versa, countries that use more recycled materials, have less stockings.

4.3. Cluster Analysis

Through cluster analysis we searched common characteristics of each state, while according to similar behavior clusters had been created. The principle of the cluster analysis is to group indicators, in our case by state, by the way that inside the cluster there was achieved maximal homogeneity of indicators and between clusters maximal variabilities were recorded.

Analysis was performed by method of hierarchical clustering. The method begins with every observation by its own cluster. In any step of the clustering process the distance between all cluster pairs are calculated and the two closest clusters are determined. The process continues until all points are included in the one cluster. Hierarchical clustering is known also as agglomerate clustering due to the use of a combined approach.

Results of the analysis in our case mean the creation of four clusters that are presented in Figure 5. In our analysis we evaluated each cluster individually.

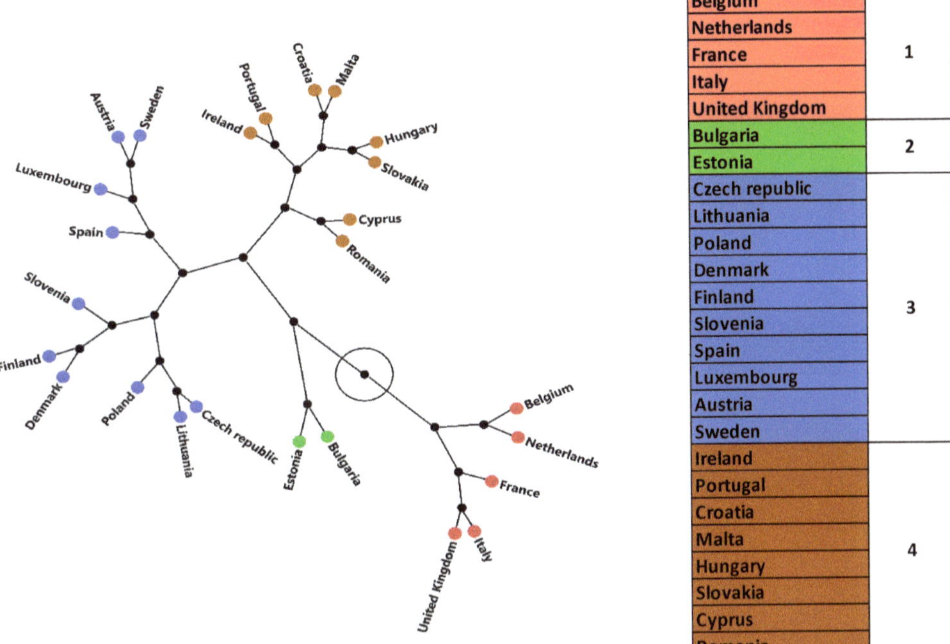

Belgium	
Netherlands	
France	1
Italy	
United Kingdom	
Bulgaria	2
Estonia	
Czech republic	
Lithuania	
Poland	
Denmark	
Finland	3
Slovenia	
Spain	
Luxembourg	
Austria	
Sweden	
Ireland	
Portugal	
Croatia	
Malta	4
Hungary	
Slovakia	
Cyprus	
Romania	

Figure 5. Results of cluster analysis in EU states according to the effectiveness of resource use in 2016. (Own processing according to ES data in software JMP, Excel).

Individual clusters were analyzed through the analysis of average cluster values of indicators. In the frame of any indicator, average achieved values were ranked from the best to the worst value according to the target values in the EU.

Indicators with the goals to growth are (Table 4):

- Productivity of sources;
- Measure of material use in a cycle;
- Index eco-innovation;
- Measure of recycling.

Indicators with the goal to decrease are:

- Waste productivity without main mineral waste on GDP;
- Measure of waste stocking.

Finally, ranked values were scored, while the first place means 1 point, second place 0.5, and third place 0.3. According to the summary of obtained points, we constructed a chart of clusters according to the ability to use resources effectively. The best results were achieved in the first cluster, which achieved the best average values in the EU, in the case of five indicators—5.5 points. The cluster is presented by Belgium, the Netherlands, France, Italy, and England. The countries have at the same time one of the lowest tax burdens in case of energy taxes and the environment. The worst results were recorded in countries from the second cluster with 0.5 points—Bulgaria and Estonia. These countries produce almost eight times more waste and they produce waste stocks five times higher, as compared to the first cluster. At the same time, there are countries that in comparing with the first cluster achieve only one-third level of indicators, a shown in the measure of recycling, index of eco-innovation, and measure of material use in a cycle. The productivity of resources is at the level 0.79 and these countries have the highest tax burden in the frame of environmental and energy policy.

Table 4. Average values of indicators in each cluster-year.

Indicator	Cluster			
	1	**2**	**3**	**4**
Productivity of Sources	3.27	0.79	1.84	1.66
Measure of Material Use in Cycle (%)	20.34	8.05	7.67	3.56
Index Eco-Innovation (%)	100.60	47.00	105.80	74.88
Measure of Recycling (%)	66.00	18.50	58.70	42.00
Measure of Waste Stocking (%)	14.20	76.50	16.90	47.38
Waste Productivity without Main Mineral Waste on GDP (kg/1000€)	66.80	533.00	74.50	77.38
Energy Taxes/GDP	0.019	0.026	0.020	0.019
Environmental Taxes (Rate of Incomes from Total Taxes) (%)	7.03	9.44	6.73	8.19
Score	5.50	0.50	3.30	1.50

(Own processing according to ES data in software Excel). (Note: red color—indicators with goal to growth, blue color—indicators with a goal to decrease. Note: Sample of cluster 1 scoring: productivity of sources—1 place/1 points + measure of material use in cycle (%)—1 place/1 points + index eco-innovation (%)—2 place/0.5 points + measure of recycling (%)—1 place/1 points + measure of waste stocking (%)—1 place/1 points + waste productivity without main mineral waste on GDP (kg/1000€)—1 place/1 points = 1 + 1 + 0.5 + 1 + 1 + 1 = 5.5).

5. Discussion

As for the comparison of the results with other studies, Haider and Bhat (2020) studied a connection between material and energetic effectiveness and total productivity of factors, finding that not all states are equal from the view of energy demands [31]. Increasing of total productivity of factors is connected with lower level of energy per output unit.

In Europe annually 18 million tons of waste could be collected in the case of installing a strategy of collection according to proven processes, which could lead to a 13% decrease of greenhouse gas production in connection with packages and package wastes. In spite of the high performance of the collection being an effective use of resources, a single improvement of the system of waste collection, separate from the resources, is not enough for recycling goals achievement or decreasing emissions production. Material loss has to be decreasing in the whole value chain, which means from selection and collection to recycling [38,39]. Kuhl et al. (2018) considered business models for circular economy (CE) that have potential environmental contributions and productivity of resources [40]. Circular business models, based on refurbishment and repeated use of materials could bring significant cost savings, as well as radically decreasing negative influences on the living environment [41]. The processing of resources has to be radically changed from the model of linear use to the sustainable, circular model. In this area, Velenturf et al. (2020) created a model that considered the complex character of resource flows [42]. Environmental sustainability has to be connected with conceptions of an ecological economy, circular economy and bio-economy [43,44].

However, the connection with sustainability is not, according to Loiseau et al. (2016), always clear [45], since there are various levels of substitution and compromises are allowed between environmental and economic advantages. Additionally, structuralized changes of the human way of life are demanded [46]. It is necessary to combine the circular economy with an approach regarding human development (HD), since one of the discussions is connecting the missing social or human dimension of the circular economy [47]. Schroder et al. (2020) included social and economic elements in the transformation from linear to circular economic models in combination with HD from studies of social science and development [48]. This could be connected with application of analysis results from area of tax burden influence for inhabitants and businessmen. Setting energy and environmental taxes at manageable levels can also increase effectiveness of productivity of resources in the countries. This approach can also be linked to the possibilities of effective use of spatial data models from GIS environment [49], that are commonly used in such research.

Due to the pandemic situation the European Union is buffeted by a range of crises since 2007, not least the economic and financial crisis. One potential victim of the economic crisis is environmental policy ambition, since during economic crises environmental policy slips down the agenda with long term consequences for environmental quality. The future environmental policy trajectory and leadership of the EU may be under threat as the Union struggles to emerge from the economic crisis. Productivity of resources could be helpful for the overcoming of such struggles.

To address the challenges of reducing impacts on the environment and of resource scarcity, it must go hand in hand with sustainable development and state policy, when material efficiency is a key element of new thinking [50,51]. Directly related to material efficiency is the concept of the circular economy, which is based on the principle of optimizing the utility embodied in materials and products throughout their life cycle.

6. Conclusions

In the EU, basic strategy results from the need to provide intelligent, sustainable, and inclusive growth, along with respect to social and economic impacts and issues surrounding the living environment and its protection, all often-discussed topics. Experts on this global theme most often refer to sustainability, considering all elements of the environment. Therefore, the analysis realized in the frame of this research states that presently, the common effort is to increase the effectiveness of Earth's resource use and protect the living environment, going hand in hand with the use of renewable energy sources [52]. It demands a follow-up of determined goals not only at the level of absolute figures for European Union and individual member states, but also at the level of each sector. The undeniable fact is that there are dependences between all sectors and processes that need higher evaluation.

Additionally, the findings of this research point to the necessity to be oriented not only to the resulting goal, but also to the secondary factors, influencing the goal achievement. It means that obtained dependencies can help countries with significant differences to orient their means to the areas that are seemingly not connected with achievement of the goal, but which finally could help speed up the goal achievement. An example could be the increasing of recycling measures that contribute to the goal achievement—a *decrease of waste stocking* can also considerably support a decrease of *domestic consumption of material on the inhabitant* or to increase *energetic productivity of the state*. An important finding of this research is that environmental tax policy has a positive influence on the meeting of environmental goals of the individual countries. Countries with high taxes on energy and environmental taxes are at the same time countries with the lowest ability to meet determined goals in the frame of Strategy Europe 2020.

Results of the analyses point to the fact that countries, oriented only to the decreasing of emission production connected with fossil fuel energy and transport, cannot meet the demanded progress if they will not intensively support innovation processes and research and development activities oriented to ecological processes. Due to the limitation of the research to the chosen indicators, the subject of further research would be single analysis of key indicators of productivity of resources, for example productivity of water, soil and energy in relation to the main indicator—productivity of resources. Each indicator should be searched in context of chosen indicators, defining standard of living, economic and financial self-sufficiency, level of education, etc.

Industry-level productivity analysis can be a useful diagnostic tool to better understand why some sectors and countries show faster overall productivity growth, and to direct research attention to parts of the economy that warrant more detailed scrutiny. A better understanding of productivity growth (or lack thereof) in industries should still be an important goal of researchers aiming to understand cross-country growth differences.

Author Contributions: Conceptualization, M.T. and K.Č.; methodology, M.T. and K.Č.; software, M.T.; validation, P.T. and L.D.; formal analysis, M.T.; resources, K.Č.; data curation, P.T.; writing—original draft preparation, M.T. and K.Č.; writing—review and editing, A.S.; visualization, P.T.; supervision, A.S.; project administration, A.S.; funding acquisition, A.S. All authors have read and agreed to the published version of the manuscript.

Funding: This research was funded of grant project VEGA 1/0797/20.

Institutional Review Board Statement: Not applicable.

Informed Consent Statement: Not applicable.

Data Availability Statement: Not applicable.

Acknowledgments: Contribution presents partial result of grant project VEGA 1/0797/20 and KEGA 006 TUKE-4/2019.

Conflicts of Interest: The authors declare no conflict of interest.

References

1. Guo, H.; Sun, Y. Research on materials classification model building and procurement strategy of coal enterprises. *J. Mines Met. Fuels* **2017**, *65*, 691–695.
2. He, M.; Wen, Y.; Jiang, W. Evaluation model of building materials suppliers considering carbon emissions. *IOP Conf. Ser. Earth Environ. Sci.* **2018**, *189*, 032067. [CrossRef]
3. Ling, W.; Cui, C.; Dong, J. Application and economic analysis of energy-saving wall materials in low-income rural areas of China. *IOP Conf. Ser. Mater. Sci. Eng.* **2018**, *423*, 012185. [CrossRef]
4. Soukupová, J.; Klimovský, D.; Ochrana, F. Key factors for public utility efficiency and effectiveness: Waste management services in the Czech Republic. *Ekon. Časopis* **2017**, *65*, 143.
5. Herciu, M.; Ogrean, C. An overview on European Union sustainable competitiveness. *Procedia Econ. Financ.* **2014**, *16*, 651–656. [CrossRef]
6. Magazzino, C.; Porrini, D.; Fusco, G.; Schneider, N. Investigating the link among ICT, electricity consumption and air pollution and economic growth in EU countires. *Energy Sources* **2021**, 1–23. [CrossRef]
7. Fry, J.; Lenzen, M.; Jin, Y.; Wakiyama, T.; Baynes, T.; Wiedmann, T.; Malik, A.; Chen, G.; Wang, Y.; Geschke, A.; et al. Assessing carbon footprints of cities under limited information. *J. Clean. Prod.* **2018**, *176*, 1254–1270. [CrossRef]
8. Hughes, G. Environmental indicators. *Ann. Tour. Res.* **2002**, *29*, 457–477. [CrossRef]
9. Sustainable Economic Development. 2019. Available online: https://www.vlada.gov.sk//sustainable-economicdevelopment/ (accessed on 15 September 2019).
10. Miller, C.; Sarewitz, D.; Light, D. Science, Technology and Sustainability. Building a Research Agenda. National Science Foundation Supported Workshop. Available online: https://www.nsf.gov/sbe/ses/sts/Science_Technology_and_Sustainability_Workshop_Rpt.pdf. (accessed on 5 June 2019).
11. Bumanis, G.; Vitola, L.; Stipniece, L.; Locs, J.; Korjakins, A.; Bajare, D. Evaluation of industrial by-products as pozzolans: A road map for use in concrete production. *Case Stud. Constr. Mater.* **2020**, *13*, e00424. [CrossRef]
12. Korhonen, J.; Honkasalo, A.; Seppala, J. Circular economy: The concept and its limitations. *Ecol. Econ.* **2018**, *143*, 37–46. [CrossRef]
13. Robaina, M.; Villar, J.; Pereira, E.T. The determinants for a circular economy in Europe. *Environ. Sci. Pollut. Res.* **2020**, *27*, 12566–12578. [CrossRef]
14. Elkington, J. Enter the triple bottom line. In *The Triple Bottom Line, Does It All Add Up? Assessing the Sustainability of Business and CSR*, 5th ed.; Henriques, A., Richardson, J., Eds.; Earthscan Publications Ltd.: London, UK, 2004; pp. 1–16.
15. Living Well, within the Limits of Our Planet. Available online: https://ec.europa.eu/environment/pubs/pdf/factsheets/7eap/en.pdf (accessed on 10 January 2020).
16. Dainiené, R.; Dagilienė, L. A TBL approach based theoretical framework for measuring social innovations. *Procedia Soc. Behav. Sci.* **2015**, *213*, 275–280. [CrossRef]
17. Kollmuss, A.; Agyeman, J. Mind the gap: Why do people act environmentally and what are the barriers to pro-environmental behavior? *Environ. Educ. Resour.* **2002**, *8*, 239–260. [CrossRef]

18. Saracoglu, B.O. Selecting industrial investment locations in master plans of countries. *Eur. J. Ind. Eng.* **2013**, *7*, 416–441. [CrossRef]
19. Joensuu, T.; Edelman, H.; Saari, A. Circular economy practices in the built environment. *J. Clean. Prod.* **2020**, *276*, 124215. [CrossRef]
20. Chen, K.; Ren, J.; Chen, C.; Xu, W.; Zhang, S. Safety and effectiveness evaluation of flexible electronic materials for next generation wearable and implantable medical devices. *Nano Today* **2020**, *35*, 100939. [CrossRef]
21. Rajca, P.; Poskart, A.; Chrubasik, M.; Sajdak, M.; Zajemska, M.; Skibiński, A.; Korombel, A. Technological and economic aspect of refuse derived fuel pyrolysis. *Renew. Energy* **2020**, *161*, 482–494. [CrossRef]
22. Chen, J.; Taylor, J.E.; Wei, H.H. Toward a building occupant network agent-based model to simulate peer induced energy conservation behavior. In Proceedings of the 2011 Winter Simulation Conference (WSC), Phoenix, AZ, USA, 11–14 December 2011; IEEE: Piscataway Township, NJ, USA, 2011; pp. 883–890.
23. Girondi, D.; Marvila, M.T.; Souza, D.; Sanchez, R.J.; Colorado, H.A.; Fontes, V.C.M. Evaluation of the application of macrophyte biomass Salvinia auriculata Aublet in red ceramics. *J. Environ. Manag.* **2020**, *275*, 111253. [CrossRef]
24. Tian, M.W.; Mihardji, L.W.W.; Moria, H.; Assadi, S.; Sadighi, D.; Khalilarya, S.; Nguyen, P.T. A comprehensive energy efficiency study of segmented annular thermoelectric generator; thermal, exergetic and economic analysis. *Appl. Therm. Eng.* **2020**, *181*, 115996. [CrossRef]
25. Stieberova, B.; Zilka, M.; Ticha, M.; Freiberg, F. Sustainability assessment of continuous-flow hydrothermal synthesis of nanomaterials in the context of other production technoologies. *J. Clean. Prod.* **2019**, *241*, 118325. [CrossRef]
26. Chepachenko, N.V.; Leontiev, A.A.; Uraev, G.A.; Ardzinov, V.D. Modeling the effect of using the innovative materials on the construction organizations economic performance. *IOP Conf. Ser. Mater. Sci. Eng.* **2019**, *698*, 077038. [CrossRef]
27. Mohamed, A.F. Economic evaluation of renewable energy systems, case study: An eco-house powered by nano-crystal PV in New Aswan, Egypt. *IOP Conf. Ser. Earth Environ. Sci.* **2019**, *397*, 012010. [CrossRef]
28. Wang, S.; Tae, S.; Kim, R. Development of a green building materials integrated platform based on materials and resources in G-SEED in South Korea. *Sustainability* **2019**, *11*, 6532. [CrossRef]
29. Long, W.J.; Li, H.D.; Wei, J.J.; Xing, F.; Han, N. Sustainable use of recycled crumb rubbers in eco-friendly alkali activated slag mortar: Dynamic mechanical properties. *J. Clean. Prod.* **2018**, *204*, 1004–1015. [CrossRef]
30. Guo, S.; Liu, Q.; Zhao, J. Mobilized thermal energy storage: Materials, containers and economic evaluation. *Energy Conserv. Manag.* **2018**, *177*, 315–329. [CrossRef]
31. Haider, S.; Bhat, J.A. Does total factor productivity affect the energy efficiency: Evidence from the Indian paper industry. *Int. J. Energy Sect. Manag.* **2020**, *14*, 108–125. [CrossRef]
32. Al-Iriani, M.A. Energy—GDP relationship revisited: An example from GCC countries using panel causality. *Energy Policy* **2006**, *34*, 3342–3350. [CrossRef]
33. Belke, A.; Dobnik, F.; Dreger, C. Energy consumption and economic growth: New insights into the cointegration relationship. *Energy Econ.* **2011**, *33*, 782–789. [CrossRef]
34. Silva, H. The ergonomics of sorting recyclable materials: A case study of a Brazilian cooperative. In *International Conference on Applied Human Factors and Ergonomics*; Springer: Cham, Germany, 2020; Volume 970, pp. 256–265.
35. Peshkov, V.V. Some aspects of construction industry development in 2018. *IOP Conf. Ser. Mater. Sci. Eng.* **2019**, *687*, 044010. [CrossRef]
36. Wang, C.C.; Sepasgozar, S.M.; Wang, M.; Sun, J.; Ning, X. Green performance evaluation system for energy efficiency based planning for construction site layout. *Energies* **2020**, *12*, 4620. [CrossRef]
37. Dini, W.; Irwan, B.; Mangara, T. Alternative selection in reducing wood scrap with green productivity approach. In *E3S Web of Conferences*; EDP Sciences: Ulis, France, 2018; Volume 73, p. 07023.
38. Ministry of Living Environment, Slovakia. *Strategy, Principles and Priorities of State Environmental Policy*; Ministry of Living Environment: Bratislava, Slovakia, 2004.
39. Agan, Y.; Kuzey, C.; Acar, M.F.; Acikgoz, A. The relationships between corporate social responsibility, environmental supplier development and firm performance. *J. Clean. Prod.* **2016**, *112*, 1872–1881. [CrossRef]
40. Kuhl, C.; Tjahjono, B.; Bourlakis, M.; Aktas, E. Implementation of circular economy principles in PSS operations. *Procedia CIRP* **2018**, *73*, 124–129. [CrossRef]
41. Linder, M.; Williander, M. Circular business model innovation: Inherent uncertainties. *Bus. Strategy Environ.* **2017**, *26*, 182–196. [CrossRef]
42. Velenturf, A.P.M.; Purnell, P.; Macaskie, L.E.; Mayes, W.M.; Sapsford, D.J. A new perspective on a global circular economy (book chapter). *RCS Green Chem.* **2020**, *62*, 3–22.
43. Barbier, E.B. The green economy post Rio+20. *Science* **2012**, *338*, 887–888. [CrossRef] [PubMed]
44. Dyllick, T.; Hockerts, K. Beyond the business case for corporate sustainability. *Bus. Strategy Environ.* **2002**, *11*, 130–141. [CrossRef]
45. Loiseau, E.; Saikku, L.; Antikainen, R.; Droste, N.; Hansjurgens, B.; Pitkänen, K.; Leskinen, P.; Kuikman, P.; Thomsen, M. Green economy and related concepts: An overview. *J. Clean. Prod.* **2016**, *139*, 361–371. [CrossRef]
46. Lieskovská, Z. A few words about the history of integration in the European area and the development of environmental policy (in Slovak). *Enviromagazín* **2004**, *9*, 14–15.

47. Jurkasová, Z.; Cehlár, M.; Khouri, S. Tools for organizational changes managing in companies with high qualified employees. In *Production Management and Engineering Sciences: International Conference on Engineering Science and Production Management, ESPM 2015, Tatranska Strba, High Tatras Mountains, 16–17 April 2015*; CRC Press: Boca Raton, FL, USA, 2016; Volume 154039, pp. 409–412.
48. Schroder, P.; Lemille, A.; Desmond, P. Making the circular economy work for human development. *Resour. Conserv. Recycl.* **2020**, *156*, 104686. [CrossRef]
49. Bindzárová Gergeľová, M.; Kuzevičová, Ž.; Kovanič, Ľ.; Kuzevič, Š. Automation of Spatial Model Creation in GIS Environment. *Inżynieria Miner.* **2014**, *33*, 15–22.
50. Cehlár, M.; Domaracká, L.; Šimko, I.; Puzder, M. Mineral resource extraction and its political risks. In *Production Management and Engineering Sciences: International Conference on Engineering Science and Production Management, ESPM 2015, Tatranska Strba, High Tatras Mountains, 16–17 April 2015*; CRC Press: Boca Raton, FL, USA, 2016; Volume 154039, pp. 39–43.
51. Dogan, E.; Inglesi-Lotz, R. The impact of economic structure to the environmental Kuznets curve (EKC) hypothesis: Evidence form European countries. *Environ. Sci. Pollut. Res.* **2020**, *27*, 12717–12724. [CrossRef] [PubMed]
52. Horodníková, J.; Khouri, S.; Rybár, R.; Kudelas, D. TESES rules as a tool of analysis for chosen OZE projects. *Acta Montan. Slovaca* **2008**, *13*, 350–356.

 *energies*

Article

The Role and Place of Traditional Chimney System Solutions in Environmental Progress and in Reducing Energy Consumption

Dariusz Bajno [1], Łukasz Bednarz [2] and Agnieszka Grzybowska [1,*]

1   Department of Building Structures, Faculty of Civil and Environmental Engineering and Architecture, University of Science and Technology, 85796 Bydgoszcz, Poland; dariusz.bajno@utp.edu.pl
2   Department of Building Structures, Faculty of Civil Engineering, Wrocław University of Science and Technology, 50370 Wrocław, Poland; lukasz.bednarz@pwr.edu.pl
*   Correspondence: agnieszka.grzybowska@utp.edu.pl; Tel.: +48-785-475-724

Abstract: Buildings, energy, and the environment are key issues facing construction around the world. The energy efficiency of buildings is a key topic when it comes to reducing the world's energy consumption, releasing harmful gases, and global climate change, as they consume about 40% of the world's energy supplies. Heat losses in buildings reduce the energy performance of buildings and are basically important to them. In the paper, the authors focus on the main problems related to heat losses generated by chimney systems, which are inseparable equipment of building structures, resulting in lower energy efficiency and, at the same time, technical efficiency and durability of the building partitions themselves. Authors present thermal imaging with its contribution to the detection of heat losses, thermal bridges, insulation problems, and other performance disturbances, and then verifications using appropriate simulation models. The mathematical apparatus of artificial neural networks was implemented to predict the temperature distributions on the surfaces of prefabricated chimney solutions. In Europe, we can often find a large building substance equipped with traditional chimneys, which disrupts the current trend of striving to reduce energy consumption, especially that derived from fossil fuels. Speaking of energy-efficient buildings, one should not ignore those that, without additional security and modern installations, are constantly used in a very wide range. Therefore, the article deals with an essential problem that is not perceived in design studies and during the operation period as having a basis in incorrect architectural solutions and which can be easily eliminated. It concerns the cooling of internal partitions of buildings on their last storeys, in places where chimneys are located, regardless of their function. The authors of the paper decided to take a closer look at this phenomenon, which may allow the limiting of its effects and at the same time reduce its impact on the energy performance of technologically older buildings.

Keywords: efficiency; operational safety; gravity; ventilation; natural ventilation

 check for
updates

Citation: Bajno, D.; Bednarz, Ł.; Grzybowska, A. The Role and Place of Traditional Chimney System Solutions in Environmental Progress and in Reducing Energy Consumption. *Energies* **2021**, *14*, 4720. https://doi.org/10.3390/en14164720

Academic Editor: Sergey Zhironkin

Received: 10 June 2021
Accepted: 25 July 2021
Published: 4 August 2021

Publisher's Note: MDPI stays neutral with regard to jurisdictional claims in published maps and institutional affiliations.

1. Introduction

Energy is one of the most essential material needs of human life. Its production, as any human activity, is accompanied by the environment, and a side effect of this is the emission of unused energy and pollution into the environment, especially when its source is coal and other fossil fuels [1]. Chimneys and combustion installations (in various forms) have accompanied man for thousands of years. Their original function was exclusively to carry away exhaust fumes from heated rooms.

Nowadays, the chimney flue and ventilation systems perform more complex functions, although the basic one—safe discharge of smoke fumes and used and humid air—remains unchanged. Modernisation of heating device construction, aiming at the highest possible efficiency of boilers, common access to various kinds of liquid and gas fuels, and changes in combustion technology of solid fuels have resulted in changes in the parameters and composition of combustion products [2–7].

Incorrectly designed chimneys, made of improper materials, functioning without inspection, maintenance, and repairs required by law are the cause of many accidents and tragedies. Every year, many users suffer from carbon monoxide poisoning, and the losses caused by fires in buildings and structures are estimated to be in the tens of millions of euros [7–10]. It is common to use chimneys without chimney sweeps and to disregard the above-mentioned obligation to periodically inspect and clean these installations or to make arbitrary changes and alterations [11]. Nowadays, they are both technological installations and at the same time responsible constructions. Therefore the European regulations place them among the construction devices or constructions subject to special supervision, regardless of their function. This is intended not only to control the technical performance of the chimneys themselves but also the extent of their impact on people and their surroundings [12–15].

Apart from the safe flue gas discharge and exchange of air in the rooms, the construction of chimneys should also take into account the uncontrolled migration of thermal energy through their walls at the chimney—building interior border–chimney—external environment border, which is connected with the constant physical and chemical processes taking place here. Chimney walls are complete partitions comparable to the external walls of buildings and constructions (including energy-saving buildings), through which heat and moisture exchange constantly occurs, not only during heating periods. Their improper design will not only cause increased CO_2 emissions into the atmosphere but also heat energy, forcing users to heat their buildings additionally. It is difficult to find innovative solutions for so-called solar chimneys in new design studies. The difference between traditional and solar chimneys lies in the fact that in the latter ones, at least one of the walls (mainly the southern wall) has a transparent layer (covering), which significantly intensifies the ventilation draught [16] so that the chimney outlet is not so intensely cooled as in traditional solutions [17–19]. The difference between traditional and solar chimneys consists in the fact that in the latter one, at least one of the walls (mainly the southern wall) has a transparent layer (casing), which significantly intensifies the ventilation draught, and the chimney outlet will not be so intensely cooled as in traditional solutions [20–22].

The authors of the article, recognising chimneys not only as devices highly responsible for the functional safety of buildings, proposed that they should also be considered as responsible building partitions, equivalent to external building partitions, which should not be routinely selected, but each time designed as a responsible building element. These barriers should effectively protect both the interiors of buildings and chimney flues from uncontrolled loss of heat, which could be used for other purposes. The issue of the efficiency of the operation of traditional flue pipes as well as the impact on ecology and the reduction of energy consumption has been addressed in numerous studies, e.g.,: [1,23–31]. In this paper, some example simulations are included to show the possibility of improving the thermal protection of the partitions described above and in this paper.

1.1. Types of Chimney Construction

Just as in the past, wood and coal are the most commonly used fuels, so the traditional materials used to make chimneys were building ceramics and, for about 120 years, silicates and cement (concrete), in the form of bricks, pipes, ceramics, silicate, and even cement and asbestos fittings. Nowadays, other materials such as stainless or acid-resistant steel or acid-resistant ceramics are increasingly used. In modern heating devices, low flue gas temperatures are the reason for condensate condensation (sometimes in considerable amounts), which, due to its composition, has a destructive influence on the chimney, lowering its technical efficiency and shortening its service life.

Mineral wool or ceramic cladding with a U-value of approximately 0.4 W/m^{2}K is used for chimney insulation. The insulation should provide the walls of such ducts with the required protection against excessive and rapid cooling, especially outside the building. There is currently a range of insulation materials available on the building market that can be used to reduce heat loss in compartments. The purpose of such insulation is the aforementioned protection against uncontrolled cooling of air and flue gases, which in the case of a negative-pressure chimney will result in the reduction of the chimney draft force effectiveness, and thus will reduce its technical efficiency and at the same time reduce the heat emission to the nearest surroundings. This is particularly important for chimneys at risk of soot ignition. To achieve the required temperature reduction on the jacket of high temperature insulated chimneys (operating temperature of 600 °C), multilayer insulation is increasingly used. The first layer of insulation (on the duct side) consists of ceramic material resistant to high temperatures (even above 1000 °C), the second one of special mats or fittings with a thermal resistance above 500 °C. A chimney made in this way is safe for the environment, both in terms of fire and to protect against the possibility of burns in case of contact with the cladding [32–34].

When analysing the operating conditions of chimney systems in Central Europe, they have to be considered as highly "specific". The condition of smoke and ventilation chimneys and flue gas installations is assessed highly critically in chimney sweeping reports. This is not helped by legal loopholes in the chimney sweeping trade, including the performance of acceptance and inspection functions, which should be carried out by authorised masters. Chimney systems, especially external chimneys, should be adapted to the climate zones in which they are installed. The climate typical of a given area influences both the type and parameters of materials used (e.g., insulation) and the chimney's operation characteristics. From the point of view of the work of heating installations, the climate of temperate zones are characterised by both extremely low (below −25 °C) and high temperatures (above 25 °C), a very long heating period (often reaching 8 months) and under such conditions, the chimney system must always work properly.

Internal ventilation chimneys, flue gas chimneys, and still smoke chimneys are an integral part of buildings and structures, and therefore their use is governed by building regulations. Therefore, for chimneys, both statutory provisions [35–40] as well as executive regulations related to the use of energy equipment, especially in relation to gas and oil installations, apply. In Europe, intensive efforts are being made against the legal background to combat environmental pollution, as exemplified by the adoption on 10 January 2007 by the European Commission of the energy and climate change package as the basis for a new energy policy for Europe (European Energy Policy, Brussels, 10.1.2007 COM(2007))). Its main strategic objective was to achieve a reduction in emissions of greenhouse gases and to increase the share of renewable energy in the total EU energy balance from less than 7% in 2006, as well as to reduce the total consumption of primary energy by 20%, compared to 2006. Energy production on an industrial scale, as with other industrial sectors, is subject to stringent environmental requirements for the introduction of the best available technologies (BAT) and the most effective means of removing pollutants from exhaust or process gases. This results in a significant reduction in pollutant emissions from industrial sources. On the other hand, the least "disciplined" sector of energy production from sources below 50 MW is the biggest shareholder of global, both worldwide and national, emissions of pollutants. Therefore, this sector has been subject to specific actions in the EU aimed at technological and legislative disciplining to increase the energy efficiency of the energy production techniques used and to reduce their environmental impact. In the Clean Air of Europe (CAFE) Thematic Strategy (COM(2005)), developed within the framework of CAFE, special attention was paid to installations with a capacity <50 MW, i.e., the so-called small combustion plants, for which there are no EU regulations, especially installations for the combustion of solid fuels producing useful energy for heating individual households and preparing domestic hot water. One of the legislative acts concerning EU measures aimed at reducing primary energy consumption, among others in households, is [41,42] establishing general principles for setting eco-product/eco-design requirements [43] for energy-using products. Due to the dominant role of energy cost in the life cycle costs of these appliances, the Directive emphasises the promotion of energy-efficient appliances. In the case of boilers, cookers, fireplaces, the main objective of the Directive is on the one hand to harmonise the regulations on appliance design (the aim is to avoid distortions in the free movement of products), and on the other hand, to induce manufacturers to produce appliances characterised by high efficiency of converting the chemical energy of fuels into useful energy—space heating, the preparation of domestic hot water, the preparation of meals, and more environmentally friendly products. Environmentally friendly products are not only from the point of view of low emission of pollutants, optimally low consumption of energy resources to produce useful energy, but also low unit consumption of materials to produce particular types of cookers/boilers/fireplaces and the possibility of recovery/recycling of materials or safe disposal at the end of operation. In 2007, the European Commission started work on establishing eco-design requirements [44] for low power combustion installations—boilers/furnaces/chimneys fired with solid fuels (coal and biomass). These actions are coherent with the assumptions of the Green Paper of the European Strategy for Sustainable, Competitive and Secure Energy and for Improving the Competitiveness of the Economy [45].

It should also be noted that irrespective of EU and national legislative measures, there is also an increase in the knowledge and awareness of the consumer—users of solid fuel combustion systems in individual buildings—which is reflected in the growing demand for high-quality equipment, especially in terms of high energy and environmental efficiency, high comfort of operation, and low operating costs. This, in turn, has resulted in the recent development of modern designs of heating devices fired by solid fuels, coal, and biomass, especially of wood origin, and both fireplaces and boilers. The greatest progress in the development of technologies used in heating devices has occurred in boiler designs fired by coal and, in recent years, also by biomass, especially in the compacted form of pellets. Considerable progress can be seen in the construction of fireplaces. In the case of industrial and professional power engineering, strict environmental protection requirements impose the use of the best available technologies for electricity and heat production (BAT technology). This results in a significant reduction in pollutant emissions from industrial and professional sources. On the other hand, the production of heat, using coal as a fuel, in dispersed sources—in the municipal sector, single-family housing, public utility buildings, agriculture, forestry, fisheries, dispersed military units and industrial plants, everywhere where it is uneconomical to apply highly efficient installations for flue gas cleaning and where outdated central heating boiler installations and household furnaces are used and where "bad practice" of burning non-sorted coals is applied, as well as burning and co-combustion of municipal waste—causes very high emissions of pollutants to the atmosphere.

This does not, however, solve the problem of improving the energy efficiency of existing buildings, which in fact require immediate intervention to improve efficiency and to provide the solutions sought by the authors during their research of historical and contemporary buildings made using older technologies. The main objective of this research is to determine the causes of corrosion degradation of building partitions related to the functioning of ventilation chimneys, as well as to assess the current extent of damage and provide methods or ways to remove the causes and remedy the effects [46].

1.2. Requirements for Flue Pipes

External chimneys, from the point of view of building regulations [16,35–40,42], are classified as either structures or equipment and are fully subject to building regulations. During chimney testing, the safety of the chimney's impact on both combustible and load-bearing elements of the building structure is checked. The manufacturer declares the minimum distance between the chimney and combustible elements under normal operating conditions, regardless of the function of the flue pipe. The temperature at the declared distance from the combustible elements (100 mm in the example) must not exceed 85 °C. Of course, safe distances of 100 or 200 mm (at a flue gas temperature of 450 °C) from combustible parts are only achievable with insulated chimneys. For single-walled uninsulated chimney liners, the distance to combustible parts should not be less than 450 mm. During the tests carried out in an accredited laboratory, other parameters declared by the manufacturer are also verified, such as: thermal resistance of chimney insulation (double-walled chimneys), resistance to atmospheric conditions, flow resistance through individual chimney elements, the durability of fixed elements (supports, brackets, consoles).

When applying for the "CE" mark for insulated chimneys, the manufacturer shall specify the parameters of the thermal insulation used. The thermal insulation of insulated chimneys must have the following characteristics:

- Be completely resistant to the stated flue gas temperatures.
- Have an adequate thermal resistance coefficient, which guarantees the maintenance of the safe temperature of the external jacket at specified flue gas temperatures, as well as the maintenance of the required operating parameters at low ambient temperatures (minimisation of condensation).

Good insulation with a thermal resistance coefficient at temperatures around 250 °C of between 0.4 and 0.6 W/m^{2}K should be resistant to short-term exposure to high temperatures, such as a soot fire, i.e., 1000 °C. The standard does not specify the thickness of insulation that should be used for chimney insulation.

According to the standard [43], the resistance of insulated chimneys is tested mainly for rainwater tightness. The tightness of the connections of individual elements is controlled by placing the chimney in a special rain chamber. The tightness is measured by examining the mass of insulation before and after the test in the rain chamber. An increase in the mass suggests a lack of jacket tightness and discredits the chimney.

Prefabricated chimney systems lose draught and cause the cooling of the top floor, which is a determinant of significant heat loss and a drop in the energy efficiency of the entire building. The consequence of this phenomenon is the condensation of moisture caused by the temperature difference, which is an excellent environment for the development of fungi and moulds, significantly affecting the comfort of use of the building [19,39,40,46–49].

2. Materials and Methods

The energy efficiency [47] of chimneys can be verified by a number of tests, i.e., thermal imaging tests, temperature measurement, pressure measurement. Thermal imaging tests are a research method, which consists of remote and noncontact evaluations of the temperature distribution on the surface of the tested object [50]. The working principle of the method is based on the registration of the distribution of infrared radiation emitted by any body whose temperature is higher than absolute zero and the transformation of this radiation in a detector into visible light. The resulting colour thermal image is a thermogram, indicating on a colour scale the level of the recorded surface temperature. If thermal images are taken from the outside during the cold season of the year (heating period), places on the external surface with an increased temperature correspond to less well-insulated areas of the partition, while a low surface temperature can usually indicate good insulating properties of the partition. The opposite situation will occur if photographs are taken of the internal side. Places with a lower surface temperature indicate deterioration of the insulation properties. This method is becoming increasingly used in the construction industry in activities classified as thermal (energy) diagnostics of buildings and their technical equipment. Using thermographic measurements, it is possible to detect not only defects in the thermal insulation of external partitions, including various types of thermal bridges, but also leaks, creating conditions for heat flow as a result of intensified air infiltration. [48]. The energy efficiency of solid fuel combustion systems, especially of traditional design, is relatively low and does not exceed 50% on average per year. It also results in excessive consumption of coal and biomass—higher than the demand for useful energy in flat, single-family houses or public utility buildings.

In general, it can be said that a neural network is a simplification of the brain structure diagram. The basic component of an artificial neural network is the processing element. It is a specific model of real cells that are part of the nervous system, responsible for processing and analysing information in the human body. The actual nerve cell can be treated as a biological information processing system. The information introduced through the inputs (dendrites) is processed inside the cell. The processed signal is sent via the axon to subsequent cells. Each neural network consists of a large number of elements with the ability to process information (neurons), which are associated with each other by connections with specific parameters (weights), which change during the learning process. The schematic diagram of the artificial neural network structure is shown in Figure 1, according to which it can be stated that nowadays, neural networks have a layered structure, where the following layers are distinguished: input, output and hidden layers [51–53]. One of the available features of STATISTICA—unidirectional MLP multilayer network is the software that is based on the backpropagation error algorithm used for the prediction in this paper.

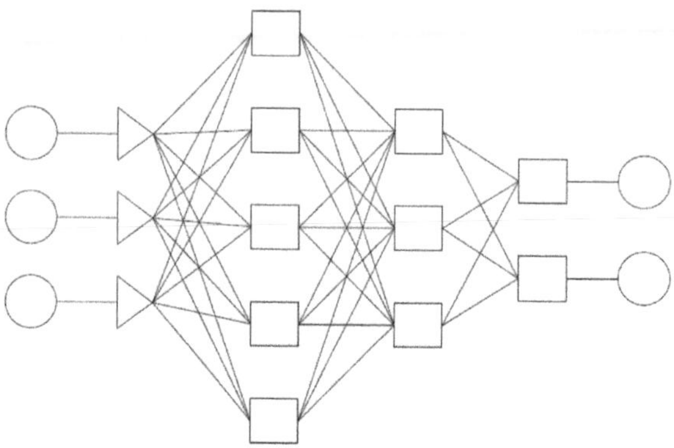

Figure 1. Diagram of an artificial neural network based on [51–53].

The neuron has a finite number of inputs, to which the inputs are given as: $x_1, x_2, \ldots, x_n$, with associated weights: $w_1, w_2, \ldots, w_n$. The process initiated by the input data is reflected by basic information occurring inside the neuron. The first is the determination of the aggregated input value and is carried out using an aggregating function, also known as the "postsynaptic potential function" (PSP function). The second process consists in determining the output value of the neuron, for which the activation function (transition function) is responsible. In general, these processes can be represented by the diagram in Figure 2 [51–54].

$$x_1 \to (w_1)$$
$$x_2 \to (w_2)$$
$$x_3 \to (w_3)$$

| s- aggregate input value | $\to$ | $y = f(x)$ | $\to (y)$ |

Figure 2. Schematic of the artificial neural network model, reflecting the processes occurring inside the neuron.

The aggregated input value of a neuron becomes the parameter of the activation function of that neuron. Common activation functions include linear, logistic, hyperbolic, exponential, sinusoidal, and Gaussian equations. The basic feature of neural networks is their ability to self-organise and adapt to changing conditions, which results in the possibility of selecting weights that can change in the learning process. SSN algorithms (artificial neural networks) are a tool perceived as informative, albeit learning, as they have the ability to adapt the stock of knowledge they possess to the possibility of its dynamic change. This is a fundamental characteristic, as this ability determines the possibility of self-learning, which significantly speeds up the execution of calculations. Learning consists of changing the coordinates of neurons so that they follow a pattern consistent with the structure of the data analysed. Therefore, the calibration of the phenomenological (abstract) model created by the network consists of an adequate selection of weights in the learning process. During learning, some connections (weights between neurons) become more important, and some do not participate in solving the problem (this could be compared to the disappearance of connections in the brain), so it is possible to determine which variables are important in terms of solving the problem. The structures presented can be used in many areas of science. Many examples can be cited, such as: electronic circuits research, psychiatric research, sales forecasting, biological research, interpretation, machinery repair, planning, production problems analysis, industrial processes control.

3. Results

The aim to achieve the highest level of energy efficiency in building envelopes and thus also in entire buildings, aimed at the development and implementation of new and modern techniques, is the result of technological progress and at the same time sustainable development. This also makes it possible to reduce the extraction and consumption of fossil fuels. The authors have analysed a very large number of types of chimney flue, and the main problems are presented in the paper.

Nowadays, we should also ask ourselves a question: when talking about energy saving, do we also mean very much cubic building resources exploited for many years, which also need such modern solutions? Unfortunately, they are still at the end of the queue for modernity. The existing housing stock, equipped with traditional brick chimneys or newly installed system solutions, is not only a problem for the environment but also for the users of these buildings. Problems are created not only by the emission of pollutants into the atmosphere but also by the cooling of building envelopes of the last storeys of buildings, such as flats and attics. The location of chimney flues on the top floors of buildings often creates very significant operational problems, including hygienic and sanitary ones. This problem initially manifests itself in the form of moisture condensing on the surfaces of external and internal partitions. In a later stage, local discolourations appear, which in the final stage turn into foci of microbiological threats due to the development of moulds on cooled fragments of thin-walled chimney ducts. Additional thermal bridges (linear and surface ones) often occur in places where chimneys are led above roofs, above heated floors. This phenomenon cannot be fully eliminated by applying common solutions, but its effects may be considerably limited. Intentional or unintentional, periodical switching off of parts of the storey, premises or only single rooms or inefficient or even inefficient ventilation may also turn out to be an important element in generating heat losses. The authors tried to present such cases on the basis of observations and own research verified by simulation models.

The cases shown in Figure 3 are quite common in construction practice, which may indicate a disregard for this problem, poor knowledge of designers, and lack of awareness of building owners and managers. Such situations not only generate higher operating costs of buildings but, above all, are the cause of significant emissions into the atmosphere due to incomplete combustion of fossil fuels. In addition, the cooling of chimney walls and adjoining walls occurs, which contributes to their slow but progressive chemical and biological degradation due to the deposition of aggressive substances on them. Such situations are also a direct source of microbiological risk for the inhabitants of the top floors of buildings.

In Europe, modern chimney systems are still being integrated into the existing walls of buildings, thus compromising their spatial rigidity. According to the manufacturers' claims, chimney risers should be independent and self-supporting structures, but in reality they are subjected to additional loads by being "glued" tightly into existing structures (Figure 4).

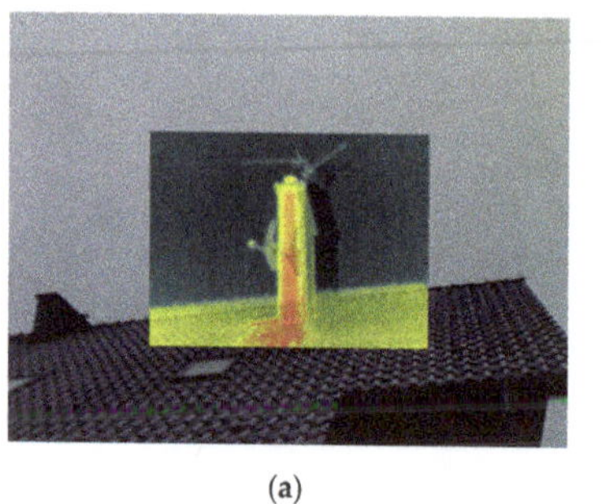
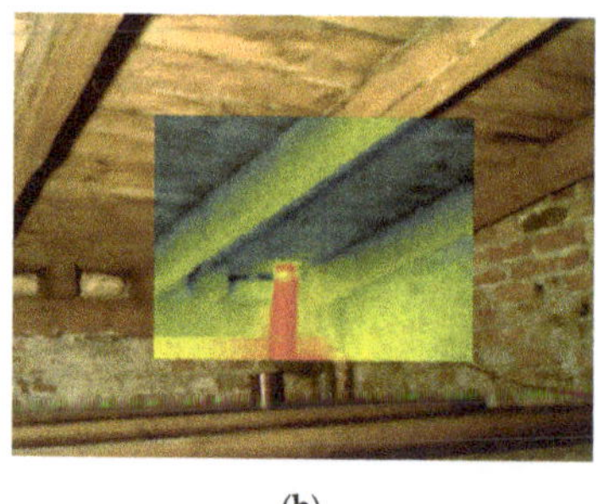

(a)   (b)

Figure 3. Non thermally insulated flue pipes: (a) in the part above the roof, (b) in the attic space.

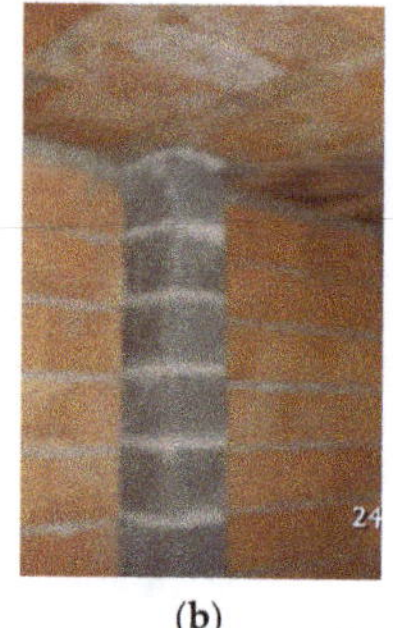

(a)   (b)   (c)

Figure 4. Non-thermally insulated chimney flues: (a) in the part above the roof, (b,c) in the attic space.

These chimneys, both in the attic space and above the roof, do not have layers protecting against excessive heat loss, and they cool down quite quickly. This is one of the reasons for chimney draught deceleration and, at the same time, the reason for the dampening of the internal partitions of buildings on the last floors and the development of harmful moulds. As already mentioned above, both walls and chimneys may be subject to faster wear and tear or damaged by corrosive degradation, which reduces their technical performance (Figure 5).

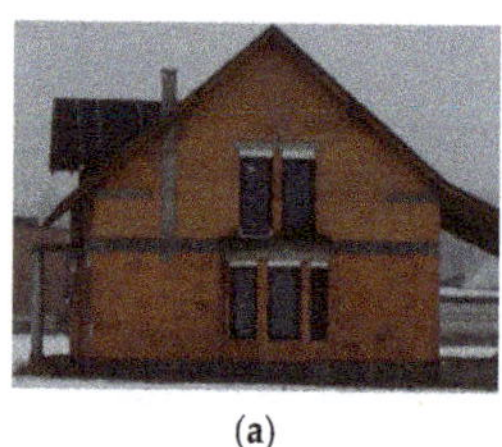

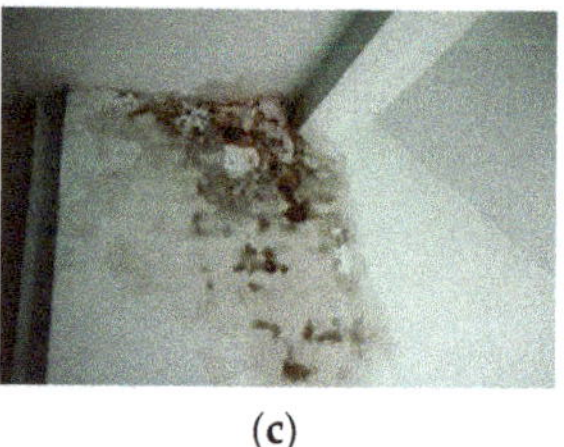

(a)   (b)   (c)

Figure 5. Consequences of inadequate protection of external and internal chimney walls (a,b) elevation, (c) inside.

This can lead to serious mechanical damage to chimneys and even entire buildings.

The variation of the temperature fields on the chimney wall is shown in Figures 6 and 7, diagrams of temperature distribution (along the line running directly under the ceiling). As the walls of the top floors are additionally heated here, and thus the rooms are heated, the heating systems require more thermal energy, which results in a significant increase in fuel consumption.

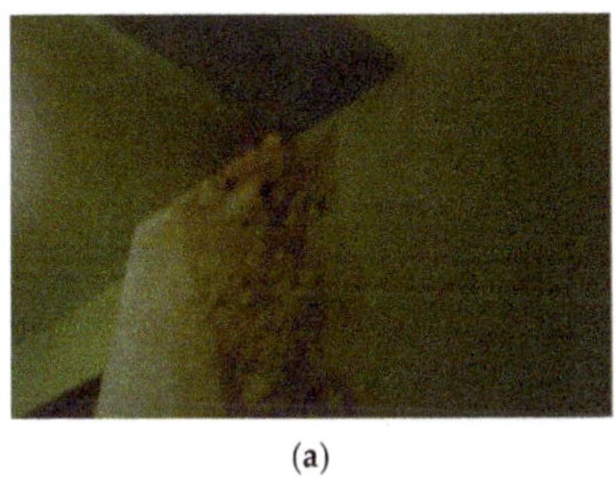

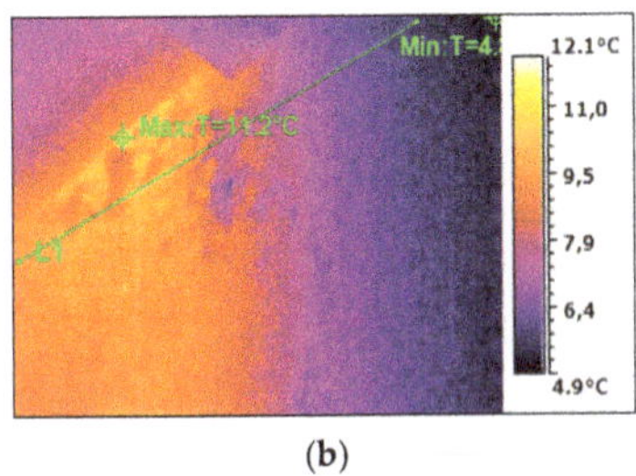

(a)　　　　　(b)

Figure 6. Thermal (**a**) image of the temperature distribution on the surface of the chimney wall in visible light (**b**).

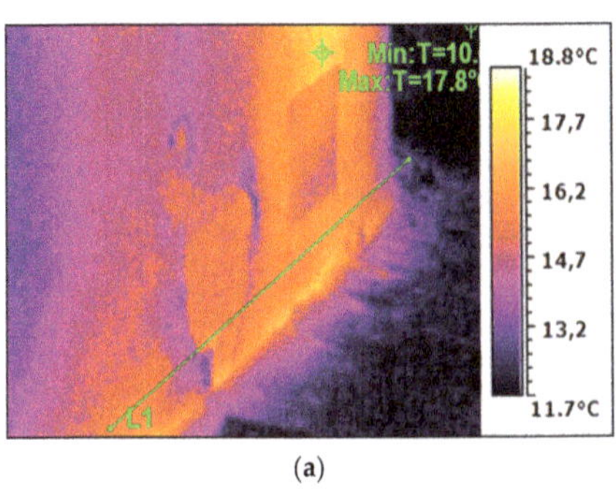

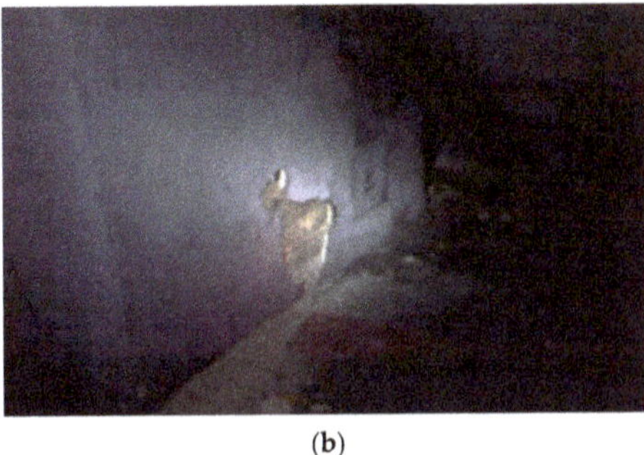

(a)　　　　　(b)

Figure 7. Thermal (**a**), and visible light (**b**) images of the chimney wall in the attic space (at the inspection grille).

Temperature distribution on the surface of a chimney located in an attic space containing smoke and ventilation ducts (Figure 7).

Figure 8 shows a certain stage in the modernisation of a single-family building through the introduction of prefabricated system chimneys. The effect of their inadequate installation has already been noticed in the winter period on the first floor of the building and in the basement, at the clean-out opening.

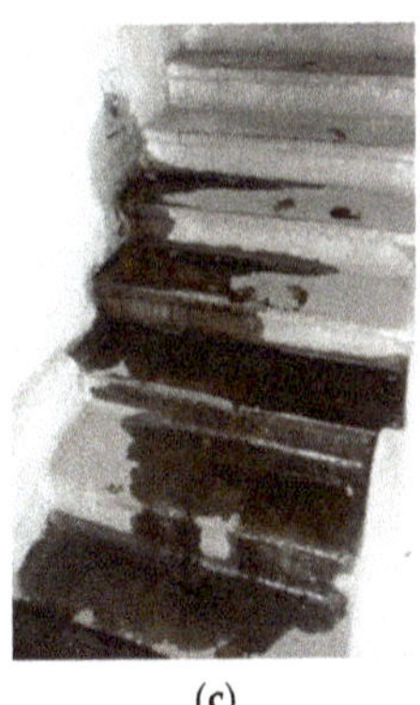

(a)　　　　　(b)　　　　　(c)

Figure 8. Prefabricated ventilation flue pipes: (**a**) inside the building, (**b**) above the roof, (**c**) condensation leakage in the basement through the clean-out opening.

It is difficult to agree with the position that our future will be based on the continuous improvement of chimney systems in modern construction (including passive construction), increasingly limiting carbon dioxide emissions into the atmosphere when a significant proportion of older buildings have not kept up with "this" modernity. Despite the incorporation of new technologies, this is done in an incompetent manner, inconsistent with the intentions of the manufacturers of these products and at the same time inconsistent with sustainable development. Therefore, it is necessary to develop modernisation and repair solutions for the existing buildings not only in the form of reinforcement of traditional masonry constructions [55] or analysis of their dampness [55], but also methods which would make it possible to ensure the highest possible level of chimney draught, prevent local heating and cooling of internal partitions (walls) on the top floors, and, at the same time, ensure efficient ventilation and comfort of rooms. This would eliminate unnecessary heating or cooling of such floors, and thus reduce energy consumption and prolong the life of the buildings and their partitions. To confirm the above thesis, a simulation has been carried out for one flue pipe, built into a wall without insulation and with insulation in several variants (Figure 9), based on the experiments presented in [55 57].

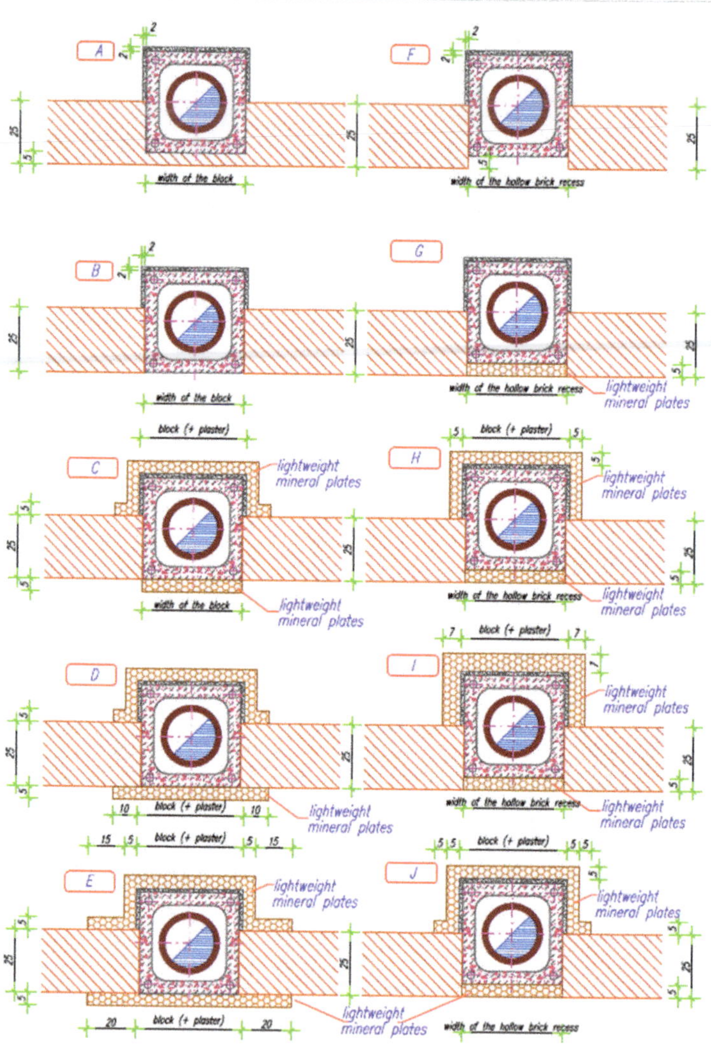

Figure 9. Design models of chimney structures adopted for simulation analyses without insulation (**A,B,F**) and with insulation (**C–E,G–J**).

3.1. Starting Assumptions for Thermal Calculations Results

For the calculation of the surface temperature [58] it is recommended to take the following values for the heat transfer resistance of internal surfaces:

- Upper part of the room: $R_{si} = 0.25 \, \text{m}^2\text{K}/\text{W}$.
- Lower part of the room: $R_{si} = 0.35 \, \text{m}^2\text{K}/\text{W}$.
- With significant shielding of the wall surface by objects such as furniture: $R_{si} = 0.50 \, \text{m}^2\text{K}/\text{W}$.

In the case studied, significant shielding of the wall surfaces was assumed: $R_{si} = 0.50 \, \text{m}^2\text{K}/\text{W}$. For the calculation of the surface temperature, the magnitude of the heat transfer resistance on the external surfaces was used: $R_{se} = 0.04 \, \text{m}^2\text{K}/\text{W}$. The design temperatures of the heated rooms have been assumed based on [35] as characteristic values for Central and Eastern Europe. For this analysis, the values of internal temperatures are summarised in Table 1.

Table 1. Values of internal temperature.

Design Temperatures	Purpose or Use of Premises	Examples of Rooms
+20 °C	— intended for permanent residence of uncovered people who do not continuously perform manual labour	living rooms, anterooms, individual kitchen with gas or electric fires, offices, meeting rooms
+24 °C	— intended to be dismantled — intended for human habitation without clothes	bathrooms, changing rooms, washrooms, showers, swimming pool hall, doctor's office, undressing, baby and nursery rooms, operating theatres

Simulation calculations have been carried out for the rooms of the last storeys of residential and office buildings using partition models, with the following assumptions:

- $t_i = + 20.0 \, °\text{C}$ (rooms)
- $t_r = + 4.0 \, °\text{C}$
- $t_w = - 5.0 \, °\text{C}$

where:
t_i means the indoor air temperature,
t_r means the temperature inside the tube (ceramic insert),
t_w means the temperature inside the ventilated space in the counterflow mode.

In the following part of the study, the temperature distribution in partitions containing chimney flues, the indication of the location of the thermal bridge with the lowest surface temperature value with its magnitude, and the layout of heat flux density lines are presented in the calculation models. To avoid surface condensation, several options were considered for insulating the locations where prefabricated system chimney flues will be installed in terms of their shape, extent, and location.

For simulation calculations of heat flow in chimneys, a square cross-section of the chimney section has been assumed instead of a round one and a square cross-section of the prefabricated protective element is often used in practice (Figure 10). This assumption is the most correct from the point of view of correctness of calculation, more so as it complies with the standard [59], where in Appendix B4 [59] it is written, "For a nonrectangular air void the thermal resistance is assumed to be equal to the resistance of a rectangular void with the same surface area and the same ratio of sides". These were the assumptions made by the authors of this paper for their calculations. The above-mentioned variants of the location of the flue pipe were analysed in terms of heat transport, taking into account the different locations of the thermal insulation layer made of lightweight mineral panels (5 and 7 cm thick), characterised by a thermal conductivity coefficient of $\lambda = 0.045 \, \text{W}/\text{mK}$.

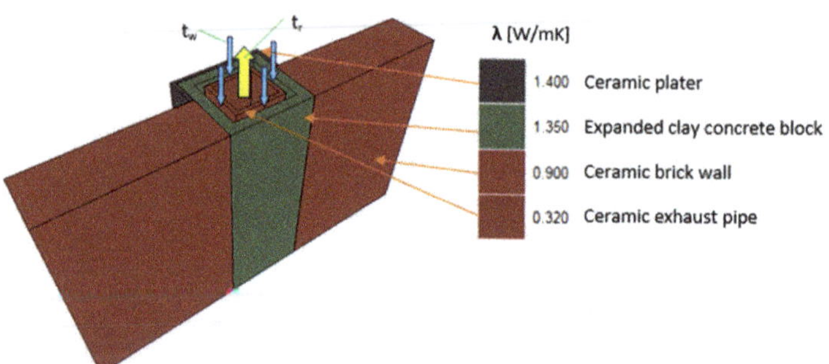

Figure 10. Calculation model.

The phenomenon of "overcooling" or "overheating" of walls on integrated traditional system chimneys is rarely recognised by architects and builders. The diagrams in Appendix A, Figures A1–A10 show how important this problem can be, especially for partitions that can lead to excessive heat loss, not taking into account the thermomodernisation of buildings. Increasing the insulation around the chimney flue, even slightly, leads to a significant energy gain.

Figures A1–A10 show the distribution of isotherms and adiabatic heat fluxes in the inner walls of the chimney stack caused by temperature differences. These models are assigned to the cross-sections shown in Figure 10 and presented as a detailed numerical analysis of the distribution of temperature and heat fluxes in the cross-section and at the height of the flue pipes. Figures A1, A2, A6 and A7 has been borrowed from construction practice as they are often used in design solutions and are also executed in this way. Such places of connection between chimney stacks and building walls generate excessive heat loss, creating a network of linear thermal bridges, which also becomes a microbiological threat to the users of the rooms through which the chimneys run. The authors of the article, using models C, D, E, H, I, and J as an example, have tried to show the possibility of limiting heat loss depending on the shape and range of the thermal insulation material.

The scales placed next to the diagrams show the temperature ranges in which particular fragments of chimney walls may be found during their operation and the adiabatic fields of heat streams assigned to them. The lowest models in Figures A1–A10 indicate the locations of the most cooling linear thermal bridges (orange colour). These are places of the smallest wall thickness in non-insulated chimneys and/or places of thermal insulation endings in single- or double-sided insulated chimneys.

This does not mean, of course, that they will consistently reach the extreme values shown above. Nevertheless, even internal partitions of buildings, which may be cooled down by introducing into them ventilation, air and fume ducts, or smoke ducts, should be additionally protected with a layer of thermal insulation protecting them against the lowering of temperature to critical values (in cold periods) or against overheating of rooms in the case of fume and smoke ducts.

Even a small layer of thermal insulation, with a low heat transfer coefficient and low diffusion resistance, is able to significantly improve the parameters of the cooled partitions, and thus the entire room. The authors of this article have used the simulation model parameters, which are characterised by mineral thermal insulation panels, ideally suited for insulating partitions from the inside. They are made from a very lightweight variety of cellular concrete and have a very low diffusion resistance ($\leq$3). Their additional advantage is their low density (115 kg/m^3); hence they will not be a significant burden on walls and chimneys, both inside buildings and those extending above the roof. Moreover, they are easy to process, have a low thermal conductivity coefficient λ = 0.043–0.045 W/mK. Therefore their thickness will be required to insulate the walls, which will result in the reduction of the usable area of the room only slightly.

The authors of the article analysed many architectural design solutions and did not find that any of the designers raised such a topic. On a global scale, the problem of unnecessary heat loss may turn out to be a very serious one, which can be very easily eliminated by raising the awareness of those responsible for building design. It will be enough to implement very simple and at the same time effective solutions, not only to reduce additional heat losses but at the same time improving the chimney draught and a more complete fuel combustion. In every European country, regulations require buildings to be properly maintained and periodically inspected. These obligations are usually incumbent on building managers and owners.

Prior to the thermomodernisation of buildings, energy audits should always be prepared, in which the issues described in the article should find their permanent place. Thermomodernisation of buildings should also include thermal insulation of such partitions inside the buildings, regardless of the adopted method of thermomodernisation of the whole building.

Table 2 summarises the results of the calculations carried out for surface temperatures in relation to the "dew point" temperatures corresponding to the indoor relative humidity of 45%, 55%, and 75%. The calculated values of the surface temperatures of partitions placed in columns 4 and 5 should be higher by at least 1 °C than those placed in column 3.

Table 2. Temperature comparison for t_i = +20 °C and different φ from 45% to 75%.

Calculation Scheme	t_s (°C)			Surface Temperature Side "A" t (°C)			Surface Temperature Side "B" t (°C)		
				t_i = +20 °C φ = 45%	t_i = +20 °C φ = 55%	t_i = +20 °C φ = 75%	t_i = +20 °C φ = 45%	t_i = +20 °C φ = 55%	t_i = +20 °C φ = 75%
A	7.7	10.7	15.4	+6.0	+6.0	+6.0	+3.0	+3.0	+3.0
B	7.7	10.7	15.4	+2.0	+2.0	+2.0	+3.0	+3.0	+3.0
C	7.7	10.7	15.4	+5.0	+5.0	+5.0	+11.0	+11.0	+11.0
D	7.7	10.7	15.4	+11.0	+11.0	+11.0	+11.0	+11.0	+11.0
E	7.7	10.7	15.4	+13.0	+13.0	+13.0	+13.0	+13.0	+13.0
F	7.7	10.7	15.4	+2.0	+2.0	+2.0	+3.0	+3.0	+3.0
G	7.7	10.7	15.4	+11.0	+11.0	+11.0	+3.0	+3.0	+3.0
H	7.7	10.7	15.4	+11.0	+11.0	+11.0	+9.0	+9.0	+9.0
I	7.7	10.7	15.4	+11.0	+11.0	+11.0	+9.0	+9.0	+9.0
J	7.7	10.7	15.4	+11.0	+11.0	+11.0	+11.0	+11.0	+11.0

The above data and examples of thermal insulation should not be treated as a form of universalisation of solutions to the example of a few examined cases but as an indication of the directions for further research and implementation. Thinner and improperly insulated walls of chimneys on the last storeys and in attics (regardless of their function) may become the cause of excessive, unpredicted heat loss and, at the same time, degradation of building partitions. The reason for this will be the intensive movement of air and exhaust fumes. For the obtained values, the relative error was calculated: humidity = 0.1%. Surface temperature sides "A" and "B" 0.2%, t_s 0.1%. Based on the obtained error values, the significant digits in Tables 3 and 4 were removed. As mentioned above, each case of thermal bridges should be anticipated and considered individually. This applies not only to the location of chimney flues on the top floors of residential and public buildings.

Table 3. Summary of data for analysis.

Humidity	Calculation Scheme	t_s (°C)	Temp. "A" Side Surface	Temp. "B" Side Surface
45	A	7.7	6	3
45	B	7.7	2	3
45	C	7.7	5	11
45	D	7.7	11	11
45	E	7.7	13	13
45	F	7.7	2	3
45	G	7.7	11	3
45	H	7.7	11	9
45	I	7.7	11	9
45	J	7.7	11	11

Table 4. Temperature values for prediction.

No.	Temp. "B" Side Surface	Prediction Spreadsheet for Temp. "B" Side Surface (Spreadsheet1)1. MLP 12-5-2
1	3.0	3.0
2	3.0	3.0
3	11.0	11.0
4	13.0	12.9
5	3.0	3.0
...		
146	11.0	10.9
147	13.0	12.9
148	3.0	3.0
149	3.0	3.0
150	9.0	9.0

In the design studies of new buildings and at the stage of designing modernisation works in the existing facilities (replacement of installations, superstructure, reconstruction, and extension), the issues of migration and excessive heat loss, even in the internal partitions of these facilities, must be taken into account, due to the introduction into all kinds of structures that are not walls, in which any form of forced or gravitational air movement may occur. The issues discussed in the article concerning local cooling of fragments of the internal walls of the last storey are presented with the example of a prefabricated system chimney (produced by one of the leading chimney systems manufacturers in the world), built into the walls of the last storey. Such situations should be anticipated, and, where necessary, preventive measures should be taken in the form of local insulation of chimney walls in the immediate vicinity with a material characterised by high thermal resistance and the lowest possible diffusion resistance. At the same time, it is advisable to inspect such walls in buildings already in use, using thermal imaging cameras, moisture meters, and for the presence of salts and other harmful compounds.

3.2. Results of Analyses Carried out Using Artificial Neural Networks

The simulation was made with the help of artificial neural networks, which allowed the prediction of the surface temperature in relation to the "dew point", temperatures corresponding to the relative humidity of indoor air, on the basis of the research results (the results presented above were extended and the prediction database was built). For the case under consideration (temperature prediction), the problem is defined as an attempt to develop a phenomenological model, determining the relationship between warming and temperature—which is a very complex issue. Neural networks are used when the derivation of the model is impossible. Complicated derived dependencies are insufficiently accurate. Then it is advantageous to use neural networks. The obtained solution will be accurate because the data in the learning set will be the real values of research. As stated above, the neural network follows the pattern given during learning. In the analysed case, the patterns are the results of temperature tests. The data subjected to analysis were divided into sets:

- Teachers (70% of case numbers).
- Test (15% of case numbers).
- Validation (15% of the number of cases).

This division of the data makes the neural network draw 85% of the data on which the learning process occurs and adjusts the weights for the input data (calculation scheme representing the way of warming, t_s temperature, and humidity). The validation set allows the evaluation (examination) of the predicted temperature with values that the program has not learned or tested. The network was built for forecasting the problem under consideration. It has the following topology:

- Characteristics of the problem: regression; this description of relationships is used to build models showing the actual relationship between the input data (explanatory) and the output variable (explained). In this case, the events are performed in the following sequence: the values of explanatory variables, neural network, the value of the explained variable.
- Number of inputs: 5.
- Network type: multilayer perceptron (unidirectional multilayer networks. MLP networks).
- Learning algorithm: BFGS (variable metric method).
- Number of neurons in the hidden layer: 4–6.
- Error function: sum of squares.
- Output function: linear.

The operation of the MPL network consists of: input data, determining the output values of hidden neurons, and determining the output value of the output neuron. The result of the network depends on: the value of the weights of hidden and output neurons and the structure of the network. Learning of unidirectional networks is performed by a "teacher". This means that the values entered as inputs to the network together with the corresponding output values enter the learning set. The aim of the learning process is, therefore, to generate such values that determine the weights for which the network outputs will coincide with the real values. The learning process is expected to result in a reduction of the network error, which is an aggregated measure of the difference between the actual output values (test results) and the calculated ones (generated during prediction). A bidirectional information flow can be identified in the network. Namely: from the input layer to the output layer. The information necessary to calculate the output variables is transported and from the output layer to the input layer. The error information used during the learning process is transmitted. Due to the orientation of the error information flow, the algorithm is called backward error propagation. Note the number of data analysed (50 results for each moisture content). A total of 150 results were obtained, which are summarised accordingly (Table 3). Figure 11 shows a view of the input and output vector selection window.

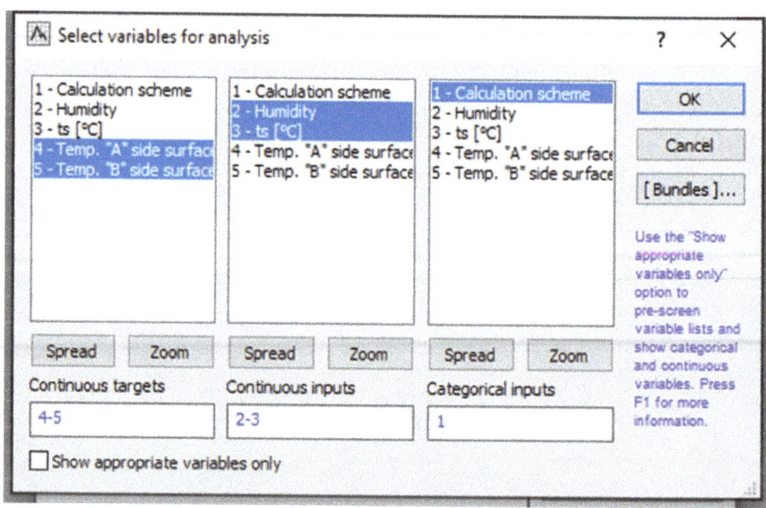

Figure 11. Selection of the input and output vector.

In solving regression problems, the explanatory variable (temperature at the partition surface) is quantitative. While the output explanatory variables can be quantitative (temperature, humidity) or/and qualitative (calculation scheme). The results were analysed according to the defined network topology. The results obtained as a result of the network operation explain the problem with high precision, as the validation (quality of forecasting is understood as the comparison of the forecasted value with the values from the validation set (Table 5), i.e., the one on which the program did not learn or test) amounted to 98%, which should be considered a correct result at this level of neural network creation. Example results of temperature value prediction (for a single network and ensemble of networks) are presented in Table 4 and graphically in Figure 12.

The graph presented in Figures 12 and 13 shows the fitting of the network results. The abscissa axis shows the output variable, i.e., the temperatures obtained from the tests, while the ordinate shows the values obtained as a result of the programme's prediction. Analysing the above results, it is possible to notice a good fit of the network to the analysed problem, as the graph does not show the occurrence of significant network errors.

Table 5. Sensitivity analysis.

Networks	Calculation Scheme	t_s	Humidity
MLP 12-5-2	2.91	24.54	24.04
MLP 12-5-2	5.37	8.13	7.26
MLP 12-9-2	6.51	1.00	1.00
MLP 12-6-2	1.06	1.01	1.02
MLP 12-3-2	2.77	30.56	31.24

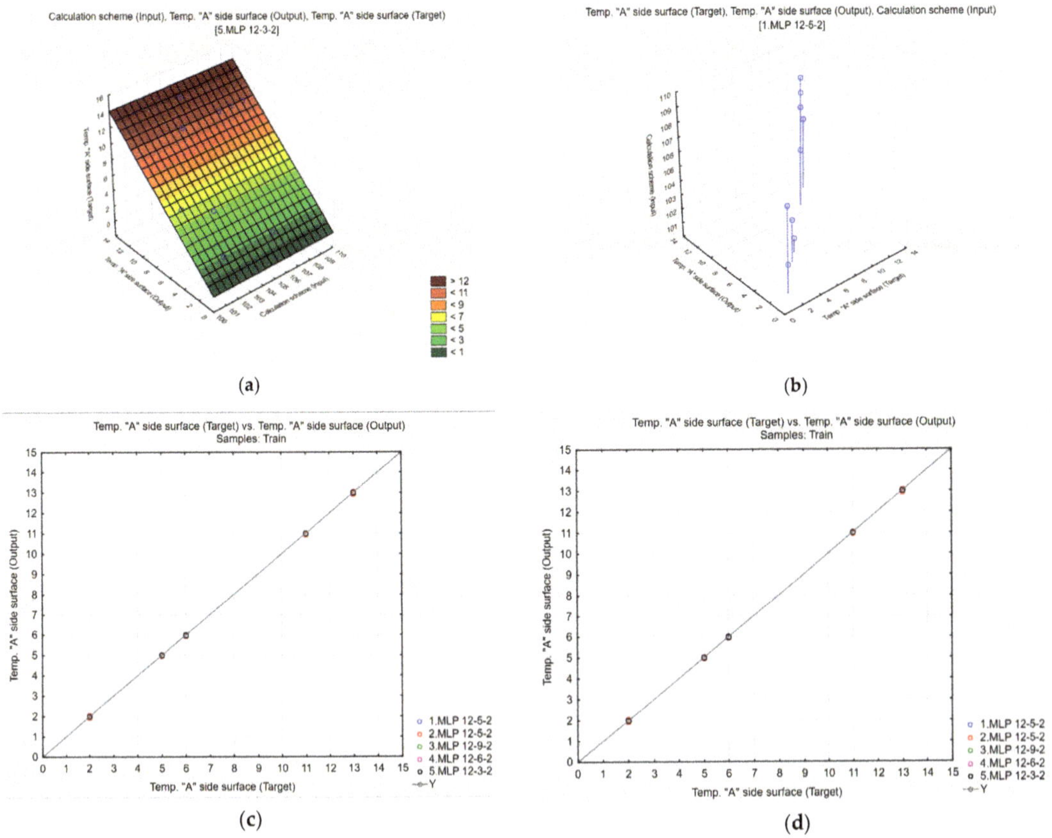

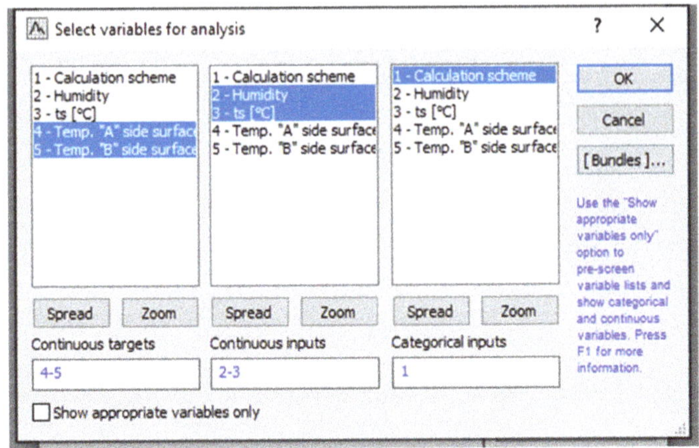

Figure 12. Graphical presentation of the network results (**a**) scheme of temp. A according to predicted value, (**b**) scheme of temp. A according to predicted value and calculation scheme, (**c**,**d**) scheme of temp. A according to predicted value for five different spreadsheets.

Figure 13. Matching the results of network runs to set input variables.

4. Discussion

The article deals with facilities that have been in operation for many years, as well as newer ones that are equipped with obsolete technological solutions, not only in terms of environmental protection but also for the users themselves. The trend towards maximum energy efficiency, which is already unstoppable today, should take into account this traditional substance that is holding it back. It is not popular nowadays to mention such buildings that shatter the statistics indicating steadily decreasing CO_2 emissions into the atmosphere by striving for energy-efficient buildings, zero-energy buildings, and passive buildings. In the presented analyses, the authors propose solutions that significantly improve the situation of indoor air exchange and are technically and financially feasible to implement in widely understood traditional construction. The results confirm the trend of recent years presented in the literature review. In the near future, the authors intend to address the model shown in Figure 14 of a multifunctional chimney: including one smokestack and a combination of ventilation for flue gas.

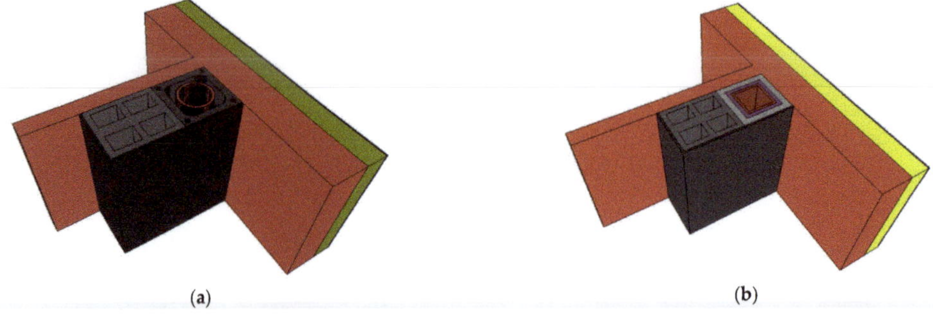

Figure 14. Actual model of the multifunctional chimney (**a**), substitute of the calculated model (**b**).

After analysing the broader context of the issue, it will be possible to answer questions such as: what can be done with technologically older and historic buildings, especially those under conservation protection? It is certainly necessary here to think about working out concrete and feasible solutions for them, which will also allow them to be on the path of technological and energy-saving progress, but without detriment to themselves. We cannot turn a blind eye and pretend that these problems do not exist. An important role of ventilation systems, including traditional ones, is the exchange of an appropriate amount of used and humid air in rooms, which should be ensured to maintain the expected microclimate in rooms, as well as to properly maintain the building envelope, and at the same time extend its lifespan.

Furthermore, the conditions that are not strongly influenced by the user in everyday life, such as building physics, heating systems, or fuel prices in the world, there is also conscious energy management. The basic problem, however, is the efficiency knowledge of the users in the field of energy-saving, even when there are no tangible benefits. e.g., monetary (when there is no possibility of individual cost accounting). Unfortunately, in buildings where the heating system is not efficient, and there is no possibility of regulation, rational heat management is very difficult.

It should be stressed that measurable effects in reducing air pollutant emissions will be brought by both the use of high-quality coal (high calorific value and low sulphur content) and the promotion of combustion of qualified coal fuels (high calorific value and low sulphur content with repeatable quality parameters) in highly efficient boilers. Replacement of high emission boilers with modern boilers (fired by gas or biofuels) or integrated heating systems based on solid fuel burning stoves combined with the use of renewable energy sources (e.g., heat pumps, solar collectors) will also contribute to reducing emissions and improving energy utilisation.

Above all, in poorly insulated buildings with old, worn-out, and low-efficiency heating systems, despite very high heat consumption, rooms can remain underheated. The most vulnerable rooms are those located at the ends of the installation. This situation not only generates high energy consumption and air pollutant emissions but also results in high costs associated with the use of energy carriers. According to all forecasts and trends in recent years, energy carriers will become more expensive in practically every form. This is due to several factors. First and foremost, the ever-increasing global demand for energy. Another major factor is the increasing dependence on supplies of energy carriers from politically unstable countries. In addition, the progressive increase in the costs of energy production and extraction (the need to search for deposits of natural resources that are increasingly difficult to access) and climate threats cause the introduction of increasing restrictions on the domestic production of pollution.

It can be seen, for example, that the use of a low-efficiency boiler results in at least a 30% loss of fuel. This is a typical value for boilers fired with solid fuel that are about 20 years old. For modern boilers, on the other hand, this loss is between 10% and 20%. All this translates, of course, into a reduction in the amount of fuel used and therefore in operating costs, but also in the amount of flue gases emitted into the air.

Another risk of poorly insulated building envelopes is the freezing of walls during frosty spells, which causes moisture from the air to condense on cold wall surfaces indoors, creating favourable conditions for mould and fungi to grow. Appearing dampness not only contributes to the deterioration of aesthetic conditions (stains, discolouration of paint coats, chipping, and falling of plaster), it is also the cause of a microclimate that negatively affects the health conditions of people staying in such rooms. Moreover, an increase in the humidity of partitions increases the heat transfer coefficient, and when the humidity turns into ice at a negative temperature, the thermal insulating power of the materials decreases further. Another example of poorly functioning heating systems can be the overheating of parts of rooms. A frequent cause in the situation of large cubature objects, such as schools and hospitals, are situations in the absence of the possibility of regulating the amount of heat supplied to various parts of the building. Part of the room is underheated despite the fact that the system operates at maximum efficiency. Then another part of the room is strongly overheated, and the only way to cope with the problem is to ventilate the room with cold external air.

To conclude on the role and place of traditional chimney system solutions in environmental progress and the reduction of energy consumption, it is necessary to state that only the appropriate insulation of these structures and conscious energy management can lead to real cost savings and climate protection.

Author Contributions: Conceptualization, D.B., A.G. and Ł.B.; methodology, D.B., A.G. and Ł.B.; software, D.B., A.G. and Ł.B.; validation, D.B., A.G., and Ł.B.; formal analysis, D.B., A.G. and Ł.B.; investigation, D.B., A.G. and Ł.B.; resources, D.B., A.G. and Ł.B.; data curation. D.B., A.G. and Ł.B.; writing—original draft preparation, D.B., A.G. and Ł.B.; writing—review and editing, D.B., A.G. and Ł.B.; visualization, D.B., A.G. and Ł.B.; supervision, D.B., A.G. and Ł.B.; project administration, D.B., A.G. and Ł.B.; funding acquisition, D.B., A.G. and Ł.B. All authors have read and agreed to the published version of the manuscript.

Funding: This research received no external funding.

Institutional Review Board Statement: Not applicable.

Informed Consent Statement: Not applicable.

Data Availability Statement: Not applicable.

Conflicts of Interest: The authors declare no conflict of interest.

Appendix A

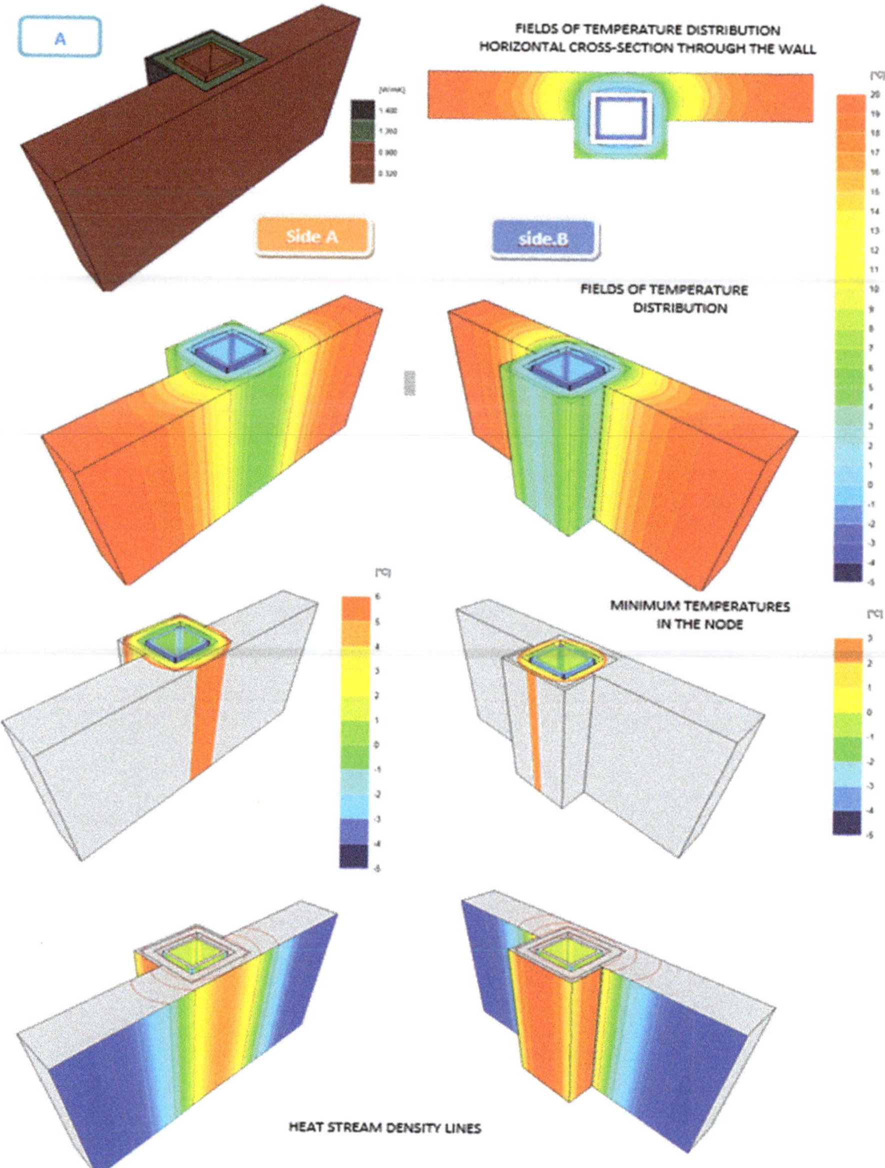

Figure A1. Scheme A: isotherm field distribution, minimum surface temperature, and heat flux distribution.

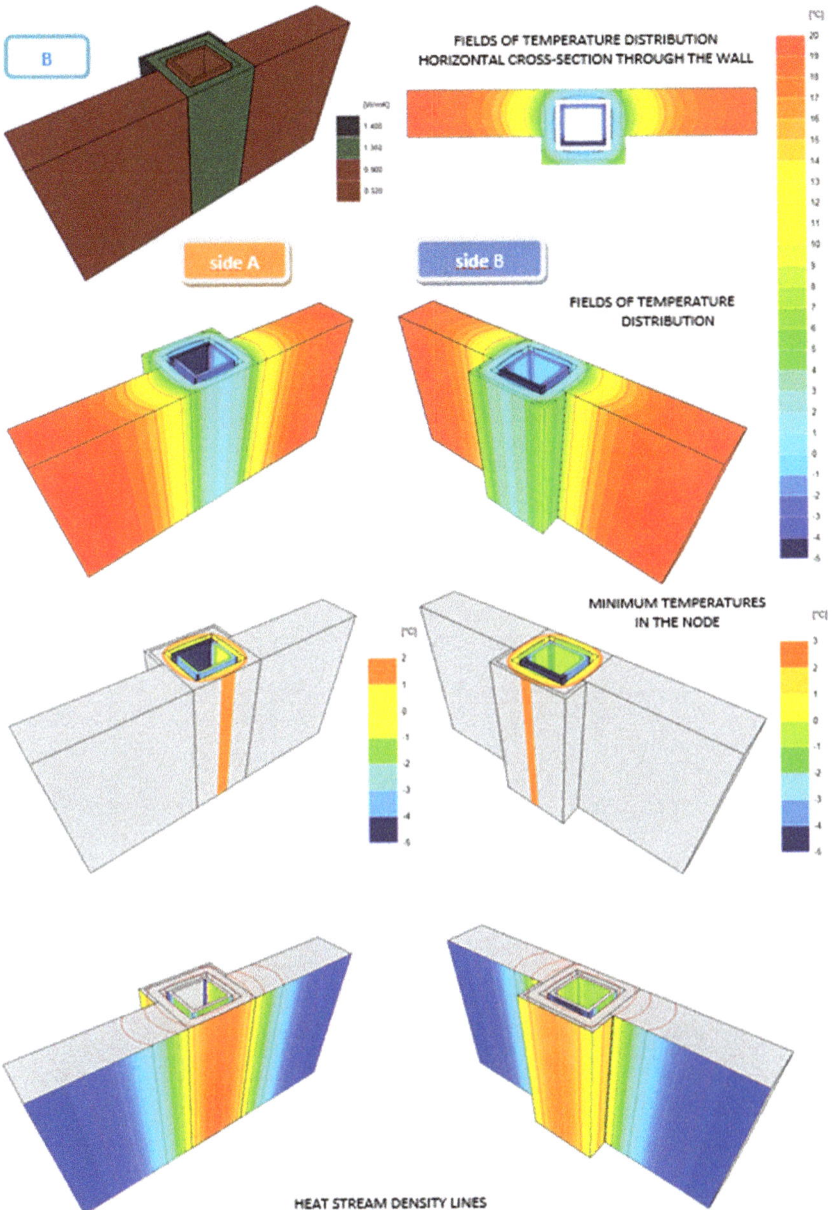

Figure A2. Scheme B: isotherm field distribution, minimum surface temperature, and heat flux distribution.

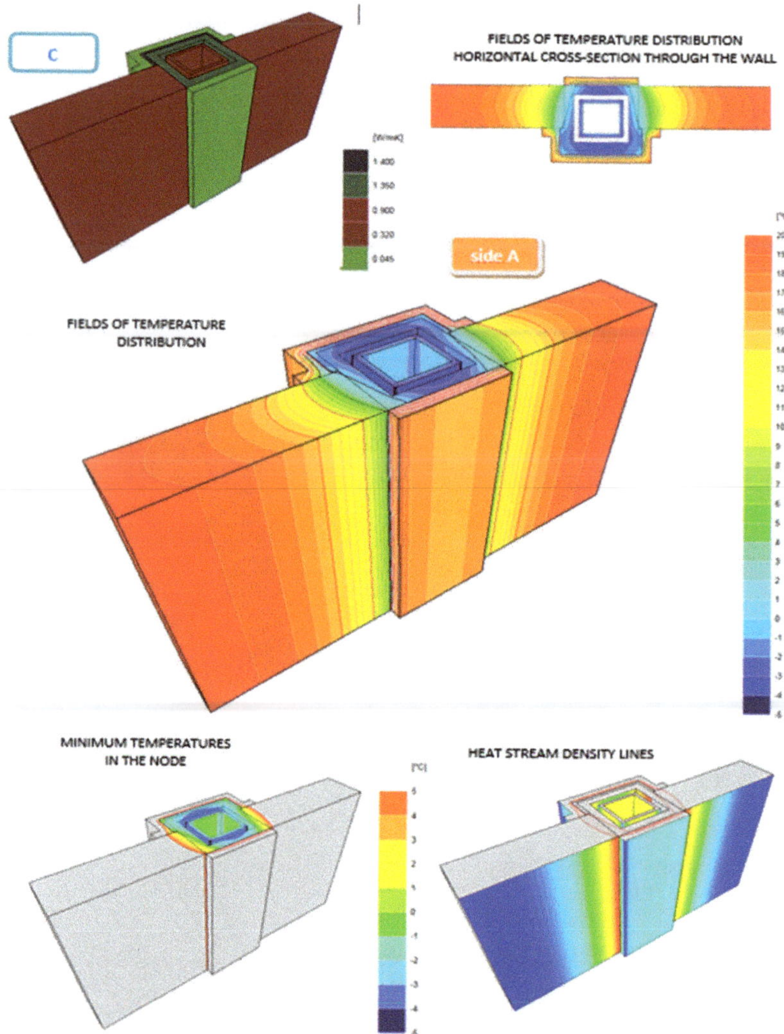

Figure A3. Scheme C: isotherm field distribution, minimum surface temperature, and heat flux distribution.

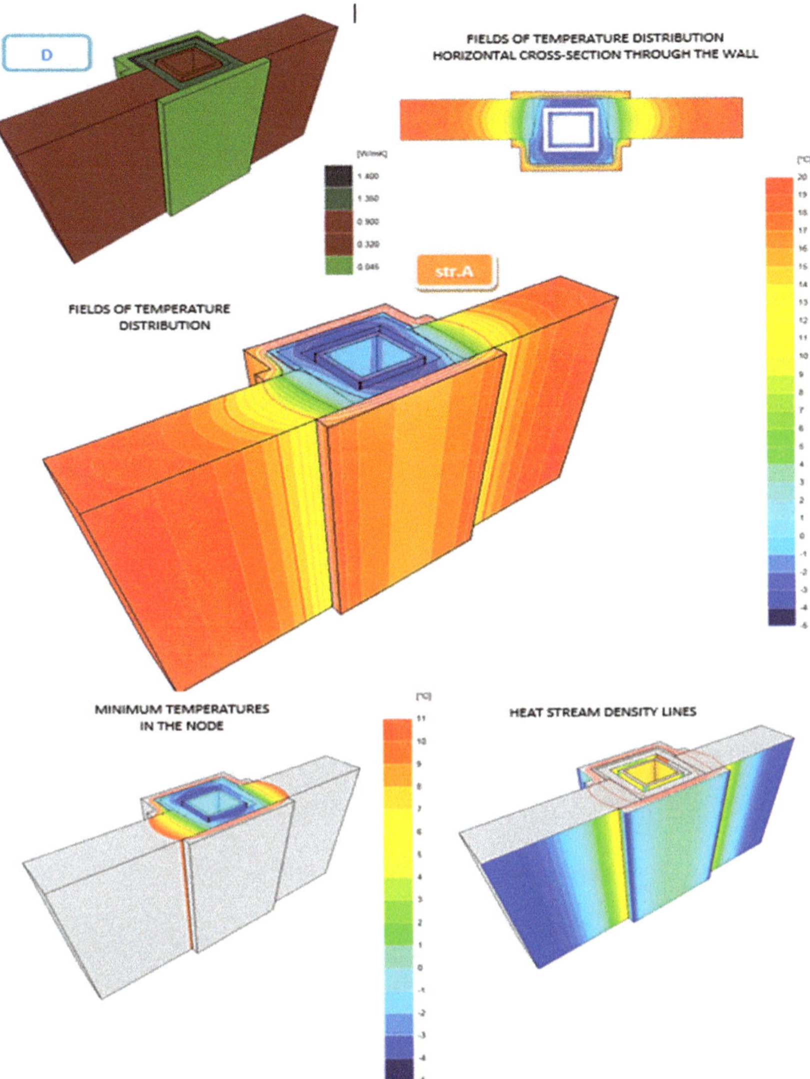

Figure A4. Scheme D: isotherm field distribution, minimum surface temperature, and heat flux distribution.

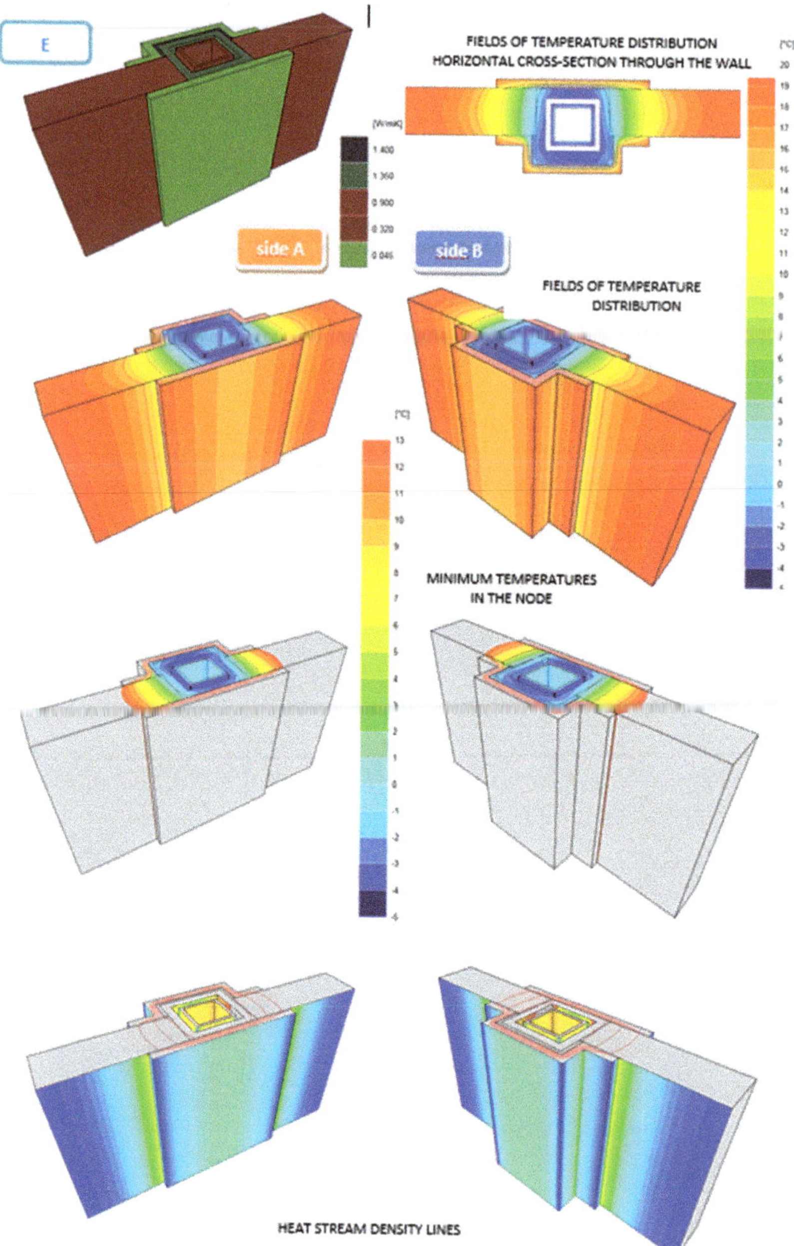

Figure A5. Scheme E: distribution of isotherms, minimum surface temperature, and heat flux distribution.

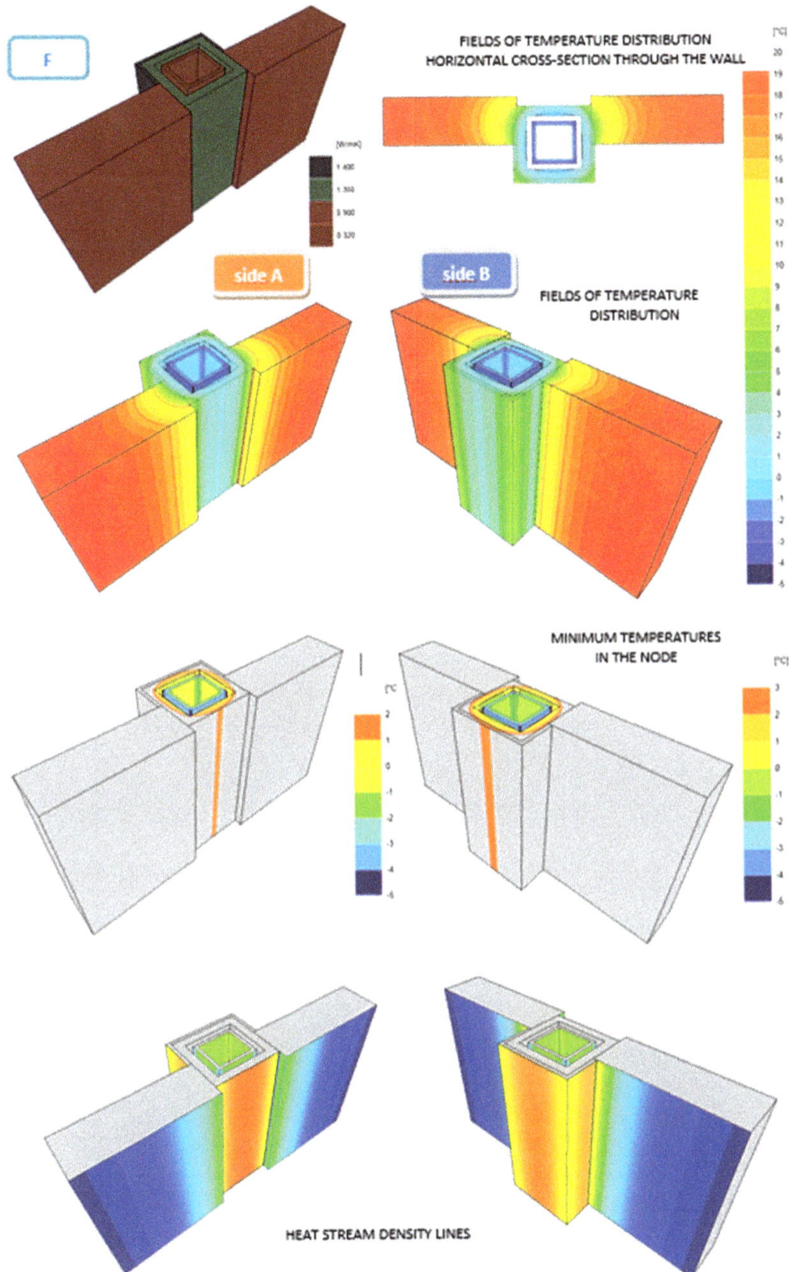

Figure A6. Scheme F: distribution of isotherm field, minimum surface temperature, and heat flux distribution.

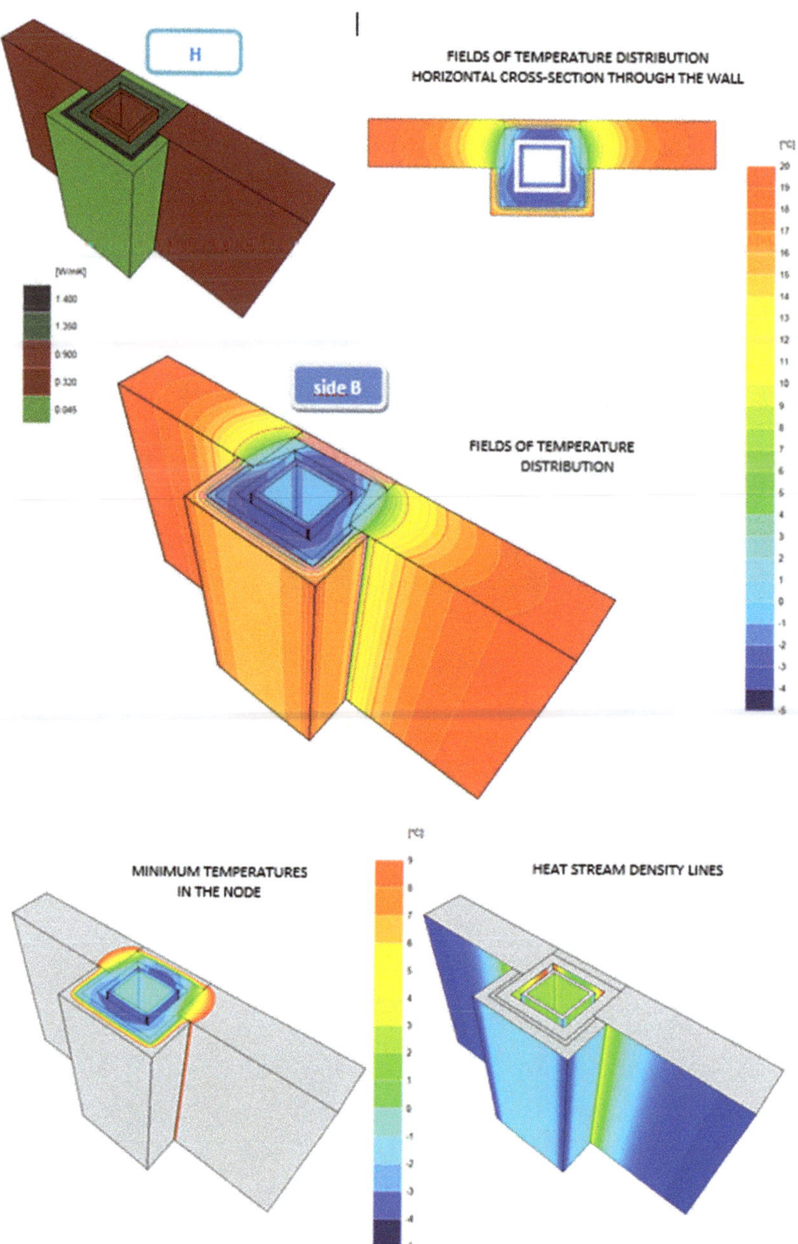

Figure A7. Scheme H: distribution of isotherm field, minimum surface temperature, and heat flux distribution.

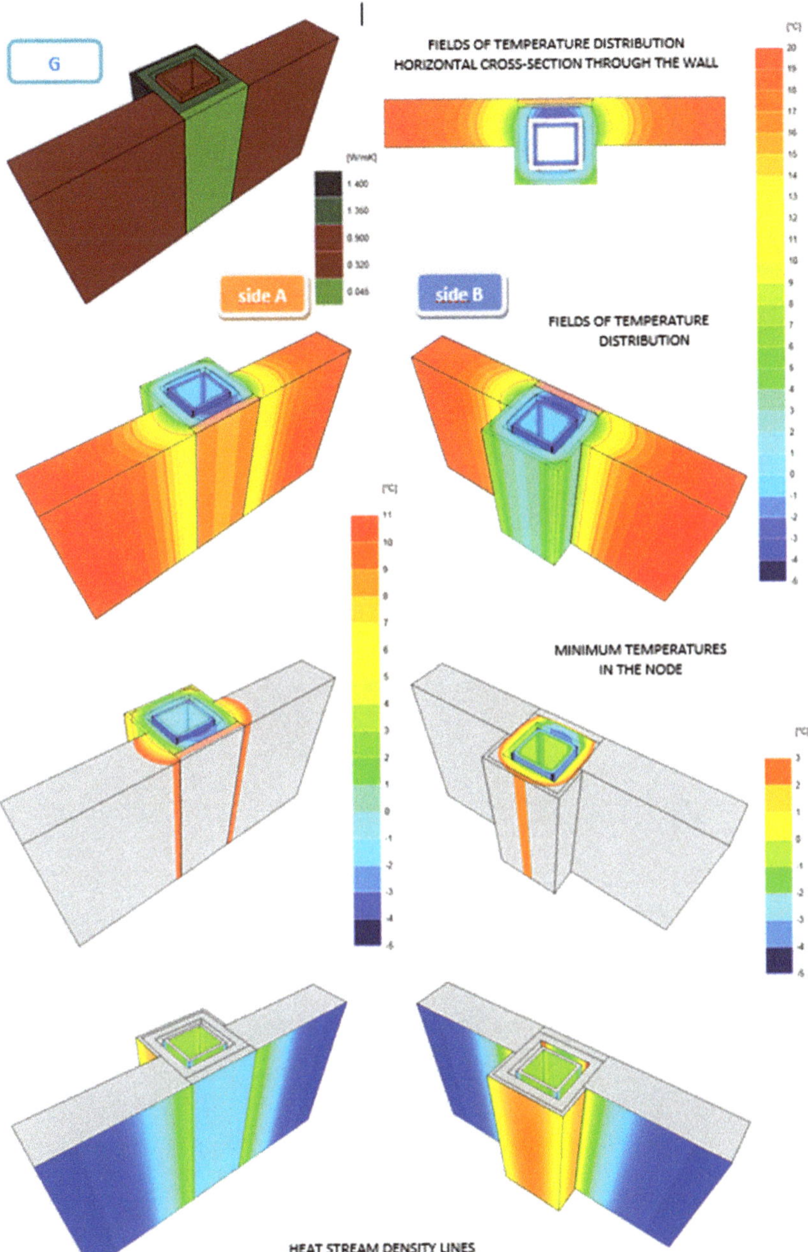

Figure A8. Scheme G: distribution of isothermal field, minimum surface temperature, and heat flux distribution.

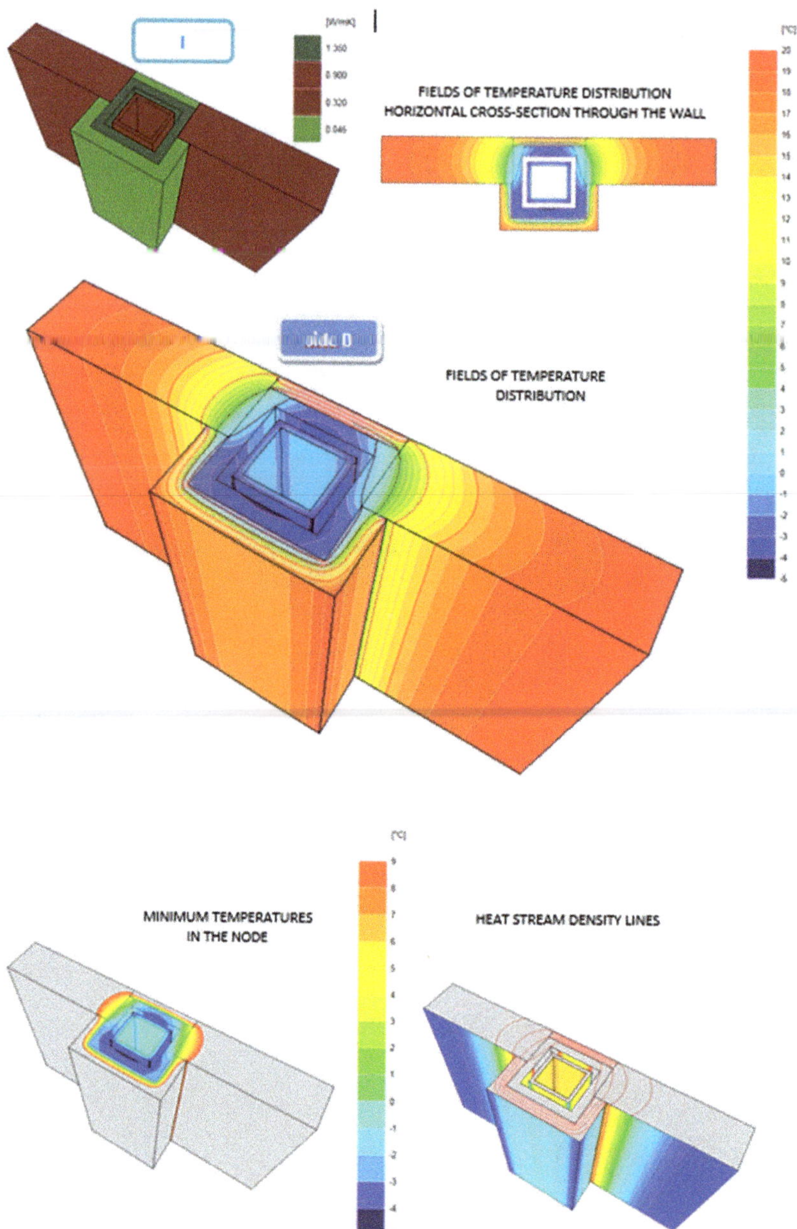

Figure A9. Scheme I: isotherm field distribution, minimum surface temperature, and heat flux distribution.

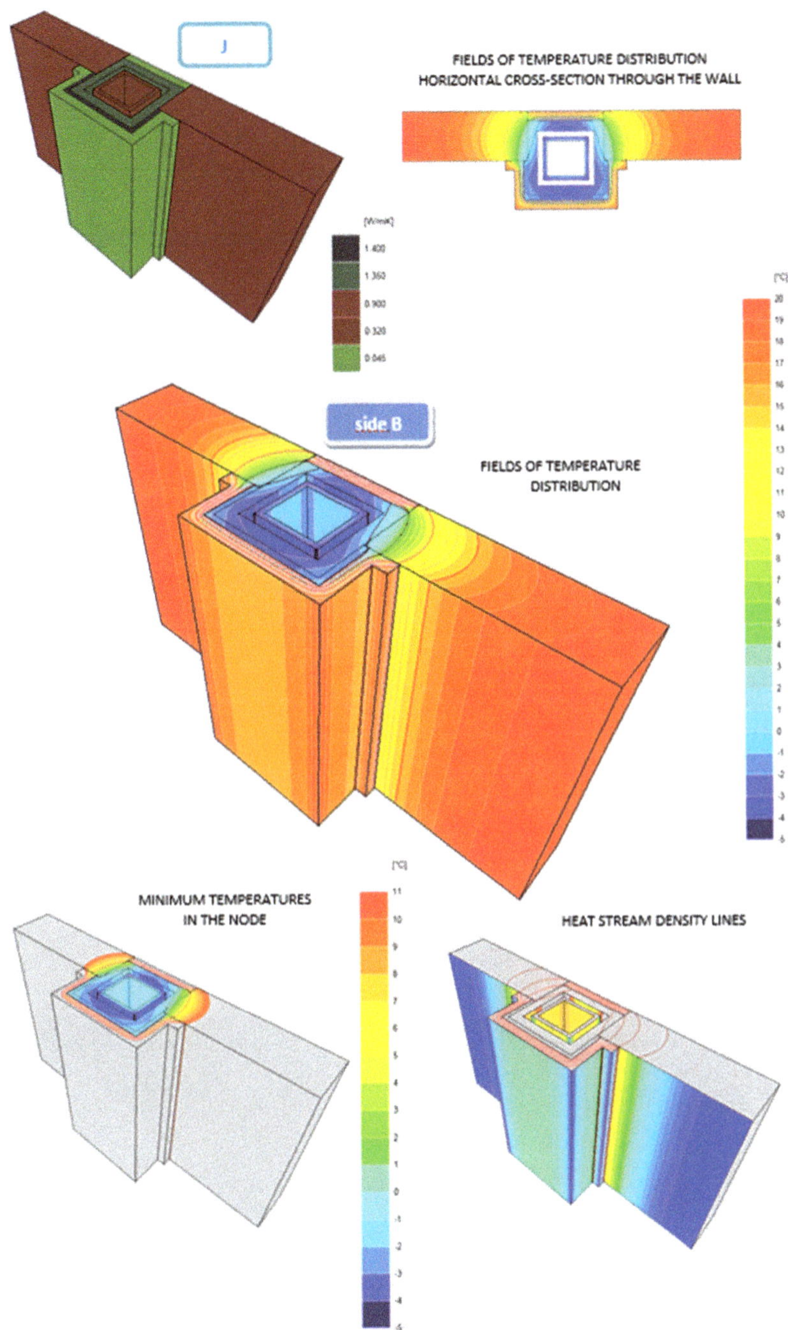

Figure A10. Scheme J: isotherm field distribution, minimum surface temperature, and heat flux distribution.

References

1. Nowak, R. Efektywność energetyczna w aspekcie innowacyjnych systemów kominowych. *GAZ WODA Tech. Sanit.* **2020**, *1*, 20–23. [CrossRef]
2. Salata, F.; Alippi, C.; Tarsitano, A.; Golasi, I.; Coppi, M. A First Approach to Natural Thermoventilation of Residential Buildings through Ventilation Chimneys Supplied by Solar Ponds. *Sustainability* **2015**, *7*, 9649–9663. [CrossRef]
3. Motyl, P.; Wikło, M.; Bukalska, J.; Piechnik, B.; Kalbarczyk, R. A New Design for Wood Stoves Based on Numerical Analysis and Experimental Research. *Energies* **2020**, *13*, 1028. [CrossRef]
4. Rezaei, A.; Samadzadegan, B.; Rasoulian, H.; Ranjbar, S.; Abolhassani, S.S.; Sanei, A.; Eicker, U. A New Modeling Approach for Low-Carbon District Energy System Planning. *Energies* **2021**, *14*, 1383. [CrossRef]
5. Lipnicki, Z.; Gortych, M.; Staszczuk, A.; Kuczyński, T.; Grabas, P. Analytical and Experimental Investigation of the Solar Chimney System. *Energies* **2019**, *12*, 2060. [CrossRef]
6. Hidaka, K.; Miyazawa, A.; Hu, H.; Mitsuji, K.; Nagai, Y.; Yoshimoto, N.; Suenaga, Y. Basic Study of Ventilation Using Semi-Transparent Organic Photovoltaic Sheets for Solar Chimney Systems. *Technologies* **2018**, *6*, 93. [CrossRef]
7. Han, C.; Li, G.; Ding, Y.; Yan, F.; Bai, L. Chimney Detection Based on Faster R-CNN and Spatial Analysis Methods in High Resolution Remote Sensing Images. *Sensors* **2020**, *20*, 4353. [CrossRef] [PubMed]
8. Daliga, K.; Kurałowicz, Z. Comparison of Different Measurement Techniques as Methodology for Surveying and Monitoring Stainless Steel Chimneys. *Geosciences* **2019**, *9*, 429. [CrossRef]
9. Manfren, M.; Sibilla, M.; Tronchin, L. Energy Modelling and Analytics in the Built Environment—A Review of Their Role for Energy Transitions in the Construction Sector. *Energies* **2021**, *14*, 679. [CrossRef]
10. Bonoli, A.; Zanni, S.; Serrano-Bernardo, F. Sustainability in Building and Construction within the Framework of Circular Cities and European New Green Deal. The Contribution of Concrete Recycling. *Sustainability* **2021**, *13*, 2139. [CrossRef]
11. Rutkowska, G.; Klepak, O. Problemy strat ciepła w istniejących budynkach jednorodzinnych w kontekście błędów wykonawczych. *Rocz. Ochr. Sr.* **2013**, *15*, 2625–2639.
12. Lee, K.H.; Strand, R.K. Enhancement of natural ventilation in buildings using a thermal chimney. *Energy Build.* **2009**, *41*, 615–621. [CrossRef]
13. Adamu, Z.A.; Price, A.D. Natural Ventilation with Heat Recovery: A Biomimetic Concept. *Buildings* **2015**, *5*, 405–423. [CrossRef]
14. Kim, J.; Kim, Y. Optimal Combination of External Wall Insulation Thickness and Surface Solar Reflectivity of Non-Residential Buildings in the Korean Peninsula. *Sustainability* **2021**, *13*, 3205. [CrossRef]
15. Zhang, D.; Yang, Y.; Pan, M.; Gao, Z. Toward a Heat Recovery Chimney. *Sustainability* **2011**, *3*, 2115–2128. [CrossRef]
16. PN-EN 1443:2019-05 Chimneys-General Requirements. Available online: https://sklep.pkn.pl/pn-en-1443-2019-05e.html (accessed on 30 July 2021).
17. Rahman, M.M.; Chu, C.-M. *Cold Inflow-Free Solar Chimney*; Rahman, M.M., Chu, C.-M., Eds.; Springer: Singapore, 2021; ISBN 978-981-33-6830-9.
18. Soto, A.; Martínez, P.J.; Tudela, J.A. Simulation and experimental study of residential building with north side wind tower assisted by solar chimneys. *J. Build. Eng.* **2021**, *43*, 102562. [CrossRef]
19. Drozdzol, K. CFD Thermal Analysis of a Three-layer Chimney Used in Residential Buildings. *J. Build. Eng.* **2021**, *44*, 102665. [CrossRef]
20. Kurpińska, M.; Karwacki, J.; Maurin, A.; Kin, M. Measurements of Thermal Conductivity of LWC Cement Composites Using Simplified Laboratory Scale Method. *Materials* **2021**, *14*, 1351. [CrossRef]
21. Antczak-Jarząbska, R.; Niedostatkiewicz, M. Wpływ nasady kominowej na poprawę sprawności użytkowej wentylacji grawitacyjnej. *Przegląd Bud.* **2017**, *88*, 48–52.
22. Pawłowski, K.; Nakielska., N. Selected design aspects of sections and joints in low energy buildings. *Bud. Zoptymalizowanym Potencjale Energ.* **2018**, *7*, 123–132. [CrossRef]
23. Saifi, N.; Settou, N.; Dokkar, B.; Negrou, B.; Chennouf, N. Experimental Study And Simulation Of Airflow In Solar Chimneys. *Energy Procedia* **2012**, *18*, 1289–1298. [CrossRef]
24. Leppänen, P.; Inha, T.; Pentti, M. An Experimental Study on the Effect of Design Flue Gas Temperature on the Fire Safety of Chimneys. *Fire Technol.* **2014**, *51*, 847–866. [CrossRef]
25. Pallarés, F.J.; Ivorra, S.; Pallarés, L.; Adam, J.M. State of the art of industrial masonry chimneys: A review from construction to strengthening. *Constr. Build. Mater.* **2011**, *25*, 4351–4361. [CrossRef]
26. Straatman, A.G.; Tarasuk, J.D.; Floryan, J.M. Heat Transfer Enhancement from a Vertical, Isothermal Channel Generated by the Chimney Effect. *J. Heat Transf.* **1993**, *115*, 395–402. [CrossRef]
27. Jörg, O.; Scorer, R. An experimental study of cold inflow into chimneys. *Atmospheric Environ.* **1967**, *1*, 645–654. [CrossRef]
28. McCloseKy, F.G.; The Northern Ireland Housing Executive. Failures in traditionaly constructed domestic chimneys. In Proceedings of the 8th International Brick/Block Masonry Conference, Dublin, Ireland, 19–21 September 1988; pp. 1732–1738.
29. Zhang, H.; Yang, D.; Tam, V.W.; Tao, Y.; Zhang, G.; Setunge, S.; Shi, L. A critical review of combined natural ventilation techniques in sustainable buildings. *Renew. Sustain. Energy Rev.* **2021**, *141*, 110795. [CrossRef]
30. Corradi, M.; DI Schino, A.; Borri, A.; Rufini, R. A review of the use of stainless steel for masonry repair and reinforcement. *Constr. Build. Mater.* **2018**, *181*, 335–346. [CrossRef]

31. Hartinger, S.M.; Commodore, A.A.; Hattendorf, J.; Lanata, C.F.; Gil, A.I.; Verastegui, H.; Aguilar-Villalobos, M.; Mäusezahl, D.; Naeher, L.P. Chimney stoves modestly improved Indoor Air Quality measurements compared with traditional open fire stoves: Results from a small-scale intervention study in rural Peru. *Indoor Air* **2013**, *23*, 342–352. [CrossRef]
32. Maj, M.; Ubysz, A.; Hammadeh, H.; Askifi, F. Non-Destructive Testing of Technical Conditions of RC Industrial Tall Chimneys Subjected to High Temperature. *Materials* **2019**, *12*, 2027. [CrossRef]
33. Al-Absi, Z.; Hafizal, M.; Ismail, M.; Ghazali, A. Towards Sustainable Development: Building's Retrofitting with PCMs to Enhance the Indoor Thermal Comfort in Tropical Climate, Malaysia. *Sustainability* **2021**, *13*, 3614. [CrossRef]
34. Leng, P.; Ling, G.H.T.; Ahmad, M.; Ossen, D.; Aminudin, E.; Chan, W.; Tawasil, D. Thermal Performance of Single-Story Air-Welled Terraced House in Malaysia: A Field Measurement Approach. *Sustainability* **2020**, *13*, 201. [CrossRef]
35. Ustawa z 10 Września 2015 r. o Zmianie Ustawy–Prawo Ochrony Środowiska. Dziennik Ustaw RP z 12 Października 2015 r. poz. 1593. Available online: https://isap.sejm.gov.pl/isap.nsf/DocDetails.xsp?id=WDU20150001593 (accessed on 30 July 2021).
36. Ustawa z Dnia 7 Lipca 1994 r. Prawo Budowlane (Dz. U. z 2018 r. poz. 1202. 1276. 1496 i 1669). Available online: http://isap.sejm.gov.pl/isap.nsf/DocDetails.xsp?id=WDU20180001202 (accessed on 30 July 2021).
37. Rozporządzanie Ministra Infrastruktury z Dnia 12 Kwietnia 2002 r. w Sprawie Warunków Technicznych. Jakim Powinny Odpowiadać Budynki i Ich Usytuowanie (Tekst Jednolity Dz. U. z 2017 r. poz. 1523). Available online: https://isap.sejm.gov.pl/isap.nsf/DocDetails.xsp?id=WDU20020750690 (accessed on 30 July 2021).
38. Rozporządzenie Ministra Transportu. Budownictwa i Gospodarki Morskiej z Dnia 5 Lipca 2013 r. Zmieniające rozporządzenie w sprawie warunków technicznych. Jakim powinny Odpowiadać Budynki i Ich Usytuowanie (Dz. U. Poz. 926). Available online: http://isap.sejm.gov.pl/isap.nsf/DocDetails.xsp?id=WDU20130000926 (accessed on 30 July 2021).
39. Rozporządzenie Parlamentu Europejskiego i Rady (UE) nr 305/2011 z Dnia 9 Marca 2011 r. Ustanawiające Zharmonizowane Warunki Wprowadzania do Obrotu Wyrobów Budowlanych i Uchylające Dyrektywę Rady 89/106/EWG. Available online: https://eur-lex.europa.eu/legal-content/PL/TXT/PDF/?uri=CELEX:02011R0305-20140616&qid=1576663491684&from=EN (accessed on 30 July 2021).
40. Dyrektywa Parlamentu Europejskiego i Rady 2010/30/UE z Dnia 19 Maja 2010 r. w Sprawie Wskazania Poprzez Etykietowanie Oraz Standardowe Informacje o Produkcie. Zużycia Energii Oraz Innych Zasobów Przez Produkty Związane z Energią. Available online: https://eur-lex.europa.eu/legal-content/PL/TXT/HTML/?uri=OJ:L:2010:153:FULL&from=NL (accessed on 30 July 2021).
41. Dyrektywa 5 Parlamentu Europejskiego i Rady 2005/32/WE z Dnia 6 Lipca 2005. Available online: https://eur-lex.europa.eu/legal-content/PL/TXT/PDF/?uri=CELEX:32005L0032&from=en (accessed on 30 July 2021).
42. Ustawa z Dnia 29 Sierpnia 2014 r. o Charakterystyce Energetycznej Budynków (Dz. U. poz. 1200 oraz z 2015 r. poz. 151). Available online: http://isap.sejm.gov.pl/isap.nsf/DocDetails.xsp?id=wdu20140001200 (accessed on 30 July 2021).
43. Dyrektywa Parlamentu Europejskiego i Rady 2009/125/WE z dnia 21 Października 2009 r. Ustanawiająca Ogólne Zasady Ustalania Wymogów Dotyczących Ekoprojektu dla Produktów Związanych z Energią. Available online: https://eur-lex.europa.eu/legal-content/PL/TXT/PDF/?uri=CELEX:32009L0125&from=ES (accessed on 30 July 2021).
44. Directive 2005/32/EC of the European Parliament and of the Council of 6 July 2005 Establishing a Framework for the Setting of Ecodesign Requirements for Energy-Using Products and Amending Council Directive 92/42/EEC and Directives 96/57/EC and 2000/55/EC of the European Parliament and of the Council. Available online: https://eur-lex.europa.eu/LexUriServ/LexUriServ.do?uri=OJ:L:2005:191:0029:0058:EN:PDF (accessed on 30 July 2021).
45. *Green Paper—A European Strategy for Sustainable, Competitive and Secure Energy*; Official Journal of the European Union: Brussels, Belgium, 2006.
46. Bajno, D. Ventilation as a particular safety element in housing development. In *Monografia Rudolf Kania Opole*; Polish Engineering Association (NOT): Vienna, Austria, 2011; pp. 173–180.
47. PN-EN 1859+A1:2013-09 Kominy-Metody Badań. Available online: https://sklep.pkn.pl/pn-en-1859-a1-2013-09p.html (accessed on 30 July 2021).
48. Qiao, M.; Xu, F.; Saha, S.C. Scaling analysis and numerical simulation of natural convection from a duct. *Numer. Heat Transfer. Part A Appl.* **2017**, *72*, 355–371. [CrossRef]
49. Fedorczak-Cisak, M. Efektywność energetyczna oraz niska emisja w krajowych i regionalnych uwarunkowaniach prawnych. *Mater. Bud.* **2016**, *1*, 22–24. [CrossRef]
50. Ujma, A.; Kysiak., A. Diagnostyka elementów konstrukcji budynków z wykorzystaniem kamery termowizyjnej. *Bud. Zoptymalizowanym Potencjale Energ.* **2015**, *1*, 182–190.
51. Knyziak, P. Sztuczne sieci neuronowe w oszacowaniu zużycia technicznego bydynków mieszkalnych. In *Theoretical Foundation of Civil Engineering*; Oficyna Wydawnicza Politechniki Warszawskiej: Warszawa, Poland, 2008; Volume 16.
52. Osowski, S. Sztuczne sieci neuronowe-Podstawowe struktury sieciowe i algorytmy uczące. *Prz. Elektrotechniczny* **2009**, *85*, 1–8.
53. Żurada, J.; Barski, M.; Jędruch, W. Sztuczne sieci neuronowe. *Ann* **2004**, *1*, 16–19.
54. Chithra, S.; Kumar, S.S.; Chinnaraju, K.; Ashmita, F.A. A comparative study on the compressive strength prediction models for High Performance Concrete containing nano silica and copper slag using regression analysis and Artificial Neural Networks. *Constr. Build. Mater.* **2016**, *114*, 528–535. [CrossRef]
55. Jasienko, J.; Bednarz, Ł.; Nowak, T. The effectiveness of strengthening historic brick vaults by contemporary methods. In Proceedings of the Protection of Historical Buildings, PROHITECH 09, Rome, Italy, 21–24 June 2009; pp. 1299–1304.

56. Bajno, D.; Bednarz, L.; Matkowski, Z.; Raszczuk, K. Monitoring of Thermal and Moisture Processes in Various Types of External Historical Walls. *Materials* **2020**, *13*, 505. [CrossRef]
57. Bajno, D. Defekty konstrukcyjno–budowlane w procesie realizacji oraz na etapie eksploatacji tradycyjnych kominów w budynkach mieszkalnych. In *Szkody Kominowe*; Rudolf Kania: Opole, Poland, 2015; pp. 103–120.
58. ISO 12011-Technical Product Documentation-Technical Drawings; Lines-Part 1: Basic Conventions. Available online: https://www.iso.org/obp/ui/#iso:std:iso:128:-1:ed-1:en (accessed on 30 July 2021).
59. PN-EN ISO 6946: 2008 Komponenty Budowlane i Elementy Budynku. Opór Cieplny i Współczynnik Przenikania Ciepła. Metoda Obliczania. Available online: http://sklep.pkn.pl/pn-en-iso-6946-2008d.html (accessed on 30 July 2021).

Article

Assessment of the Development Potential of Post-Industrial Areas in Terms of Social, Economic and Environmental Aspects: The Case of Wałbrzych Region (Poland)

Beata Raszka, Halina Dzieżyc and Maria Hełdak *

Institute of Spatial Management, Wrocław University of Environmental and Life Sciences, ul. Grunwaldzka 55, 50-357 Wrocław, Poland; beata.raszka@upwr.edu.pl (B.R.); halina.dziezyc@upwr.edu.pl (H.D.)
* Correspondence: maria.heldak@upwr.edu.pl

Abstract: The purpose of this research was to assess the development potential of cities and municipalities in Wałbrzych County, approached from the perspective of social, economic and environmental potential. Comparisons were performed for the periods 2002–2004 and 2016–2018. The following problem questions were formulated in the study: Which strategic goals can currently act as incentives for taking action in the analysed post-industrial areas, and for which ones should development be particularly strengthened? What development direction should be taken by individual municipalities in Wałbrzych County? The research covering the development potential of the municipalities was conducted using the multidimensional analysis method, and the similarities in municipal potential were analysed taking into account a comparison of distances between individual diagnostic variables. Among the analysed municipalities, in terms of social, economic and environmental potential, Szczawno-Zdrój and Czarny Bór achieved the best results, and Boguszów Gorce ranked the worst. In some municipalities, a noticeable increase in social potential (Jedlina-Zdrój, Mieroszów, Walim) or economic potential (Jedlina-Zdrój) was observed, and a significant decline in economic potential in Stare Bogaczowice was seen. As a result of the research, the following was established: the policies of municipal authorities have to focus on improving the living conditions of residents; the crucial factors are opening new jobs, providing appropriate living conditions and services and offering diverse sports and tourism options. Efforts should be made to take advantage of the inherent potential in the area, also by highlighting the preserved post-industrial buildings and constructions.

Keywords: environmental potential; development potential; social potential; municipal development indicators; Wałbrzych region

Citation: Raszka, B.; Dzieżyc, H.; Hełdak, M. Assessment of the Development Potential of Post-Industrial Areas in Terms of Social, Economic and Environmental Aspects: The Case of Wałbrzych Region (Poland). *Energies* **2021**, *14*, 4562. https://doi.org/10.3390/en14154562

Academic Editors: Sergey Zhironkin and Michal Cehlar

Received: 25 May 2021
Accepted: 18 July 2021
Published: 28 July 2021

Publisher's Note: MDPI stays neutral with regard to jurisdictional claims in published maps and institutional affiliations.

1. Introduction

1.1. Theoretical Foundations of Research Concept

Integrating social, economic and environmental phenomena remains the basis for developing an integrated order to create a system of strategic goals related to social, economic, environmental, institutional and political issues. That can be implemented through social, economic and cultural capital, as well as protection of the natural environment. When managing local development, local governments follow the statutory legal regulations that provide tools for influencing the local environment. Management focused on meeting the needs and aspirations of society requires formulating economic, environmental and social goals. The latter are related to the desire to improve living conditions, which will determine the standards of residents' current life and their individual development perspectives [1].

The interdisciplinary approach to research is widely used in relation to making investment or political decisions. In numerous instances, policymakers and funding agencies call for researchers to address the grand societal challenges of globalized economic, ecological and demographic changes [2,3]. According to Reed et al., focusing efforts on only a single challenge may result in negative social or environmental outcomes [4].

Therefore, it is important to consider the problem of economic transformations much more extensively, as dictated by the contemporary problems of environmental protection in the context of energy technology, in addition to social and economic development.

As noted by Hełdak and Raszka [5], after Kozłowski [6] and Pęski [7], planning is one of the most important tools in the management process, is also used as a tool to form the future image of a country or a region, and, in the case of Poland, is decisive in terms of the quality of people's lives and the functioning of the natural environment.

The decisions made locally and at higher levels of government should consider at least the environmental, social and economic aspects that are addressed in the presented research.

1.2. Effects of State Policy

The systemic transformation initiated in Poland in 1989 created new development opportunities; however, in some regions of the country, including the Wałbrzych region (part of the Lower Silesia Voivodship), it contributed to the decline of the leading industries [8]. The economic recession in the Wałbrzych region was aggravated by the closure of hard coal mines, which were the primary places of employment for the local community. The last mine was closed in 1997, and the entire restructuring process took only five years.

According to Buczak [9], the transformations that took place in Wałbrzych after the mines closed caused chaos in their initial phase; both the society and city authorities found it difficult to function in the new reality, while the city, with a growing unemployment rate and social dissatisfaction, was perceived as a black spot on the map of Poland, the centre of social pathology.

Economic transformations, including high levels of unemployment, are directly related to the deterioration of the quality of life of local residents. Along with the collapse of the economy, many people were forced to migrate in order to search for work and many families required financial support from the state, and the consequences of political transformation are still experienced today. The influence of the extractive industry is well described in the source literature. Based on their research, Karakaya and Nuur [10] stated that the environmental effects of mining have been a key topic in academic discourse for decades [11,12].

There is a noticeable trend towards encouraging mining companies to take the necessary remedial actions in connection with the closure of mines and to monitor the situations in local communities. It is important to focus on whether the existing restructuring plans fully address the environmental, social and economic conditions in terms of providing care and opportunities for relocated workers, and regarding the implications for government and other actors at all levels [13].

According to Solomon et al. [14], the social dimension of the extractive industry is increasingly recognized as essential to business success; however, it remains the least understood aspect of the business concept of sustainable development, the "triple bottom line" of economy, environment and society.

1.3. Quality of Life in Research

Research covering the differentiation of living conditions makes a distinction between the standard of living, which is a reflection of the degree of fulfilment of material needs, and quality of life, which results from the level of satisfaction obtained from various spheres of life or areas of activity [15–19]. Quality of life is an interdisciplinary concept referring to many different aspects, including the condition of the natural environment and owning material and non-material goods [20–22].

Research addressing quality of life not only can be useful for exploration-oriented purposes in terms of infrastructure, but also can be used for better planning and assessment of effectiveness regarding implemented social and economic policies, e.g., monitoring the use of financial resources. The multifaceted nature of the quality of life concept can be reflected in statistical measurements [23] using various indicators, including material living conditions (e.g., indicators of poor sanitary conditions) and the environment quality in residential areas (e.g., exposure to excessive noise and pollution, satisfaction with recreational and green areas). Many of these indicators fall within the scope of those used to assess sustainable development [24].

Various diagnostic variables defined by the indicators of social, economic and environmental development should be taken into account within the framework of broadly understood living conditions.

The purpose of this research was to assess the development potential of cities and municipalities of Wałbrzych County, approached from the perspective of social, economic and environmental potential. Poland faces the challenge of more mines being closed, and this research and the point of view presented in the discussion highlight the problems accompanying the economic transformations. Changes in energy industry policies will affect other regions of Poland where the energy industry is based on coal mining in the same way. In this context, this research could have national impact. An important element of the research is highlighting the residents' quality of life in areas affected by the economic turndown resulting from the country's political transformation and entering the free market economic system.

2. Materials and Methods

The spatial scope of the research covers an area in Lower Silesia, Poland, in the region of the former Wałbrzych Voivodship. All municipalities located in Wałbrzych County were selected for analysis: urban: Boguszów-Gorce, Jedlina-Zdrój and Szczawno-Zdrój; urban–rural: Głuszyca and Mieroszów; and rural: Czarny Bór, Stare Bogaczowice and Walim. Before the period of systemic transformation, this region was dominated by production, much of it dating back to the period before the Second World War. At the end of the 1980s, the period of industrial development in Poland came to an end.

In the studied area, excessively developed industry, characterised by inadequate sector structure and technological underdevelopment, ceased to be the leading development factor [25]. In addition, the closure of hard coal mines resulted in high unemployment rates and the depopulation of cities around Wałbrzych. The conducted analyses show the negative socio-economic effects of the collapse of many industries in the region. In general terms, the restructuring process brought about catastrophic economic consequences, although they could be perceived as beneficial for the natural environment [26–28].

The reclamation of post-mining areas, including the management of mining waste dumps, remains a significant problem in the Wałbrzych region, and the return of some municipalities to the pre-transformation level of development turned out to be a long-term process and still has not been accomplished.

The authors of the study put forward the following problem questions:

- What is the current social, economic and environmental potential of Wałbrzych County?
- Which strategic goals can currently act as incentives for action in the studied post-industrial areas, and for which ones should development be particularly strengthened?
- What development direction should be taken by individual municipalities in Wałbrzych County?

The research addressing the development potential of the municipalities was carried out using a multidimensional analysis method, zero unitarization. It allows comparisons of objects described by diagnostic variables of different nature. The method normalizes these variables and creates a synthetic variable, which enables the analysed objects to be ordered according to it.

The normalization of diagnostic variables is performed in line with the following formulas: For diagnostic variables that have a positive impact on the analysed phenomenon (stimulants):

$$z_{ij} = \frac{x_{ij} - \min_{i} x_{ij}}{\max_{i} x_{ij} - \min_{i} x_{ij}} \quad \text{for } j \text{ corresponding to stimulants} \tag{1}$$

For variables having a negative impact on the analysed phenomenon (destimulants):

$$z_{ij} = \frac{\max_{i} x_{ij} - x_{ij}}{\max_{i} x_{ij} - \min_{i} x_{ij}} \quad \text{for } j \text{ corresponding to destimulants} \tag{2}$$

where $i = 1, 2, \ldots, r$ and $j = 1, 2, \ldots, s$; r is the number of objects and s is the number of diagnostic variables.

The synthetic variable (indicator) for the ith object is determined according to the following formula:

$$Q_i = \frac{1}{s} \sum_{j=1}^{s} z_{ij} \tag{3}$$

Next, the value of U can be determined:

$$U = \frac{\max_{i} Q_i - \min_{i} Q_i}{3} \tag{4}$$

Further, the categories of objects can be established as follows:
I. Objects with the highest indicator values:

$$Q_i \epsilon \left(\max_{i} Q_i - U, \ \max_{i} Q_i \right]$$

II. Objects with average indicator values:

$$Q_i \epsilon \left(\max_{i} Q_i - 2U, \ \max_{i} Q_i - U \right]$$

III. Objects with the lowest indicator values:

$$Q_i \epsilon \left[\min_{i} Q_i, \ \max_{i} Q_i - 2U \right]$$

In addition to determining the total potential indicator, the similarities in municipal potential were also examined by comparing the distances of individual diagnostic variables. The agglomeration method was used, which creates a tree diagram that allows grouping the analysed objects into clusters. Euclidean distance was applied as the distance measure and Ward's algorithm as the clustering method.

In determining the measures and distances, reference was made to the source literature [29,30] and subsequent research using the above-mentioned methods [31].

Most of the variables were taken directly from the Local Data Bank of Statistics Poland [32], which confirms their credibility, and only a few were calculated based on the same data parameters, e.g., some variables were defined per capita. The zero unitarization method was presented in an Excel spreadsheet.

The calculations were performed for standardized variables (TIBCO Software Inc.) [33].

The diagnostic variables that were included in the research are shown in Table 1.

Table 1. Diagnostic variables adopted in the research.

Diagnostic Variable Groups	Symbol	Diagnostic Variables
Social potential	S1	birth rate per 1000 people (person)
	S2	migration balance per 1000 residents (person)
	S3	share of pre-working age population as % of total population (%)
	S4	share of working age population as % of total population (%)
	S5	working population per 1000 residents (person)
	S6	share of registered unemployed in working age population (%)
	S7	expenditure on health care per capita in municipal budget (PLN)
	S8	expenditure on education per capita in municipal budget (PLN)
	S9	expenditure on culture and protection of national heritage per capita in municipal budget (PLN)
	S10	health care facilities (outpatient entities) per 10,000 residents (pcs)
Economic potential	EC1	revenues of municipal budgets per capita (PLN)
	EC2	expenditures of municipal budgets per capita (PLN)
	EC3	funding and co-funding EU programmes and projects per capita (PLN)
	EC4	share of municipal investment expenditure in total expenditure of municipal budget (%)
	EC5	expenditure on residential housing management per capita (PLN)
	EC6	expenditure on transport and communication per capita (PLN)
	EC7	expenditure on tourism per capita (PLN)
	EC8	expenditure on agriculture per capita (PLN)
	EC9	length of bike paths (km)
	EC10	business entities (from REGON (statistical business identification number) register) per 10,000 residents (pcs)
Environmental potential	EN1	expenditure on municipal management and environmental protection per capita (PLN)
	EN2	water consumption per capita (m^3)
	EN3	length of active sewage network (km)
	EN4	population using water supply network vs. total population (%)
	EN5	population using sewage network vs. total population (%)
	EN6	population using sewage treatment plants vs. total population (%)
	EN7	share of forest land in total area of municipality (%)
	EN8	population using gas vs. total population (%)
	EN9	share of parks, lawns and housing estate green areas in total area (%)

Source: authors' compilation.

The share of registered unemployed in the working age population was referred to as a destimulant, and the remaining variables as stimulants.

The research was carried out for the present period, taking into account the average diagnostic parameters from 2016–2018 and the beginning of the 21st century and the average parameters from 2002–2004. The averaging was performed because of certain variability over the years, primarily regarding the financial parameters. For the period 2002–2004, because there were no data, the following variables were not taken into account: employed per 1000 residents, expenditure on tourism per 1 resident and the length of bike paths. The characteristics of the variables used in the calculations covering the period 2016–2018 compared to the data of the Lower Silesia Voivodship and the entire country are presented in Table 2.

Table 2. Characteristics of variables included in research methods against the background of data from Lower Silesia Voivodship and the entire country.

Average for 2016–2018	Average Value for Municipalities of Wałbrzych County	Minimum Value for Municipalities of Wałbrzych County	Maximum Value for Municipalities of Wałbrzych County	Value for Lower Silesia Voivodship	Value for Country
S1	−5.2	−9.9	0.1	−1.4	0.5
S2	0.7	−3.2	4.7	1.1	−1.1
S3	15.9	13.5	18.9	16.9	18.0
S4	61.9	59.2	63.5	61.2	61.2
S5	118.2	43.3	280.3	389.3	407.3
S6	5.4	3.8	7.7	4.1	4.9
S7	23.5	13.0	64.9	26.7	16.2
S8	938.6	696.9	1253.2	877.4	903.9
S9	177.7	106.3	312.3	135.8	93.6
S10	4.6	1.9	8.8	5.3	5.6
EC1	2008.6	1318.1	3417.7	1603.3	1264.3
EC2	4492.3	3905.6	5677.7	3076.7	2925.9
EC3	162.5	74.7	318.3	162.7	
EC4	18.9	11.5	30.9	15.6	
EC5	392.6	77.9	672.8	121.9	71.8
EC6	371.5	126.9	918.4	150.7	246.9
EC7	65.2	0.0	209.2		
EC8	17.8	0.7	50.9	45.1	65.4
EC9	0.9	0.0	3.9	780.2	14,714.7
EC10	1161.0	828.0	2004.0	1269.0	1120.0
EN1	510.5	405.0	821.5		
EN2	23.5	14.0	36.9	35.6	32.4
EN3	25.1	18.1	48.1	11,485.2	157,166.7
EN4	88.3	61.4	100.0	94.9	
EN5	56.4	20.3	95.9	76.5	
EN6	62.2	8.2	97.3	80.9	73.7
EN7	3.9	0.9	9.8	30.9	30.9
EN8	50.1	0.4	91.0	61.2	
EN9	0.5	0.0	2.5		0.2

Source: Authors' calculations based on bdl.stat.gov.pl (accessed on 30 October 2020) [32].

A certain limitation in the research was the lack of access to some data referring to the whole country; these data were not published at the time of the research. It was not possible to determine the position of the analysed municipalities against the background of Poland in relation to some of the variables, but this did not affect the essence of the research.

In order to better illustrate which municipalities recorded strengthened development potential in relation to other municipalities and in which ones this potential declined, scatter plots were prepared showing the indicator of potential for 2002–2004 vs. 2016–2018.

3. Results

In order to identify municipal development potential, the research findings were analysed. For each diagnostic variable, its least favourable state (for stimulants it is min x'' and for destimulants it is max x'') has a value of zero. The state considered the most favourable (for stimulants it is max x' and for destimulants it is min x') has the highest value in the variability range of the normalized variables, i.e., unity [34–37].

The established indicators of social development potential in the municipalities of Wałbrzych County are presented in Table 3.

Table 3. Indicators of social development potential in municipalities of Wałbrzych County calculated using zero unitarization method.

Municipality	Years 2002–2004			Years 2016–2018		
	Indicator	Ranking Position	Category	Indicator	Ranking Position	Category
Boguszów-Gorce	0.253	8	III	0.184	8	III
Jedlina-Zdrój	0.320	6	III	0.372	7	II
Szczawno-Zdrój	0.500	2	I	0.560	2	I
Czarny Bór	0.539	1	I	0.604	1	I
Głuszyca	0.469	3	I	0.373	6	II
Mieroszów	0.324	5	III	0.389	5	II
Stare Bogaczowice	0.422	4	II	0.538	3	I
Walim	0.316	7	III	0.417	4	II

Source: Authors' calculations based on bdl.stat.gov.pl (accessed on 30 October 2020).

Among the diagnostic variables, the following were analysed: birth rate; migration; share of pre-working age population; share of working age population; working population; unemployment rate; and expenditure on culture, health and education, and health care facilities. In terms of social potential, the two analysed periods 2002–2004 and 2016–2018, Boguszów-Gorce municipality performed very poorly, ranking last with values of 0.253 and 0.189, respectively. Considering the possible numerical value in the range [0, 1], the result achieved here is alarmingly low. Significantly worse indicators were recorded in the Głuszyca municipality, which dropped from third position in 2002–2004 (0.469) to sixth position in 2016–2018 (0.373). At the same time, two municipalities remained in the lead and continuously increased their social potential: Czarny Bór, with values of 0.539 and 0.604 in the two analysed periods, and Szczawno-Zdrój, with 0.5000 and 0.560.

In the next part of the research, the economic potential of Wałbrzych municipalities was analysed (Table 4).

Table 4. Indicators of economic development potential for Wałbrzych County municipalities calculated using zero unitarization method.

Municipality	2002–2004			2016–2018		
	Indicator	Ranking	Category	Indicator	Ranking	Category
Boguszów-Gorce	0.192	4	III	0.154	8	III
Jedlina-Zdrój	0.176	5	III	0.489	3	II
Szczawno-Zdrój	0.614	1	I	0.515	2	I
Czarny Bór	0.430	3	II	0.660	1	I
Głuszyca	0.158	8	III	0.178	7	III
Mieroszów	0.172	6	III	0.278	4	III
Stare Bogaczowice	0.467	2	I	0.214	6	III
Walim	0.169	7	III	0.231	5	III

Source: Authors' calculations based on bdl.stat.gov.pl (accessed on 30 October 2020).

The diagnostic variables adopted to assess economic development potential included expenditure by selected sectors of the municipal economy (e.g., residential housing management, transport and communication, tourism, and agriculture), revenues of municipal budgets, expenditure of municipal budgets, funding and co-funding EU programmes and projects, share of the municipal investment in the total expenditure of the municipal budget, the length of bike paths and economic entities. In this case, in both analysed periods, the municipalities of Szczawno-Zdrój and Czarny Bór achieved the best results, while Głuszyca and Boguszów-Gorce were rated the worst. The established indicators, and therefore also the ranking positions, changed in the analysed years.

The least significant changes when comparing 2002–2004 and 2016–2018 took place in terms of the environmental potential (Table 5). The highest indicator in both analysed periods was recorded in Szczawno Zdrój and Boguszów-Gorce.

Table 5. Indicators of environmental development potential for Wałbrzych County municipalities calculated using zero unitarization method.

Municipality	2002–2004			2016–2018		
	Indicator	Ranking	Category	Indicator	Ranking	Category
Boguszów-Gorce	0.649	2	I	0.711	2	I
Jedlina-Zdrój	0.576	3	II	0.510	3	II
Szczawno-Zdrój	0.948	1	I	0.814	1	I
Czarny Bór	0.333	7	II	0.294	6	III
Głuszyca	0.554	4	II	0.360	4	II
Mieroszów	0.469	5	II	0.322	5	II
Stare Bogaczowice	0.024	8	III	0.067	8	III
Walim	0.416	6	II	0.262	7	III

Source: Authors' calculations based on bdl.stat.gov.pl (accessed on 30 October 2020).

There was a certain stability in the assessment of environmental potential indicator results from, among others, variables that involve considerable resources and time to expand, e.g., sewage systems, water supply systems, gas pipelines and sewage treatment plants, and also from the implicit constant nature of some of the adopted features (e.g., share of forest area). The diagnostic variables adopted to assess environmental potential included expenditure on municipal management and environmental protection, water consumption, length of active sewage network, population using water supply network, population using sewage treatment plants, share of forest land in the total area of the municipality, population using gas vs. total population, and share of parks, lawns and housing estate green areas in the total area.

In the interval between the analysed periods, 2002–2004 and 2016–2018, no significant change in the ranking of individual municipalities was observed; however, a decline in potential was noticed in most of them.

The graphs in Figures 1–3 show the dispersion of social, economic and environmental potential of Wałbrzych municipalities for the periods 2002–2004 and 2016–2018. In terms of the social potential indicator, the most stable and best situation occurred in Czarny Bór and Szczawno-Zdrój, while Walim, Mieroszów and Jedlina-Zdrój remained close to each other.

The municipalities presenting the highest economic potential in 2016–2018 were, in the following order, Czarny Bór, Szczawno Zdrój and Jedlina Zdrój; the potential significantly increased since the beginning of the century in Czarny Bór and Jedlina Zdrój (Figure 2). Boguszów-Gorce, Głuszyca, Mieroszów and Walim, whose economic potential was relatively low, show a clear concentration in both analysed periods.

The dispersion indicator of environmental development potential ranks Szczawno-Zdrój in the best position, and shows improvement in Boguszów-Gorce and a lower assessment in Jedlina-Zdrój, Głuszyca, Mieroszów and Walim.

Further analyses using the agglomeration method confirmed the previously obtained results (Figures 4–6).

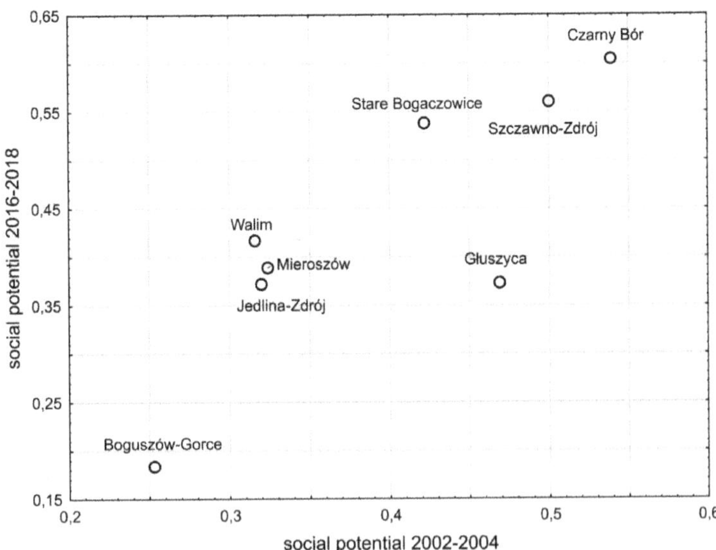

Figure 1. Scatter plot of social development potential in 2002–2004 vs. 2016–2018. Source: authors' compilation based on bdl.stat.gov.pl (accessed on 30 October 2020) [32].

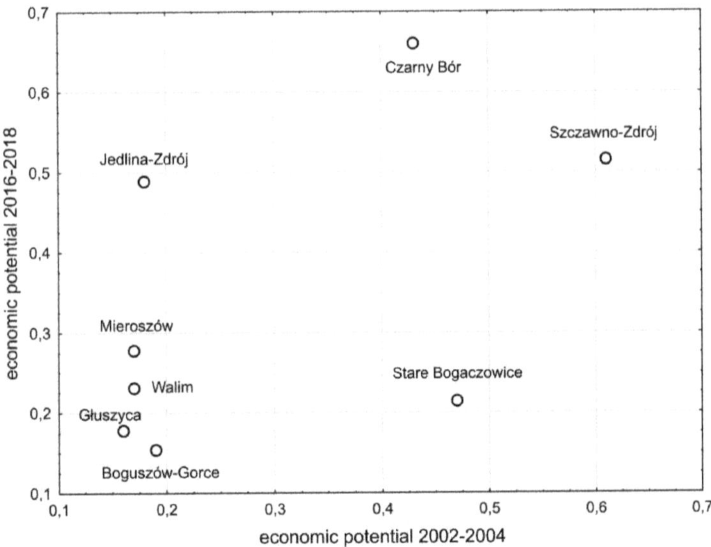

Figure 2. Scatter plot of economic development potential in 2002–2004 vs. 2016–2018. Source: authors' compilation based on bdl.stat.gov.pl (accessed on 30 October 2020) [32].

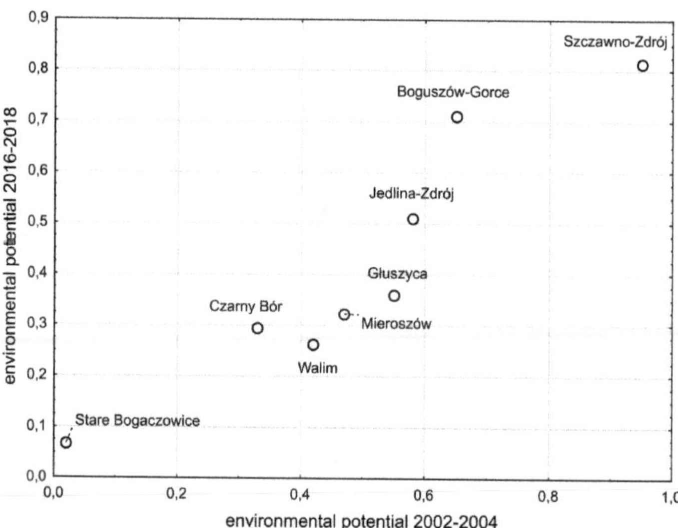

Figure 3. Scatter plot of environmental development potential in 2002–2004 vs. 2016–2018. Source: authors' compilation based on bdl.stat.gov.pl (accessed on 30 October 2020) [32].

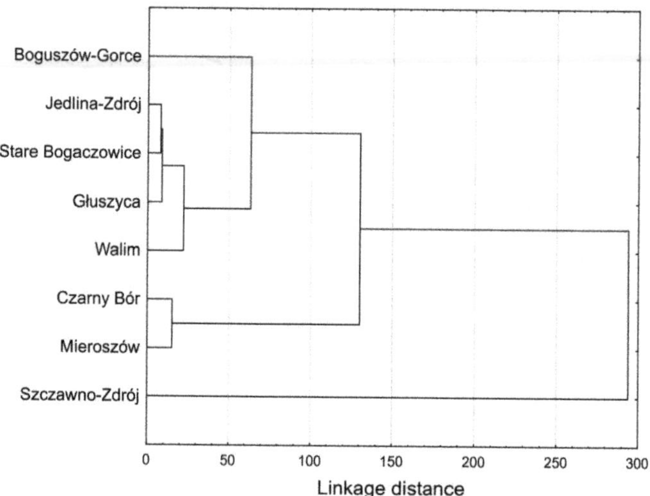

Figure 4. Tree diagram of social potential 2016–2018. Source: authors' compilation based on bdl.stat. gov.pl (accessed on 30 October 2020) [32].

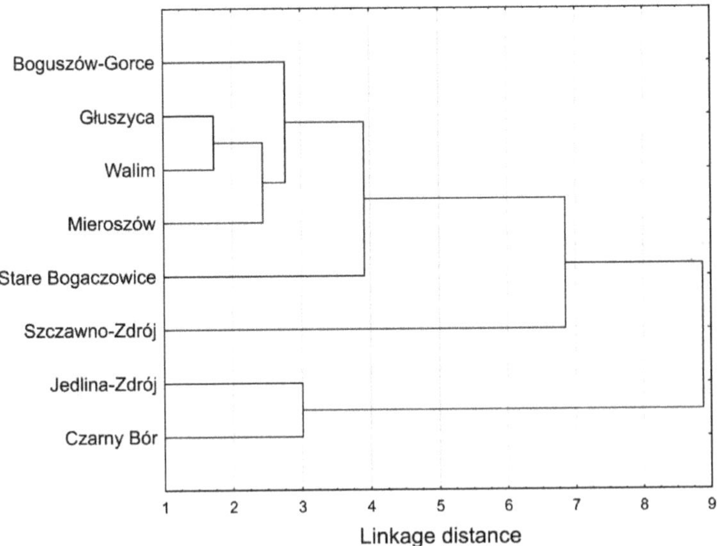

Figure 5. Tree diagram of economic potential 2016–2018. Source: authors' compilation based on bdl.stat.gov.pl (accessed on 30 October 2020) [32].

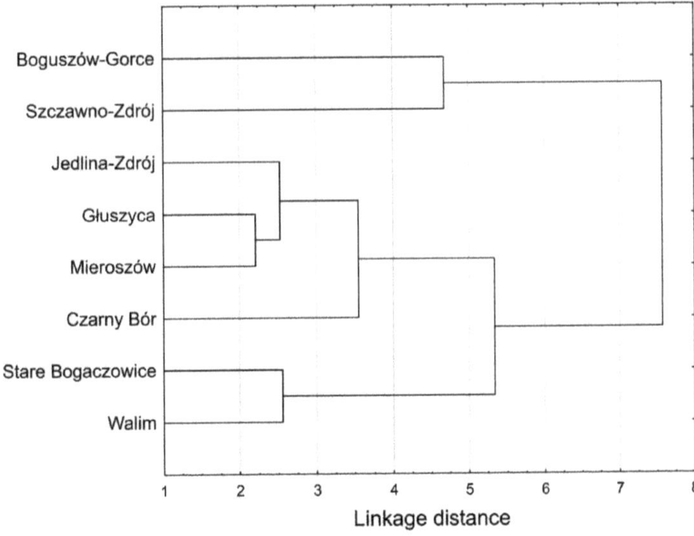

Figure 6. Tree diagram of environmental potential 2016–2018. Source: authors' compilation based on bdl.stat.gov.pl (accessed on 30 October 2020) [32].

The agglomeration method (Figure 4) shows that the social potential parameters of Szczawno-Zdrój included in the research differ significantly from the parameters of other municipalities. The municipalities most similar in this respect are Jedlina Zdrój, Stare Bogaczowice and Głuszyca. Taking into account the cluster analysis tree in terms of economic potential (Figure 5), it was found that Czarny Bór and Jedlina Zdrój constitute a separate group. As regards the environmental potential indicator, the highest value in both analysed periods was achieved by Szczawno Zdrój and Boguszów-Gorace. They also constitute a separate cluster in the agglomeration method (Figure 6). Stare Bogaczowice and Walim were similar in terms of environmental parameters, characterised by the lowest Q indicator.

4. Discussion

4.1. Current Social, Economic and Environmental Potential of Wałbrzych County

Having analysed the diagnostic variables, it can be concluded that the position of Wałbrzych County municipalities is usually worse compared to the average values achieved for the entire Lower Silesia Voivodship area. The shutdown of some production plants and the complete closure of hard coal mines resulted in the aggravation of social problems in these municipalities. In recent years, a lower average birth rate per 1000 people (-5.2) has been observed in this area compared to Lower Silesia and the entire country (-1.4 and 0.5). Other diagnostic features confirm the unfavourable demographic structure of the discussed municipalities and the small share of employed people in the total population. The number of employed people per 1000 inhabitants raises concerns and suggests the need to take up remedial measures; on average, only 118 people per 1000 residents work in the municipalities surrounding the city of Wałbrzych, compared to 389 people per 1000 residents in Lower Silesia.

Detailed research findings confirm that the social development potential of the municipalities is alarmingly low. Boguszów-Gorce remains the worst-ranked municipality in this respect. It is also the only municipality which received the lowest third category ranking. Together with the low economic potential, this assessment translates into a low quality of life for residents of this municipality. However, Boguszów Gorce has an advantage in the form of environmental potential (rank I), which can be used in further functional transformations.

The next municipalities in the group featuring low social potential are Jedlina Zdrój, Mieroszów, Walim and Głuszyca.

The municipalities characterised by low economic potential also include Głuszyca, Mieroszów, Stare Bogaczowice and Walim. Thus, some municipalities that present unfavourable social potential are also characterized by unfavourable economic potential. Further analysis also revealed that the natural potential of municipalities such as Stare Bogaczowice, Walim, Mieroszów and Głuszyca is not highly rated.

Concerns are raised by the fact that the same municipalities appear to be among the worst assessed in terms of social, economic and environmental potential.

4.2. Identifying Main Municipal Strategic Goals and Tasks Faced by Local Authorities Based on the Conducted Research

In accordance with the research conducted by Stacherzak et al. [38] on functional transformations, since 1996, municipalities have been strengthening the function of industrialization developed for tourism purposes, almost without agriculture (where agriculture is not of great importance).

The process of transformations is slow and difficult. Most likely, in the next several dozen years, it will not be possible to restore the population back to the level prior to the transformation period (before 1991). It is crucial to focus on taking action and improving the quality of life of the population at the current level. The actions for individual municipalities are listed below.

Municipalities: Boguszów Gorce, Jedlina-Zdrój, Mieroszów, Stare Bogaczowice, Walim, Głuszyca:

- Retain young people entering the labour market locally by creating jobs, providing facilities for families with children and developing sports and recreational offers for various age groups.
- Introduce the principles of planned housing management in the municipal housing stock through the flat replacement programme, and renovation and demolition of decapitalised buildings.

- Increase municipal activity of obtaining EU funds as well as funds from external sources in order to stimulate the economic activity of residents in the region.
- Promote and advertise the entire Wałbrzych region in Poland along with developing tourist and accommodation facilities, and create opportunities to participate in specific events and make unique tourist attractions available for sightseeing.

The municipalities of Szczawno-Zdrój and Czarny Bór are ranked best in terms of social and economic potential. Additionally, Szczawno Zdrój has high environmental potential. The main strategic goals include the following:

- Strive to maintain the municipal development potential.
- Further develop services for the population and care for municipal economic development, including job retention.

4.3. Further Directions for the Development of Municipalities Based on Their Potential

Economic processes and sustainability remain crucial from the perspective of broader development processes taking place in the economy [25,39,40], including the aspect of providing the population with an appropriate standard of living and preventing excessive migration. These municipalities, according to our assessment, do not present significant potential for environmental development; however, in their area, there are underground military buildings from the Second World War (Osówka Complex, Underground City, Walim Adits) and monuments of architecture and technology (Grodno Castle, a water dam on the Bystrzyca River, the Bystrzyca Railway). The municipalities have many attractions that allow the development of tourism; they are surrounded by the beautiful landscape of the Sowie, Wałbrzyskie and Kmienne Mountains, but the activity of the local community is at a lower level than the rest of the Lower Silesia region. Local resources play a significant role in theoretical thinking about development; however, in practice they are difficult to identify [41].

Although the main determinant of quality of life is in the form of psychological factors [42], it is also closely related to other factors. Quality of life includes, among other things, economic conditions [43], individual expectations, participation in social life [44], place and conditions of residence [45], quality of the environment in which people live [46,47] and the degree of satisfaction of various human needs [48,49].

The good social, economic and environmental conditions of the Szczawno Zdrój municipality result from its location, environmental value, including medicinal waters used in spa treatment, and access to green areas, as well as caring for the city's image. The good ranking of Czarny Bór municipality is explained by the location of the Special Economic Zone within its area and the investments carried out there.

It can be suggested that using the inherent potential in the discussed area should determine the direction of this region's development; apart from tourism and recreation, the focus should also be on investing in rebuilding and developing industrial production. The mining industry used to be the driving force behind the economic development of Wałbrzych County, and today it should be industry based on new technologies in order to develop the potential of the local community. The establishment of the Wałbrzych Special Economic Zone allowed for the passivity and dependence on coal monoculture to be partially overcome [50]. This trend is developing simultaneously with the sector of small and medium-sized enterprises.

4.4. Limitations and Further Development of the Research

Certain limitations in reaching the diagnostic variables covering the entire area of the country appeared in the study. The authors plan to continue the research, focused on updating the analysis regarding the development strategies of individual municipalities and of Wałbrzych County. The planned research will refer to the observations and findings on the status and potential of the municipalities covered by this study. In further work, the authors would like to show compliance with strategic development goals included in the development strategies of individual municipalities with the current social, economic and environmental situation, and to identify the areas (strategic goals) which require changes as a result of the research findings.

5. Conclusions

The conducted research allowed us to answer the problem questions and formulate the following final conclusions.

The research revealed that the social potential of some municipalities in Wałbrzych County is alarmingly low, and a return to the level of development before the transformation has proved to be a long-term process and has not yet been reached.

Among the analysed municipalities, in terms of social, economic and environmental potential, Szczawno-Zdrój and Czarny Bór achieved the best results, and Boguszów Gorce was ranked the worst. In some municipalities, a noticeable increase in social potential (Jedlina-Zdrój, Mieroszów, Walim) and economic potential (Jedlina-Zdrój) was observed, and a significant decline in economic potential in Stare Bogaczowice was also noticeable.

The policies of municipal authorities have to focus on improving the living conditions of residents. Monitoring alone, without taking any specific measures, could result in significant depopulation of the area.

Municipal authorities should support the development of local initiatives that take advantage of the beauty of nature and technical monuments.

In the analysed municipalities of Wałbrzych County, special care should be taken to improve the living conditions of residents (to strengthen economic and social potential), which may have a significant impact on whether or not young people decide to stay there after graduation. The crucial elements are to open new jobs provide appropriate living conditions and services to allow taking up employment while caring for children, and to offer diverse sports and tourist options.

The analysis shows that the vast majority of municipalities offer great environmental potential. Efforts should be made to use the inherent potential in the area, also by highlighting preserved post-industrial buildings and constructions.

Beyond any doubt, the Wałbrzych region requires systemic support from the state to stop the depopulation of municipalities.

Author Contributions: Conceptualization, B.R. and H.D.; methodology, B.R. and M.H.; software, B.R.; validation, B.R., M.H. and H.D.; formal analysis, B.R., M.H. and H.D.; investigation, B.R; resources, B.R. and M.H.; data curation, H.D.; writing—original draft preparation, B.R. and M.H.; writing—review and editing, B.R and M.H.; visualization, H.D.; supervision, B.R.; project administration, M.H.; funding acquisition, B.R and M.H. All authors have read and agreed to the published version of the manuscript.

Funding: The research and the publication were financed under the Leading Research Groups support project from a subsidy increased for the period 2020–2025 in the amount of 2% of the subsidy referred to Art. 387 (3) of the Law of 20 July 2018 on Higher Education and Science, obtained in 2019. The open-access publication fee was financed under the Leading Research Groups support project from the subsidy increased for the period 2020–2025 in the amount of 2% of the subsidy referred to Art. 387 (3) of the Law of 20 July 2018 on Higher Education and Science, obtained in 2019.

Institutional Review Board Statement: Not applicable.

Informed Consent Statement: Not applicable.

Data Availability Statement: The data analysed in this study are subject to the following licenses/restrict Some of the datasets presented in this study are included in the paper. Some of the data can be obtained by contacting the authors. Requests for access to these datasets should be directed to Maria Hełdak: maria.heldak@upwr.edu.pl.

Conflicts of Interest: The authors declare no conflict of interest.

References

1. Oleszko-Kurzyna, B. Jakość życia a procesy zarządzania rozwojem gmin wiejskich [Quality of life vs. management processes of rural municipalities development]. In *Polityka Społeczna Wobec Problemu Bezpieczeństwa Socjalnego w Dobie Przeobrażeń Społeczno-Gospodarczych [Social Policy towards the Problem of Social Security in the Age of Socio-Economic Transformations]*; Rączaszek, A., Koczur, W., Eds.; University of Economics in Katowice: Katowice, Poland, 2014; Volume 179, pp. 85–93.
2. Langfeldt, L.; Godø, H.; Gornitzka, A.; Kaloudis, A. Integration modes in EU research: Centrifugality versus coordination of national research policies. *Sci. Public Policy* **2012**, *39*, 88–98. [CrossRef]
3. Pedersen, D. Integrating social sciences and humanities in interdisciplinary research. *Palgrave Commun.* **2016**, *2*, 16036. [CrossRef]
4. Reed, J.; Van Vianen, J.; Deakin, E.L.; Barlow, J.; Sunderland, T. Integrated landscape approaches to managing social and environmental issues in the tropics: Learning from the past to guide the future. *Glob. Chang. Biol.* **2016**, *22*, 2540–2554. [CrossRef]
5. Hełdak, M.; Raszka, B. Evaluation of the Local Spatial Policy in Poland with Regard to Sustainable Development. *Pol. J. Environ. Stud.* **2013**, *22*, 395–402.
6. Kozłowski, S. *The Future of Eco-Development*; KUL Publishing: Lublin, Poland, 2005; Volume 197, p. 586. (In Polish)
7. Pęski, W. *Zarządzanie Zrównoważonym Rozwojem Miast [Managing the Sustainable Development of a City]*; Wydawnictwo Arkady: Warszawa, Poland, 1999; Volume 20, p. 294. (In Polish)
8. Hełdak, M. Przemiany funkcjonalne obszarów wiejskich Sudetów po integracji z Unią Europejską [Functional Transformations of Rural Areas of the Sudeten after the European Union Integration]. In *Infrastruktura i Ekologia Terenów Wiejskich [Infrastructure and Ecology of Rural Areas]*; No. 8/2008; The Kraków Branch of the Polish Academy of Sciences: Kraków, Poland, 2008; pp. 91–102.
9. Buczak, A. *Centrum Nauki i Sztuki "Stara Kopalnia" w Wałbrzychu jako Przykład Rewitalizacji Obiektu Poprzemysłowego, Samorząd Terytorialny w Polsce z Perspektywy 25-lecia jego Funkcjonowania [The Science and Art Centre "Old Mine" in Wałbrzych as an Example of the Post-Industrial Facility Revitalization, Local Government in Poland from the Perspective of the 25th Anniversary of Its Functioning]*; Laskowski, P., Ed.; Research Papers of Wałbrzych Higher Scholl of Management and Enterprise: Wałbrzych, Poland, 2015; Volume 33, pp. 33–52. Available online: http://www.pracenaukowe.wwszip.pl/prace/prace-naukowe-33.pdf (accessed on 15 April 2021).
10. Karakaya, E.; Nuur, C. Social sciences and the mining sector: Some insights into recent research trends. *Resour. Policy* **2018**, *58*, 257–267. [CrossRef]
11. Cappuyns, V.; Swennen, R.; Vandamme, A.; Niclaes, M. Environmental impact of the former Pb-Zn mining and smelting in East Belgium. *J. Geochem. Explor.* **2006**, *88*, 6–9. [CrossRef]
12. Dudka, S.; Adriano, D.C. Environmental impacts of metal ore mining and processing: A review. *J. Environ. Qual.* **1997**, *26*, 590. [CrossRef]
13. IIED. Breaking New Ground: Mining, Minerals and Sustainable Development (MMSD). 2002. Available online: http://pubs.iied.org/9084IIED/ (accessed on 15 April 2021).
14. Solomon, F.; Katz, E.; Lovell, R. The Social Dimensions of Mining: Research, Policy and Practice Challenges for the Minerals Industry in Australia. *Resour. Policy* **2008**, *33*, 142–149. [CrossRef]
15. Borys, T. Propozycja Siedmiu Typologii Jakości Życia [A Proposal of Seven Typologies of the Quality of Life]. In *Gospodarka a Środowisko [Economy vs. Environment]*; Research Papers of Wrocław University of Economics: Wrocław, Poland, 2008; Volume 9, pp. 125–134.
16. Gill, M.; Feinstein, A.R. A Critical Appraisal of the Quality of Quality-of-Life Measurements. *JAMA* **1994**, *272*, 619–626. [CrossRef]
17. Keles, R. The Quality of Life and the Environment. *Procedia Soc. Behav. Sci.* **2012**, *35*, 23–32. [CrossRef]
18. Malina, A.; Zeliaś, A. Taksonomiczna analiza przestrzennego zróżnicowania jakości życia ludności w Polsce w 1994 r. [Taxonomic analysis of spatial differentiation in the quality of life of the population in Poland in 1994.]. *Stat. Rev.* **1997**, *44*, 11–27.
19. Uzzell, D.; Moser, G. Environment and quality of life. *Rev. Eur. Psychol. Appl.* **2006**, *56*. [CrossRef]
20. Kusterka-Jefmańska, M. Wysoka jakość życia jako cel nadrzędny lokalnych strategii zrównoważonego rozwoju [High quality of life as the primary goal of local sustainable development strategies]. *Public Manag.* **2010**, *4*, 115–123.
21. Burckhardt, C.S.; Anderson, K.L. The Quality of Life Scale (QOLS): Reliability, Validity, and Utilization. *Health Qual. Life Outcomes* **2003**. Available online: http://www.hqlo.com/content/1/1/60 (accessed on 4 April 2021).
22. Kucher, A.; Hełdak, M.; Kucher, L.; Raszka, B. Factors Forming the Consumers' Willingness to Pay a Price Premium for Ecological Goods in Ukraine. *Int. J. Environ. Res. Public Health* **2019**, *16*, 859. [CrossRef]
23. Zenka-Zganiacz, A. Jakość życia w badaniach ekonomicznych, społecznych i w statystyce publicznej [Quality of life in economic and social research and in public statistics]. In *Pomiar Jakości Życia w Układach Regionalnych i Krajowych. Dylematy i Wyzwania [Measurement of the Quality of Life in Regional and National Systems. Dilemmas and Challenges]*; Studies and Materials of the Faculty of Management and Administration of the Jan Kochanowski University: Kielce, Poland, 2017; Volume 21/3 (1), pp. 93–105.
24. The Indicators of Sustainable Development for Poland. Team of Employees of the Statistical Office in Katowice with the Co-Participation of: Statistical Office in Kraków Statistical Office in Szczecin Analysis and Comprehensive Studies Department of CSO under the Management of Edmund Czarski. Ph.D. Thesis, Central Statistical Office, Katowice, Poland, 2011. Available online: https://stat.gov.pl/cps/rde/xbcr/gus/as_Sustainable_Development_Indicators_for_Poland.pdf (accessed on 12 March 2021).

25. Parysek, J. *Wewnętrzne i Zewnętrzne Uwarunkowania Transformacji Przestrzenno—Strukturalnej i Rozwoju miast Polski w końcu XX wieku. Przemiany bazy Ekonomicznej i Struktury Przestrzennej Miast pod red. Naukową Janusza Słodyczka [Internal and External Determinants of Spatial and Structural Transformation and the Development of Polish Cities at the End of the 20th Century. Changes in the Economic Base and Spatial Structure of Cities, ed. Janusz Słodyczka]*; University of Opole: Opole, Poland, 2002.

26. Borówka, A. *Stan i Perspektywy Wykorzystania Obiektów Poprzemysłowych w Wałbrzychu i Boguszowie-Gorcach [The Condition and Perspectives for Using Post-Industrial Facilities in Wałbrzych and Boguszów-Gorce]*; Institute of Geography and Regional Development, University of Wrocław: Wrocław, Poland, 2010.

27. Kosmaty, J. Wałbrzyskie Tereny Pogórnicze po 15 Latach od Zakończenia Eksploatacji Węgla [Wałbrzych Post-Mining Areas 15 Years after the Closure of Coal Mining]. *Coal Min. Ecol.* **2011**, *6*. Available online: https://www.polsl.pl/Wydzialy/RG/Wydawnictwa/Documents/kwartal/6_1_13.pdf (accessed on 10 May 2021).

28. Wójcik, J. Przemiany Wybranych Komponentów Środowiska Przyrodniczego Rejonu Wałbrzyskiego w Latach 1975–2000 w Warunkach Antropopresji, ze Szczególnym Uwzględnieniem Wpływu Przemysłu [Transformations of the Selected Components of the Wałbrzych Region Natural Environment in the Years 1975–2000 in the Conditions of Anthropopressure, with Particular Emphasis on the Impact of Industry]. Scientific Thesis, The Institute of Geography and Regional Development of the University of Wrocław 21 [Habilitation Degree], University of Wrocław, Wrocław, Poland, 2011.

29. Ward, J.H. Hierarchical grouping to optimize an objective function. *J. Am. Stat. Assoc.* **1963**, *58*, 236. [CrossRef]

30. Szkutnik, W.; Sączewska-Piotrowska, A.; Hadaś-Dyduch, M. *Metody Taksonomiczne z Programem STATISTICA*; Wydawnictwo Uniwersytetu Ekonomicznego w Katowicach: Katowicach, Poland, 2015; p. 144.

31. Basiura, B. Metoda Warda w zastosowaniu klasyfikacji województw Polski z różnymi miarami odległości [The Ward method in the application for classification of Polish voivodeships with different distances]. In *Prace Naukowe Uniwersytetu Ekonomicznego we Wrocławiu [Research Papers of Wrocław University of Economics]*; Wydawnictwo Uniwersytetu Ekonomicznego we Wrocławiu: Wrocławiu, Poland, 2013; pp. 209–216.

32. Statistics Poland, Local Data Bank. Available online: https://bdl.stat.gov.pl/BDL/start (accessed on 4 September 2019).

33. TIBCO Software Inc. Statistica (Data Analysis Software System), Version 13. 2017. Available online: http://statistica.io (accessed on 10 February 2021).

34. Kukuła, K.; Bogocz, D. Zero Unitarization Method and its Application in Ranking Research in Agriculture. *Econ. Reg. Stud.* **2014**, *7*, 5–13.

35. Kukuła, K. Metoda unitaryzacji zerowej na tle wybranych metod normowania cech diagnostycznych [Zero unitarization method at the background of selected methods for the diagnostic features normalization]. *Acta Sci. Acad. Ostroviensis* **1999**, 5–31. Available online: http://bazhum.muzhp.pl/ (accessed on 4 March 2021).

36. Kukula, K. *Metoda Unitaryzacji Zerowanej*; PWN: Warszawa, Poland, 2000; Volume 71.

37. Kukuła, K. *Badania Operacyjne w Przykładach i Zadaniach*; PWN: Warszawa, Poland, 2004; Volume 289.

38. Stacherzak, A.; Hełdak, M. Borough Development Dependent on Agricultural, Tourism, and Economy Levels. *Sustainability* **2019**, *11*, 415. [CrossRef]

39. Przybyła, K. Employment structure transformations in large polish cities. In *Hradec Economic Days*; Jedlička, P., Ed.; University of Hradec Kralove: Hradec Kralove, Czech Republic, 2018; Volume 8, pp. 196–205.

40. Przybyła, K.; Kachniarz, M.; Hełdak, M. The Impact of Administrative Reform on Labour Market Transformations in Large Polish Cities. *Sustainability* **2018**, *10*, 2860. [CrossRef]

41. Surnacz, M. Zasoby lokalne jako potencjały rozwoju lokalnego w dokumentach strategicznych na przykładzie wybranych gmin wiejskich województwa śląskiego [Local resources as the potential for local development in the strategic documents based on selected rural municipalities of the Silesian Voivodship]. *Stud. Rural. Areas* **2017**, *46*, 99–115. [CrossRef]

42. Smith, K.W.; Avis, N.E.; Assman, S.F. Distinguishing between quality of life and health status in quality of life research: A meta-analysis. *Qual. Life Res.* **1999**, *8*, 447–459. [CrossRef]

43. Sompolska-Rzechuła, A. Jakość życia jako kategoria ekonomiczna [Quality of life as an economic category]. *Folia Pomer. Univ. Technol. Stetin Oeconomica* **2013**, *301*, 127–140. Available online: http://foliaoe.zut.edu.pl/pdf/files/magazines/2/38/457.pdf (accessed on 4 March 2021).

44. Papuć, E. Jakość życia—definicje i sposoby jej ujmowania [Quality of life—definitions and presentation methods]. *Curr. Probl. Psychiatry* **2011**, *12*, 141–145. Available online: https://docplayer.pl/8578848-Jakosc-zycia-definicje-i-sposoby-jej-ujmowania-ewa-papuc.html (accessed on 4 March 2021).

45. Campbell, A.; Converse, P.E.; Rogers, W.L. *The Quality of American Life: Perception, Evaluation, and Satisfaction*; Rasel Sage Foundation: New York, NY, USA, 1976.

46. Kusterka, M. "Audyt miejski" jako narzędzie pomiaru jakości życia w miastach europejskich ["City audit" as a tool to measure the quality of life in European cities]. In *Gospodarka a Środowisko [Economy vs. Environment]*; Borys, T., Ed.; Research Papers of Wrocław University of Economics: Wrocław, Poland, 2008; Volume 22, pp. 135–142.

47. Saxena, S.; Orley, J. Quality of life assessment. The World Health Organization perspective. *Eur. Psychiatry* **1997**, *12* (Suppl. S3), 263–266. [CrossRef]

48. Polak, E. Jakość życia w polskich województwach—analiza porównawcza wybranych regionów [Quality of life in Polish voivodships—Comparative analysis of selected regions]. *Soc. Inequal. Econ. Growth* **2016**, *48*, 66–89. [CrossRef]

49. Tobiasz-Adamczyk, B. Jakość życia w naukach społecznych i medycynie [Quality of life in social sciences and medicine]. *Art Treat.* **1996**, *2*, 33–40.
50. Zakrzewska-Półtorak, A. Problemy rozwoju Wałbrzycha i ich znaczenie dla rozwoju województwa dolnośląskiego [Problems of Wałbrzych development and their impact on Lower Silesia development]. *Bibl. Reg.* **2010**, *10*. Available online: https://dbc.wroc.pl/Content/35006/Zakrzewska-Poltorak_Problemy_Rozwojowe_Walbrzycha_i_Ich.pdf (accessed on 8 April 2021).

Article

Numerical Analysis of Concentrated Solar Heaters for Segmented Heat Accumulators

Martin Beer , Radim Rybár, Jana Rybárová, Andrea Seňová * and Vojtech Ferencz

Institute of Earth Sources, Faculty of Mining, Ecology, Process Technologies and Geotechnology, Technical University of Košice, Letná 9, 042 00 Košice, Slovakia; martin.beer@tuke.sk (M.B.); radim.rybar@tuke.sk (R.R.); jana.rybarova@tuke.sk (J.R.); vojtech.ferencz@tuke.sk (V.F.)
* Correspondence: andrea.senova@tuke.sk

Abstract: This presented paper focuses on the design and evaluation of the concept of concentrated solar heaters for segmental heat accumulators, which are designed to cover the energy needs of selected communities in terms of food preparation without the need for fossil fuels, which have a negative impact not only on the climate but especially on health. The proposed device is based on the traditional method of food preparation in the so-called earth oven; however, the fire-heated stones are replaced with heat accumulators heated by solar radiation. This approach eliminates the need to change common and long-term habits of food preparation for selected communities. The device connects solar vacuum heat pipes, a solar radiation concentrator, and heat accumulators. The concept was evaluated based on computational fluid dynamics (CFD) analysis with the use of a transient simulation of selected operating situations in three geographical locations. The results showed a significant temperature increase of the heat accumulators, where in the most effective case the temperature increased up to 227.23 °C. The concept was also evaluated based on a calorimetric analysis of the system consisting of heat accumulators and food. The resulting temperature in the considered case reached the pasteurization temperature necessary for safe and healthy food preparation.

Keywords: solar energy; solar oven; concentrating; CFD analysis

Citation: Beer, M.; Rybár, R.; Rybárová, J.; Seňová, A.; Ferencz, V. Numerical Analysis of Concentrated Solar Heaters for Segmented Heat Accumulators. *Energies* **2021**, *14*, 4350. https://doi.org/10.3390/en14144350

Academic Editor: Carlo Renno

Received: 22 June 2021
Accepted: 15 July 2021
Published: 19 July 2021

Publisher's Note: MDPI stays neutral with regard to jurisdictional claims in published maps and institutional affiliations.

1. Introduction

One of the most significant characteristics of the last decade is the fight against climate change, the negative impact of which is gradually manifested in a whole range of natural as well as human processes and phenomena [1–5]. Consumption of primary energy sources on a global scale is constantly growing [6–9]; over the last decade it is possible to record an increase of 14.6% [10]. However, this growth is uneven from a global perspective, with many developed countries in Western Europe and North America showing stagnation or even decline in this parameter for a long time. The increase in consumption is mainly due to the region of Africa, the Middle East, and Southeast Asia [11]. When studying the projections for the consumption of primary energy sources for the coming decades, it is clear that the gap between developed and developing countries will increase.

With current trends, the so-called developing countries will account for 67% of total energy consumption in 2040 [12], which is mainly covered by fossil fuels, the extraction, processing and use of which have many negative effects [13,14]. In developing countries, this increase is caused by the rapid growth of the population, which puts pressure on the provision of basic energy needs in the housing sector, i.e., especially cooking, heating, and preparation of hot water.

In developing countries, food preparation is largely covered using coal, solid biomass, and only to a lesser extent are gaseous fuels used in the form of propane-butane [15,16]. According to a study [17] about 3 billion people use for food preparation the so-called open fire, which burns solid biomass or coal. Depending on the country, in addition to coal, a whole range of solid biomass is used as an energy carrier, starting with animal excrement, wood biomass, or waste from wood processing activities [18]. The use of solid biomass brings several negative factors [19–21]. The most fundamental is the extensive use of forests beyond their sustainability. The second, no less serious, is the impact of imperfect combustion and particulate matter on human health [22–25]. This impact is multiplied by the fact that there are mainly women and small children in the vicinity of the open fire, for whom solid pollutants or increased CO_2 levels are more significant threats. According to a study [26], 1 million people, mainly women and children, lose their lives each year because of the safety and health impacts of open fires.

The availability of energy as well as the reduction of negative environmental impacts in the food preparation process is currently addressed by many technologies [27–32]. Among the simplest and most promising is the use of solar energy. It is mainly the use of concentration technology, increasing the power input from solar radiation, the overall efficiency, and thus also the resulting temperatures influencing the time or ability to prepare food. According to Solar Cookers International, more than 4 million solar cookers of various technologies are currently used by more than 11 million people [33]. Solar cookers can be divided into three main categories: solar box/panel cookers, parabolic solar cookers, and evacuated tube solar cookers [34]. The principle of operation of the first category is based on the capture of solar radiation in a thermally insulated box, the inner walls of which show a high degree of reflectivity of solar radiation. This guarantees an increase in the temperature throughout the inner volume, allowing the preparation of food, which, however, usually takes a longer period of time compared to other methods.

By contrast, parabolic solar cookers use a reflective-concentration element (dish), which concentrates sunlight into a small space and thus ensures the achievement of high temperatures and the speed of food preparation. The most effective way of using solar energy for food preparation is the use of evacuated tube solar cookers, where food is placed in a thermally insulated glass vacuum tube, which is in the focal point of the concentrating device. This method is comparable in time and temperature to the preparation of food on gas cookers, but the disadvantage is the space limitation given by the diameter of the tube. Another way to increase the thermal efficiency of solar cookers is to implement heat accumulators into their construction. Solar Cookers are prominently represented in the field of scientific research, and efforts to increase the efficiency of existing installations can be observed, as well as proposals for new technical solutions for solar cookers.

Morales et al. [35] designed a multifunctional solar system, consisting of two modules for food preparation and water heating. The design combines the features of a solar radiation collector that utilizes a revolving compound parabolic concentrator and channel. In the presented tests, the solar cooker part of the device achieved a heat output of 150 W. Zafar et al. [36] presented a new design of a solar cooker using the upper, lower, and front reflective surface. In the experiment, the new device reached a temperature of 100 °C–135 °C in 1 to 1.5 h. Ayub et al. [37] focused on increasing the power of a solar cooker. The authors supplemented the solar oven with a heat accumulator based on metal hydride. The results of the numerical analysis confirmed the expected increase in the energy efficiency of the modified solar oven from 6% to 42%.

Bhave and Kale [38] have developed a solar cooker with a heat accumulation function for high-temperature applications. A eutectic mixture of sodium nitrate and potassium nitrate was used as a heat accumulator, allowing the authors to reach a cooking temperature of 170 °C–180 °C. Kumaresan et al. [39] proposed a flat plate cooking unit for an indirect mode of solar cooking, using heat accumulation via phase change material (PCM) from the sugar alcohol group D-Mannitol. A numerical assessment of the operation showed an increase in the efficiency of the flat plate cooking unit to 41%. A change in the internal arrangement of the solar cooker was discussed by Sagade et al. [40]. The authors proposed a transparent modification of the top cover of the interior space intended for food preparation, which increased the heat flow directly into the prepared food and thus the total preparation temperature at the level of 120 °C–130 °C.

Mawire et al. [41], in their work, compared two devices designed for cooking using solar radiation. Both devices used heat accumulation usable at a time of lower solar radiation intensity. Sunflower oil and erythritol have been used as a heat accumulator. In terms of the use of accumulated energy, the authors confirmed greater efficiency in favor of erythritol. Chaudhary and Yadav [42] proposed a device enabling the simultaneous preparation of two separate meals, using a vacuum tube collector, which heats heat transfer medium and subsequently two in-series connected vessels containing the prepared food. The presented simultaneous operation increases the usability of the solar cooker in terms of repeated cooking during the day, as well as cooking over a longer period. Hosseinzadeh et al. [43] described a solar cooker consisting of a vacuum tube in which a steel vessel is placed for preparing food or heating water. The heat flow was increased by the concentration of solar energy using parabolic concentration equipment. However, for this type of device, the limiting element is the diameter of the vacuum tube, which limits the volume as well as the dimensions of the prepared dish.

This presented paper deals with the design and evaluation of an innovative technical design of cooking with the use of solar energy. The essential feature is the connection of the solar vacuum tube collector, the concentrating device, and the heat accumulation element. The resulting device, by changing the concept, solves two problems of solar cookers using vacuum tube-limited capacity of the space for food preparation and the need to change traditional cooking procedures. The proposed concept implements modern technologies into centuries-old methods of food preparation, which can increase the acceptance of the device in selected communities for everyday use. The innovativeness of the proposed device lies in the use of heat accumulators, which are heated by solar radiation in proposed device. The heated heat accumulators are then used outside the body of the solar heater for food preparation, heating, or disinfection of water. Verification of the function and evaluation in terms of the achieved parameters was performed through numerical CFD analysis.

2. Materials and Methods

2.1. Concentrated Solar Heater for Segmental Heat Accumulators

The technical solution concerns a solar heater designed for heating heat accumulators for their further use mainly for the heat treatment of food (meals) in the so-called earth oven. The method of preparing food in the earth oven consists of placing the food and a source of heat (most often fire-heated stones) in the prepared pit and its subsequent covering. The slowly released heat, thanks to thermal insulation in the form of soil, will then heat-treat the desired food.

The proposed device eliminates the need for a fire while using solar radiation to heat the heat accumulators, which replace the stones heated by fire. The solar heater of heat accumulators is intended primarily for use in warmer climatic zones with predominantly sunny weather, in regions with a low standard of living, or low availability of fuels or other heat sources. The proposed solar heater consists of three main (1–3) and four supporting parts (4–7), which are shown in the partial section view in Figure 1. The functional part of the design consists of the double-layered solar vacuum glass tube with open endings and without selective coating, in which heat accumulators made of cast iron are located. In the gap between the layers of glass is a secured vacuum, which increases the thermal resistance of the tube and thus improves its thermal insulation properties.

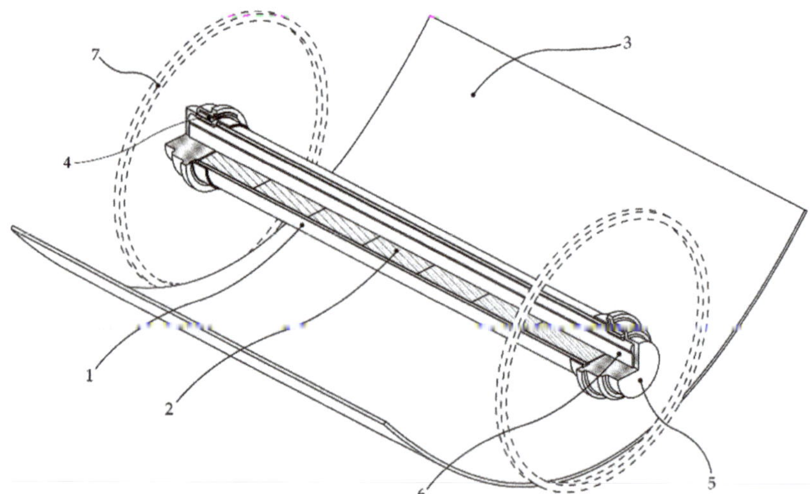

Figure 1. Concentrated solar heater for segmental heat accumulators (1—double-layered solar vacuum tube, 2—heat accumulator, 3—solar radiation parabolic concentrator, 4—heat accumulator support system, 5—thermal insulation block, 6—central beam, 7—chassis (not part of the design).

Solar radiation is focused by means of a concentration device attached to the chassis and wheels, respectively, which, however, are not essential from the point of view of the heat accumulation function and have therefore not been detailed in this phase of the design. The sides of the vacuum tube, the supporting system, and the heat storage elements are covered by a heat insulation block eliminating heat losses.

The heat accumulators are slid onto a beam which is self-supporting, attached to the hubs of the chassis and the glass tube, as shown in Figure 2. Thanks to this solution, the heat accumulators do not directly load the glass vacuum tube with their weight and the position of their center of gravity is still at one point, which is achieved by the independent mounting of the central beam with respect to the chassis hubs.

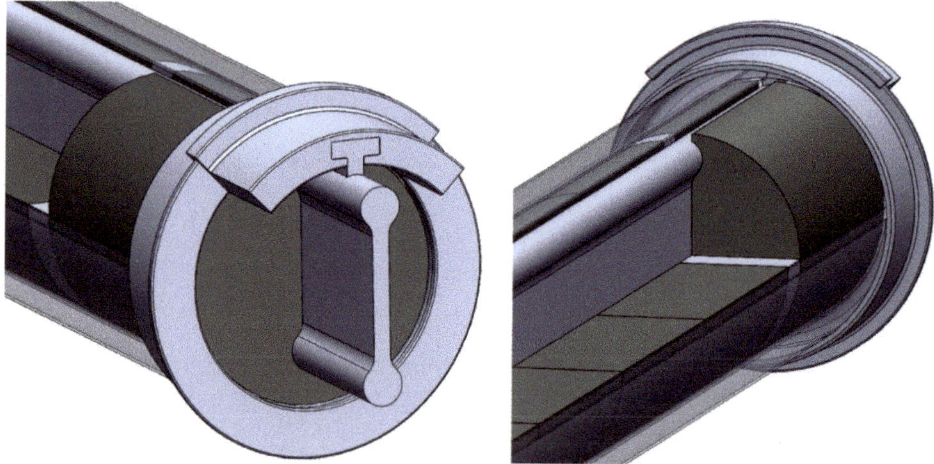

Figure 2. Detailed view of the 3D model of the supporting system of the heat accumulators.

Due to the technical and operational properties of the vacuum tube, it is advantageous that the concentrated solar heater of segmental heat accumulators can also be used to store heat accumulators in a charged state for their later use (for the time without sufficient solar radiation). Another advantageous fact is that the absorption surface where the photothermal conversion takes place is the surface of the segmental heat accumulators themselves, and not the inner surface of the glass tube as is the case with conventional solar vacuum collector designs, which helps with the heat transfer process. It is also positive that the vacuum tube of a transparent heating chamber allows for a visual inspection of the heating process of the individual segmental heat accumulators, which can be provided with a temperature indicator for this purpose. From the point of view of service life, it is advantageous that segment heat accumulators, unlike PCM and other heat accumulators, have an unlimited service life and a high resistance to damage.

Positioning the concentrator in the required position relative to the transparent heating chamber, depending on the position of the sun, is manual, by simultaneously turning both support wheels of the chassis. The working position of the solar heater with respect to the direction of incidence of solar radiation, during its operation, is such that the longitudinal axis of the horizontally placed heating chamber is at a right angle to the direction of incidence of the sun's rays. The basic principle of the proposed solar concentrated device is depicted in Figure 3.

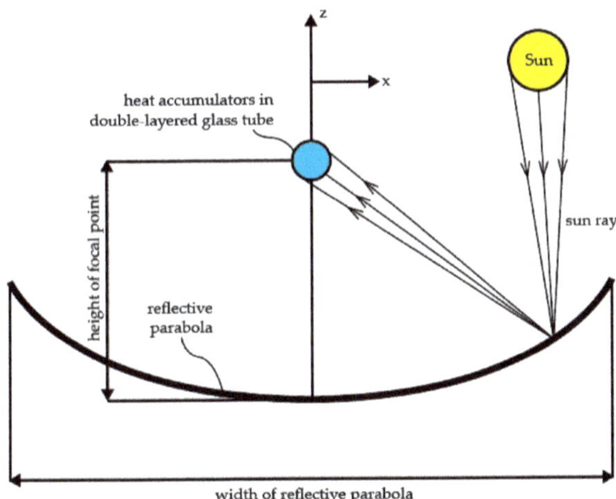

Figure 3. Basic principle of the proposed concentrated solar device.

2.2. Methodology

The concept of the concentrated solar heater of segmental heat accumulators was evaluated from the point of view of achieving selected values of operating parameters: average and daily temperature course of the heat accumulator temperature and the average temperature of the solar glass vacuum tube. The proposed design was assessed through numerical CFD analysis, which saves time, costs, working material, and with which the authors have many years of experience [44]. The analyzed geometry was simplified by eliminating the support system and chassis, which do not occur in the process of heat accumulation. Thus, only the concentration-reflective element, the glass vacuum tube, the heat accumulators, and the thermal insulation blocks closing the openings of the vacuum tube were featured in the simulations. This simplification has resulted in a significant reduction in the demands on the computing apparatus and hence the computing time. Figure 4 shows the basic layout, boundary conditions, and description of the calculation model.

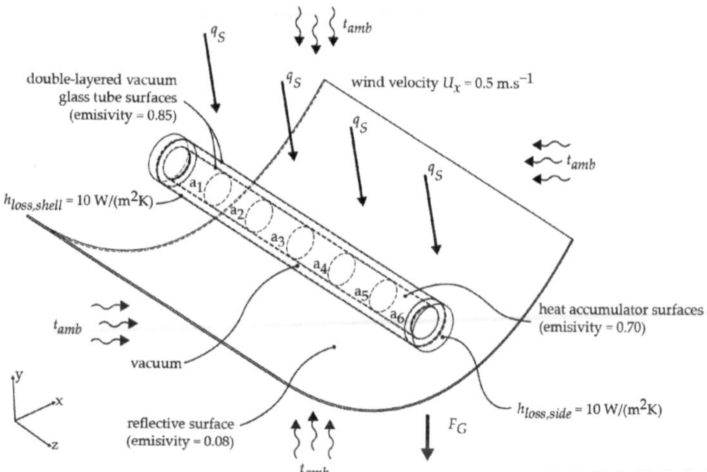

Figure 4. Basic layout and description of the calculation model.

The basic part of the model is heat accumulation blocks made of gray cast iron of a cylindrical shape with a diameter of 100 mm and a length of 200 mm (marked as a_1 to a_6). The model contains six pieces of heat accumulators, which were placed in a double-layered borosilicate vacuum tube, when a vacuum was defined between the layers, serving as thermal insulation. The thickness of the glass is 2 mm, the gap between the layers is 20 mm. The length of the tube corresponds to the length of the six heat accumulators and the length of the edge fillet, i.e., 1224 mm. At both ends, thermal insulating blocks are closing the heat accumulators inside the vacuum glass tube, their thickness is 40 mm and they exactly copy the shape of the end of the tube. The last part is a concentration-reflective element in the shape of a parabola with a width of 1100 mm, a total length of 1200 mm. The focal point at which the assembly of heat accumulators, glass vacuum tube, and insulators are located is 250 mm from the top of the parabola. Boundary conditions (BC) and computational domain are depicted in Figure 5.

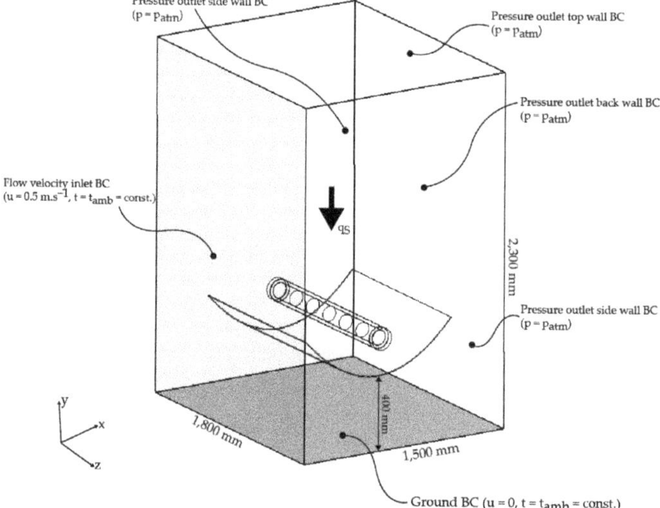

Figure 5. Computational domain and boundary conditions (BC).

In modeling, heat losses of 10 W/(m^{2}K) from the outer shell and sides of the tube were considered. The airflow was set in the x-axis direction at 0.5 m/s. Ambient temperature, intensity, and angle of incidence of solar radiation were modeled based on a specific geographical location. Table 1 summarizes other material, physical and thermal properties of each part of the model. Numerical CFD analysis of the monitored parameters was carried out in Solidworks Flow Simulation version 2017 software (published by Dassault Systemes). The Discrete Ordinates radiation model was used to simulate solar heating, which also includes the absorption of thermal radiation by solids, due to the simple situation from the point of view of the medium flow in the model without expecting the formation of significant turbulent regions; the standard k-ε turbulence model was used. A module for monitoring the intensity of solar radiation based on latitude, day number of the year, and time was also used. This computing apparatus is long and extensively used [45–47].

Table 1. Material, physical and thermal properties of each part of the model.

	Material	Density [kg/m^3]	Total Weight [kg]	Thermal Conductivity [W/(mK)]	Specific Heat Capacity [J/(kgK)]
Heat accumulator	Cast iron	7200	67.86	52	460
Vacuum glass tube	Borosilicate glass	2200	4.232	1.4	670
Concentration-reflective parabola	Polished aluminum foil	2700	0.120	237	900
Thermal isolation block	Mineral glass wool	130	0.156	0.035	1030

In fluid regions, SOLIDWORKS Flow Simulation solves the Navier–Stokes equations, which are formulations of mass, momentum and energy conservation laws:

$$\frac{\partial \rho}{\partial t} + \frac{\partial (\rho u_i)}{\partial x_i} = 0 \tag{1}$$

$$\frac{\partial (\rho u_i)}{\partial t} + \frac{\partial}{\partial x_j}(\rho u_i u_j) + \frac{\partial p}{\partial x_i} = \frac{\partial}{\partial x_j}\left(\tau_{ij} + \tau_{ij}^R\right) + S_i \tag{2}$$

$$\frac{\partial \rho H}{\partial t} + \frac{\partial \rho u_i H}{\partial x_i} = \frac{\partial}{\partial x_i}\left(u_j\left(\tau_{ij} + \tau_{ij}^R\right) + q_i\right) + \frac{\partial p}{\partial t} - \tau_{ij}^R \frac{\partial u_i}{\partial x_j} + \rho \varepsilon + S_i u_i + Q_H \tag{3}$$

where u is fluid velocity, ρ is density, S_i is the mass-distributed external force (gravity or forces of rotation), H = h + u^2/2, h is enthalpy, Q_H is the heat source per unit volume, τ_{ij} is viscous stress tensor, $\tau^R{}_{ij}$ is Reynolds viscous stress tensor, and q_i is diffusive heat flux. Indices indicate the summation in three coordinate directions. Heat transfer in solids and fluids with energy exchange between them (conjugate heat transfer) is an essential and implicit element of CAD-embedded CFD software. The phenomenon of heat conduction in solid media is described by the following equation:

$$\frac{\partial \rho e}{\partial t} = \frac{\partial}{\partial x_i}\left(\lambda_i \frac{\partial T}{\partial x_i}\right) + Q_H \tag{4}$$

where e is the specific internal energy, Q_H is specific heat release (or absorption) rate per unit volume and λ_i are the eigenvalues of the thermal conductivity tensor. In the Discrete Ordinates model, the whole 4π directional domain at any location within the computational domain is discretized into the specified number of equal solid angles. A radiation governing equation can be written as follows:

$$\frac{dI\left(\vec{s},\vec{r}\right)}{ds} = -(\kappa + \sigma_s)\cdot I\left(\vec{s},\vec{r}\right) + \kappa \cdot n^2 \cdot I_b\left(\vec{r}\right) + \frac{\sigma_s}{4\pi}\int_0^{4\pi} \phi\left(\vec{s'},\vec{s}\right)\cdot I\left(\vec{s'},\vec{r}\right)\cdot d\Omega' \tag{5}$$

where $\vec{r}$ is position vector, $\vec{s}$ is direction vector, $\vec{s}\,'$ is scattering direction vector, s is path length, κ is absorption coefficient, n is refractive index, I_b is black body radiation intensity, σ_s is scattering coefficient, I is radiation intensity, which depends on position $(\vec{r})$ and $(\vec{s})$, ϕ is phase function, Ω' is solid angle. A more detailed description can be found in the supplied technical documentation [48].

The concentrated solar heater model for segmented heat accumulators was transformed into a computational mesh for simulation purposes. The Solidworks Flow Simulations software uses a unique mesh creating methodology, which consists of using immersed-body mesh. With this approach, the creation of a mesh begins independently of the geometry itself, and each mesh cell can arbitrarily intersect the boundary between solid and liquid. This allows using Cartesian-based mesh. Such a mesh can be defined as a set of cuboids (rectangular cells) that are adjacent to each other and to the outer boundary of the computational space and at the same time are oriented along with Cartesian coordinates. These cells can be divided into solid cells, fluid cells, and partial cells (cells intersected the immersed boundary). On partial cells, which intersect the surface separating the solid and fluid region, the two-scale wall function is then applied, which consists of two methods for thin boundary layer treatment or thick boundary layer treatment. This approach at the fluid/solid interface allows the solution of Navier–Stokes equations in these defined volumes. A complete mathematical apparatus for grid formation methods is provided in the available technical documentation of the software [49]. The result of this procedure was a computational mesh that had 876,294 cells, of which 659,257 were fluid cells and 217,037 solid cells.

The quality of the mesh and its effect on the result was tested using the grid-independent study, in which the gradual development of the parameter of the average temperature of heat accumulators was monitored depending on the change of the quality of the mesh, respective to the number of cells. The results of the grid-independent study are shown in Figure 6, where the stabilization of the monitored value at the level of approximately 875,000 cells can be seen; the overall mesh in section-view is depicted in Figure 7.

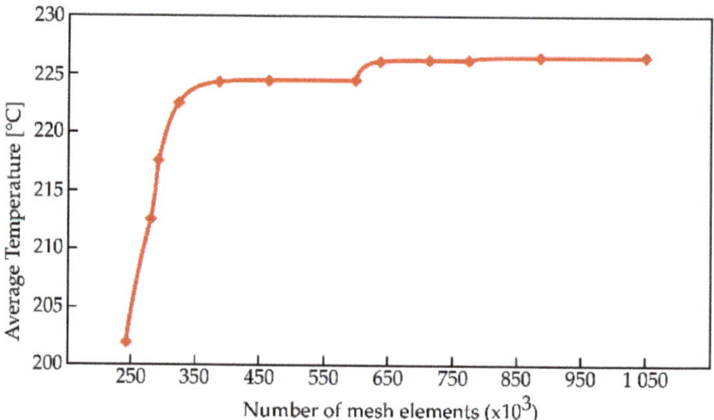

Figure 6. Results of the grid-independent study.

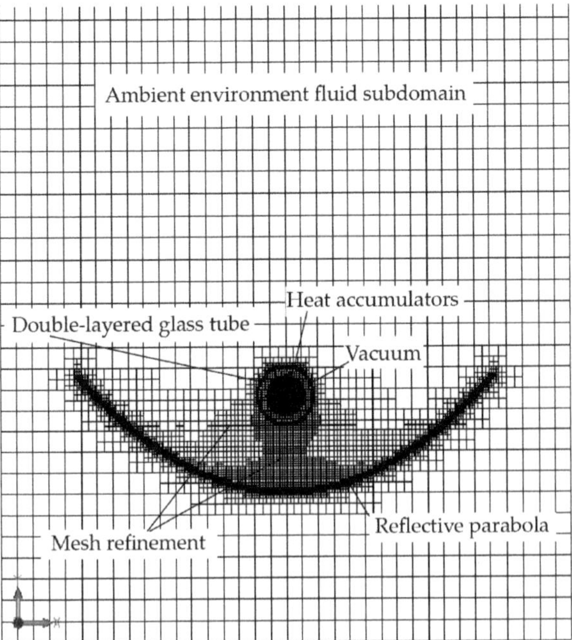

Figure 7. Section-view of meshed model.

3. Results

The evaluation of the presented design was carried out using transient simulations of the selected operation cases. For better evaluation of operating parameters, three geographical locations were selected, which represent different conditions in terms of solar intensity and ambient temperature, but at the same time, all three represent areas culturally suitable for implementing the proposed food preparation method. These areas are North Africa, South China, and Indonesia. From the point of view of solar radiation intensity, these areas were selected to cover the southern parts of the northern hemisphere in the range from 33° N, through the Tropic of Cancer, i.e., 23.5° N to 1° N, where suitable climatic and insolation conditions can be expected for the operation of the proposed device. These regions will hereinafter be referred to as regions A, B, and C. The reference date was chosen on June 21 and simulations followed the entire 24-h period. For a combination of location B (23.5° N), date June 21, and time 12:00, we obtained the ideal combination when the sun was directly above the concentrator and thus the device was attacked by the maximum intensity of solar radiation. It should be emphasized that in the simulations, static operation of the proposed concentrator was considered without gradual positioning of the device with respect to the changing position of the sun in the sky. In real operation, it is assumed that the device will be gradually positioned at approximately 30-min intervals.

The results of the analysis presented in Table 2 show the average temperature of all heat accumulation elements and the temperature of the outer surface of the vacuum glass tube at 12:00. The presented values quantified the expected results, i.e., the highest average temperature of the six heat accumulators, is reached by the device in locality B, followed by localities A and C. These results copy the angle of incidence of solar radiation and its intensity, respectively. For locality A, a maximum intensity value of 1030 W/m^2 was calculated, for locality B it is 1045 W/m^2 and for locality C it is 960 W/m^2.

Table 2. Average temperature of the heat accumulators and outer surface of the vacuum glass tube at 12:00.

	The Maximum Intensity of Solar Radiation [W/m^2]	Ambient Temperature [°C]	Latitude	The Average Temperature of the Heat Accumulators [°C]	The Average OuterSurface Temperature of the Vacuum Glass Tube [°C]
Locality A (North Africa)	1030	25.1	33° N	206.83	94.55
Locality B (South China)	1045	25.6	23.5° N	227.23	106.29
Locality C (Indonesia)	960	25.9	1° N	137.65	62.99

Figure 8 shows the temperature field in the longitudinal section view of the device and a detailed view of the top and bottom of the model. The figure shows that the six heat accumulators do not reach the same temperature, while the heat accumulators that are located at the ends of the vacuum glass tube are most affected by heat loss through the heat insulation block. At higher temperatures (Figure 8A,B) the temperature stratification is more pronounced, with the highest temperature being in the middle of the device. At lower temperatures, this temperature stratification is less pronounced. A closer look at the top and bottom of the device shows the apparent heat transfer through conduction from the inner layer of the glass tube through the rounding to the outer layer, as well as the heat transfer by radiation from the cast iron heat accumulators. For better mutual comparison, all cases are visualized in the same color spectrum and using the same scale.

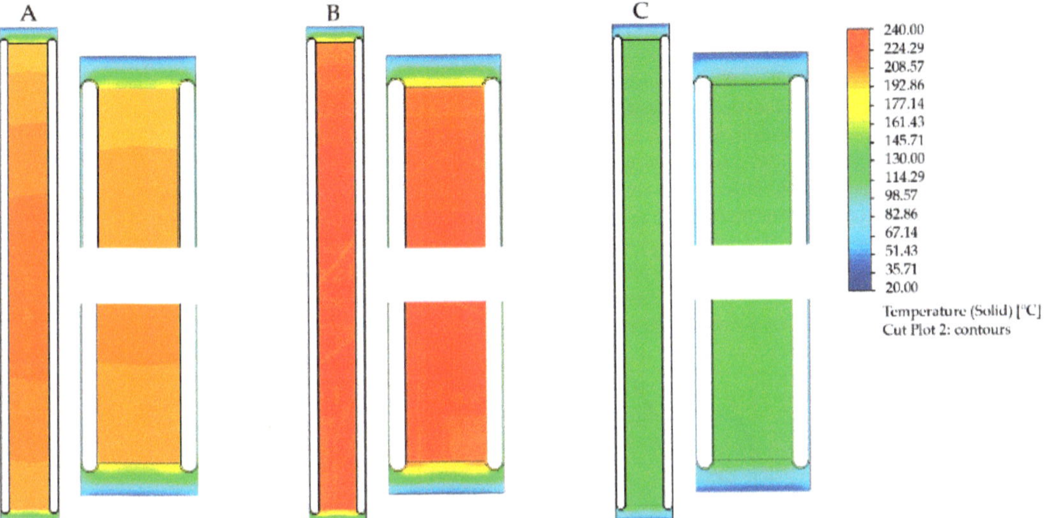

Figure 8. Temperature field in the longitudinal section view of the device and a detailed view of the top and bottom of the model for the North Africa (**A**), South China (**B**) and Indonesia (**C**).

A much more interesting result was obtained by comparing the average surface temperature of the outer layer of the vacuum glass tube, which in the case of sites A and B oscillates around 100 °C. The reason for this high temperature is the mentioned radiation of cast iron cylinders, which reach a temperature of over 200 °C, as well as heat transfer by conduction from the inner layer (which is in direct contact with the heat accumulators) through rounding to the outer layer.

From Figure 9, where the heat accumulators are shown in an isometric view, in addition to the temperature distribution, it is also possible to see the temperature non-uniformity in the Z-axis and the Y-axis. From the point of view of the Y-axis, a higher temperature can be seen in the lower part, which is caused by a higher heat flux of the reflected solar radiation from the parabolic concentrator (this fact is best visible in the case of location B).

Two scenarios were simulated in the next phase of the analysis. The first focused on a 24-h period in which the device was not positioned depending on the direction of the sun's rays and the heat accumulators were not removed. It is therefore a simulation of operation, where only the accumulation of thermal energy occurs without the direct input of the operator. Figure 10 shows the time course of the change in the average temperature of all heat accumulators (solid line) as well as the intensity of the solar radiation (dashed line) for each of the three locations.

solar heater, where the change in temperature of individual heat accumulators can be seen. The figures show the first hour of charging. Over time, it is clear to see a decrease in the temperature of the heat accumulators that were in the device, and thus initially had a higher temperature, as well as a further increase in replenished heat accumulators, both due to heat transfer from charged heat accumulators and also due to solar radiation.

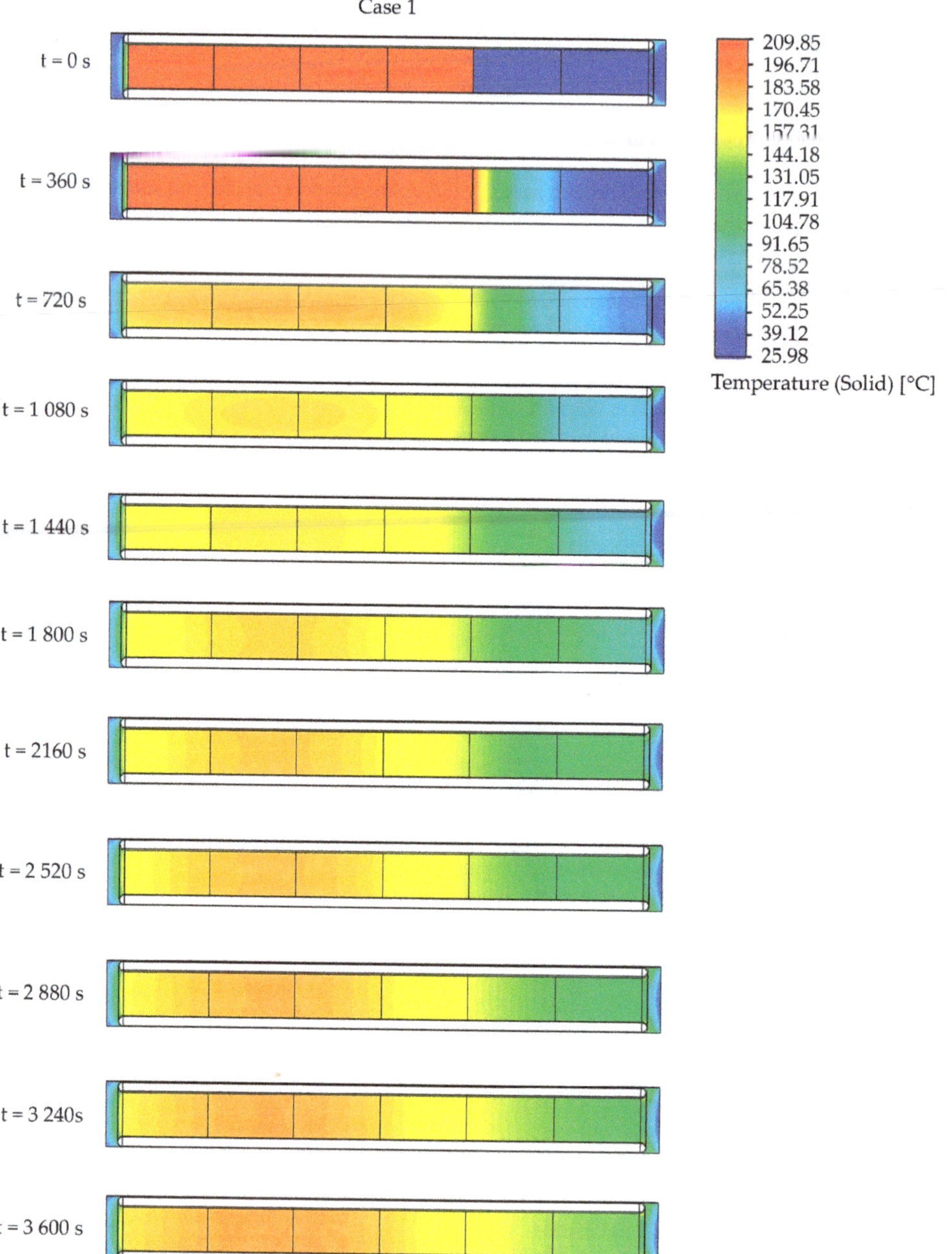

Figure 12. Temperature field in the longitudinal section view of heat accumulator in process of charging in Case 1.

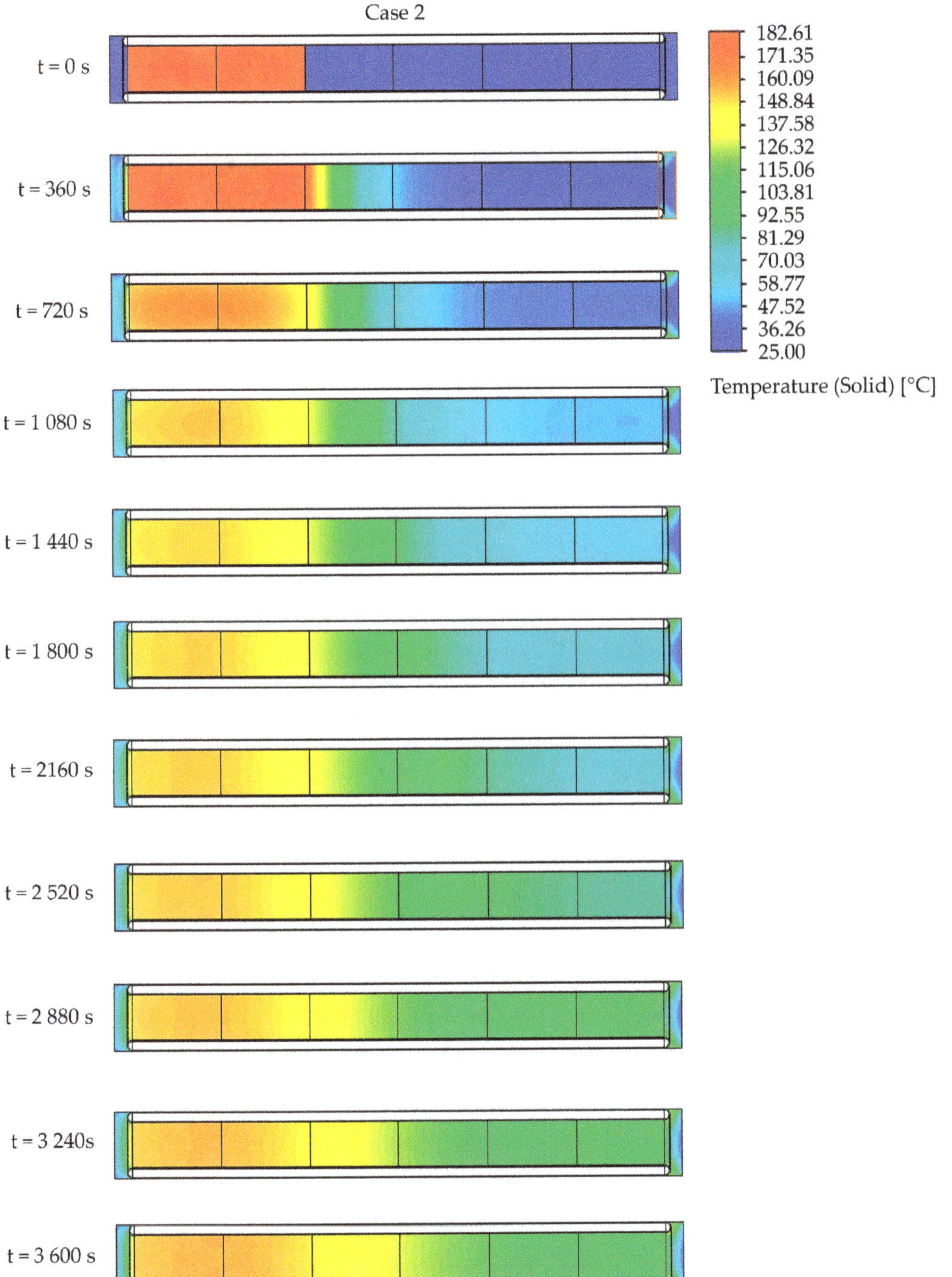

Figure 13. Temperature field in the longitudinal section view of heat accumulator in process of charging in Case 2.

Time course of the whole simulation is shown in Figure 14, where Case 1 on the left and Case 2 on the right can be seen. In both cases there are visible initial temperature decreases of charging heat accumulators; in Case 1, heat accumulator number 3 drops to the temperature of 154.1 °C, in Case 2 heat accumulator number 5 drops to the temperature of 142.8 °C. This decrease is caused by the heat conduction into the added heat accumulators (especially the adjacent ones). With increasing time, it is possible to observe the gradual heating of the added heat accumulators, while there is no longer heat removal from the originally charged heat accumulators, but heating caused by the solar radiation. During the entire heating period, the temperatures of the heat accumulators do not significantly match, with the largest difference occurring at the time of maximum temperature, being 23.7 °C for Case 01 and 25.8 °C for Case 2. The temperature of the heat accumulators only equalizes when they fall below 100 °C.

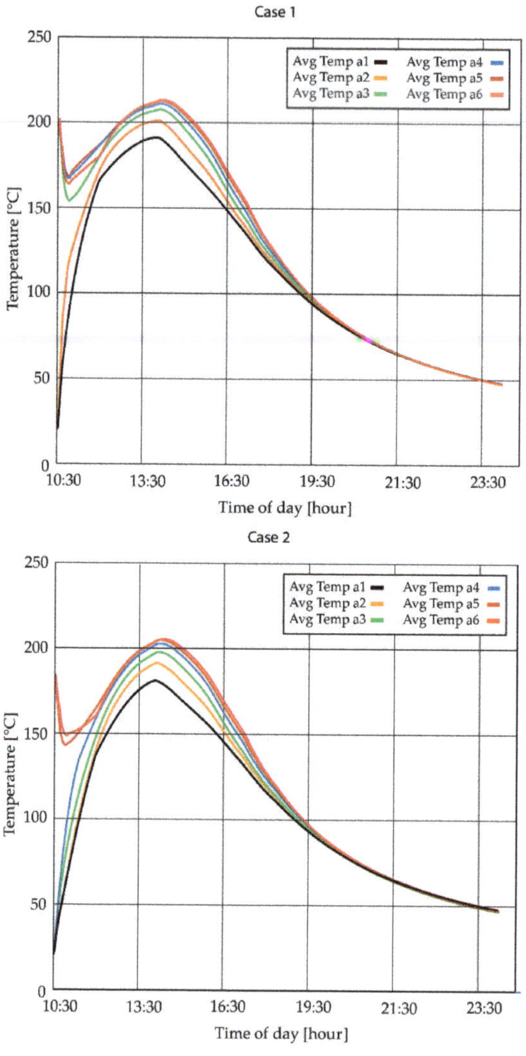

Figure 14. The course of the temperature of heat accumulators for Case 1 and Case 2.

The usability of the proposed device was finally assessed by calorimetric analysis of accumulated heat. This analysis assumes the need to exceed a certain temperature of the prepared food; this increase in temperature is provided by the heat accumulators, with which the food is in contact and there is heat exchange between them in terms of the second law of thermodynamics. The resulting temperature of the heat accumulator/food system can be determined using a calorimetric equation (Equations (6) and (7)) [50].

$$m_a \cdot c_a \cdot (t_a - t) = (m_F \cdot c_F \cdot (t - t_F)) \tag{6}$$

$$t = \frac{(m_a \cdot c_a \cdot t_a) + (m_F \cdot c_F \cdot t_F)}{m_a \cdot c_a + m_F \cdot c_F} \tag{7}$$

where m is the weight, c is the specific heat capacity and t is the temperature. Index a applies to quantities and values related to the heat accumulators, F to food. The resulting temperature of heat accumulators/food system is denoted as t. This analysis was based on known parameters of the proposed device, i.e., six heat accumulators made of gray cast iron (specific heat capacity 460 J/(kg.K) and a total weight of 67.86 kg) and properties of the prepared food, which in this case is represented with chicken meat (specific heat capacity 3100 J/(kg.K) and a total weight of 3.5 kg). Using Equation (2), the resulting temperature is 175 °C, which is 10 °C higher than the so-called pasteurization temperature required for chicken meat. Thanks to the thermal insulation of the food preparation site with a layer of soil cover, it is then possible to assume longer maintenance of the required temperature using the slow cooking method of food preparation.

The overall analysis of the results pointed to three areas of design that need to be addressed accordingly. The first area concerns the effort to eliminate the increased temperature of the outer surface of the glass vacuum tube. The achieved temperatures at the highest performances may pose a threat to the health of the operator of this equipment in careless handling. If we consider the heat transfer in the earth oven from the heat accumulators to the food through heat transfer by conduction and convection, then we would be able to eliminate this increased temperature of the outer glass vacuum tube by the selective coating of cast iron cylinders, thereby reducing the amount of radiation from their surface. The second area is the design of the chassis, which must not significantly cast a shadow on the concentration-reflective surface, which could reduce the overall performance of the device. In solving this problem, it is possible to build on existing technical wheel solutions with a low number of spokes. The third area concerns the proposal of further use of heat accumulators, which can be used for water heating and disinfection. At the achieved maximum temperatures of the heat accumulators, it is possible to reach water temperatures above 100 °C, with a water volume in the range of 8 L to 10 L. Here, however, it is necessary to change the approach from an earth oven, where water heating is out of the question, and propose the equipment enabling thermal connection of the vessel with water and heat accumulators. A possible solution could be a container with an increased heat exchange area through fins and with openings in the casing for accommodating heat accumulators. However, this issue will be addressed in the next phase of research.

5. Conclusions

The presented paper deals with the design and evaluation of the concept of a concentrated solar heater for segmented heat accumulators. The purpose of this design is to cover the energy needs of selected communities in terms of food preparation without the need to use fossil fuels, which have a negative impact in the form of extensive use of forests beyond their sustainability, endangerment of human health by combustion products, and accidents arising from the use of open flames. The proposed device is based on the traditional method of food preparation in the so-called earth oven; however, the fire-heated stones are replaced by heat accumulators heated by solar radiation. This procedure eliminates the need to change the usual and long-term methods of food preparation of selected communities. The device combines a parabolic concentrator, heat accumulators, and solar glass vacuum tubes, which reduces heat loss and increases the performance of the device. For better evaluation of operating parameters, three geographical locations were selected, which represent different conditions in terms of solar intensity and ambient temperature, but at the same time, all three represent areas culturally suitable for implementing the proposed food preparation method. These areas are North Africa, South China, and Indonesia. From the point of view of solar radiation intensity, these areas were selected to cover the southern parts of the northern hemisphere in the range from 33 °N, through the Tropic of Cancer, i.e., 23.5° N to 1° N, where suitable climatic and insolation conditions can be expected for the operation of the proposed device. The results showed an increase in the temperature of the heat accumulators up to 206.83 °C; 227.23 °C and 137.65 °C for each of the locations. In terms of time, the heat accumulators reached their maximum temperature in about eight hours, but without the active approach of the operator to position the entire device in terms of the movement of the sun in the sky, which would significantly shorten this time. Based on a calorimetric analysis of the system, which consists of food and heat accumulators, it was found that the resulting temperature exceeds 175 °C, which ensures the so-called pasteurization temperature necessary for safe and healthy food preparation. The results of simulations demonstrate the overall functionality of the device, even in operating conditions, which bring non-ideal insolation conditions.

Author Contributions: Conceptualization, M.B. and R.R.; methodology, M.B.; validation, J.R.; formal analysis, M.B.; resources, R.R.; data curation, M.B., A.S., V.F.; writing—original draft preparation, M.B.; writing—review and editing, M.B., A.S.; visualization, M.B.; supervision, R.R. All authors have read and agreed to the published version of the manuscript.

Funding: This paper was created in connection with the project VEGA 1/0290/21 Study of the behavior of heterogeneous structures based on PCM and metal foams as heat accumulators with application potential in technologies for obtaining and processing earth resources. This research was funded by the Scientific Grant Agency of the Ministry of Education, Science, Research and Sport of the Slovak Republic (VEGA). This paper was created in connection with the project 048TUKE-4/2021 Universal educational-competitive platform.

Institutional Review Board Statement: Not applicable.

Informed Consent Statement: Not applicable.

Conflicts of Interest: The authors declare no conflict of interest.

References

1. Liu, W.; Hou, Y.; Liu, W.; Yang, M.; Yan, Y.; Peng, C.; Yu, Z. Global estimation of the climate change impact of logging residue utilization for biofuels. *For. Ecol. Manag.* **2020**, *462*, 118000. [CrossRef]
2. Wang, X.; Xu, L.; Cui, S.; Wang, C.-H. Reflections on coastal inundation, climate change impact, and adaptation in built environment: Progresses and constraints. *Adv. Clim. Chang. Res.* **2020**, *11*, 317–331. [CrossRef]
3. Zhao, X.; Huang, G.; Lu, C.; Zhou, X.; Li, Y. Impacts of climate change on photovoltaic energy potential: A case study of China. *Appl. Energy* **2020**, *280*, 115888. [CrossRef]
4. Yang, Y.C.E.; Son, K.; Hung, F.; Tidwell, V. Impact of climate change on adaptive management decisions in the face of water scarcity. *J. Hydrol.* **2020**, *588*, 125015. [CrossRef]
5. Macusi, E.D.; Macusi, E.S.; Jimenez, L.A.; Catam-isan, J.P. Climate change vulnerability and perceived impacts on small-scale fisheries in eastern Mindanao. *Ocean. Coast. Manag.* **2020**, *189*, 105143. [CrossRef]
6. Li, R.; Leung, G.C.K. The relationship between energy prices, economic growth and renewable energy consumption: Evidence from Europe. *Energy Rep.* **2021**, *7*, 1712–1719. [CrossRef]
7. Jun, W.; Mughal, N.; Zhao, J.; Shabbir, M.S.; Niedbała, G.; Jain, V.; Anwar, A. Does globalization matter for environmental degradation? Nexus among energy consumption, economic growth, and carbon dioxide emission. *Energy Policy* **2021**, *153*, 112230. [CrossRef]
8. Meng, J.; Hu, X.; Chen, P.; Coffman, D.; Han, M. The unequal contribution to global energy consumption along the supply chain. *J. Environ. Manag.* **2020**, *268*, 110701. [CrossRef] [PubMed]
9. Inchauspe, J.; Li, J.; Park, J. Seasonal patterns of global oil consumption: Implications for long term energy policy. *J. Policy Model.* **2020**, *42*, 536–556. [CrossRef]
10. BP Plc. *BP Statistical Review of World Energy 2020*; BP Plc: London, UK, 2020; pp. 8–11.

11. U.S. Energy Information Administration. *The Middle East, Africa, and Asia Now Drive Nearly All Global Energy Consumption Growth—International Energy Statistics 2020*; U.S. Energy Information Administration: Washington, DC, USA, 2020; pp. 1–10.

12. U.S. Energy Information Administration. *International Energy Outlook 2020 (IEO2020)*; U.S. Energy Information Administration: Washington, DC, USA, 2020; pp. 1–7.

13. Cehlár, M.; Teplická, K.; Seňová, A. Risk management as instrument for financing projects in mining industry. In Proceedings of the 11th International Multidisciplinary Scientific Geoconference and EXPO—Modern Management of Mine Producing, Geology and Environmental Protection, SGEM 2011, Varna, Bulgaria, 20–25 June 2011; Volume 1, pp. 913–920.

14. Hrehová, D.; Cehlár, M.; Rybár, R.; Mitterpachová, N. Mining technology with drilling-blasting operations. In Proceedings of the 12th International Multidisciplinary Scientific GeoConference and EXPO—Modern Management of Mine Producing, Geology and Environmental Protection, SGEM 2012, Varna, Bulgaria, 17–23 June 2012; Volume 1, pp. 675–682.

15. Batchelor, S.; Talukder, M.A.R.; Uddin, M.R.; Mondal, S.K.; Islam, S.; Redoy, R.K.; Hanlin, R.; Khan, M.R. Solar e-Cooking: A Proposition for Solar Home System Integrated Clean Cooking. *Energies* **2018**, *11*, 2933. [CrossRef]

16. Cravioto, J.; Ohgaki, H.; Che, H.S.; Tan, C.; Kobayashi, S.; Toe, H.; Long, B.; Oudaya, E.; Rahim, N.A.; Farzeneh, H. The Effects of Rural Electrification on Quality of Life: A Southeast Asian Perspective. *Energies* **2020**, *13*, 2410. [CrossRef]

17. World Health Organization. *Burning Opportunity: Clean Household Energy for Health, Sustainable development, and Wellbeing of Women and Child*; World Health Organization: Geneva, Switzerland, 2016; p. 9.

18. Malla, S.; Timilsina, G.R. *Household Cooking Fuel Choice and Adoption of Improved Cookstoves in Developing Countries*; The World Bank Development Research Group: Washington, DC, USA, 2014.

19. Anwar, M.A.; Nasreen, S.; Tiwari, A.K. Forestation, renewable energy and environmental quality: Empirical evidence from Belt and Road Initiative economies. *J. Environ. Manag.* **2021**, *291*, 112684. [CrossRef] [PubMed]

20. Manolis, E.N.; Zagas, T.D.; Karetsos, G.K.; Poravou, C.A. Ecological restrictions in forest biomass extraction for a sustainable renewable energy production. *Renew. Sustain. Energy Rev.* **2019**, *110*, 290–297. [CrossRef]

21. Galik, C.S.; Benedum, M.E.; Kauffman, M.; Becker, D.R. Opportunities and barriers to forest biomass energy: A case study of four U.S. states. *Biomass Bioenergy* **2021**, *148*, 106035. [CrossRef]

22. Balmes, J.R. Household air pollution from domestic combustion of solid fuels and health. *J. Allergy Clin. Immunol.* **2019**, *143*, 1979–1987. [CrossRef]

23. Li, Y.; Xu, H.; Wang, J.; Hang Ho, S.S.; He, K.; Shen, Z.; Ning, Z.; Sun, J.; Li, L.; Lei, R.; et al. Personal exposure to PM2.5-bound organic species from domestic solid fuel combustion in rural Guanzhong Basin, China: Characteristics and health implication. *Chemosphere* **2019**, *227*, 53–62. [CrossRef] [PubMed]

24. Wang, D.; Li, Q.; Shen, G.; Deng, J.; Zhou, W.; Hao, J.; Jiang, J. Significant ultrafine particle emissions from residential solid fuel combustion. *Sci. Total. Environ.* **2020**, *715*, 136992. [CrossRef]

25. Du, W.; Yun, X.; Fu, N.; Qi, M.; Wang, W.; Wang, L.; Chen, Y.; Shen, G. Variation of indoor and outdoor carbonaceous aerosols in rural homes with strong internal solid fuel combustion sources. *Atmos. Pollut. Res.* **2020**, *11*, 992–999. [CrossRef]

26. Ezzati, M.; Kammen, D.M. The Health Impacts of Exposure to Indoor Air Pollution from Solid Fuels in Developing Countries: Knowledge, Gaps, and Data Needs. *Environ. Health Perspect.* **2002**, *110*, 1057–1068. [CrossRef]

27. Ray, I.; Smith, K.R. Towards safe drinking water and clean cooking for all. *Lancet Glob. Health* **2021**, *9*, 361–365. [CrossRef]

28. Quinn, A.K.; Bruce, N.; Puzzolo, E.; Dickinson, K.; Sturke, R.; Jack, D.W.; Mehta, S.; Shankar, A.; Sherr, K.; Rosenthal, J.P. An analysis of efforts to scale up clean household energy for cooking around the world. *Energy Sustain. Dev.* **2018**, *46*, 1–10. [CrossRef]

29. Rosenthal, J.; Quinn, A.; Grieshop, A.P.; Pillarisetti, A.; Glass, R.I. Clean cooking and the SDGs: Integrated analytical approaches to guide energy interventions for health and environment goals. *Energy Sustain. Dev.* **2018**, *42*, 152–159. [CrossRef]

30. Neto-Bradley, A.P.; Rangarajan, R.; Choudhary, R.; Bazaz, A. A clustering approach to clean cooking transition pathways for low-income households in Bangalore. *Sustain. Cities Soc.* **2021**, *66*, 102697. [CrossRef]

31. Aemro, Y.B.; Moura, P.; de Almeida, A.T. Experimental evaluation of electric clean cooking options for rural areas of developing countries. *Sustain. Energy Technol. Assess.* **2021**, *43*, 100954.

32. Kajumba, P.K.; Okello, D.; Nyeinga, K.; Nydal, O.J. Experimental investigation of a cooking unit integrated with thermal energy storage system. *J. Energy Storage* **2020**, *32*, 101949. [CrossRef]

33. Solar Cooker Distribution—S.C.I. Available online: http://www.solarcookers.org/our-work/solar-cooker-distribution/ (accessed on 3 June 2020).

34. Arunachala, U.C.; Kundapur, A. Cost-effective solar cookers: A global review. *Sol. Energy* **2020**, *207*, 903–916. [CrossRef]

35. Rodríguez Morales, J.Á.; González-Avilés, M.; Servín Campuzano, H.; Masera, O. T'imani a Multifunctional Solar System to Provide Cooking and Water Heating Rural Energy Needs. *Energies* **2020**, *13*, 3429. [CrossRef]

36. Zafar, H.A.; Badar, A.W.; Butt, F.S.; Khan, M.Y.; Siddiqui, M.S. Numerical modeling and parametric study of an innovative solar oven. *Sol. Energy* **2019**, *87*, 411–426. [CrossRef]

37. Ayub, I.; Nasir, M.S.; Liu, Y.; Munir, A.; Yang, F.; Zhang, Z. Performance improvement of solar bakery unit by integrating with metal hydride based solar thermal energy storage reactor. *Renew. Energy* **2020**, *161*, 1011–1024. [CrossRef]

38. Bhave, A.G.; Kale, C.K. Development of a thermal storage type solar cooker for high temperature cooking using solar salt. *Sol. Energy Mater. Sol. Cells* **2020**, *208*, 110394. [CrossRef]

39. Kumaresan, G.; Santosh, R.; Raju, G.; Velraj, R. Experimental and numerical investigation of solar flat plate cooking unit for domestic applications. *Energy* **2018**, *157*, 436–447. [CrossRef]

40. Sagade, A.A.; Samdarshi, S.K.; Lahkar, P.J.; Sagade, N.A. Experimental determination of the thermal performance of a solar box cooker with a modified cooking pot. *Renew. Energy* **2020**, *150*, 1001–1009. [CrossRef]
41. Mawire, A.; Lentswe, K.; Owusu, P.; Shobo, A.; Darkwa, J.; Calautit, J.; Worall, M. Performance comparison of two solar cooking storage pots combined with wonderbag slow cookers for off-sunshine cooking. *Sol. Energy* **2020**, *208*, 1166–1180. [CrossRef]
42. Chaudhary, R.; Yadav, A. Twin vessel solar cook stove for the simultaneous cooking of two different cooking articles. *Sol. Energy* **2020**, *208*, 688–696. [CrossRef]
43. Hosseinzadeh, M.; Faezian, A.; Mirzababaee, S.M.; Zamani, H. Parametric analysis and optimization of a portable evacuated tube solar cooker. *Energy* **2020**, *197*, 116816. [CrossRef]
44. Radim Rybár, R.; Kudelas, D.; Beer, M.; Horodníková, J. Elimination of Thermal Bridges in the Construction of a Flat Low-Pressure Solar Collector by Means of a Vacuum Thermal Insulation Bushing. *J. Sol. Energy Eng.* **2015**, *137*, 054501. [CrossRef]
45. Iqbal, M. *An Introduction to Solar Radiation*; Academic Press: Toronto, ON, Canada, 1983; pp. 303–334.
46. Bird, R.E.; Hulstrom, R.L. *A Simplified Clear Sky Model for Direct and Diffuse Insolation on Horizontal Surface*; Solar Energy Research Institute: Golden, CO, USA, 1981; p. 46.
47. Badescu, V. *Modeling Solar Radiation at the Earth's Surface*; Springer: Berlin, Germany, 2008; pp. 93–113.
48. Dassault Systemes. *Solidworks Flow Simulations Technical Reference*; Dassault Systemes: Vélizy-Villacoublay, France, 2017.
49. Sobachkin, A.; Dumnov, G. *Numerical Basis of CAD-Embedded CFD*; Solidworks Technical Reference; Dassault Systems: Vélizy-Villacoublay, France, 2017.
50. Cengel, Y.; Boles, M. *Thermodynamics: An Engineering Approach*, 8th ed.; McGraw-Hill Education: New York, NY, USA, 2014.

Article

The Relationship between ROP Funds and Sustainable Development—A Case Study for Poland

Łukasz Mach [1,*] , Karina Bedrunka [2], Ireneusz Dąbrowski [3] and Paweł Frącz [4]

[1] Faculty of Economics and Management, Opole University of Technology, Luboszycka 7, 45-036 Opole, Poland
[2] Department of Coordination of Operational Programs, Marshal Office of Opolskie Voivodship, Krakowska 38, 45-075 Opole, Poland; karbed74@gmail.com
[3] Department of Applied Economics, SGH Warsaw School of Economics, Madalińskiego 6/8, 02-513 Warsaw, Poland; ireneusz.dabrowski@sgh.waw.pl
[4] Faculty of Electrical Engineering Automatic Control and Informatics, Opole University of Technology, Prószkowska 76, 45-758 Opole, Poland; p.fracz@gmail.com
* Correspondence: l.mach@po.edu.pl; Tel.: +48-505848287

Abstract: The aim of this research is to analyse the correlation between public intervention in Poland within the Regional Operational Programmes and the key macroeconomic variables for the sustainable development of regions, i.e., the labour market, with particular emphasis on the unemployment rate and the level of employment; the average monthly remuneration; the residential construction market, with particular emphasis on the number of permits issued for the construction of apartments and the number of apartments under construction. The research and the analyses carried out on the basis of the above-mentioned aspects made it possible to indicate the relations between the studied macroeconomic indicators and the EU funds spending in Polish provinces, which will enable the implementation of the sustainable development policy. The capitals used in the research process are very important components of the region's and country's sustainable development. In the research, a calculation methodology was applied based on the analysis of time variability of the examined determinants, their correlation and regression relationships. The tools and methods of data analysis used allowed the quantification of the relationship between the macroeconomic determinants studied and the pace and value of payments made. The conducted analyses have shown a positive influence of the payments made in Poland within the framework of Regional Operational Programmes on selected macroeconomic indicators, i.e., regional economic and social-institutional capitals. The research results obtained may have a practical decision-making aspect for regional and national authorities responsible for the disbursement of EU funds.

Keywords: sustainable development; determinants of sustainable development; regional operational programmes; European Union funds

Citation: Mach, Ł.; Bedrunka, K.; Dąbrowski, I.; Frącz, P. The Relationship between ROP Funds and Sustainable Development—A Case Study for Poland. *Energies* **2021**, *14*, 2677. https://doi.org/10.3390/en14092677

Academic Editor: Sergey Zhironkin

Received: 23 March 2021
Accepted: 26 April 2021
Published: 6 May 2021

Publisher's Note: MDPI stays neutral with regard to jurisdictional claims in published maps and institutional affiliations.

1. Introduction

The weakening position of the European Union as the dominant global economy, including the deteriorating social, economic [1], and demographic situation [2] in Europe, forces the application of a new approach to programming activities whose basic goal is to strengthen the EU competitiveness. It is necessary that the concentration of intervention covered correctly diagnosed areas of development, which will result in the macroeconomic development of the European Union as a whole, as well as its individual member states and regions [3]. Economic growth is highly influenced by certain macroeconomic indicators [4]. Thus, the sustainable development of the EU, a country, or a region is a complex process influenced by numerous internal and external factors [5]. The shaping of development policy requires a thorough analysis of the factors as well as great courage, willingness, and readiness to actively and effectively minimise the negative impact of unfavourable trends [6].

In the national dimension, the Polish economy is an integral part of the global economic system; thus, it should be remembered that one of the key factors affecting Poland's economic development is the global economic situation [7,8]. The dynamics of the Polish economy and its dependence on cycles and development trends of the global system are influenced both by direct economic relations and global markets conditions, including the consequences of participation in the common market of the European Union and in multilateral free trade agreements [6]. It should also be borne in mind that the development of the country as a whole is a component of the development of its individual regions, whose growth potential has been and continues to be stimulated thanks to the support of the European Union structural funds. In Poland, this support has been in place since 1999 and significantly increased after accession in 2004. At the moment, the third perspective of the EU funds spending is coming to an end; these are the years 2004–2006, 2007–2013, and 2014–2020. Poland is the largest recipient of the EU assistance among all EU member states. In 2004, the Polish GDP per capita equalled 51% of the EU's average, and over a period of 15 years, it has grown by 15%, i.e., to 73% in 2019 [9]. Between 2014 and 2020, the amount reached over 87 billion euros. Over 35.9% of this allocation, i.e., 31.3 million euros, is assigned for 16 Polish regions under regional operational programmes. These funds are managed by regional authorities and are earmarked for increasing socio-economic competitiveness, including sustainable development [9].

Taking the above into account, the aim of the subject research is to analyse the relationship between the public intervention under Regional Operational Programmes (ROP) and macroeconomic variables, crucial for the development of the regions, i.e.,

- The labour market, with particular reference to the unemployment rate and the employment level;
- Average monthly remuneration;
- The residential construction market, with particular attention to the number of construction permits and the number of apartments under construction.

The research, and the analyses carried out on its basis, will indicate the relations between the studied development determinants and the EU funds spending in the 16 Polish provinces. In the research, the following hypotheses were made:

Hypothesis 1. *There is observed a seasonality in the payments made in relation to the analysed variables.*

Hypothesis 2. *The correlation between the payment and macroeconomic variables depends on the amount of financial support.*

Hypothesis 3. *There is a time correlation between the payments and the analysed macroeconomic areas.*

Hypothesis 4. *The analysed macroeconomic variables are characterized by a different flexibility with relation to the payments made.*

The implementation of the research process was performed in several key stages. The first provides an in-depth analysis of strategic development planning from a transnational, national, and regional perspective. The second one is the literature analysis of the studied issue, and the third one contains calculations showing the influence, expressed quantitatively, between the EU funds spending and selected determinants of regional development. The conducted research will show the relationship caused by spending EU funds on selected regional macroeconomic variables. As a result of the literature analysis, the selection of variables for the analysis was narrowed down to the identification of the correlation between the EU spending and regional economic, social and institutional capitals of sustainable development. The selected capitals are very important economic and social components of sustainable development. It should also be strongly emphasised that the conducted research is based on created simulation models, the aim of which is to show—in a simplified and generalised way—the researched interdependencies.

Undoubtedly, EU project-related payments are significant for the variables under study. This article focuses only on the regional dimension of EU funds. At the same time, it should be noted that in Poland, EU funds are also allocated within the framework of national programmes. Thus, the macroeconomic variables under study are influenced by other public funds and private capital on the market. It should also be noted that the effectiveness and structure of the obtained funds within the ROP programme in relation to the total value of implemented projects is usually only 30–40%. This creates problems in obtaining the remaining funds for the implementation of projects [10]. ROP programmes are sometimes criticized for supporting urban regions or rich regions that have their own sources of funding to implement projects. This, instead of balanced development of the regions, may cause even greater polarisation and increased economic disparities between regions [11]. An important issue is also the analysis of differences in the amount of funds absorbed by individual countries and regions [12].

2. Cascading Strategy of Development Planning—Established Assumptions

While presenting a cascading strategy description for economic and social development planning, first, the adopted assumptions for development at the transnational level were described, i.e., first, the European Union; then, the national level, i.e., Poland; and finally, the regional level, i.e., Polish provinces. Planning documents are shaped at the EU, national, and regional levels and are the basis for negotiations on structural funds for specific years.

The crisis has wiped out the results of many years of economic and social progress and exposed structural weaknesses in the European economy. At the same time, the world has been changing rapidly, and long-term issues, such as globalisation, increasing demands on limited resources, and ageing populations, are becoming more and more pressing. Europe must take care of its future. Europe can succeed if it acts together as a Union. Appropriate strategic planning is needed, including planning documents. The main one is the Strategy for smart, sustainable, and inclusive growth—Europe 2020 (hereinafter Europe 2020), which will make the EU economy smart, sustainable, and inclusive, with high rates of employment, productivity, and social cohesion. Europe 2020 is a vision of a social market economy for Europe in the 21st century. It comprises three interrelated priorities:

- Smart growth: economy development based on knowledge and innovation;
- Sustainable development: promotion of a more resource-efficient, greener, and more competitive economy;
- Inclusive growth: fostering of a high-employment economy delivering social and territorial cohesion.

The European Union had to define where it wanted to be at the end of the described planning periods. To this end, several overarching and measurable targets have been proposed, relating to the following: the employment rates for people aged of 20–64 years (should be 75%); 3% of the Union's GDP should be invested in research and development; the "20/20/20" climate and energy targets should be met (including a reduction in carbon emissions of up to 30%, if conditions allow); the number of early school leavers should be reduced to 10%, and at least 40% of the younger generation should pursue higher education; the number of people at risk of poverty should be reduced by 20 million [13].

While presenting the adopted strategic planning process in the national perspective, it should be noted that for the financial perspective 2014–2020, European funds for Poland have been recognised as the main, although not the only, source of funding for investments ensuring dynamic, sustainable, and balanced development. Thus, the programming logic was based on the correlation of European expectations as regards the concentration on the objectives of Europe 2020 with the national objectives indicated in the medium-term national development strategy, i.e., the National Development Strategy 2020—Active society, competitive economy, efficient state, and operationalised in the integrated strategies. This makes it necessary to look at Poland's development more broadly than just in the context of using the EU funds [14]. It contains recommendations for public policies, providing a basis for changes in the development management system, including the existing strategic documents with a long- and medium-term perspective (strategies, policies, and programmes); however, it also requires verification of other implementation instruments [14]. This means that the assumptions at the level of regions with regard to the implementation of the EU funds had their basis in it. While presenting specific measures, it should be reminded that in 2017 the Polish Government adopted the Responsible Development Strategy by 2020 (with the perspective by 2030). The main objective of the development measures designed in the Strategy is the creation of conditions for growth of income of the inhabitants of Poland with a simultaneous increase of cohesion in the social, economic, environmental, and territorial dimension. The strategy is geared towards inclusive socio-economic development. The document assumes that social cohesion is the main driver of development and a public priority. The strategy subordinates the activities in the economic sphere to achieving objectives related to the standard and the quality of life of the Polish citizens and puts emphasis on making the citizens, and the areas so far neglected in the development policy, benefit from the economic development to the extent greater than so far. The strategy presents a new development model—responsible development, i.e., the development that, while building competitive strength using new development factors, ensures participation and benefits to all social groups living in different parts of our country. This will be done by focusing legal, institutional, and investment actions on three objectives, i.e., sustainable economic growth based increasingly on knowledge; data and organisational excellence; socially responsive and territorially balanced development; effective state and institutions for growth and social and economic inclusion [15].

3. Research Evolution and Development: Region, Development, Sustainable and Smart Growth, and Smart Specialisations

Poland's accession to the European Union structures gave rise to intensified academic discussion on development, regions and regional development, sustainable and intelligent development, and the methods for their evaluation. There are many publications in Poland and Europe which describe these issues directly or indirectly. The following theoretical analysis of the issue has been carried out according to the procedure shown in Figure 1.

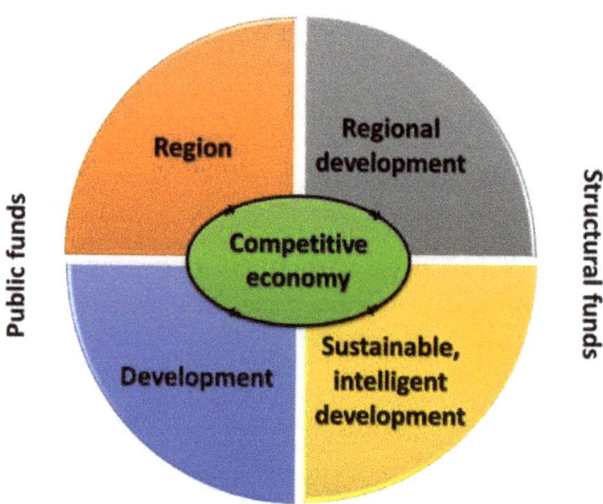

Figure 1. Diagram of relationship between EU Structural Funds and competitiveness of economy.

The notion of development should be understood as any change in the economic, social, and environmental system; however, the attribute of such a change is its irreversibility [16]. Development refers to the desired positive transformations of quantitative, qualitative, and structural properties of a given system [17]. Thus, spatial and temporal irreversibility is not a single attribute of development. It is worth adding another property to it, i.e., a positive evaluation of the changes taking place from the point of view of a particular system.

In the context of social and economic development, a region should be considered in terms of the relationship between the changes occurring at the local and global level [18] (p. 4). In the global economy, which is dominated by the process of globalisation, the area can become competitive only when it takes advantage of its individual characteristics while adapting to the conditions and requirements of the global environment [19]. Currently, the region is classified mainly in terms of economy, where it is possible to identify areas coherent by the role of a particular branch of services or industry [20]. Thus, the events and processes occurring in the region most often determine whether the region is developing or not. The region is identified as an element of developmental policy in terms of economic, institutional, demographic, natural, infrastructural, spatial, potential, and living conditions of inhabitants [21]. In the economic aspect, the region can be considered in relation to the functioning and mutual interaction of the private and the public sectors. In turn, taking into account the logic of market economy, regions treated as public sector entities function in a multi-level system. In this aspect, the national and transnational levels are most significant since the regions which receive financial support from central authorities and transnational institutions, and in which high-level institutions and infrastructure are located, have a chance to strengthen their competitiveness [5].

In this context, the concept of regional development should be investigated. It is more and more frequently defined as a holistic, structural, and strategic process by which a region's resources and conditions, its technological and cultural potential, and the opportunities identified in regional, national, and global markets are exploited by companies [5]. J. Regional development is influenced by both internal (endogenous) and external (exogenous) factors. At the same time, regional development models, which define a comprehensive and coherent way of explaining the mechanisms of regional development, are concerned with identifying only the key (priority) potentials that are important for development; they mainly revolve around economic growth [5,22]. Regional development can be considered to be a systematic improvement of competitiveness of entities and living standards of inhabitants as well as an increase in the economic potential of regions, contributing to the social and economic development of the country [23].

Among researchers, there is no coherent approach to the concept of regional development because, due to the changing environment, it is subject to constant modification. Sustainable development is one of the concepts of regional development. It is defined as a process of changes in the states of dynamic balance among local economic, social, as well as ecological and spatial development. On the basis of respect for natural resources, the ultimate goal of this process is to improve the quality of life in a broad sense [24]. The development policy objectives formulated by public authorities are characterised by certain principles which make it possible to operationalise them [25]. Sustainable development paradigm in regional development policy contains some conceptual features and operating principles; these are the development maintenance and sustainability [26]. Sustainable development as a concept of development policy defines the process of changing the states of dynamic balance between regional social, economic, as well as environmental and spatial development. There are two integrated pillars of this concept, i.e., balancing social (including political), economic, and environmental governance as well as the sustainability of development capitals achieved through the creation and diffusion of innovations [27]. Capitals are identified by, inter alia, human capital [28], social and institutional capital [29], and by physical and natural (ecological) capital [30].

Sustainable development should be viewed very broadly, including the way companies operate in the global economy. Currently, research is carried out in sustainable production [31]. Research is carried out to evaluate the progress in the implementation of the concept of sustainable development in the social aspect of the European Union between 2014 and 2018, with particular emphasis on Poland [32]. It is posited that increased emphasis on knowledge and economy factors increases country's competitiveness, which contributes to its sustainable development [33]. Sustainable development is also influenced by fiscal issues [34], climate protection [35], sustainable product life-cycle management, big databases and the use of artificial intelligence in businesses [36,37], networked and integrated urban technologies and sustainable smart energy systems, as well as sensor-based big data applications and computational urban network in a smart energy management system [38,39].

Sustainable development in all EU Member States is regarded as an integral factor in the economic and social policy of the state [40]. At the same time, this approach promotes the growth model proposed in Europe 2020, which is based on three priorities: smart growth, sustainable growth, and inclusive growth [14]. Smart growth means increasing the role of knowledge and innovation as drivers of our future development. This involves raising the quality of education, improving research performance, promoting innovation and knowledge transfer throughout the Union, making full use of information and communication technologies, and ensuring that innovative ideas can be turned into new products and services that create growth, jobs, and solutions for societal problems in Europe and globally. Entrepreneurship, financial resources, consideration of users' needs, and market opportunities are also necessary [14]. Inclusive growth is understood as a set of actions meant to promote a high-employment economy that ensures social and territorial cohesion. It is implemented through the Agenda for new skills and jobs and the European Platform against Poverty [41]. Research shows the effects of the implementation of the Europe 2020 from the point of view of the objectives on poverty and social exclusion [42].

A number of evaluation methods and tools are available in the literature that can be used to evaluate elements of regional development policy, including sustainable development. In the conducted research, attention is primarily paid to its usefulness within the framework of multidimensional processes of regional development, in which it is necessary to take into account the social, economic, and environmental dimensions of sustainable development [25]. The method of ratio analysis can be used to evaluate the effectiveness of sustainable development [43]. The same method should also be used to study the effectiveness of strategies and programmes based on the capitals as well as on the orders of sustainability and regional development (the so-called integrated strategic effectiveness) from the point of view of effectiveness, efficiency, and feasibility [44].

In the ratio analysis of integrated performance evaluation and in the analysis of effectiveness and efficiency of sustainable development for orders and capitals, static, dynamic, and criterion analyses are taken into account, including the integrating orders criteria, and in the capital—the spatial and temporal criteria [45]. The complexity of the category of sustainable development—in the concept of a set of features, objectives, principles, and integration of orders—entails attempts to operationalise this concept and the size of the cross-section of the ratio analyses. Criteria for the classification of indicators include, among others, the extent to which the characteristics, objectives and principles as well as governance of sustainable and balanced development have been achieved [43].

The new EU financial perspective for 2014–2020 and the closely related strategic vision of Europe 2020 clearly define the approach to the environment and its natural resources; it is clearly based on a strong principle of sustainable development [46]. The concept of smart specialisation is conducive to directing regions towards the creation of eco-innovations, which are understood as intentional activity, characterised by entrepreneurship and which encompasses a product design phase and its integrated management throughout its life cycle, that contributes to the pro-ecological modernisation of societies by taking into account environmental concerns in the development of products and related processes [47,48]. Eco-innovation reflects the concept of a clear focus on reducing environmental impacts, where such effects may or may not occur without limitations to product, process, marketing, or organisational innovation but also including innovation in social structures [49].

Intelligent and sustainable development is closely related. The process of arriving at smart specialisations in Regional Innovation Strategy of Podkarpackie Province was fully of an entrepreneurial discovery process. It basically covered two years, i.e., 2012 and 2013, although it used a number of documents and research findings. The methodology of creating of the Regional Innovation Strategy, including the methodology used for the evaluation of all stakeholders, as well as the criteria for the selection of smart specialisations, was of a uniform nature, showing the continuity and cohesion of individual stages. While preparing the document, a triangulation of methods was made, so that the final result was not derived from only one method used but was adopted when all methods used gave the same or similar result. The basic methods used in the process of creating the Strategy were the following: the analysis of strategic documents and other available sources of knowledge; the analysis of foresight projects carried out for the region; SWOT analysis in terms of social and economic potential of Podkarpackie; the analysis of stakeholders—also performed to identify the most important stakeholders; various forms of meetings and discussions, practiced on a continuous basis; the analysis of the potential and opportunities for development of clusters; and performing primary research with a very wide economic spectrum [50].

In Poland, in the Opolskie province, there has been developed an original model for the selection of regional smart specialisations and the creation of the Regional Innovation Strategy by 2020. It was based on the following methods and tools: content analysis; industry analysis; desk research; time series/trend forecasting; stakeholder consultation; Delphi method; creative imaging; impact assessment; PEST (Political, Economic, Socio-cultural, Technological); logic diagram; environmental scan; visioning; and workshops on future occurrences [51,52]). At the same time, it broadly describes the monitoring process of the Strategy on the basis of the Action Plan and selected indicators, which—to a lesser extent—is visible in the works carried out for Podkarpackie Province.

Another interesting example of research in this area is the analysis of higher education institutions from the point of view of their role as innovation brokers in the context of smart specialisations [53], or analyses of the whole regional innovation system in the context of the backwardness of European regions [54].

4. Analysis of the Correlation between the Regional Operational Programme and Selected Macroeconomic Determinants of Development—Research Perspective for Poland

When analysing the correlation between payments made under the ROP for Poland and the selected macroeconomic determinants, three selected areas of the economy were described. In terms of macroeconomic theory, these areas are crucial for the country's and regions' sustainable development. The first is the area of the labour market (employment); the second is the average monthly remuneration, while the third area of analysis is the housing market, which was described by two dimensions, i.e., the number of building permits issued and the number of current construction projects in the housing market. The choice of the indicated areas of analysis results from the analysis of issues concerning capitals and governance described by the authors of that publication in Chapter 2. It should be reminded that the selected areas of analysis belong to economic, social, and institutional capitals and their analysis is aimed at showing the relationship between public intervention under regional operational programmes and the creation and consolidation of these capitals.

Each of the described areas was scrutinised according to the methodological scheme taking into account the following:

- The analysis of the time variability of the examined determinants against the background of the payments made under ROP for Poland in general. The variability analysis was performed by assessing the nature of developmental trends for the relationships studied.
- The correlation (and cross-correlation) analysis between the examined determinants and payments made within ROP for Poland in total. The correlation analysis was performed using Pearson's linear correlation, assuming that x and y are the random

variables analysed with discrete distributions. x_i and y_i denote random sample values of these variables ($i = 1, 2, \ldots, n$), while $\bar{x}$ and $\bar{y}$ are the mean values of these samples. Then, the estimator of the linear correlation coefficient was determined according to Equation (1).

$$r_{xy} = \frac{\sum_{i=1}^{n}(x_i - \bar{x})(y_i - \bar{y})}{\sqrt{\sum_{i=1}^{n}(x_i - \bar{x})^2}\sqrt{\sum_{i=1}^{n}(y_i - \bar{y})^2}} \tag{1}$$

- The functional multiple regression analysis based on the estimation of structural parameters of the analysed models. The estimation was made according to Equation (2).

$$\begin{bmatrix} y_1 \\ y_2 \\ \ldots \\ y_n \end{bmatrix} = \begin{bmatrix} 1 & x_{11} & x_{12} & \ldots & x_{1k} \\ 1 & x_{21} & x_{22} & \ldots & x_{2k} \\ \ldots & \ldots & \ldots & \ldots & \ldots \\ 1 & x_{n1} & x_{n2} & \ldots & x_{nk} \end{bmatrix} \begin{bmatrix} a_0 \\ a_1 \\ \ldots \\ a_k \end{bmatrix} + \begin{bmatrix} \varepsilon_1 \\ \varepsilon_2 \\ \ldots \\ \varepsilon_n \end{bmatrix} \tag{2}$$

where vectors $y = \begin{bmatrix} y_1 \\ y_2 \\ \ldots \\ y_n \end{bmatrix}\begin{bmatrix} y_1 \\ y_2 \\ \ldots \\ y_n \end{bmatrix}$, $a = \begin{bmatrix} a_0 \\ a_1 \\ \ldots \\ a_k \end{bmatrix}$, $\varepsilon = \begin{bmatrix} \varepsilon_1 \\ \varepsilon_2 \\ \ldots \\ \varepsilon_n \end{bmatrix}$

and matrix $X = \begin{bmatrix} 1 & x_{11} & x_{12} & \ldots & x_{1k} \\ 1 & x_{21} & x_{22} & \ldots & x_{2k} \\ \ldots & \ldots & \ldots & \ldots & \ldots \\ 1 & x_{n1} & x_{n2} & \ldots & x_{nk} \end{bmatrix}$

contain the following values:

y—dependent variable, a—model parameters, ε—residual value, X—independent variable.

For each correlation under analysis, there were estimated parameters of two functional correlations, i.e., linear (cf. 3) and logarithmic (cf. 4)

$$y = ax + b \tag{3}$$

$$y = a * e^{\frac{-x}{b}} + c \tag{4}$$

It should also be added that the extent to which the model fits the data, i.e., quality of the developed model was assessed based on the coefficient of determination R^2 (Formula (5)).

$$R^2 = \frac{\sum_{i=1}^{n}(\hat{y}_i - \bar{y})^2}{\sum_{i=1}^{n}(y_i - \bar{y})^2} \tag{5}$$

where:

y_i—real value of variable Y in moment/period i.

$\hat{y}_i$—model-based value in moment/period i.

$\bar{y}$— arithmetic average of empirical values of the dependent variable.

The data was obtained from the databases of the Central Statistical Office and from the databases of individual Marshal Offices. It was collected with a time interval of one month. The data used for analysis was collected from January 2015 to July 2020 (in aggregate, 66 values for each studied variable were collected). The collected data include:

x_1—independent variable—total payments under the ROP (data acquired from the Marshal Office of Opolskie Voivodship).

y_1—dependent variable—employment rate (data obtained from the Main Office of Statistics).

y_2—dependent variable—average employment rate (data obtained from the Main Office of Statistics).

y_3—dependent variable—number of apartment construction permits (data obtained from the Main Office of Statistics).

y_4—dependent variable—number of apartments under construction (data obtained from the Main Office of Statistics).

On the basis of the data presented above, a total of 4 groups of models have calculated (for each y_i) for which the cross-correlation, functional correlation relationship and parameters of the regression function have been described in detail. By computing 4 different models, knowledge was gained about the relationship between each dependent variable (y_i) analysed and payments made. It should also be noted that another approach in the conducted analyses may also be the simultaneous inclusion in the model of multiple dependent variables and conducting multiple regression analysis.

When analysing the first of the listed variables, i.e., payments made for Poland in total, it should be noted that the increased payments occurred annually at the end of each of the surveyed years. It should also be noted that there was an upward trend in the payments made during the period in question (see Figure 2). The seasonal character of the payments made under the ROP in the years covered by the study is caused, among other things, by the annual financial settlement of the programme-related activities. By subjecting to the analysis of the development of the employment rate in the studied period, we can also notice the tendency of its increase, especially visible in 2016–2018. Comparing the data in Figure 2 on the two horizontal diagrams on it, one can argue about the relationship between the payments made under ROP and the level of employment in Poland. It should also be noted that there are practically no time lags in the described variables in the relationships studied. This demonstrates the high elasticity of change between the studied variables.

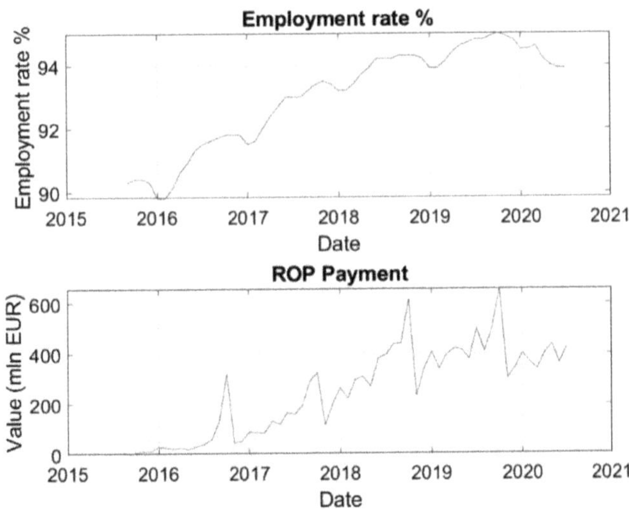

Figure 2. Employment and payments made under ROP—Poland.

In order to confirm the occurrence of significant elasticity of changes between the examined variables, cross-correlation diagrams were prepared for the variable employment to the ROP payments made. From the presented Figure 3, from the obtained decreasing correlation values over time, it can be concluded that the highest correlation relationship is for periods 0, 1, and 2. The results obtained for the calculated cross-correlation confirm the almost immediate effect of payments on employment. Of course, one should be aware that the described correlations illustrate a simplified case based on the cross-correlation model. It should be remembered that this correlation does not explain the cause-effect relationship, which should be the investigated into by experts dealing with sustainable development of regions.

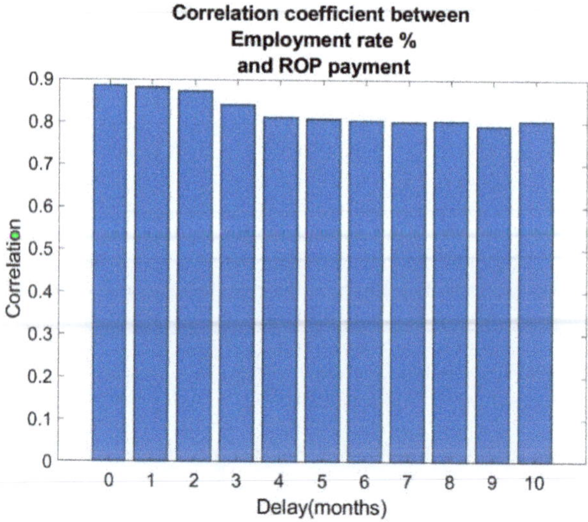

Figure 3. Cross correlation between employment and made payments.

In the next step of the research, the relationship between the quantitative impact of the payments made on the labour market was analysed because the employment rate is one of key economic barometers. Figure 4 shows the graph of the estimated regression function along with the calculated value of determination coefficients. From the relationship obtained, it can be concluded that the nature of the relationships studied is in the form of a logarithmic function (higher value of the coefficient of determination). The obtained results of the research allow us to put forward a thesis that the effectiveness of the payments made and their impact on the labour market is high, up to payments of about 400 million euros. Above this value, employment growth is of a slowing nature.

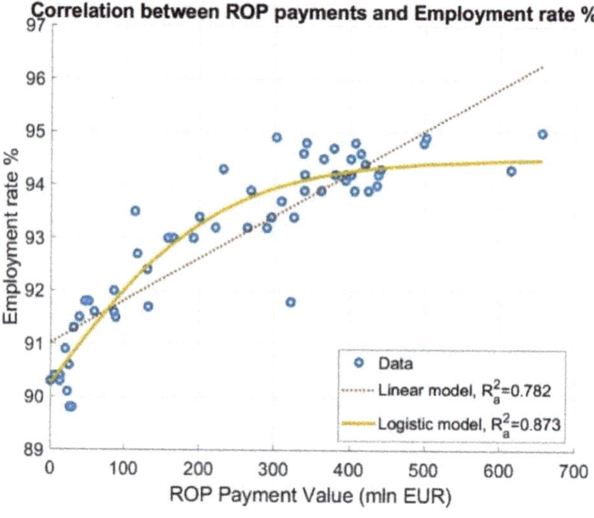

Figure 4. Correlation function between ROP payments and employment—Poland.

The second variable analysed is remuneration. For this variable, in the first step, the course of its variability was checked, and its course was compared with the variability of made payments within ROP (Figure 5). The significance of lags between comparable variables was also examined (Figure 6) and, after functional estimation for the described relationships, their regression dependence was assessed (Figure 7).

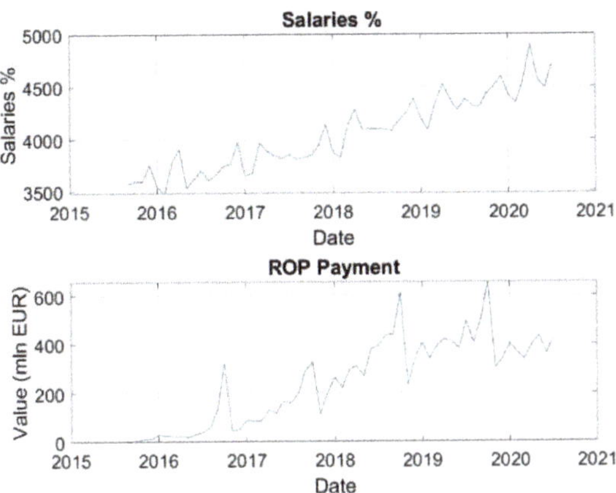

Figure 5. Salaries and payments made under ROP—Poland.

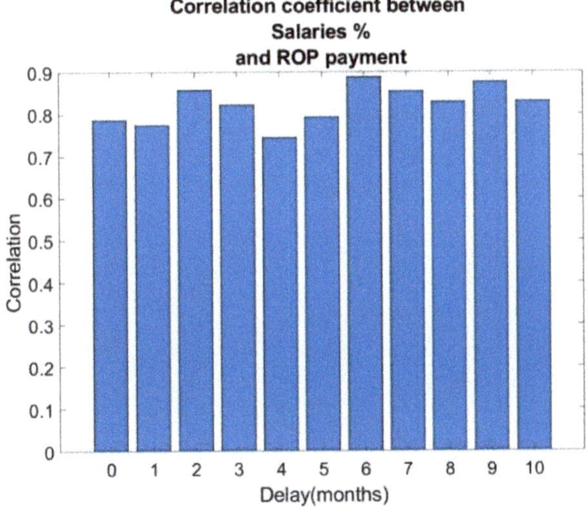

Figure 6. Cross correlation between employment and made payments.

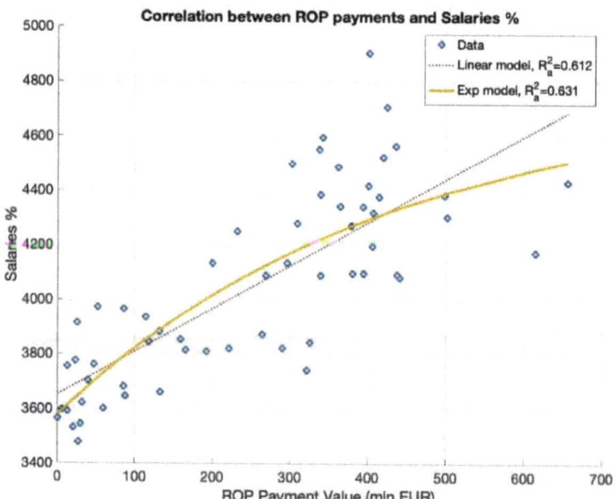

Figure 7. Correlation function between ROP payments and remuneration—Poland.

Comparing the time courses for the remuneration variable and made ROP payments, we can see an increasing trend for both variables studied. It should be noted that the remuneration variable is characterised by seasonality in the beginning/end of the year. This periodicity is due to the nature of the remuneration paid in Poland, where it is customary to pay additional remuneration at the end of the year, such as awards or annual bonuses. On the other hand, at the beginning of the year, in many companies, the so-called thirteenth salary is paid (see Figure 5). In the attempt to check the relationship between the remuneration variable and made payments, we can see (cf. Figure 6) that a six-month delay is the one with the highest value. This indicates that the time shift between the variables under study is characterised by a six-month period.

While interpreting the results obtained for the calculated autocorrelation functions (cf. Figure 7), we can see that the values of determination coefficients for the linear function and the logarithmic one are close to each other. Based on the results of the linear trend function, it can be noticed that the salaries grow along with the increased level of payments under the ROP in Poland. The average statistic increase in the average salary in relation to the increase amount of payments is expressed by formula $y_t = 1.579t + 3651$.

The last area examined is the residential construction market. The research examined the relationship between made payments and the first two stages of the housing construction process, which included the number of permits issued for the construction of new housing units and the number of units under construction. In Figures 8 and 9, we can see that for both examined variables there is a dynamic increase of values at the beginning of each examined year. It may also be noted that the execution of a construction project, as described by commenced construction, has a visible periodic component, which results from a strong dependence of the execution of construction projects on seasonal variability in the housing market. These phenomena result, among other things, from the differences between the climatic seasons in Poland.

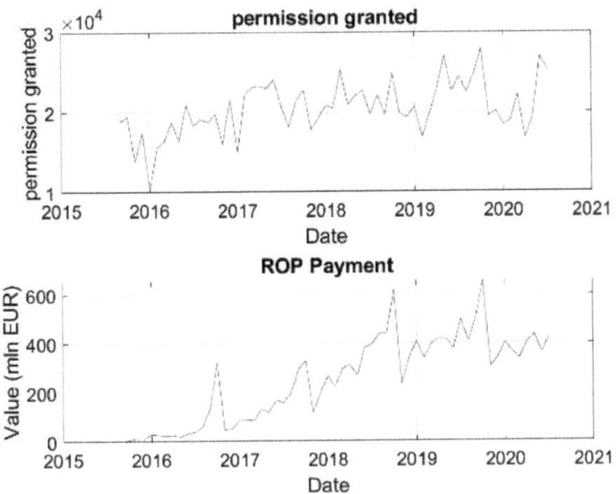

Figure 8. Permits issued for the construction of new apartments and payments made under ROP.

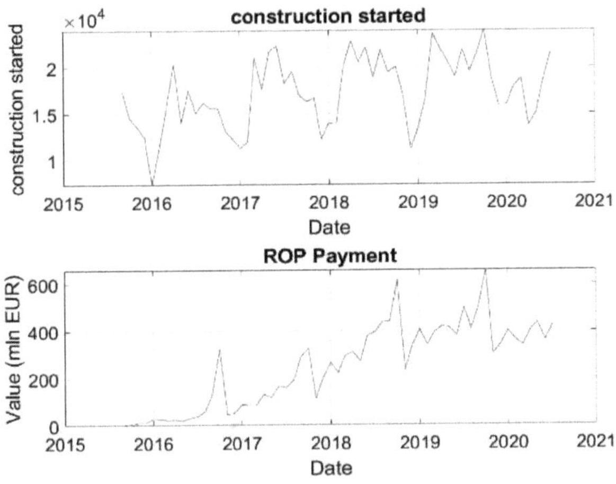

Figure 9. Apartments under construction and ROP payments made.

Making an attempt at a quantitative analysis of the examined dependencies, taking into account the linear regression models constructed, it may be noticed that if we increase the payments within the framework of ROP in Poland by 1 million euros, we will obtain an average increase of 110,500 building permits issued, while in the case of apartments under construction, an increase in payments by 1 million euros will result in an average increase of 114,300 units under construction. Comparable values of estimated parameters of regression functions for the analysed variables prove similar sensitivity of changes (cf. Figures 10 and 11).

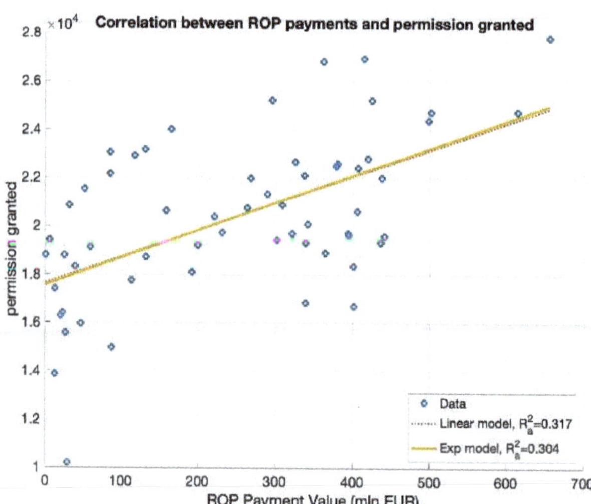

Figure 10. Correlation function between ROP Opolskie Province payments and permits issued for the construction of new apartments.

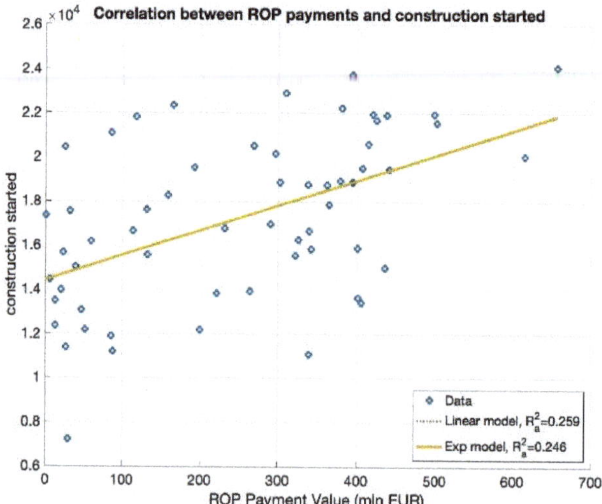

Figure 11. Correlation function between ROP Opolskie Province payments and housing units whose construction has begun.

Trying to assess significant time lags between the examined variables, it should be noted (cf. Figures 12 and 13) that no single significant time lag can be unambiguously identified for the examined relationships. This may testify to the fact of cyclical changes on the residential property market, and this study covered the analysis of seasonal fluctuations over a period of 10 months.

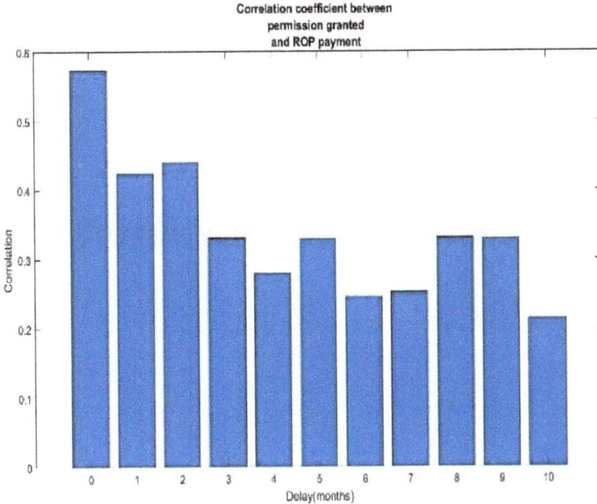

Figure 12. Cross correlation between completed payments and permits issued for new residential construction.

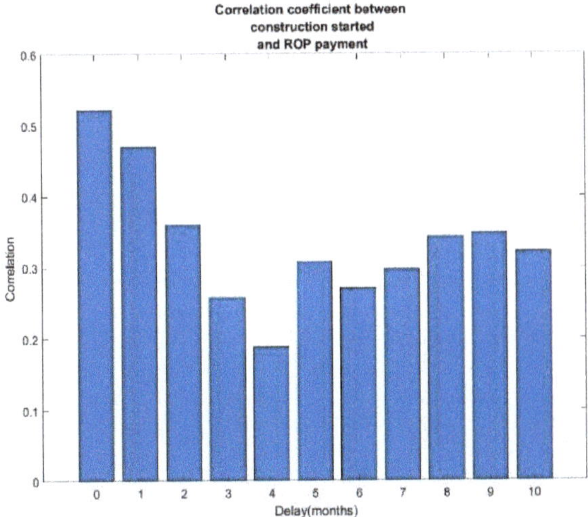

Figure 13. Cross correlation between payments made and dwellings started.

5. Results

The research carried out facilitated the identification and analysis of the relationship between the payments made throughout Poland under the regional operational programmes and selected macroeconomic variables. The analysis covered key macroeconomic aspects directly affecting, inter alia, the creation of the sustainable development potential of 16 Polish provinces and, consequently, the creation of Poland's competitive position in Europe and worldwide.

In order to perform a possibly systematic research inference, the literature analysis of the issue was conducted in the first stage. This analysis presents the validity of the cascading strategy planning process from the perspective of spending the European Union funds. It comprehensively describes the evolution of the perception of development of individual countries and regions, i.e., sustainable development, smart development, and inclusive growth, including the importance of smart specialisations. It is important from the point of view of shaping the directions of spending structural funds in particular periods of European Union programming.

Discussing the utilitarian research dimension of the conducted analyses, it is stated as follows:

- During the period considered, there are annually increased payment values at the end of each calendar year, which evidences the seasonal fluctuations.
- Analysis of the development of the employment rate shows that over the period considered, there is a tendency for it to increase, which was particularly noticeable between 2016 and 2018.
- The effectiveness of the payments made and their impact on the labour market is high up to payments of around 400 million euros. Above this value, employment growth is of a slowing nature.
- When comparing the time courses for the variables employment and ROP payments made, there is an increasing tendency. The employment variable is characterised by seasonality in the beginning/end of the year. In this analysis, the six-month lag is the one with the largest value.
- In the relationship between the made payments and the first two stages of the housing construction process, i.e., the number of permits issued for the construction of new apartments and the number of apartments under construction, a dynamic increase of the value at the beginning of each analysed year for both variables is noticeable.
- When attempting a quantitative analysis of the examined dependencies, it can be seen that if we increase the payments under ROP in Poland by 1 million euros, we will obtain an average increase of 110,500 building permits issued, while in the case of dwellings units under construction, an increase in payments by 1 million euros will result in an average increase of 114,000 units under construction.
- When attempting to assess significant time lags between the researched variables, i.e., made payments and the first two stages of the housing construction process, it should be noted that no single significant time lag can be unambiguously identified for the examined relationships within a period of 10 months.

It is worth underlining that it is not easy to assess the impact of structural funds on the economy. The authors are aware that in practice, it is often difficult to indicate how much public intervention is a direct cause of a region's social and economic development, including its sustainable development. When interpreting the results described in this study, one should be aware of their model approach, i.e., an approach showing a simplified picture of the reality which, by rule, focuses only on correlative and regressive relationships, without allowing for the cause-and-effect relationships present in economy.

The analyses carried out showed a positive relationship between the payments made under ROP and the selected macroeconomic indicators. The results obtained from the research may have a practical aspect for decision-making for both regional and national authorities responsible for disbursement of the EU funds. Referring to the practical goal of the research, recommendations for authorities at all levels of the EU spending are as follows:

- Determining the principles for the implementation of regional operational programmes in individual provinces should be the subject of permanent discussions and meetings of regional authorities with programme stakeholders. This approach guarantees greater efficiency in the disbursement of funds in a given province.
- The assessment of the rate of disbursement of the EU funds, i.e., payments from the operational programme, should be continuously monitored by the regional authorities so that corrective measures can be introduced in due time. The remedial models used should be based on past experience. They should be discussed with programme stakeholders.
- At the level of implementers of project initiatives financed from the EU funds, information activities should be carried out to promote the fastest possible implementation of projects and, at the same time, the importance of this approach from the point of view

of macroeconomic indicators, important for the development of economy, including job creation, the level of wages, or the pace of housing construction.

In conclusion, the economic and social position of the European Union on the global stage is a determinant for coordinated actions by regional and national authorities in the twenty-eight individual Member States. For Poland, it is of particular importance because it was the largest recipient of the EU funds between 2014 and 2020 as well as in the new EU perspective for 2021–2027. Structural Funds are the main vehicle for project initiatives and have a positive impact on the country's macroeconomic indicators. This situation leads to the emergence of new barriers that need to be eliminated in the short term to achieve the best possible results in the disbursement of the EU funds.

The authors have analysed ROP payments and their correlation with the selected macroeconomic indicators. It should be emphasized that it is worthwhile to continue the research that would focus on the influence of other public funds on selected indicators.

Author Contributions: Conceptualization, K.B. and Ł.M.; methodology, K.B. and Ł.M.; software, Ł.M.; validation, P.F. and I.D.; formal analysis, Ł.M. and I.D.; investigation, K.B. and P.F.; resources, Ł.M.; data curation, K.B.; writing—original draft preparation, K.B.; writing—review and editing, Ł.M.; visualization, K.B. and Ł.M.; supervision, I.D.; project administration, P.F.; funding acquisition, P.F. All authors have read and agreed to the published version of the manuscript.

Funding: The APC was funded by Opole University of Technology.

Acknowledgments: The paper presents the personal opinions of the authors and does not necessarily reflect the official position of Marshal Office of Opolskie Voivodship.

Conflicts of Interest: The authors declare no conflict of interest.

References

1. Radulescu, M.; Fedajev, A.; Sinisi, C.I.; Popescu, C.; Iacob, S.E. Europe 2020 Implementation as Driver of Economic Performance and Competitiveness. Panel Analysis of CEE Countries. *Sustainability* **2018**, *10*, 566. [CrossRef]
2. Ogólski, M. Wyzwania demograficzne Eropy i Polski. In *Europejskie Wyzwania dla Polski i Jej Regionów*; Ministerstwo Rozwoju Regionalnego: Warsaw, Poland, 2010.
3. Bedrunka, K.; Malik, K. Zintegrowana efektywność polityki rozwoju regionalnego w kolejnym okresie programowania 2014–2020. In *Handel Wewnętrzny:Trendy i Wyzwania Zrównoważonego Rozowoju w XXI Wieku*; Numer Specjalny (Lipiec-Sierpień), Tom 2; Institute for Market, Consumption and Business Cycles Research: Warsaw, Poland, 2012; pp. 7–18.
4. Iuga, I.C.; Mihalciuc, A. Major Crises of the XXIst Century and Impact on Economic Growth. *Sustainability* **2020**, *12*, 9373. [CrossRef]
5. Klasik, A. Przedsiębiorczość i konkurencyjność a rozwój regionalny -wprowadzenie. In *Przedsiębiorczość i konkurencyjność a rozwój regionalny*; Klasik, A., Ed.; Akademia Ekonomiczna im. Karola Adamieckiego w Katowicach: Katowice, Poland, 2006; p. 598.
6. *Strategia Rozwoju Kraju 2020—Aktywne Społeczeństwo, Konkurencyjna Gospodarka, Sprawne Państwo*; In Załącznik do Uchwały nr 157 Rady Ministrów z dnia 25 Września 2012; Warsaw, Poland, 2012.
7. Mobarak, H.; Murshed, N. *Antoine Augustin Cournot: The Pioneer of Modern Economic Ideas*; MPRA Paper; University Library of Munich: Munich, Germany, 2018.
8. Kuczmarka, M.; Pietryka, I. (Eds.) *Problemy Gospodarki Światowej*; Polskie Towarzystwo Ekonomiczne, Oddział w Toruniu: Toruń, Poland, 2010.
9. *Programowanie Perspektywy Finnasowej 2014–2020—Umowa Partnerska*; Styczeń 2020; Ministerstwo Infrastruktury i Rozwoju: Warsaw, Poland, 2020.
10. Mrozińska, A. Struktura oraz skuteczność pozyskiwania środków na projekty unijne realizowane w miastach wojewódzkich w perspektywie 2007-2013/Structure and effectiveness of obtaining funds on the EU projects implemented in Polish voivodeship cities in the perspective. *Prace Naukowe Uniwersytetu Ekonomicznego we Wrocławiu* **2017**, *467*. [CrossRef]
11. Jańczuk, L. Finanse i nieruchomości w rozwoju lokalnym i regionalnym. *Prace Naukowe Uniwersytetu Ekonomicznego we Wrocławiu* **2013**, *280*, 84–93.
12. Zoppi, C.; Lai, S. Differentials in the regional operational program expenditure for public services and infrastructure in the coastal cities of Sardinia (Italy) analyzed in the ruling context of the Regional Landscape Plan. *Land Use Policy* **2013**, *30*, 286–304. [CrossRef]
13. Europa 2020. Strategia na Rzecz Inteligentnego i Zrównoważonego Rozowju Sprzyjającego Włączeniu Społecznemu. Communication from th European Union: Brussels, Belgium, 2010.
14. *Programowanie Perspektywy Finnasowej 2014–2020 – Umowa Partnerska*; Sierpień 2017; Ministerstwo Infrastruktury i Rozwoju: Warsaw, Poland, 2020.

15. *Strategia Na Rzecz Odpowiedzialnego Rozwoju do 2020 (z Perspektywą do 2030)*; Dokument Przyjęty Uchwałą Rady Ministrów w Dniu 14 Lutego 2017 r; Council of Ministers, Rady Ministrów: Warsaw, Poland, 2017.
16. Czaja, S. Etropijno-energetyczna analiza procesów gospodarczych. Kierunki rozwoju analizy enegetycznej. *Ekonomia i Środowisko* **1998**, *2*, 9–20.
17. Sztado, A. Oddziaływanie samorządu loklanego na rozwój lokalny w świetle ewolucji modeli ustrojowych gmin. *Samorząd Terytorialny* **1998**, *11*, 12–29.
18. Chądzyński, J.; Nowakowska, A.; Przygodzki, Z. *Region i Jego Rozwój w Warunkach Globalizacji*; Publisher CeDeWu: Warsaw, Poland, 2007.
19. Di Mauro, F.; Dees, F.; McKibbin, W. *Globalisation, Regionalism and Economic Interdependence*; Cambridge University Press: Cambridge, NY, USA, 2008.
20. Bedrunka, K.; Dymek, Ł. Pojecie regionu i znaczenie jego rozwoju. In *Ekonomia: Przewodnik dla Studentw i Doktorantów Kierunków Technicznych*; Malik, K., Ed.; Politechnika Opolska: Opole, Poland, 2016; p. 308.
21. Stawasz, D. Gospodarka regionalna—Teoria i praktyka. In *Ekonomiczne i Organizacyjne Uwarunkowania Rozwoju Regionu—Teoria i Praktyka*; Wydawnictwo Uniwersytetu Łódzkiego: Lodz, Poland, 2004; p. 58.
22. Mayer-Stamer, J. Systematic Competitiveness and Local Economic Development. In *Large Systematic Change: Teories*; Bodhanya, S., Ed.; Modelling and Practices; Mesopartner: Duisburg, Germany, 2008; p. 7.
23. Szlachta, J. Główne problemy polityki rozwoju regionalnego Polski na przełomie XX i XXI wieku. In *Strategiczne Wyzwania dla Polityki Rozwoju Regionalnego Polski*; Broszkiewicz, R., Ed.; Friedrich-Ebert-Stiftung: Warsaw, Poland, 1996.
24. Borys, T. Jakość życia a zrównoważony rozwój. Relacja i pomiar. In *Ekonomia a Rozwój Zrównoważony, Tom 1*; Piontek, F., Ed.; Ekonomia i Środowisko: Białystok, Poland, 2001; p. 89.
25. Malik, K. *Ewaluacja Polityki Rozwoju Regionu. Metody, Konteksty i Wymiary Rozowju Zrównoważonego; Tom 135*; Polska Akademia Nauk, Komitet Przestrzennego Zagospodarowania Kraju: Warsaw, Poland, 2011.
26. Borys, T. *Wskaźniki Zrównoważonego Rozwoju*; Ekonomia i Środowisko: Warsaw, Poland; Białystok, Poland, 2005.
27. Barbier, B.; Markandya, E.A.; Pearce, W. Environmental sustainability and cost-benefit analysis. *Environ. Plan. A* **1990**, *22*, 1259–1266. [CrossRef]
28. Biela, A. Metodologia wyceny kapitału ludzkiego w kontekscie koncepcji zrównoważonego rozwoju. In *Ekonomia a Rozwój Zrównoważony, Tom 1*; Wydawnictwo Ekonomia i Środowisko: Bialystok, Poland, 2001; pp. 207–228.
29. Tomer, J. *Organizatorial Capital: The Path to Higher Productivnity and Well-Being*; Praeger: Westport, NY, USA, 1987.
30. El Serafy, S. The Environment as Capital. In *Economical Economics: The science and Managment of Sustainability*; Costanza, R., Ed.; Columbia University Press: New York, NY, USA, 1991; pp. 168–175.
31. Królczyk, G.; Król, A.; Kochan, O.; Su, J.; Kacprzyk, J. (Eds.) *Sustainable Production: Novel Trends in Energy, Environmental and Materials Systems*; Springer: Cham, Switzerland, 2020.
32. Barska, A.; Jędrzejczak-Gas, J.; Wyrwa, J.; Kononowicz, K. Multidimensional Assessment of the Social Development of EU Countries in the Context of Implementing the Concept of Sustainable Development. *Sustainability* **2020**, *12*, 7821. [CrossRef]
33. Širá, E.; Vavrek, R.; Vozárová, I.K.; Kotulič, R. Knowledge Economy Indicators and Their Impact on the Sustainable Competitiveness of the EU Countries. *Sustainability* **2020**, *12*, 4172. [CrossRef]
34. Ionescu, L. The economics of teh carbon Tax: Enviromental Performance, Sustainable Energy, and Green Financial Behavior. *Geopolit. Hist. Int. Relat.* **2020**, *12*, 101–107. [CrossRef]
35. Pflugmann, F.; De Blasio, N. The Geopolitics of Renewable Hydrogen in Low-Carbon Energy Markets. *Geopolit. Hist. Int. Relat.* **2020**, *12*, 7. [CrossRef]
36. Tacker, G. Sustainable Product Lifecycle Management, Industrial Big Data, and Internet of Things Sensing Networks in Cyber-Physical System-based Smart Factories. *J. Self-Governance Manag. Econ.* **2021**, *6*, 9–19. [CrossRef]
37. Lazaroiu, G.; Klistik, T.; Novak, A. Internet of Things Smart Devices, Industrial Artificial Intelligence, and Real-Time Sensor Networks in Sustainable Cyber-Physical Production Systems. *J. Self-Governance Manag. Econ.* **2021**, *6*, 20–30. [CrossRef]
38. Harrower, K. Networked and Integrated Urban Technologies in Sustainable Smart Energy Systems. *Geopolit. Hist. Int. Relat.* **2020**, *12*, 45. [CrossRef]
39. Eysenck, G. Sensor-based Big Data Applications and Computationally Networked Urbanism in Smart Energy Management Systems. *Geopolit. Hist. Int. Relat.* **2020**, *12*, 52. [CrossRef]
40. Antonescu, D. Regional Development Policy in Context of Europe 2020 Strategy. *Procedia Econ. Finance* **2014**, *15*, 1091–1097. [CrossRef]
41. Mazur-Wierzbicka, E. Wzrost sprzyjający włączeniu społecznemu jako wyzwanie dla Polski w świetle Strategii Europa 2020. In *Studia Ekonomiczne*; Zeszyty Naukowe Uniwersytetu Ekonomicznego w Katowicach: Katowice, Poland, 2015.
42. Lafuente, J.Á.; Marco, A.; Monfort, M.; Ordóñez, J. Social Exclusion and Convergence in the EU: An Assessment of the Europe 2020 Strategy. *Sustainability* **2020**, *12*, 1843. [CrossRef]
43. Malik, K. Analiza wskaźnikowa zintegrowanej efektywności strategii rozowju loklanego i regionalnego. In *Efektywność Zrównoważonego i Trwałego Rozwoju*; Wydawnictwo Instytut Śląski: Opole, Poland, 2004; p. 237.
44. Bedrunka, K.; Malik, K. (Eds.) Sustainable development as a criterion for eveluating the integrated implementation of the regional development policy. In *Multi-Dimensional Effectiveness of Regional Development Policy. Implementation: Evaluation Scheme for the Opole Voivodeship*; Polish Academy for Sciences Committee for Spatial Economy and Regional Planning: Warsaw, Poland, 2014; p. 168.

45. Malik, K. Monitorowanie efektywności zrównoważonego i trwałego rozwoju. In *Rola Małych Regionów w Rozowju Społeczno-Gospodarczym Kraju i Integracji Europejskiej*; Heffner, K., Ed.; Polska Akademia Nauk, Komitet Przestrzennego Zagospodarowania Kraju: Warsaw, Poland, 2004.

46. Borys, T. Nowe kierunki ekonomii środowiska i zasobów naturalnych w aspekcie nowej perspektywy finansowej Unii Europejskiej. *Ekonomia i Środowisko* **2013**, *1*, 12.

47. Carley, M.; Spapens, P. *Dzielenie Się Światem*; Instytut na rzecz Ekorozwoju: Warsaw/Białystok, Poland, 2000.

48. Del Castillo, J.; Barroeta, B.; Paton, J. Converting Smart Specialisation into a Regional Strategy. *INFYDE Working Paper* **2011**, *2/1*, 103–110.

49. *Sustainable Manufacturing and Eco-Innovation. Fremork, Practice and Measurement*; Synthesis Raport; OECD Directorate for Science, Technology and Industry: Paris, France, 2009.

50. Dziedzic, S.; Wozniak, L.; Chrzanowski, M. Inteligentne specjalizacje jako droga dla zrównoważonego rozwoju. In *Zrównoważony Rozwój Organizacji—Odpowiedzialność Środowiskowa*; Borys, T., Bortniczak, B., Ptak, M., Eds.; Uniwesytet Ekonomiczny we Wrocławiu: Wrocław, Poland, 2015.

51. Bedrunka, K. *Specjalizacje regionalne jako baza do podnoszenia poziomu innowacyjności województwa opolskiego In Efektywny Transfer Wiedzy z Nauki do Przemysłu w Województwie Opolskim*; Malik, K., Dymek, Ł., Eds.; Difin: Warsaw, Poland, 2015; pp. 70–79.

52. *Regionalna Straregia Innowacji Województwa Opolskiego do 2020 roku*; Załączniknr 1 do Uchwały Nr 5250 ZarząduWojewództwa Opolskiego z Dnia 1 Lipca 2014r; Opole, Poland, 2014.

53. Kangas, R.; Aarrevaara, T. Higher Education Institutions as Knowledge Brokers in Smart Specialisation. *Sustainability* **2020**, *12*, 3044. [CrossRef]

54. Ponsiglione, C.; Quinto, I.; Zollo, G. Regional Innovation Systems as Complex Adaptive Systems: The Case of Lagging European Regions. *Sustainability* **2018**, *10*, 2862. [CrossRef]

 *energies*

Article

"Green Energy" and the Standard of Living of the EU Residents

Mariusz Malinowski 

Department of Economics, Faculty of Economics, Poznan University of Life Sciences, 60-639 Poznan, Poland; mariusz.malinowski@up.poznan.pl

Abstract: The author intended to present the relationship between the standard of living of EU citizens and the level of the development of renewable energy. It is particularly important in the context of the implementation of the sustainable development idea, by ensuring a high standard of living for both current and future generations, with rational use of available natural resources. The first, theoretical part of the article presents the problem related to the impact of renewable energy on the standard of living in a synthetic way. The second part involves empirical research conducted in all countries of the EU. To evaluate the level of renewable energy development and the standard of living, the author constructed original measures based on the TOPSIS method. Variables were selected on the basis of substantive, statistical and formal criteria (primarily the completeness and availability of data in 2019). Within the framework of the conducted study, the author obtained, among other things, a relatively high value of Spearman's rank correlation coefficient between the constructed synthetic measures (0.47). Canonical analysis was used to identify the relationship between them. Numerous indicators, including canonical correlations, complete redundancy and extracted variances, were determined with the use of canonical analysis. Seven statistically significant canonical variables were identified. The value of the greatest and most statistically significant canonical correlation exceeded 0.94, and for the last statistically significant canonical variable, the value reached over 0.31. Statistical data were primarily obtained from the publicly available EUROSTAT database.

Keywords: standard of living; renewable energy; linear ordering; canonical analysis; sustainable development

Citation: Malinowski, M. "Green Energy" and the Standard of Living of the EU Residents. *Energies* **2021**, *14*, 2186. https://doi.org/10.3390/en14082186

Academic Editor: Sergey Zhironkin

Received: 26 March 2021
Accepted: 12 April 2021
Published: 14 April 2021

Publisher's Note: MDPI stays neutral with regard to jurisdictional claims in published maps and institutional affiliations.

1. Introduction

It is hard to disagree that energy, regardless of its form, constitutes the basis of any economic activity. Availability of energy is currently the basis of human existence and society will become more and more dependent on energy sources, mostly since all kinds of everyday devices are powered with electricity. The development of electricity (the increase in uninterrupted energy supply) has played a special role in health care, decreasing infant mortality, development of agricultural production and industry, thus affecting the standard of living of the population. Electric energy will remain an important determinant of an adequate standard of living and the importance of energy security in each country (broadly understood as the ability to satisfy the demand for energy in terms of quantity and quality, at the lowest price possible and maintaining environmental protection) will grow. Due to these demands, new sources of energy are needed. The demand for energy increasing along with civilizational progress [1], as well as the simultaneous depletion of discovered and accessible conventional energy sources (mainly fossil fuels) and progressive degradation of the environment, makes renewable energy sources (the so-called "green energy") the desired direction for energy generation. As a result, renewable energy sources (the so-called "green energy") has become desired methods of energy production. Various solutions, adapted to regional and local conditions, are adopted for this purpose. Energy is obtained with the use of the wind, the sun, tides and ocean currents, river drops, or energy obtained from biomass (most often hay from crop cultivation and firewood). Landfill biogas and biogas obtained from sewage disposal processes or decomposition of components of plant or animal remains are also listed among such sources. The increase in the importance of such energy sources stems from, among other things, their positive impact on the environment or increasing of energy security (apart from the effective way of storing the produced energy). It is particularly important in the context of the implementation of the sustainable development idea, by ensuring a high standard of living for both current and future generations, with rational use of available natural resources. For this reason, increasing the share of energy obtained from renewable sources in energy balances of different countries constitutes an important aspect of the adopted sustainable development strategy. At this point, it must also be noted that a sort of coexistence of renewable energy sources and modern ways of utilising conventional energy sources (e.g., lignite) sometimes occurs as well. An example is Germany, one of the leaders in the European Union [1] (and in the world in general) in terms of lignite mining; at the same time, in the first quarter of 2019, the share of renewable energy in the amount of electricity supplied in Germany surpassed the share of electricity generated there using fossil fuels. [2].

Although the Europe 2020 Strategy is about to expire, this approach is in line with the "TOWARDS A SUSTAINABLE EUROPE BY 2030" document, which envisions a European Union-level transition to a low-carbon, climate-neutral, resource-efficient and biodiverse economy [3]. In turn, according to the European Green Deal strategic programme established by the European Commission, the goal is to make Europe a climate-neutral continent by 2050. The most important initiatives included in the programme are achieving global CO_2 emissions neutrality, incorporating a wide range of renewable energy sources into the energy system, establishing a circular economy and achieving zero emissions of pollutants [4].

In the subject literature, the impact of renewable energy on the standard of living was analysed in the context of energy poverty (by such scholars as C.W. Njiru and S.C. Letem [5], E.M. Getie [6], or M.A. Hussein and W.L. Filho [7].

In the literature, the problem of the impact of renewable energy on the standard of living has been analysed in the context of energy poverty. Less frequently renewable energy is strictly treated as a livelihood enabler.

This article aims to determine the correlation between the level of renewable energy development and the standard of living of EU citizens. An additional goal assumed by the author was to popularize the relatively rarely used canonical analysis. The aim, however, was not to scrupulously explain this calculation method in detail, but rather demonstrate its usefulness. To create a measurement tool for the analysed phenomena, the author used a multi-dimensional comparative analysis, which was a set of methods of the construction of synthetic measurements and linear ordering of objects described by a large number of variables (including the TOPSIS method). A correlation analysis was conducted to determine the direction and the strength of the correlation between the constructed, original synthetic measures of the standard of living and the level of "green energy" development. On the other hand, the author used canonical analysis, one of the advanced methods of multidimensional statistical analysis, to identify multidimensional correlations between the standard of living and the level of renewable energy development. The study involved 27 member states of the EU. Statistical data obtained from international statistical institutions, mainly EUROSTAT [8], were used in the paper. The data concerned the year 2019.

2. The Standard of Living and Renewable Energy—The Theoretical Aspect

The notion of the standard of living is commonly applied in everyday life while its meaning varies, which results primarily from the fact that it constitutes the subject of research in many scientific fields (including sociology, philosophy, economics, physiology, or psychology). On the one hand, there is a broad interdisciplinary research perspective—both problem-related and methodological—while on the other hand, there is a problem of operationalisation of this research category. What is missing in the numerous and constantly evolving literature on the subject is one commonly accepted definition.

In 1954, the UN expert committee defined the standard of living of the inhabitants as the entirety of their actual living conditions, as well as the level of material and cultural satisfaction of needs through numerous goods and paid services, as well as those derived from the social funds (quoted after [9]). This interpretation has become the basis for many other definitions, formulated in the future. A. Luszniewicz (1982) defined the standard of living as the degree to which material and cultural needs of households are met (i.e., its security) through the streams of paid goods and services and funds for collective consumption [10]. U. Grzega considered the standard of living to be "the degree of satisfaction of human needs, resulting from the consumption of material goods and services, as well as the use of the values of the natural and social environment" [11]. According to One Global Economy, the standard of living is, first and foremost, defined by three categories [12]: income (changes in annual income, savings, employment and career, entrepreneurship); education (finishing high school, university admission); health (availability of the health care system, disease management programmes, preventive medicine (including prenatal care, sanitation services, vaccinations).

The category if the standard of living is largely based on the theory of needs (see: Table 1). If human needs are satisfied to a large extent, it means that the standard of living is high. A need is defined as a perceived state of absence of something, while social needs are needs whose fulfilment requires the existence and action of various social institutions for the intended purposes. A characteristic feature of needs is their variability over time, which is less related to basic needs (e.g., food, shelter) and more to the higher (luxurious) ones [13]. According to A. Chabior et al., a need is a state of a deficiency, absence, imbalance in the qualities of the organism or environment that are important to a person. Therefore, a need means a lack of something that motives one to initiate action aimed at compensating for such lack and restoring the disturbed optimum in life [14]. According to K.W. Frieske and P. Poławski needs are conditions that must be met to make people able to cooperate, take autonomous decisions and in general participate in collective life [15]. In T. Tomaszewski's approach, the most detailed needs mentioned in the literature can be classified into three categories: biological (or elementary) needs, social needs and cultural needs. Social or elementary needs are understood as needs related to the structure of the human organism, the satisfying of which is necessary to keep a given organism or species alive (the need for food, water, oxygen). In turn, based on the fact that humans are in a certain sense dependent on others, we can distinguish social needs (such as the need to be accepted by others, the need to be loved or recognized). The third group of needs constitutes an expression of human dependency on the creations of human culture and the operation of social institutions. Such needs can be called cultural needs, which include products of material culture (accommodation, radio, fridge, car) and creations of spiritual culture (books, the cinema and the theatre, social conversation) [16].

T. Słaby observes that in terms of economy the need to have or feel for any purpose constitutes the basis for the production and exchange of goods. The need for goods and services constitutes the basic component of rational environmental management towards sustainable development [20].

As a multidimensional phenomenon, the diversity of the standard of living is conditioned by several factors, which generally may be divided into external and internal. External factors influence the variability of the said standards in time, while internal factors- in space. Internal factors are directly or indirectly influenced by the inhabitants of a given region. This group of factors includes, among others, the manner of managing the local self-government unit, which is reflected in the organisation of life (spatial plans and their implementation, the location of jobs, housing construction, provision of social, educational, health and other services) and the level of economic development. External factors include the dynamics of population development and its structure, as well as equipment with the technical infrastructure [21]. According to J. Piasny, in socio-economic terms, the standard of living includes all circumstances characterizing material and cultural as well as social conditions of the life of society. According to the author, the determinants include work conditions (e.g., whether it was easy to obtain it, its difficulty, the length of the working week), wage or income level, consumption, housing situation, the possessed durable consumer goods, the level of health and social care, the condition of education and culture, the provision of water to households, gas, electricity, sewage systems etc. [22].

Table 1. Categories of human needs.

Max-Neef [10]	1. Subsistence; 2. Protection; 3. Affection, 4. Understanding; 5. Participation; 6. Idleness; 7. Creation; 8. Identity; 9. Freedom.
Słaby [6]	1. Biological condition (food, housing, health, natural environment, leisure); 2 Professional status (having a job, working hours, wages); 3. Material status (savings, prices, durable goods); 4. Educational status (education of children and youth, education of adults, culture and art); 5. Social status (social security, income egalitarianism, social pathology, family and social bonds, politics).
Murray, Pauw, Holm [11]	1. Subsistence; 2. intactness, arrangement, intake, waste, movement, temperature, receptivity); 2. Protection (e.g., maintain physical subsistence);3. Affection (e.g., pleasure, trust, loyalty, respect); 4. Participation (receiving, giving); 5. Understanding (perception, cognition, emotion, reflex); 6. Creation (transform matter, transform symbols, procreate). 7. Idleness (catharsis, revitalization); 8. Identity (e.g., physical disposition and appearance, personality); 9. Freedom (choice, value); 10. Transcendence (affirmation of life, overcome meaninglessness).
Ding, Jiang, Riloff [12]	1. Physiological Needs; 2. Physical Health and Safety Needs; 3. Leisure and Aesthetic Needs; 4. Social, Self-Worth, and Self-Esteem Needs; 5. Finances, Possessions, and Job Needs; 6. Cognition and Education Needs; 7. Freedom of Movement and Accessibility Needs.

Source: own elaboration based on [13,17–19].

Disparities in the level of life and social inequalities (like the use of renewable energy sources) have recently become a persistent focus of economic analyses. A. Zieliaś [23] explains the growing interest of researchers in the economic category that is the standard of living with the transition from the stage of fascination with the pace of technical and economic progress into the stage of reflection on the benefits and dangers of the civilizational progress. The adverse phenomena that economic brings include [24,25]: accelerated degradation of the natural environment, threatening the lives of humans and animals; a significant increase in morbidity and premature deaths due to certain diseases, mainly the so-called "civilization diseases" (such as cardiovascular diseases, cancer); an increase in various social pathologies (such as frustration, crime, alcoholism); a rapid increase in the number of traffic accidents and accidents at work; disorientation in the value system, destruction of old systems without the creation of new ones instead; an increase in social disparities in various social cross-sections and so on; excessive consumption of products and goods, resulting in environmental pollution; inability to produce substitutes for non-renewable resources without increasing ecological risk for the environment and humans.

In recent decades, amidst the transformation of many EU economies, the standard of living was determined by many factors as well as socio-economic and political processes. Nowadays, environmental issues, including the use of renewable energy sources, are becoming particularly important.

The contemporary economic system is characterized by cost externalization and "non-payment of bills" as well as the "grow first and clean up later" strategy. As a result of this phenomenon, there is an imbalance in the economic system, which is reflected in environmental pollution of water, air and land. Increasing environmental pollution has an impact on the standard of living and production capacity—production costs are growing because technological processes require the use of clean water and clean air. In the long term, it may lead to irreversible changes in the natural environment [26]. As it has already been mentioned, ensuring a high standard of living to both current and future generations with rational use of the available natural resources constitutes a basic prerequisite for sustainable development. Such an approach dominates international economic relations, and in recent decades we have been observing a concentration of actions oriented towards the transformation of socio-economic systems into the so-called green economy.

The concept of a "green economy" is relatively new. The term itself first appeared in 1989 in the report entitled Blueprint for a Green Economy, prepared for the government of Great Britain (to support the introduction of the sustainable development idea) by D. Pearce, A. Markandya and E.B. Barbier [27]. From the perspective of these considerations, it is worth quoting the definition of "green economy" adopted by UNEP, which provides that it is an economy that improves welfare and social equality (justice) while reducing environmental risk and the use of natural resources [28]. It can be assumed that an important role in the initiation thereof was played by the United Nations Environmental Protection Programme of 2008, which called for the conclusion of an agreement, the so-called Global Green New Deal, within the framework of which opportunities and chances to recover from the global economic crisis through the development of "green" economy sectors were indicated. The Global Green New Deal. Policy Brief report, published in 2009, recommended investing in environmentally important areas with the greatest potential for transformation, headed towards "green economy", such as, among other things: renewable energy, clean technologies, energy-efficient construction, recycling and waste management, sustainable use of land, green coal mining techniques (including the underground coal gasification process) [29]. Energetic self-sufficiency, defined by the ratio of the amount of energy obtained to the amount of energy used in a given country/region, became one of the most important indicators characterizing sustainable development [30]. The green economy thus becomes a path to achieve sustainable development.

The concept of sustainable development is supposed to contribute to solving the main challenges of contemporary economies, which comprise fears related to climate changes and excessive air pollution, but also depletion of natural resources and energy security. In the face of these challenges, the development of renewable energy seems to be an important element in the implementation of the new energy policy. Two main areas can by confirmed by the effects of implementing sustainable development:

- production and consumption, where it is expected that natural resources will be used in a rational way and energy will be used effectively, waste and poverty will be reduced and economic competitiveness will be strengthened.
- energy use and production, where it is expected that air quality will be improved, the energy intensity of production will be reduced, energy efficiency will be improved and that energy security and general access to clean energy will be increased. The area assumes that it is critically important to increase the use of renewable energy sources and slowly become less dependent on traditional energy sources (including oil and natural gas) [31].

The processes of transitioning to a low-carbon fossil-fuel-free economy to a certain extent stem from the obligations imposed on governments and national economies by international organizations, mainly within the framework of the United Nations Framework Convention on Climate Change (UNFCCC or FCCC) and the Kyoto Protocol, and partially result from the emergence of new technologies and innovations that are climate-friendly and provide a competitive advantage [32]. Although the guidelines stemming from the concept of the green economy should not be ignored by any country, the condition and implementation thereof in most of them is still not satisfactory. The basic problem in many countries is not the amount of energy produced and used, but the fact that the source thereof are fossil fuels. It is estimated that the total technical potential of renewable energy can be 100 times higher than the current global demand for energy [33].

Nowadays, the energy needs of consumers are impacted by a wide variety of factors, including the level of development of a society to which a consumer belongs, economic and cultural traditions; geographical location; mobility and development aspirations and so on [34].

Until 2050, the technological revolution will take place not only with a share of an increase in energy efficiency and renewable energy sources but also with a growing role of energy prosumers in the economy [35]. Therefore, the involvement of the main stakeholders-citizens of individual countries and regions, who must participate in the creation of their (as well as local/regional) energy security—is necessary to make the energy revolution possible. Already in 1947, M. McLuhan and B. Nevitt [36] formulated a thesis that with the development of new electrical technologies, the consumer will be increasingly often becoming the producer. Businesses (including households) using new technological solutions for the production of electricity (and/or heat energy) that satisfies their needs are referred to as prosumers in the energy market. A prosumer is a participant in the prosumption process, in other words—both a user and a consumer, producing a product to consume it on its own. A prosumer is an active consumer-customer, who not only buys electricity (and/or heat energy) from traditional suppliers but also engages in active purchase and sale relations with them. It produces energy with the use of distributed power generation equipment (DPGE) and sales the surplus. It also sells system services, such as demand reduction. It is equipped with storage DPGE technologies which provide it with back up power, especially electricity, in the case of grid failures [37]. Pro-consumer energy represents the lowest level of distributed energy. It includes such microgeneration installations as solar collectors, photovoltaic microsystems, biomass boilers, micro wind turbines, biogas-powered cogeneration microsystems, heat pumps. The most important prosumer groups include households, small and medium-sized farms as well as small and medium-sized enterprises (located mainly in rural and suburban areas).

The development of prosumer energy has many benefits for prosumers (it impacts the standard of living in the case of households), national economies and the environment. The benefits are as follows [38]: a prosumer lives in a household that generates profit by selling the surplus of produced energy; with the use of energy storage devices together with the use of microgeneration devices, a prosumer has constant access to the electricity grid, even during grid failure; by using smart power intake management capabilities, the prosumer can create their own energy tariffs, which leads to a decrease in the cost of electricity use; a prosumer becomes independent of increases in the prices of conventional energy carriers.

The main reasons for an energy policy including renewable sources are benefits (Table 2), which can be synthetically divided into three groups [39–41]:

- economic-better use of production resources-land, labour, capital; economic activation of local communities; lower energy production cost; reduced fuel imports.
- social-the creation of new workplaces; improvement of residents' standard of living.
- environmental-improvement in the condition of the natural environment and reduction of greenhouse gas emissions, which will have an impact on limiting the extraction of fossil fuels and, as a result, decreasing environmental burden related to the exploitation of deposits and will reduce the risk of natural disasters.

M. Wozniak and B. Saj point out that energy generated from renewable sources makes it possible to [39]: diversify the available energy sources, create active prosumer attitudes concerning the use of renewable energy in the general energy system of a given country, manage fossil fuels in a reasonable way, obtain relatively cheap, renewable electricity; significantly reduce the negative impact on the environment; reduce the costs related to energy transfer; increase social awareness in terms of ecology; improve the stability of energy supplies (especially in rural areas); expand the entrepreneurial attitudes of residents.

Table 2. Advantages and disadvantages of the impact of renewable energy sources.

RES	Advantages	Disadvantages
Solar energy	<ul><li>its availability related to the presence of sunlight</li><li>it produces no noise, harmful emissions or pollutions,</li><li>it is environmentally friendly as it does not exacerbate the greenhouse effect,</li><li>it constitutes a free and inexhaustible source of energy</li></ul>	<ul><li>its cyclicity associated with seasons and the time of day,</li><li>the necessity to create a proper</li><li>infrastructure that will keep up with the movement of the Sun and expenditures for the construction of assistive devices related thereto</li></ul>
Wind energy	<ul><li>it is a virtually free and clean source of energy,</li><li>it does not pollute the environment,</li><li>it can be obtained virtually everywhere,</li><li>alternative workplaces,</li><li>no waste</li></ul>	<ul><li>high cost of equipment installation,</li><li>lack of continuity due to windless periods,</li><li>it is a threat to birds,</li><li>vibrations producing infrasound,</li><li>noise emission,</li><li>limitations related to location,</li><li>visual interference</li></ul>
Water energy	<ul><li>it supports local development,</li><li>it supports rural electrification,</li><li>it enables saving other natural resources,</li><li>it reduces environmental pollution,</li><li>electricity production with the use of this method is relatively stable,</li><li>damming up of water has a positive impact on the hydrological balance</li></ul>	<ul><li>negative impact of the dam infrastructure</li><li>on the quality of the environment,</li><li>artificial reservoirs cause changes in the hydrological conditions of rivers,</li><li>siltation of river bottoms,</li><li>changes in the physical and chemical properties of waters,</li><li>changes in the landscape,</li><li>affecting the aquatic environment by interfering with the world of fish living therein and limiting their freedom of migration,</li><li>changes in fish population,</li><li>generation of large amounts of methane</li><li>in dammed reservoirs,</li><li>they become a major threat to</li><li>the environment, mainly due to a large amount of the stored organic matter and significant methane emissions, contributing to greenhouse gas emissions,</li><li>they constitute obstacles in animal migration,</li><li>cumbersome noise,</li><li>the local populace often needs to be resettled; this may lead to intergenerational trauma in the resettled people</li></ul>
Biomass	<ul><li>a large, untapped production potential for agriculture,</li><li>adequately developed animal production unused animal manure,</li><li>reduction of emissions related to the use of conventional fuels</li></ul>	<ul><li>production of biomass as competition for food production,</li><li>dispersion of agricultural resources,</li><li>obtaining of biomass depends on the season,</li><li>high cost of preparation and transport</li></ul>

Table 2. *Cont.*

RES	Advantages	Disadvantages
Geothermal energy	• independent of weather conditions, • steady power level throughout the entire year, • with proper use, exploiting geothermal energy should not generate any pollution	• the potential side effects of using geothermal energy include the discharge of harmful gases and minerals into the atmosphere or surface waters and groundwaters, • not all areas have access to equally effective geothermal sources (ones with the appropriate temperature), • expensive generation technology (boring is required; the entire infrastructure must be constructed), • geothermal sources may shift, which may result in their unavailability in the areas where the infrastructure is set up

Source: Own study based on [39–41].

In the context of identifying economic benefits, it is important to assess the profitability of investments in renewable energy sources, for which mainly methods based on cash flow (static and dynamic) are used. Statistical methods used in this regard are pay-back period and rate of return. The pay-back period of the investment in RES is the time after which the revenues cover the expenses. For any investor, a higher level of investment evaluation is when the time is as short as possible. On the other hand, the rate of return of investment outlays in RES is determined based on the relation of revenues obtained as a result of the investment and the amount of the capital employed. The most often used dynamic methods to assess the cost-efficiency of the investment include the net present value (NVP), the internal return rate (IRR) and the modified internal return rate (MIRR). It is often difficult to assess the profitability of investments in the market of energy obtained from renewable sources, due to limitations associated with the lack of complete information on the current and, especially, the future operation of this market, and, in particular, the lack of sufficient information on financial forecasts, future CO2 emission costs and, most importantly, the direction of changes or stability of the existing legislation [42].

Without much reservation, it can be assumed that the main argument for the use of green energy is the prevention of environmental changes and that it is of particular importance in the context of air pollution reduction and prevention of global warming. Global warming changes the environment, increasing the frequency and intensity of extreme weather events.

In 2017, weather-related disasters caused record economic losses, amounting to EUR 283 billion. The European Commission predicts that by the year 2100, such a disaster could affect approximately two-thirds of the population of Europe, while only 5% of the population is affected by them these days. For example, annual damage from river flooding in Europe could rise to EUR 112 billion from the current EUR 5 billion. It is estimated that reduction in predicted food availability with global warming by 2 °C is more significant than with global warming of 1.5 °C (as per the "aim of the Paris Agreement"), including the regions of key importance for the safety of the EU, such as Northern Africa and the rest of the Mediterranean. It could undermine security and prosperity in the broadest sense, harming the economic, food, aquatic and energy system, which would, in turn, cause conflicts and migration pressures [43].

From the perspective of these considerations, it is worth mentioning at this point that the most important negative health effects resulting from air pollution (reflected in the standard of living) include [44]: problems with memory and concentration, depression, anatomical changes in the brain, accelerated ageing of the nervous system, stroke; breathing problems, cough, runny nose, sinusitis; myocardial infarction, hypertension, ischaemic heart disease, cardiac arrhythmias, heart failure; exacerbation of asthma, lung cancer, exacerbation of chronic obstructive pulmonary heart disease, more frequent respiratory infections; low birth weight, missed abortion, premature birth.

At this point, it is also worth mentioning the results of studies conducted by M. Szyszkowicz et al. [45] which show that there is a statistically significant correlation between air pollution with carbon monoxide, sulphur dioxide, nitrogen dioxide and the number of suicide attempts in Vancouver. It has been empirically proven that air pollution is associated not only with respiratory problems but also with depression, anatomical changes in the brain, accelerated ageing of the nervous system, the likelihood of stroke, missed abortion. According to the report by the Jagiellonian University, approx. 44,000 in Poland, one of the infamous European leaders in the level of air pollution, die due to air pollution every year [46]. In its 2015 report Economic Cost of the Health Impact of Air Pollution in Europe, the World Health Organisation states that the estimated economic cost of deaths occurring in Europe as a result of environmental pollution is approximately USD 1.6 trillion per year [47].

According to the BP Energy Outlook 2020 report, there is a clear upward trend in energy consumption all over the world, which leads to many environmental threats. The report shows that oil and coal continue to be the basic source of energy (the trend, however, is downward), and the share of renewable energy sources has been increasing for another year in a row [1].

As M. Tomala [48] correctly observes, renewable energy frequently still turns out to be more expensive than the traditional one, but the use of ecological energy must be based on a high level of social awareness consisting in the understanding of the fact that the wealth of a country does not depend solely on the accumulated capital, but also on the standard of living (as it is in the Nordic countries).

It seems that the cost of investment and operating costs of an energy or heat/cold source remains the most important criterion for the selection of energy production technology. Technologies using renewable energy sources require a relatively large initial investment, but the cost of their ongoing operation is low (solar, wind, water energy). From the perspective of these considerations, an important thing is that lower prices of renewable energy will increase the fund of free decisions of consumers, who will be able to spend the money saved on energy bills on products and services. For this reason, apart from the direct costs of financial investments in RES, we should also take into account the benefits stemming from the use of renewable resources, which to a greater or lesser extent also affect the standard of living of residents (Figure 1).

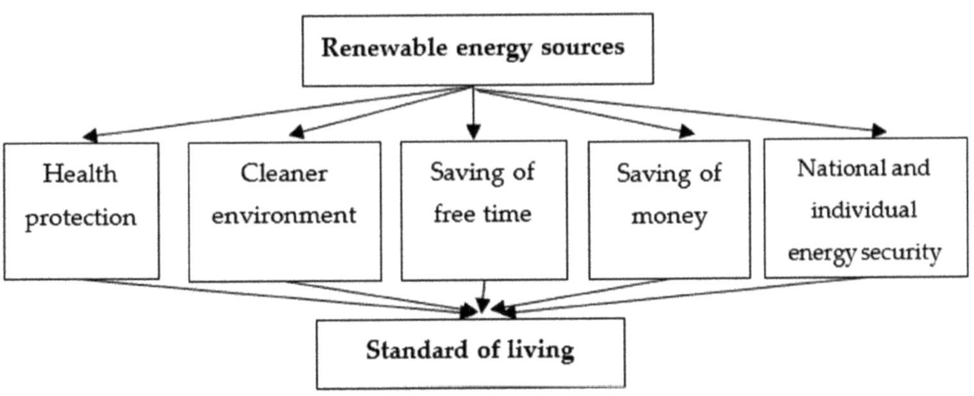

Figure 1. The impact of RES on the standard of living.

The amount of the initial capital investment has a large impact on the economic competitiveness of a given technology, especially when estimating the LCOE coefficient (levelized cost of energy). Due to the need for large expenditures at the beginning of the operation, renewable energy sources, especially those using solar technologies, frequently have a higher LCOE coefficient than conventional technologies. This argument is frequently put forward to prove that the use of renewable energy sources is too expensive. Even though the comparison of costs of energy production from renewable and conventional sources depends on the location and conditions in a given country, the LCOE coefficient suggests that using renewable energy sources we can already offer energy services at a competitive price. Moreover, it is believed that technological advancement and greater efficiency will further increase the price competitiveness of all types of RES, compared to conventional technologies. For example, it is assumed that the LCOE coefficient for wind and solar energy will have decreased by 35% (by 2030) and 50% (by 2050), respectively [33].

M. Iganrska observes that the development of the RES sector has a positive impact on the labour market, thus determining the standard of living of residents. This is especially true for the industries manufacturing renewable energy and construction equipment, due to an increase in the demand for construction services related to the construction of power generating facilities and modernization of the existing ones. Employment can also be expected to increase in the banking sector, due to the expansion of programmes financing green energy investments. There will be a shift of human capital from traditional to highly innovative sectors, which will consequently contribute to the development of the knowledge-based economy [49].

Areas allocated for RES investments do not lose their tourist and utility value. Such investments can also result in other types of investments. The areas where RES investments are implemented are perceived as investor-friendly regions, fostering the development of new technologies and protecting the environment, which makes them worth investing in [50].

In the context of the use of renewable energy sources and their impact on the standard of living of residents, it is worth mentioning the change in the time needed for the operation of thermal energy source. In general, taking into account the operation time, we can distinguish two groups of thermal energy sources unmanned sources (such as oil and gas boilers, heat pumps, solar collectors and electric heating) and sources requiring constant maintenance (all kinds of solid fuel and biomass boilers). Their time-consuming nature is dependent and can vary. In the case of changing the power source from group one to group two, the time needed for service will increase, thus generating costs of lost free time. Switching from group one to group two, however, will bring benefits associated with saving free time. Changing an individual source of thermal energy production may be associated with changes in the time needed for the operation of the source. The time needed for the following should be particularly taken into account: operation of the source producing thermal energy (supplying the source with fuel as well as other operation-related activities) during and outside the heating season; fuel management on the property (for example reloading the imported fuel from the storage area) [51].

3. Materials and Methods

All 27 member states of the European Union were included in the empirical analysis. The aim of the analysis was not to compare national fuel resources in individual countries, the structure of energy production by the source, or the share of renewable energy in electricity production as it is the subject of many widely available scientific analyses and reports (more in REN21 [52], bp Statistical Review of World Energy [53], IRENA—RENEWABLE CAPACITY STATISTICS [54]). The author aimed to construct original, synthetic measures of the standard of living and the level of renewable energy development and, on the basis thereof, conduct an assessment of development diversification in the countries of the EU as well as use an advanced, multivariate exploratory technique, i.e., canonical analysis, to assess the correlations between them.

In the literature, the problem of the impact of renewable energy on the life situation of residents was most often analysed in the context of energy poverty—a situation when a household cannot afford energy or energy services to satisfy their basic, daily needs."

Less frequently renewable energy is strictly treated as a livelihood enabler. The use of renewable energy as a tool that eliminates energy poverty and, consequently, contributes to improving the standard of living was emphasized by, among others, C.W. Njiru and S.C. Letema [5] who demonstrated that energy poverty has an impact on the physical health, well-being and welfare of the people living in the Kirinyaga province, Kenya. Similar studies were conducted by E.M. Getie [6] conducted similar studies which revealed that energy poverty has a direct or indirect impact on the standard of living of people in Ethiopia (including human resource development, residents' health and agricultural automation which improve the standard of living). M.A. Hussein and W.L. Filho [7] analysed the quantitative relationship between energy availability and improved living conditions and elimination of poverty in sub-Saharan Africa. According to the authors, renewable energy technologies play a special role in the improvement of access to energy services for poor people and isolated rural communities. At this point, it is also worth mentioning a study that investigated the relationship between the level of economic development and renewable energy. M. Simionescu et al. [55] showed that there was a potential relationship between the share of renewable energy sources in total electricity and the real GDP per capita. Based on the constructed models for panel data, the authors indicated that GDP per capita had a positive, but very low impact on the share of RES in electricity in the years 2007–2017 in EU countries, except for Luxembourg. On the other hand, based on data from 1992–2018, J. Grabara et al. [56] analysed the relationships between economic growth and the use of renewable energy consumption and direct foreign investments in Kazakhstan and Uzbekistan. Based on the results of the Granger causality test it was, among other things, demonstrated that there is a relationship between direct foreign investments and the use of renewable energy in the countries under consideration. Based on a very extensive literature review of the problem, C. Llamosas and B.K. Sovacool [57] indicate numerous benefits resulting from the construction and operation of transboundary dams. According to them, the economic benefits include the possibility of exporting electricity, effectively generating income for hydropower exporters, while the catalytic benefits include knowledge transfer, as well as building confidence and industry experience.

An initial determination of variables describing the analysed objects (EU countries) constituted the first stage of constructing synthetic measures of EU residents' standard of living and the level of the development of renewable energy sources (a kind of "green energy index"), according to the multivariate analysis. Both in the case of the quantification of the standard of living and the level of renewable energy development, the variables should be selected in such a way so as to reflect various aspects of these multivariate categories. As M. Walesiak correctly observes, the selection of variables is among the most important and most difficult issues. This is because the quality of variables determines the reliability of the final classification results and the accuracy of the decisions made on their basis. Only the variables which can discriminate a set of objects should be included in the classification procedure [58].

On the one hand, issues related to the assessment of the standard of living (and the quantification of environmental aspects, including those related to renewable energy) enjoy a great interest of researchers, but on the other—they are controversial. This is largely due to the multifaceted and interdisciplinary nature of this research category, which fosters the emergence of various measures and indicators relating to this phenomenon. No single, universally accepted method of measuring the standard of living of residents has been developed to date. To illustrate, at least partially, the results of works on the construction of synthetic measures obtained so far, listed below are those which in the author's opinion are most popular in the literature in the context of the assessment of the living standard (and related categories, such as the quality of life or living conditions) and the degree of sustainability of socio-economic development ("greening of the economy" (including the use of renewable energy sources), the level of sustainable development):

- Index of the Economic Aspects of *Welfare* (EAW) is one of the first measures of economic welfare to more broadly incorporate the ecological aspect and a broad spectrum of qualitative factors. It was applied for the first time by X. Zolotas in 1981. Its structure is focused on the current flow of goods and services. It takes into account expenditures on public buildings, the value of household works, expenditures on durable consumer goods, advertising, the value of free time, the value of public sector services, corrected by the expenditures related to health care and education, costs of environmental pollution and the depletion of natural resources [59,60].

- Index of Sustainable Welfare (ISEW), developed in 1989 by H. Daly and J. Cobb. The first step in the construction of this measure is the correction of personal expenditures of a population by the income index spread. The obtained values are then modified by adding or subtracting monetary values of a predetermined set of factors (of social, economic and environmental nature), depending on whether a given factor has a positive or negative impact on welfare. Expenditures related to, among other things, public education and health care, or the value of services from domestic work, increase the base value and decrease, among other things, the costs of commuting, unemployment and natural environment exploitation as well as crime-related costs [61,62].

- Human Development Index (HDI), which is a complex measure based on the geometric mean of three (normalized) indicators relating to basic dimensions of human life: the ability to live a long and healthy life, measured by life expectancy at birth; the ability to acquire knowledge, measured by average years of education and the expected years of education; and the ability to attain a decent standard of living, measured by gross income per capita. Since 1900, the index has been recurrently published as part of the Human Development Report prepared by the UN Development Programme [63].

- Multidimensional Poverty Index (MPI), which replaced the HPI (Human Poverty Index) index applied since 1997. It comprises 10 elements aggregated into 3 dimensions [64]: I. Education (1. no household member has been receiving education for at least 6 years); 2. no school-age child is attending school); II. Health (1. at least one household member is malnourished; 2. child mortality); III. Living conditions (1. lack of access to electricity; 2. lack of access to clean drinking water; 3. lack of access to sanitary facilities; 4. use of "dirty cooking fuel" (e.g., charcoal); 5. disorder in the household; 6. owning at least one asset related to access to information (radio, television, telephone), mobility or supporting the household (fridge, arable land, livestock).

- Quality of Life Index (QoL), an indicator developed by "The Economist" and used for the first time in 2005 for 111 countries. The indicator is based on a unique methodology combining the results of subjective satisfaction with life and an examination of objective determinants of the quality of life. The following parameters are taken into account while calculating the index: financial situation, political stability and safety, family life (divorce rate per 1000 residents), community life (the rate of church attendance or trade union membership), climate and geography, job security (unemployment rate), political freedom, gender equality (the ratio of average earnings of men and women) [65].

- Better Life Index (BLI), created in 2011 by OECD for international comparison of social welfare. The index consists of 11 parameters: income and wealth (corrected net disposable income of a household, net household assets), work and remuneration (employment rate, long-term unemployment rate, average gross earnings of full-time employees, uncertainty in the labour market), housing (number of rooms per person, houses without basic amenities, housing expenditures), health (average life expectancy at birth, self-reported health), life and work (employees spending long hours at work (more than 50 h per week), time for leisure and personal care), education and skills (level of education, cognitive skills of students, expected years of education), social bonds (support in social media), civic involvement (participation in the development of legislation, voter turnout), quality of the natural environment (air pollution, satisfaction with water quality), personal security (homicide rates, the sense of safety while walking alone), satisfaction with life [66].

- The Happy Planet Index (HPI) was developed by the New Economics Foundation. It is used to measure the well-being in specific countries and is the product of the life expectancy and the citizens' general satisfaction with life, as well as a measure reflecting the uneven distribution of life expectancy and the well-being experienced in a given country, divided by the so-called ecological footprint, i.e., the use of the natural environment [67].
- The Social Progress Index (SPI) is constructed based on 50 variables concerning 12 components, which are grouped into three categories: basic human needs, foundations of well-being and opportunity: nutrition and basic medical care, air, water and sanitation, shelter and personal safety (included in the 1st category), access to basic knowledge, access to information and communication, health and well-being, environmental quality (included in the 2nd category), personal rights, personal freedom and choice, inclusiveness and access to advanced education (included in the 3rd category) [68].

Among synthetic measures more focused on the environmental aspects, the following are worth mentioning:

- Living Planet Index (LPI), an index promoted by the Word Wildlife Foundation, measuring biological diversity based on data on various species of vertebrae and calculating the average change in their number over time. The measurement aims to identify biodiversity threatened by human activity. The situation of the analysed populations is compared with the situation observed in 1970 [69].
- Ecological Footprint, used as a measure of human demand for broadly defined natural capital. The "Ecological Footprint" determines how many biologically productive land and sea areas are necessary to provide resources for consumption and absorb the generated waste, based on the existing technological solutions combined with specific resource management practices [70].
- Global Green Economy Index (GGEI)—developed in 2010 r. by Dual Citizens Inc. as a complex analytical tool offering stakeholders a system for improving their operation and image within the framework of "green economy". In 2018, its structure was based on 20 partial indicators relating to four main thematic areas: leadership in green economy implementation (actions of public entities, managements, creation of institutions, international cooperation) and climate changes; effectiveness sectors (including construction, transport and energy); markets and investments; environment [71].

According to J. Piasny [22], synthetic measures, rather than partial indicators, are a more appropriate measure of the inhabitants' standard of living. However, some limitations on the use of synthetic measures should be kept in mind [72]: subjective selection of diagnostic variables used for the construction of a synthetics measure; subjective selection of weights for individual variables in the aggregation formula.

Considering the limited availability of statistical data, it is difficult to conduct a complex measurement of the standard of living, because the level of this multifaceted phenomenon is determined by the level of satisfaction of the above-described needs, both material and non-material. In the context of multidimensional phenomena, K. Nermend [73] indicates what, in terms of content and form, the analysed variables should be characterised by, which includes capturing the most important properties of studied phenomena, being precisely defined, logically interrelated, measurable (directly or indirectly), expressed in natural units (in the form of intensity indicators), containing a significant information load, and being characterised by high spatial variability.

The procedure of diagnostic variable selection consisted of two stages (both in the case of variables related to the standard of living and renewable energy). In the first stage, a set of potential diagnostic variables was proposed based on the above-listed formal and substantive premises (the data was obtained from the EUROSTAT database—Tables A1–A3). The selection primary of sets of variables, apart from the substantive and formal criteria, was largely determined by the availability, completeness and the fact that this data is up-to-date. It was assumed that the included partial variables will be indicative and will not be absolute values, which was aimed at eliminating the distortion related to the fact that some objects (EU countries) have some characteristic traits (eg. a significantly larger area than other objects).

As a result, 29 potential diagnostic variables related to the standard of living were proposed. The variables were then divided according to substantive criteria into seven thematic groups [9,13,74,75]:

- demography: S1—Average life expectancy at birth; S2—The birth rate; S3—Population density; S4—Infant mortality rate; S5—Total birth rate. S6—Average age of mothers at birth.
- education: S7—Students enrolled in early childhood education (pre-school education) per 1000 inhabitants; S8—Percentage of young people not working or studying (aged 15–24); S9—Percentage of university students in relation to the population S10—Percentage of people with higher education in the age group 25–64 (variables S9 and S10 are similar, however, it is assumed that not every student completes his/her studies and obtains higher education).
- economy and labour market: S11—average remuneration; S12—gross domestic product in market prices per person; S13—long-term unemployment rate (12 months and more); S14—professional activity rate (age 15–74); S15—unemployment rate, S16—the percentage of people at risk of poverty and social exclusion.
- health: S17—available beds in hospitals (per 100,000 residents); S18—doctors (per 100,000 residents), S19—the percentage of people with chronic illnesses or health problems,
- tourism: S20—nights spent at tourist accommodation establishments (per 1000 residents); arrivals at tourist accommodation establishments (per 1,000 residents); S22—net occupancy of beds and rooms in hotels and similar establishments;
- transport: S23—cars per 1000 residents; S24—length of highways per 100 km²,
- housing conditions: S25—level of severe housing deprivation; S26—the average number of rooms per person,
- environmental protection: S27—forestation rate, S28—dangerous waste production (in tonnes per person); S29—the percentage of people exposed to pollution or other environmental problems.
- In turn, a set of 11 diagnostic variables, divided into two thematic groups, was used to determine the level of renewable energy development in individual EU countries.
- RES production: R1—production of electricity and derived heat based on hydroenergy (in TOE per 1000 residents); R2—production of electricity and derived heat based on wind energy (in TOE per 1000 residents); R3—production of electricity and derived heat based on photovoltaic energy (in TOE per 1000 residents), R4—production of electricity and derived heat based on biogas fuels (in TOE per 1000 residents) R5—production of electricity and derived heat based on communal waste (in TOE per 1000 residents), R6—share of energy from renewable sources in final electricity consumption, R7—share of energy from renewable energy sources in energy used for heating and cooling, R8—share of energy from renewable sources in energy used in transport,
- RES infrastructure: R9—installed heat pump capacity (in megawatts per 1000 residents), R10—solar collector surface (in km^2 per 1000 residents), R11—"autoproducers" of electricity from renewable energy sources per 1000 residents.

The multitude of variables (being carriers of various information) describing the analysed objects in multidimensional comparative analyses makes it necessary to choose the most important ones from the research point of view. Therefore, at the second stage of variable selection, to limit the number of potential diagnostic variables, statistical procedures were used so that the selected variables characterized the studied objects as fully as possible and, simultaneously, created a set as small as possible. As M. Walesiak [58] correctly observes, the approach which is to take account of as many variables as possible is unsubstantiated as adding one or several irrelevant variables to the set makes it impossible to detect the proper structure in the set of objects.

Studying variability and the degree of correlation of potential diagnostic data (information criterion) is of particular importance in the process of variable selection.

During variable selection, it is required that individual observations exhibit adequate variability (discriminatory ability) as poorly varied variables provide little analytical value. It was assumed that those traits will be eliminated from both primary sets for which the absolute value of classic variability coefficient will be below the arbitrarily determined, critical threshold value of this coefficient at the level of 10% (the traits were considered quasi-constant, not bringing significant information on the studied phenomenon). Subsequently, the degree of variable correlation (information capacity) was investigated, since it is assumed that two highly correlated variables are carriers of similar information, which makes one of them redundant. For this reason, in order to conduct information value assessment, one of the methods of feature discrimination depending on the value of correlation matrix—the so-called method of inverse correlation matrix—was used (more in: [76]). Based on the correlation matrix, the inverse correlation matrix was calculated for each thematic variable subgroup

$$R^{-1} = [\tilde{r}_{jj'}]\, j, j' = 1, 2, \ldots, m, \text{ where :} \tag{1}$$

$\tilde{r}_{jj'} = \frac{(-1)^{j+j'}|R_{jj'}|}{|R|}$, where: $R_{jj'}$—the matrix reduced by deleting the j-th row and the j'-th column; R, $|R_{jj'}|$—determinants of the R and Rjj' matrices respectively.

In features that are overly correlated with other diagonal elements of the inverse correlation matrix are much greater than 1 (which means that the matrix is ill-conditioned). According to this method, the overly correlated feature (which corresponds to the diagonal element of the inverse correlation matrix characterized by the value exceeding the arbitrarily determined threshold value (most often r* = 10) is eliminated from the primary set of features. Then the inverse correlation matrix is calculated again, and it is analysed whether the diagonal values do not exceed the determined threshold value. The procedure is continued until all diagonal values not exceeding the determined threshold are obtained (i.e., until the stability of the inverse correlation matrix is achieved).

As a consequence, taking into account the discriminatory criterion, 4 variables (S1, S6, S9, S14) should be eliminated from the set of variables relating to the standard of living. On the other hand, all variables in the set of variables relating to the level of renewable energy development were characterized by a variable coefficient greater than the adopted critical threshold of 10%. For this reason, all variables in this set were further analysed. After conducting the assessment of the information potential (based on the results obtained with the use of the inverse correlation method) variables R4, R7 and R11 were eliminated from the set describing the level of renewable energy development, while variables S12 and S29 were eliminated from the set relating to the standard of living.

In studies aimed at linear ordering of a set of objects, variable classification by preferences among variables is of particular significance. In this context, we distinguish stimuli (high values desired from the perspective of the essence of the phenomenon under consideration), dampers (desired low values) and neutral variables (where certain nominal values constitute the optimal value, and deviations from the value worsen the evaluation of the analysed phenomenon). In the set of data relating to the standard of living, the following variables were included in the damper set: S3 (population density); S4 (infant mortality rate); S8 (young people not in employment and not attending any school (age 15–24)); S13 long-term unemployment rate (12 months and above) and S15 (unemployment rate); S16 (percentage of people at risk of poverty and social exclusion); S19 (percentage of people with chronic diseases or health problems); S25 (degree of significant housing deprivation); S28 (production of dangerous waste). The other variables were treated as stimuli. This also applies to the variables describing the sphere of tourism, assuming that the higher values of considered variables, the greater the tourist attractiveness and the more opportunities for leisure activities. In the set of variables relating to the level of renewable energy development, all variables under consideration were classified as stimuli.

In studies using linear ordering and classification methods, there is a need to de-value variables and standardise orders of magnitude to render them comparable. This operation is known as normalization transformation. The most common methods of data normalization include standardization, unitization and quotient transformation. Normalisation was performed by standardizing the value of a variable. The purpose of standardisation is to obtain variables with a distribution with an average of 0 and a standard deviation of 1. On the other hand, the most popular [58,77] standardization formula has the form:

$$z_{ij} = \frac{x_{ij} - \bar{x}_j}{S_j} \tag{2}$$

where: $\bar{x}_j$—arithmetic mean; S_j—standard deviation of variable x_{ij}; $i = 1, 2, ..., n$; $j = 1, 2, ..., m$.

The so selected and initially transformed set of values, became the basis for linear ordering of the considered objects (as well as further analyses). The classic TOPSIS (the Technique for Order of Preference by Similarity to Ideal Solution) method, which is considered to be a model method, was used to facilitate linear ordering of the selected countries of the European Union in terms of the standard of living of the residents and the level of renewable energy development. This method makes it possible to determine the hierarchy of objects in accordance with the adopted criteria. It is a certain modification of Hellwig's method, which is popular among scientists. In this method, the synthetic measure is constructed taking into account the Euclidean distance of observation not only to the pattern (the most desired variant) but also from the anti-pattern, known as the anti-deal reference solution (in contrast to Hellwig's method, in the case of which only the distance from the patter is considered). The following stages in the construction of a synthetic measure can be distinguished in this method [78]:

- Creating a standardized decision matrix based on the quotient transformation.

$$z_{ij} = \frac{x_{ij}}{\sqrt{\sum_{i=1}^{m} x_{ij}^2}} \text{ for } i = 1, 2, \dots \ m \text{ and } j = 1, 2, \dots, m \tag{3}$$

where: x_{ij} —the observation of the j-th variable in the i-th object.

- Constructing a matrix of weights using weighing of variables and subsequently creating a weighted standardized decision matrix (as a result of multiplying the standardized values by the weights):

$$v_{ij} = w_j \cdot z_{ij} \tag{4}$$

- Based on the standardized decision matrix, the value vector for the pattern (A^+) and the anti-pattern (A^-) is determined:

$$A^+ = \left(\max_i(v_{i1}), \ \max_i(v_{i2}), \dots, \ \max_i(v_{iN}) \right) = \left(v_1^+, v_2^+, \dots, v_n^+ \right), \tag{5}$$

$$A^- = \left(\min_i(v_{i1}), \ \min_i(v_{i2}), \dots, \ \min_i(v_{iN}) \right) = \left(v_1^-, v_2^-, \dots, v_n^- \right), \tag{6}$$

- Indicating the distance from the pattern and the anti-pattern for each analysed object based on the Euclidean metric:

$$s_i^+ = \sqrt{\sum_{j=1}^{N} \left(v_{ij} - v_j^+ \right)^2}; \ s_i^- = \sqrt{\sum_{j=1}^{N} \left(v_{ij} - v_j^- \right)^2}, \ i = 1, 2, \dots, M, \ j = 1, 2, \dots, N \tag{7}$$

- Determining the value of the synthetic variable which defines the similarity of objects to the "model" solution, in accordance with the following formula:

$$C_i = \frac{s_i^-}{s_i^+ - s_i^-}, \text{ where } 0 \leq C_i \leq 1 \tag{8}$$

The smaller the distance of a given object from the model unit and, therefore, the greater the development anti-pattern, the closer the value of the synthetic feature to 1.

A classification of the EU countries by the standard of living of their residents and the level of renewable energy development was conducted with the use of the PAM (Partitioning Around Medoids) method to determine groups of objects similar to each other in terms of variables describing them. As correctly observed by J. Korol and P. Szczuciński [79], classification methods make it possible to divide the analysed object sets into adequate subsets (classes) in such a way that the objects belonging to the same subset are most similar to each other, while those belonging to different subsets are least similar to each other. The idea behind clustering methods is to delimit a set of objects into homogenous subgroups, enabling better description thereof from the perspective of the purpose of structural comparisons.

The PAM method is among relatively new classification methods (the mechanism of clustering around medoids was proposed by Kaufman and Rousseeuw in 1987) (more in: [80,81]). The idea behind this method is to search for k objects, so-called representatives, which are centrally located in clusters (so-called medoids). An object in which the mean dissimilarity (distance to the representative) of all objects in the cluster is the smallest is considered to be the representative of the cluster. In reality, the algorithm minimizes the sum of dissimilarities instead of the mean dissimilarity. The selection of k medoids is a process consisting of two steps. First, an initial clustering is obtained by successive selection of "representative" objects until k objects are checked. The first object is the one for which the sum of dissimilarities for all other object is as small as possible (it is a certain kind of a "multidimensional median" of N objects, hence the term "medoid). In each subsequent step, the object which to the greatest extent contributes to the decreasing of the objective function (the sum of dissimilarity) is selected. The next stage is a sort of attempt to correct the set of representatives. It is done by including all pairs of objects (i, h) for which i object was selected for the set of representatives, and an h object does not belong to the set of representatives, checking whether after changing the i into the h, the objective function decreases. The final mean distance (dissimilarity), which is interpreted as the measure of the "goodness" of the final clustering of the considered objects, is expressed by the following formula:

$$F = \frac{\sum_{i=1}^{N} d_{i,m(i)}}{N} \tag{9}$$

where: $m(i)$ is the representative (medoid) of the closest i object.

Two types of isolated clusters can be distinguished in the algorithm of this method: the *L-cluster* and the *L*-cluster*. The C cluster is an *L-cluster* if the following condition is met for each i object belonging to C:

$$\max_{j \in C} d_{ij} < \min_{h \notin C} d_{ih}. \tag{10}$$

In turn, the C cluster is an *L*-cluster* if:

$$\max_{i,j \in C} d_{ij} < \min_{l \in C, h \notin C} d_{lh}. \tag{11}$$

The diameter of the C cluster, defined as the greatest dissimilarity (distance) between the objects belonging to C, is determined in this method:

$$D_C = \max_{i,j \in C} d_{ij} \tag{12}$$

When j is the medoid of the C cluster, the mean distance from the considered C objects to j is calculated as follows:

$$\bar{d}_j - \frac{\sum_{i \in C} d_{ij}}{N_j} \tag{13}$$

In addition, the maximum distance of all C objects to j is calculated as follows:

$$DIST_{max} = \max_{i \in C} d_{ij} \tag{14}$$

It was arbitrarily assumed that there would be 4 clusters.

Subsequently, a canonical analysis was conducted to present multidimensional correlations between the sets of variables relating to the standard of living of EU residents and the level of renewable energy development. Canonical analysis can be viewed as a generalization of the linear multiple regression (in which the variability of one endogenous variable may be explained by the variability of a set of exogenous variables) into two sets of variables (endogenous and exogenous). If the set of response features consists of only one feature, the method is equivalent to multiple regression. Therefore, canonical analysis searches for an answer to the question of what is the extent of the simultaneous influence of the whole set of endogenous variables on the whole set of exogenous variables. Analysis of correlations between two sets of variables comes down to analysing the relationships between two new types of variables (so-called canonical variables, also known as canonical roots). Canonical roots constitute the weighted sums of the first and second primary data sets. The weights are generated in a manner that maximizes the mutual correlation of the weighted sums. If this condition is met, it means that pairs of canonical variables are considered to be good representations of the initial data within the framework of the adopted model. Maximum correlation is sought with the use of the Lagrange multiplier method (see: [82–87]). Considering the arrangement of two random variables $[x, y]$, where: $x = \left[x_1, x_2, \ldots, x_p\right]^T$ is a vector of exogenous variables, $y = \left[y_1, y_2, \ldots, y_q\right]^T$ is a vector of endogenous variables, the aim is to maximize the value of the expression:

$$r_l = \frac{\left(w_x^T R_{xy} w_y\right)}{\sqrt{\left(w_x^T R_{xx} w_x w_y^T R_{yy} w_y\right)}}, \tag{15}$$

where: R_{xx}—the correlation matrix of exogenous variables, R_{yy}—the correlation matrix of endogenous variables, R_{xy}—the correlation matrix of both types of variables, w_x, w_y—weights for the first and second type of canonical variables; r_l—the canonical correlation coefficient.

The literature review concerning the application of canonical analysis indicates that this technique remains one of the least commonly used statistical methods in social sciences. Thus, it is also a rarely used tool in the context of the issues of living standards and factors determining the level of this phenomenon. Here, we can mention the studies of O.R. Ebenezer who conducted a canonical analysis which showed that there is a positive correlation between the levels of poverty and literacy concerning data from the state of Ekiti in central Nigeria [88]. In contrast, K. Chin-Tsai [89] conducted a canonical analysis to evaluate correlations between quality of life and professional satisfaction among cyclists. In the context of energy sources, a canonical analysis has been used, for instance, by T. Saeed and G.A. Tularam [90] to identify relationships between fossil fuel prices (oil, natural gas and coal) and climate change (expressed by variables related to green energy, carbon emissions, temperature and precipitation indexes). F.J. Santos-Allamilos et al. [91] used a canonical analysis to assess the spatio-temporal balance between regional resources of solar and wind energy. The conducted analysis indicated the optimal distribution of wind farms and solar power plants throughout the analysed territory (south of the Iberian Peninsula) to minimise the variability of total energy contribution in the power system.

It appears that the relatively low popularity of this tool among researchers and in economic analyses (compared to, for example, classical correlation analysis or regression analysis) may result from at least two reasons. Firstly, the method itself is quite complicated (it requires knowledge of, e.g., multiple regression). Secondly, there are some difficulties in interpreting the obtained results (e.g., a high number of determined indicators). Considering the multifaceted nature of both the standard of living and the level of renewable energy development, the use of this multidimensional exploratory technique to assess interactions occurring between the two appears substantiated. In the context of the multifaceted phenomena analysis, the use of, e.g., multiple regression models and separate analysis of each endogenous variable (in this case, concerning the standard of living), could entail certain "information noise" and thus the risk of narrowing down and distorting the results of conducted analyses. This may result in the loss of crucial information concerning the correlation occurring in the set of endogenous variables. Furthermore, it appears insufficient to conduct only a classical correlation analysis (e.g., Pearson's) for pairs of studied variables, as it does not take into account the correlations occurring within the analysed sets of variables.

At this point, it should be noted that to obtain reliable results of the analysis, it is necessary to operate on a sufficiently large sample. T. Panek and J. Zwierzchowski [92] believe that a sample size of 50 observations may be considered sufficient. Therefore, to increase the reliability of results obtained based on the canonical analysis, the following assumptions were made:

- to describe a pre-selected set of variables characterising the standard of living, data aggregated at the level of NUTS-2 regions were used. Due to frequent data gaps for some regions: Guadeloupe (French overseas department in Central America), Martinique (French overseas department in the Caribbean Islands), Guyane (French overseas department located in the north-eastern part of South America on the Atlantic Ocean), Mayotte (French overseas department in the Indian Ocean) and La Réunion (French overseas department in the Indian Ocean)—those regions were excluded from further analyses. Ultimately, 235 EU regions were included in the canonical analysis,
- in the case of variables concerning the renewable energy sources, it was assumed that their values were distributed proportionally to the number of residents of those regions. It resulted from the lack of statistical data aggregated at the NUTS-2 regional level,
- due to lack of regional data for variables: S11, S16, S19, S25, S26, S27 and S28, it was assumed that values of these variables are identical across the country,
- if no data were available for 2019, data for 2018 were included.

The results of the canonical analysis are sensitive to outliers. For this reason, it was preceded by the analysis of the internal structure of studied variables to identify outlier observations that may arise, for example, from transcription errors. For this purpose, the "three-sigma" rule was applied (see [93,94]), according to which, observations outside the range [mean3*standard deviation; mean+3*standard deviation] are eliminated. If outliers were identified, they were replaced with average values calculated for all regions of a given country, in which there where units characterised by partial variables exceeding the limit values. Such a necessity occurred 17 times for the set of variables concerning the standard of living (14 times as a result of exceeding the upper limit of the aforementioned range and 3 times the lower limit) and 7 times for the level of renewable energy development (as a result of exceeding the upper limit of the range).

In the canonical analysis, the key issue is to determine (by testing the statistical significance) how many pairs of canonical variables should undergo an in-depth evaluation. In significance tests in the canonical correlation analysis, the null hypothesis assumes that there are no correlations between two sets of input variables. The null hypothesis was verified using the Λ-Wilks canonical correlation significance test (Wilks' lambda). The test statistic for the *s-k* set of variables adopts the following form [95,96]:

$$\Lambda_k = \prod_{l=k}^{s}\left(1 - r_l^2\right) \tag{16}$$

where: *s*—number of canonical elements, *k*—number of removed canonical elements, r_l^2 a squared canonical correlation coefficient for the *l*-th canonical variable.

This statistic is characterised by the Λ-Wilks' probability distribution, assuming the truth of the null hypothesis with parameters $n-1, p, q$.

To facilitate the interpretation of canonical weights, the literature on the subjects includes recommendations for using a standardised data matrix [92]. For this reason (as mentioned earlier), both analysed sets of variables were subjected to a standardisation process.

As part of the conducted analyses, values of extracted variance were determined for each generated canonical variable. Such an indicator provides information about the percentage of the input variable variance explained by the said canonical variables. It is determined by adding the squares of canonical factor loadings located by a given variable in the set for a particular canonical root and subsequently dividing the result by the number of variables. The analytical form of the indicator may be presented as follows:

$$\overline{R_{u_l}^2} = \frac{1}{q}\sum_{j=1}^{q}c_{jl}^2 \text{ or} \overline{R_{v_l}^2} = \frac{1}{m-q}\sum_{j=q+1}^{m}d_{jl}^2, \, l = 1, 2, \ldots, s, \tag{17}$$

where: *q*—the number of input variables; c_{jl}—is the canonical factor loading of the *j*-th basic variable and the *l*-th canonical variable of the first type; d_{jl}—is the canonical factor loading of the *j*-th basic variable and the *l*-th canonical variable of the second type.

The product of the said mean and the square of the canonical correlations, referred to as the redundancy index, was subsequently determined (more information in [97]). Its value indicates how much of the average variance in one set is explained by a particular canonical variable, with another given set of variables. The analytical form of this index may be presented as follows:

$$R_{u_l,x2}^2 = \overline{R_{u_l}^2} \cdot \lambda_l \text{ or } R_{v_l,x1}^2 = \overline{R_{v_l}^2} \cdot \lambda_l, \, l = 1, 2, \ldots, s, \tag{18}$$

where: λ_l—the characteristic element of the square matrix of canonical correlation.

The said index is also referred to as the composite coefficient of determination or composite determination.

A single significance level of $\alpha = 0.05$ was assumed throughout the analysis, which covered only those "categories", for which the *p*-value was less than the assumed significance level.

4. Study Results: A Multivariate Analysis of Correlations Between the Standard of Living of the EU Residents and the Level of Renewable Energy Development

Countries of the European Union are significantly differentiated and disproportionate in terms of infrastructure and equipment supporting the development of RES, the system of obtaining energy from renewable sources and the share of RES in energy consumption. As a result, the constructed synthetic measures of the renewable energy development level exhibited relatively high variation. Significant discrepancies between energy systems (and energy intensity of economies) of the EU countries are influenced by numerous factors, among which the most important are [30]: geographic location and endowment with natural resources; energy, transport and housing infrastructure facilities; human capital; equipment with capital and access to assistance programmes and support funds; interest in and acceptance of solutions applying RES; innovativeness of the economy (enterprises) and equipment in R&D facilities; historical and political conditions of local, regional and national scale.

The highest value of the TOPSIS-based measure of renewable energy development level (over 0.56) was recorded in Sweden (see Table 3). In recent years, residents of Sweden have dramatically reduced the use of fossil fuels (Sweden pays the highest carbon tax in the world) and became significantly involved in the development of renewable energy. They adopted a policy of developing energy clusters based on modern technology and RES, and aim to generate energy exclusively from renewable energy sources by 2040. Austria (where the synthetic measure of the renewable energy development level was over 0.43) and Denmark (0.42) were placed at further positions in the created ranking. Austrian energy is based primarily on hydropower, and it is assumed that it will be completely derived from renewable energy sources by 2030. In turn, Denmark is the global energy leader in the wind sector, assumed to completely transition to renewable energy by 2050.

Table 3. Values of the synthetic measures of the standard of living and renewable energy development level.

Standard of Living		Renewable Energy	
Country	**I**	**Country**	**II**
Sweden	0.5432	Sweden	0.5628
France	0.5200	Austria	0.4312
Finland	0.5114	Denmark	0.4189
Germany	0.5005	Italy	0.3869
Netherlands	0.4921	Germany	0.3644
Malta	0.4900	Finland	0.3549
Belgium	0.4879	Cyprus	0.3365
Estonia	0.4859	Portugal	0.3285
Austria	0.4819	Greece	0.3079
Cyprus	0.4788	Netherlands	0.2819
Ireland	0.4787	Ireland	0.2735
Denmark	0.4722	Belgium	0.2543
Spain	0.4680	Spain	0.2510
Luxembourg	0.4652	France	0.2345
Czechia	0.4558	Malta	0.2278
Lithuania	0.4495	Luxembourg	0.2081
Poland	0.4397	Croatia	0.1937
Slovenia	0.4261	Slovenia	0.1918
Slovakia	0.4260	Latvia	0.1884
Portugal	0.4179	Romania	0.1725
Bulgaria	0.4146	Bulgaria	0.1521
Greece	0.3999	Czechia	0.1440
Romania	0.3855	Slovakia	0.1249
Croatia	0.3830	Estonia	0.1214
Latvia	0.3826	Hungary	0.1129
Italy	0.3815	Lithuania	0.0877
Hungary	0.3727	Poland	0.0747
Variation			
AA	0.4522	AA	0.2514
Vs [in %]	10.5792%	Vs (in %)	47.3466%
SD	0.0478	SD	0.1190
MED	0.4652	MED	0.2345
Q1	0.4162	Q1	0.1623
Q3	0.4869	Q3	0.3325

I—Synthetic measure of the standard of living, II—Synthetic measure of the renewable energy development level, AA—arithmetic average, V_s—coefficient of variation, SD—standard deviation, MED—median, Q1—first quartile, Q3—third quartile. Source: Own study based on [8].

Sweden, Austria and Denmark assumed leading positions in the world's 2020 energy transition ranking compiled by the World Economic Forum (Energy Transition Index—ETI). The ranking classifies countries according to the performance of their current energy systems and readiness for the energy transition. Sweden assumed the top global position in this respect; Denmark placed fourth while Austria ranked sixth (more information: [98]).

High values of the constructed synthetic measures of the renewable energy development level in these countries are derived from high or very high values of partial variables concerning primarily the share of renewable energy in final electricity consumption, the production of electricity and derived heat based on renewable municipal waste, as well as the production of electricity and derived heat based on wind energy.

What draws attention is that ten countries rated lowest in terms of the renewable energy development level are relatively new members of the European Union (all of these countries joined the Community in 2004 or later). These countries recorded low or very low values of the analysed partial variables used to construct the synthetic measure. In particular, they included variables concerning the production of electricity and derived heat based on renewable municipal waste and wind energy, as well as the solar collector surface.

Poland was the lowest-rated country in this respect. According to the data from Statistics Poland, the share of energy from renewable sources in the acquisition of total primary energy increased from 13.25% in 2015 to 15.96% in 2019. In 2019, the share of renewable energy in gross final energy consumption in power engineering increased by 1.31 percentage points compared to the previous year. Factors contributing to the increase in this measure consisted of a rise in gross final renewable electricity consumption (by 9.52%) and a decrease in gross final renewable electricity consumption (by 0.81%). Energy derived from renewable sources in Poland in 2019 comes primarily from solid biofuels (65.56%), wind energy (13.72%) and liquid biofuels (10.36%) [99].

For the partial variables included in the study, the coefficient of variation for the constructed measure of the level of renewable energy development was over 47%, while the standard deviation was nearly 0.12 (with a mean value of nearly 0.25). This confirms the significant variation in the level of renewable energy development in the European Union. The said measure was characterised by right-side asymmetry (the classical asymmetry coefficient was 0.66), which indicates the prevalence of values not exceeding the arithmetic mean.

In turn, the assessment of the standard of living of the European Union population based on the synthetic measures constructed shows that the highest level of this phenomenon (for the partial variables considered) occurred in the case of Sweden (where the synthetic measure of the standard of living was over 0.54), France (0.52) and Finland (0.51). High values for variables related to such things as the number of students enrolled in early childhood education, average wages, nights spent in tourist accommodation, and the average number of rooms per person were reported in the case of these countries.

In turn, for the included partial variables, the lowest values of the synthetic measure of the standard of living were characteristic of Hungary (0.37), Italy and Latvia (slightly over 0.38 each). These countries recorded relatively low values of partial variables concerning, for example, the number of doctors per 1000 inhabitants, the average number of rooms per person and the percentage of people with higher education.

The synthetic measure of the standard of living of the European Union inhabitants was characterised by left-side asymmetry, which indicates that most countries recorded values above the level of the arithmetic mean. The classical asymmetry coefficient was −0.06, which allowed assessing the degree of the asymmetry as weak. The classical variation coefficient was less than 10.5%, which indicates a relatively weak differentiation of the analysed phenomenon (for the analysed set of partial variables). In the case of three-quarters of the EU countries, the value of the synthetic measure did not exceed 0.49 (with a minimum value of 0.37 and a maximum value of 0.54).

To conduct more in-depth analyses, the EU countries were classified using the previously discussed PAM method (Table 4). The identified groups incorporate countries with similar standards of living and levels of renewable energy development, yet the composition of a given group does not provide information about the degree of development of the analysed phenomenon. The PAM method (or non-linear ordering methods in general) does not allow determining the hierarchy of the analysed multifaceted objects. The obtained grouping results may be compared with results of the linear ordering (in this case, the results obtained using the TOPSIS method), although they may not completely coincide. To facilitate interpretation, the results of the classification procedure were presented in a tabular form and numbered in descending order according to the arithmetic means of synthetic measures (obtained using the TOPSIS method) within a given group.

Table 4. Results of grouping the EU countries according to the standard of living and the level of renewable energy development.

Standard of Living			
I	II	III	IV
Estonia, Finland, Sweden	Belgium, Denmark, Germany, Ireland, France, Cyprus, Luxembourg, Malta, Netherlands, Austria	Greece, Spain, Italy	Bulgaria, Czechia, Croatia, Latvia, Lithuania, Hungary, Poland, Portugal, Romania, Slovenia, Slovakia
Renewable Energy			
Sweden	Denmark, Germany, Ireland, Greece, Spain, France, Italy, Austria, Portugal, Finland	Cyprus	Belgium, Bulgaria, Czechia, Estonia Croatia, Latvia, Lithuania Luxembourg, Hungary Malta, Netherlands Poland, Romania Slovenia, Slovakia

Source: Own study based on [8].

With regard to the results of grouping EU countries according to the standard of living of their inhabitants using the PAM method, the last group was the most numerous (11 countries). Group IV included the highest number of countries (15) also in terms of the level of renewable energy development. In the case of both the standard of living of the inhabitants and the level of renewable energy development, group I, characterised by the highest level of analysed phenomena, was relatively small (3 countries and 1 country, respectively). The application of the PAM method allowed identifying certain regularities, including:

- the last group distinguished due to the standard of living of the inhabitants included the majority of countries that joined the EU in 2004 and later (Croatia),
- Sweden was placed in the first group due to the standard of living and the level of renewable energy development.
- as many as 10 countries ranked in the lowest-rated group in terms of the standard of living and the level of RES development (Bulgaria, Czech Republic, Croatia, Latvia, Lithuania, Hungary, Poland, Romania, Slovenia and Slovakia).
- Estonia was the only country ranked in the highest-rated group due to the standard of living and simultaneously in the lowest-rated due to the level of renewable energy development.

In the next step, a correlation analysis was conducted to examine the relationship between the standard of living of the EU residents and the level of renewable energy development (measured using the author's synthetic measures). For this purpose, the non-parametric Spearman's rank correlation coefficient was applied.

Spearman's rank correlation coefficient is not only more resistant to outliers than the commonly used Pearson's correlation coefficient, but it is also recommended if the sample distribution does not meet the assumption of a normal distribution [100]. The value of Spearman's rank correlation coefficient between the synthetic measure of the standard of living of the EU residents and the level of renewable energy development was 0.4652, which allows assessing the strength of this correlation as average. The determined correlation coefficient was statistically significant at the significance level $p < 0.05$. For the purpose of the in-depth study, a canonical analysis was conducted, which determines the interdependencies between two sets of variables (rather than individual variables).

In the conducted canonical analysis, the number of generated canonical roots is equal to the minimum number of variables included in one of the analysed sets. In this case, it is eight canonical variables (roots), since this is the size of the reduced set of variables describing the level of renewable energy development. The first generated pair of canonical variables, which synthetically illustrates correlations between the analysed sets of variables, explains the majority of relationships between these sets. In practice, most attention is paid to the correlation for the first canonical variable. However, it is necessary to bear in mind that the first pair of canonical variables does not completely explain the relationships between the variables under study. Therefore, it becomes necessary to determine successive pairs of canonical roots that explain relationships in other (less significant) dimensions. Calculations are conducted until all canonical variables are determined—their number is equal to the minimum number of variables in any of the sets. Only statistically significant canonical variables should be subject to in-depth analysis. To identify such variables, the previously described Wilks' lambda test was conducted (Table 5).

Table 5. Wilks' lambda test results.

Root Removed	Canonical Correlation	χ^2 Test Value	Number of Degrees of Freedom for χ^2 Test	Probability Level p for χ^2 Test	Value of Wilks' Lambda Statistics
0	0.9400	1290.3840	184	0.0000	0.0027
1	0.8354	821.9887	154	0.0000	0.0231
2	0.7428	560.6475	126	0.0000	0.0764
3	0.6642	385.7014	100	0.0000	0.1705
4	0.6197	258.8623	76	0.0000	0.3050
5	0.5317	153.2339	54	0.0000	0.4951
6	0.4853	80.7912	34	0.0000	0.6903
7	0.3115	22.2584	16	0.1351	0.9029

Source: Own study based on [8].

In the last isolated pair, canonical variables do not correlate with each other in a statistically significant manner, therefore they were omitted in further description and interpretation.

In the first stage of the study, canonical weights were determined (as part of the canonical analysis) for the first pair of canonical variables that has the greatest contribution in explaining correlations between the analysed phenomena. Then, the magnitudes of weights for subsequent statistically significant canonical variables were determined. The weights created for standardised sets of variables are equivalent to *beta* coefficients in multiple regression. They inform about the contribution of each variable to the generated weighted sum. As their absolute value increases, so does the contribution (positive or negative) in the generation of the canonical variable.

Since the variables used for canonical analysis were subjected to the standardisation process, it is possible to directly compare the absolute values of the canonical weights determined (Table 6). Based on the conducted calculations, it is possible to conclude that for the first canonical variable, the S11 (0.7501) and R6 (0.4637) variables exhibit the greatest (absolute) weight values. Therefore, it can be assumed that the first canonical variable was the most significantly influenced by the correlation between the average wage and the share of renewable energy in final electricity consumption. For the assumed partial variables, the greatest contribution in the determination of the second canonical variable was made by variables S27 (-0.8422), describing the forestation rate, and R8 (0.7086), referring to the share of energy from renewable sources in energy used in transport. Variables S16 (the percentage of people at risk of poverty and social exclusion) and R3 (production of electricity and derived heat based on photovoltaic energy) had the greatest contribution in the generation of the third canonical variable, while the fourth canonical variable was influenced mostly by variables S10 (the percentage of people with higher education) and R5 (production of electricity and derived heat based on renewable municipal waste). Variables S18 and R1, S26 and R5, as well as S13 and R5 had the greatest contribution in the determination of the fifth, sixth and seventh canonical variables, respectively. Due to the large number of variables used and statistically significant, canonical variables generated, results of the canonical analysis were presented in a tabular form rather than using canonical models.

Table 6. Canonical weights and factor loadings.

	Canonical Weights							Factor Loadings						
	Variables Related to the Standard of Living of Residents													
	I	II	III	IV	V	VI	VII	1	II	III	IV	V	VI	VII
S2	0.01	−0.07	−0.31	−0.48	−0.37	0.29	−0.01	0.17	0.33	−0.35	−0.14	−0.35	0.15	0.22
S3	0.03	0.01	0.03	−0.25	−0.08	0.10	0.11	−0.02	−0.02	−0.15	−0.15	−0.09	0.05	−0.04
S4	0.09	−0.11	−0.23	−0.12	−0.15	0.05	0.09	−0.28	−0.04	−0.13	−0.12	−0.22	−0.08	0.09
S5	0.06	−0.07	0.30	0.34	0.06	−0.04	0.30	0.15	0.29	0.18	0.02	−0.06	−0.08	0.37
S7	−0.23	0.36	0.09	−0.11	0.17	0.16	−0.57	0.17	0.35	0.30	0.37	−0.30	0.16	−0.24
S8	−0.07	0.22	−0.10	0.21	0.27	−0.32	0.32	−0.22	−0.24	−0.02	0.17	0.20	0.11	−0.25
S10	−0.13	0.30	0.16	0.77	−0.10	−0.52	0.08	0.29	*0.50*	−0.12	0.18	−0.17	−0.17	−0.06
S11	0.75	−0.08	−0.40	−0.29	−0.03	−1.22	−0.70	*0.48*	0.25	−0.09	−0.39	0.32	−0.28	−0.12
S13	−0.64	−0.24	0.11	−0.22	−0.76	1.19	−1.64	−0.20	−0.33	−0.10	0.15	0.23	0.15	−0.40
S15	0.50	−0.02	−0.07	0.28	0.50	−1.20	1.31	−0.06	−0.24	−0.06	0.20	0.28	0.17	−0.36
S16	−0.54	−0.12	0.77	−0.43	0.30	0.15	−0.52	−0.22	−0.34	0.18	−0.30	*0.56*	0.01	−0.09
S17	−0.20	−0.30	−0.20	−0.06	−0.71	−0.21	0.27	−0.28	−0.28	0.04	−0.20	−0.35	−0.30	0.46
S18	0.15	0.00 *	−0.46	−0.11	0.78	−0.11	0.29	0.15	−0.34	−0.37	−0.19	0.38	−0.10	0.20
S19	−0.24	0.41	0.45	0.01	0.23	−0.60	0.08	0.16	0.17	0.17	0.00 *	−0.13	−0.23	*0.45*
S20	−0.19	−0.13	0.16	−0.32	−0.06	−0.39	−0.11	0.01	−0.20	−0.18	−0.08	0.02	−0.08	0.12
S21	0.06	0.03	0.34	0.02	0.35	−0.02	0.58	0.19	0.04	0.35	−0.19	0.37	0.07	0.35
S22	0.20	−0.01	−0.49	0.49	−0.07	0.70	−0.13	0.10	−0.06	−0.23	0.23	0.24	0.25	0.09
S23	−0.15	0.09	−0.12	−0.03	−0.09	0.20	0.11	−0.01	−0.01	0.11	−0.17	0.13	0.08	−0.06
S24	0.08	−0.22	0.28	−0.10	−0.21	−0.18	−0.25	0.03	0.20	−0.02	−0.15	−0.14	−0.13	−0.36
S25	0.23	−0.4	−0.28	0.21	0.00 *	0.29	−0.05	−0.18	−0.40	0.06	0.36	−0.22	−0.20	−0.10
S26	−0.06	−0.08	−0.61	0.04	−0.11	1.62	0.32	0.23	0.43	−0.14	−0.33	0.25	0.01	0.06
S27	0.73	−0.84	0.17	0.2	−0.59	0.53	0.03	*0.68*	−0.32	0.40	0.10	−0.16	0.21	0.08
S28	−0.02	0.46	0.25	−0.63	0.08	0.28	−0.17	*0.49*	0.41	0.29	−0.32	−0.21	0.09	0.01
	Variables Related to the Level of Renewable Energy Development													
	I	II	III	IV	V	VI	VII	I	II	III	IV	V	VI	VII
R1	0.33	−0.4	0.15	0.64	−1.02	−0.01	−0.01	*0.79*	−0.12	−0.01	0.24	*−0.51*	0.06	−0.14
R2	0.01	0.56	−0.59	1.23	0.03	0.50	1.14	*0.52*	0.43	*−0.51*	0.39	0.14	−0.22	−0.06
R3	−0.01	0.06	−0.75	−0.60	−0.32	0.83	0.25	−0.02	0.23	*−0.72*	−0.25	−0.32	0.38	−0.22
R5	0.25	0.10	0.22	−1.01	0.18	−1.09	−1.53	*0.63*	*0.51*	−0.25	−0.01	−0.07	−0.26	−0.38
R6	0.46	−0.67	−0.40	−0.66	0.51	−0.26	0.19	*0.78*	−0.40	−0.18	−0.06	0.35	−0.10	0.23
R8	0.30	0.71	0.46	−0.06	0.29	0.80	0.42	*0.76*	0.34	0.40	−0.09	0.12	0.35	0.04
R9	−0.04	−0.28	0.04	0.34	0.45	0.31	−0.81	0.12	−0.28	−0.16	0.40	0.38	*0.49*	*−0.59*
R10	−0.03	0.22	0.08	0.19	0.29	−0.53	−0.23	−0.02	0.26	*−0.48*	0.13	0.01	−0.11	−0.02

* The value 0.00 results from the adopted format of data presentation. In reality, a value greater than 0. Source: Own study based on [8].

In the subsequent step, canonical factor loadings and redundancy values were determined (see Table 7). Factor loadings are interpreted as values of correlation between canonical and interchangeable variables in each analysed set. The greater they are (in terms of an absolute value), the more emphasis should be placed on this variable during the interpreting. T. Panek and J. Zwierzchowski [92] recommend interpreting variables for which the square of this correlation coefficient is greater than 0.50. In turn, G. Więcek and

A. Sękowski [101] propose analysing only those variables for which the value of loads (not their squares) is greater than 0.30. For this analysis, it was assumed that the critical value of the correlation coefficient square is 0.20 (to facilitate the interpretation, those values were presented in Table 6 in bold and italics).

Table 7. Isolated variances and redundancies.

Specification	Set of Variables Related to the Level of RES Development		Set of Variables Reflecting the Standard of Living of the EU Inhabitants	
	Isolated Variance	Redundancy	Isolated Variance	Redundancy
First canonical variable	0.3112	0.2749	0.0685	0.0605
Second canonical variable	0.1163	0.0812	0.0821	0.0573
Third canonical variable	0.1613	0.0890	0.0437	0.0241
Fourth canonical variable	0.0571	0.0252	0.0491	0.0216
Fifth canonical variable	0.0831	0.0319	0.0689	0.0265
Sixth canonical variable	0.0803	0.0227	0.0250	0.0071
Seventh canonical variable	0.0771	0.0182	0.0601	0.0142

Source: Own study based on [8].

In the set of variables concerning the standard of living of the residents, for the first canonical root, the highest factor loading is presented by variable S27 (0.6762); for the second canonical variable, the highest factor load is shown by variable S10 (0.4852), for the third—variable S27 (0.3987), for the fourth—variable S11 (−0.3931), and the fifth—variable S17 (−0.3477). For the last two significantly static canonical roots, the greatest factor loadings were exhibited by variables S17 and S13 (-0.3031 and -0.3986, respectively). With regard to the set of variables concerning the renewable energy development level, for the first canonical variable, the highest factor loading is carried by variable R1 (0.7932), for the second—variable R5 (0.5088), for the third—variable R3 (−0.7187), for the fourth—variable R9 (0.3963), for the fifth—variable R1 (−0.5072), for the sixth and seventh—variable R9 (0.4900 and −0.5868, respectively).

In the literature on the topic, there are opinions, which claim that, in the interpretation of results obtained based on the canonical analysis, the interpretation of individual variables should be conducted using values of canonical factor loadings. This is substantiated by the fact that they are easy to understand intuitively. However, it is necessary to note that the values of such coefficients indicate correlations of individual input variables with canonical variables and, unlike the canonical weights, do not take into account the effects of covariance within a given set of input variables. As a result, the interpretation of canonical roots based on the values of correlation coefficients may lead to different conclusions than a more complete "multidimensional" interpretation based on canonical weights [93].

Based on the values of canonical weights and factor loadings, it can be concluded that the first statistically significant canonical root explained the following relationships:

- the higher the production of electricity and derived heat based on hydroenergy (R1), wind energy (R2) and renewable municipal waste (R5), the higher the average wage of workers (S11);
- the higher the production of electricity and derived heat based on hydroenergy (R1), wind energy (R2) and renewable municipal waste (R5), the lower the generation of hazardous waste (S28),
- as the share of renewable energy in final electricity consumption (R6) and the share of renewable energy in energy consumption in transport (R8) increase, so does the forestation rate. Therefore, it is possible to presume that activities related to the "greening" of the economy go hand in hand.

Based on the values of canonical weights and factor loadings for the second canonical root, it can be concluded that there is a positive correlation between the production of electricity and derived heat based on renewable municipal waste and the percentage of people with higher education in the age group 25–64 (S10). In turn, based on the fourth canonical root (or, more precisely, its weights and canonical loads), it can be concluded that the percentage of people at risk of poverty and social exclusion (S16) decreases as the production of electricity and derived heat based on hydroenergy (R1) increases. However, based on the values of canonical weights and factor loadings for the last necessary significant variable, it may be concluded that, as the installed heat pump capacity (R9) increases, the percentage of people with chronic diseases or health problems (S19) decreases.

When analysing factor loading values for the third, fourth and sixth canonical roots, it is possible to notice that in at least one analysed set, the square of the correlation coefficient between canonical variables and partial variables was less than 0.2 for all considered variables. For this reason, such canonical variables were not interpreted in terms of factor loadings and canonical weights.

In the subsequent step, the mean of the factor loading squares of every analysed set was determined for each statistically significant canonical variable, thus obtaining an indicator referred to as the extracted variance. In turn, by multiplying the said mean by the square of the canonical correlation, the redundancy value was obtained. The table below presents the values of the extracted (isolated) variance and redundancies (Table 7).

The most statistically significant canonical variable isolates more than 31% of the variance in the set of variables relating to the level of renewable energy development and nearly 7% in the second set (concerning the standard of living of EU residents). In turn, the second and third canonical variables isolate respectively 11.6% and 16% of the variance in the set describing RES, as well as 8% and 4% in the set of variables related to the standard of living. In the case of subsequent canonical variables, the degree of variance isolation in the set concerning the level of renewable energy development is significantly smaller; in the second set, the extracted (isolated) variance did not exceed 7%. The last statistically significant canonical variable isolates nearly 8% of the variance in the first set and 6% in the second set.

For the set of variables related to the standard of living of the EU residents, it is possible to explain 6.1%, 5.7%, 2.4%, 2.2%, 2.7%, 0.7%, and 1.4% of the variance of the set of variables describing the level of renewable energy development, respectively. In turn, with regard to the set of primary variables concerning the level of renewable energy development, it is possible to explain 27.5%, 8.1%, 8.9%, 2.5%, 3.2%, 2.3%, and 1.8% of the variance based on the first seven statistically significant canonical variables, respectively. Therefore, the fourth and subsequent statistically significant canonical variables already make a small specific contribution to explaining this variation. When comparing the amount of explained variance in both analysed sets of variables, it can be seen that in terms of each generated canonical root, the group of variables describing the level of RES development has a greater contribution to explaining the standard of living of the residents. This results in a research-relevant conclusion: variables describing the RES development level constitute a better predictor of variation in the standard of living that vice versa.

The next step involves the calculation of the total redundancy, interpreted as the average percentage of the variance explained in one set of variables for a second given set, based on all canonical variables. The conducted calculations show that the knowledge of values of variables describing the renewable energy development level allows explaining nearly 55.5% of the variance of variables from the set describing the standard of living of the EU residents. This value may be assessed as relatively high, and to obtain even better results, further research should be conducted in the future, using a different set of input variables and a changed number of such variables.

When studying multidimensional correlations between the standard of living and level of renewable energy development, it is worth noting the high and, more importantly, statistically significant (see Table 5) canonical correlation values. However, at this point, it should be emphasised that canonical correlation cannot be interpreted in the same manner as classical correlation (e.g., Pearson's). These values are interpreted as correlations between the weighted sum values in each set and the weights calculated for subsequent canonical variables. The value of the greatest and most statistically significant canonical correlation was 0.94. For the last (seventh) statistically significant canonical variable, this value was nearly 0.49. The square of these canonical correlations constitutes a measure of the degree of explanation, through linear relationships, of the variability of one set of variables by the other input set, through successive pairs of canonical variables. For the first statistically significant canonical variable, the square of the canonical correlation is over 0.88, while for the second one, it is nearly 0.70. For the last statistically significant canonical variable (seventh), this coefficient is close to 0.24. It can be assumed that this generated model describes the analysed data sets relatively well.

The figure below presents the scatter plots of the first and last statistically significant canonical variable (Figure 2). The OX axis refers to the set of variables concerning the level of renewable energy development while the OY axis refers to the standard of living of the EU inhabitants.

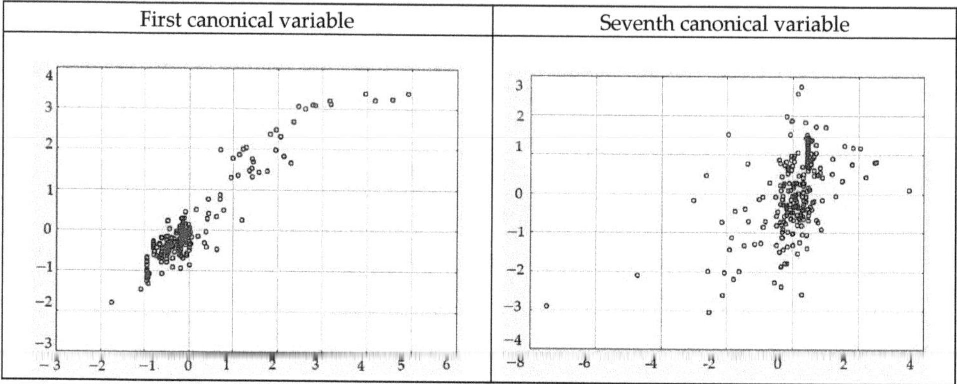

Figure 2. Scatter plot of the first and last statistically significant canonical variable.

For the first canonical variable, the scatter plot does not present any strong scatter of points representing the analysed objects. These points are arranged along a straight line. This indicates that the generated pairs of canonical variables carry a significant amount of information about the covariance of the two analysed sets of input variables. The proximity of most points (which in the case of canonical analysis represent the European Union regions) may indicate a similar structure of input variables. In the scatter plot compiled for the last statistically significant canonical variable, points representing the analysed objects are also arranged along an upwardly inclined line, yet are more scattered relative to the said line. This indicates that such a pair of canonical variables carries a significantly smaller amount of information about the covariance of two analysed variables than the first pair of canonical variables.

5. Conclusions

Initially, renewable energy was marginalised due to very high investment costs. However, their progressive decline indicates that renewable energy is currently perceived not only as a source of energy but also as an instrument that facilitates resolving many other global problems. Among other things, it is crucial for ensuring energy security, reducing the effects of environmental contamination and mitigating the influence of excessive greenhouse gas emissions, which is particularly important in the context of implementing the concept of sustainable development. One of the basic tasks of the state consists in ensuring energy security (especially important nowadays, from the perspective of the standard of living) which should not, however, take place at the cost of environmental degradation. Renewable energy sources generate zero to small amounts of pollution, which in the face of a deteriorating condition of the natural environment is an undeniable advantage.

The conducted studies aimed to detect correlations between the sets of variables describing the standard of living of the EU inhabitants and the level of renewable energy development. The canonical analysis appears to be the most appropriate statistical tool, which may be used to study multidimensional interactions between two sets of variables. Due to the multifaceted character of the analysed categories, the sole use of classical correlation analysis or multiple regression analysis seems to be inadequate. According to the author, for the analysis of socio-economic phenomena, it is important to popularise the use of multidimensional exploratory methods (such as canonical analysis) to identify correlations between compiled, multidimensional categories.

In the conducted studies, the canonical analysis was preceded by the creation of the author's synthetic measures and the determination of the correlation coefficient between them. For the partial variables included in the study, the coefficient of variation for the constructed measure of the level of renewable energy development was over 47%, while the standard deviation was nearly 0.12 (with a mean value of nearly 0.25). This confirms the significant variation in the level of renewable energy development in the European Union. In turn, for the constructed synthetic measure of the standard of living, the classical variation coefficient was less than 10.5%, which indicates a relatively weak differentiation of the analysed phenomenon (for the analysed set of partial variables). In the case of three-quarters of the EU countries, the value of the synthetic measure did not exceed 0.49 (with a minimum value of 0.37 and a maximum value of 0.54). Based on the conducted correlation analysis, it can be concluded that there is a positive, moderate and statistically significant correlation between the standard of living of the EU residents and the level of renewable energy development (measured by synthetic measures constructed using the TOPSIS method) (Spearman's rank correlation coefficient was nearly 0.47). Seven statistically significant canonical variables were identified using canonical analysis. Based on the value of the redundancy coefficient determined as part of the canonical analysis, it can be concluded that, with the knowledge of the considered variables describing the level of renewable energy development in the EU, it is possible to explain nearly 55.5% of the variance of variables from the set concerning the standard of living of the residents. In other words, more than half of the variation associated with the standard of living of the EU residents is determined by partial variables relating to the level of renewable energy development that were taken into account. It should also be mentioned that high values of canonical correlation coefficients were identified for statistically significant canonical variables. For the most statistically significant canonical variable, this coefficient was 0.94, while for the least statistically significant, it was 0.31.

The results of conducted studies (i.a. the ranking of countries in terms of the standard of living of the residents), can be used indirectly by, for example, central and self-government authorities responsible for local and regional development (including undertaking pro-social and pro-environmental actions) in the context of selecting the direction of socio-economic restructuring of individual countries and self-government units (taking into account the financial capacity of the society). Furthermore, the results of the studies may indirectly prompt self-government authorities to undertake actions directed towards more efficient use of the capacity for funding investment projects aimed to develop renewable energy technologies. Quantification of such important economic categories as the standard of living and the renewable energy sources development level in comparison with other areas may be conducive to initiating activities (e.g., of local authorities in shaping sustainable development strategies) directed towards stimulation of development aimed to achieve the highest standard of living while respecting natural resources. The conducted studies and obtained results may constitute the starting point for further analyses using different statistical methods (e.g., Granger causality analysis) and/or diagnostic variables, or to encourage similar studies at the level of other countries and self-government units. In further analyses, it would also be worth it to examine the spatial interactions between the phenomena analysed, including analysing autocorrelation and spatial heterogeneity, as well as constructing spatial regression models (including SEM and SLM models). It is possible to consider weighting the applied diagnostic variables based on expert opinions and/or statistical methods, which, however, in the case of spatial unit analyses, is a controversial solution.

Funding: This research received no external funding.

Institutional Review Board Statement: Not applicable.

Informed Consent Statement: Not applicable.

Data Availability Statement: Data available on the Eurostat website (https://ec.europa.eu/eurostat/web/main/data/database, accessed on 12 April 2021).

Conflicts of Interest: The funders had no role in the design of the study; in the collection, analyses, or interpretation of data; in the writing of the manuscript, or in the decision to publish the results.

Appendix A

Table A1. Data on the standard of living—part 1.

	S1	S2	S3	S4	S5	S6	S7	S8	S9	S10	S11	S12	S13	S14	S15	S16
Belgium	81.70	5.80	377.30	3.80	1.62	29.00	39.29	12.80	37.42	1.00	47.50	24193.00	41450.00	2.30	69.00	5.40
Bulgaria	75.00	−7.00	63.40	5.80	1.56	26.20	31.55	17.50	28.65	1.00	32.50	7051.00	8780.00	2.40	73.20	4.20
Czechia	79.10	4.10	138.20	2.60	1.71	28.40	34.40	13.20	37.41	1.00	35.10	11996.00	20990.00	0.60	76.70	2.00
Denmark	81.00	2.90	138.50	3.70	1.73	29.50	30.68	10.00	41.41	1.00	49.00	33932.00	53760.00	0.80	79.10	5.00
Germany	81.00	1.80	235.20	3.20	1.57	29.70	28.41	9.30	54.50	1.00	35.50	26128.00	41510.00	1.20	79.20	3.10
Estonia	78.50	3.10	30.50	1.60	1.67	27.70	50.60	11.60	29.65	1.00	46.20	13193.00	21220.00	0.90	78.90	4.40
Ireland	82.30	12.20	71.90	2.90	1.75	30.50	25.06	12.60	48.24	1.00	55.40	32644.00	72260.00	1.60	73.30	5.00
Greece	81.90	−0.60	82.40	3.50	1.35	30.40	14.16	20.70	30.15	1.00	43.10	10060.00	17110.00	12.10	66.10	17.30
Spain	83.50	8.40	93.80	2.70	1.26	31.00	27.61	16.00	35.42	1.00	44.70	19135.00	26430.00	5.30	73.80	14.10
France	82.90	2.10	106.10	3.80	1.88	28.70	37.86	14.00	50.30	1.00	47.50	27062.00	35960.00	3.40	71.70	8.50
Croatia	78.20	−4.40	72.80	4.20	1.47	28.80	28.10	15.00	42.81	1.00	33.10	9227.00	13340.00	2.40	66.50	6.60
Italy	83.40	−2.90	201.50	2.80	1.29	31.20	24.71	23.80	29.42	1.00	27.60	20570.00	29660.00	5.60	65.70	10.00
Cyprus	82.90	13.70	95.70	2.40	1.32	29.80	27.97	14.60	30.63	1.00	58.80	21492.00	25310.00	2.10	76.00	7.10
Latvia	75.10	−6.40	30.20	3.20	1.60	27.20	40.31	12.00	29.43	1.00	45.70	10852.00	15920.00	2.40	77.30	6.30
Lithuania	76.00	0.00	44.60	3.40	1.63	27.80	37.01	11.60	58.82	1.00	57.80	10871.00	17460.00	1.90	78.00	6.30
Luxembourg	82.30	19.70	239.80	4.30	1.38	30.90	28.64	6.90	36.58	1.00	56.20	48452.00	102200.00	1.30	72.00	5.60
Hungary	76.20	−0.30	107.10	3.30	1.55	28.20	31.83	14.60	40.09	1.00	33.40	7458.00	14950.00	1.10	72.60	3.40
Malta	82.50	41.70	1595.10	5.60	1.23	29.20	19.10	9.60	25.89	1.00	38.10	17290.00	26670.00	0.90	75.90	3.60
Netherlands	81.90	7.20	507.30	3.50	1.59	30.00	27.49	7.00	46.43	1.00	51.40	27213.00	46710.00	1.00	80.90	3.40
Austria	81.80	4.80	107.60	2.70	1.47	29.50	29.04	9.20	38.52	1.00	42.40	28094.00	44780.00	1.10	77.10	4.50
Poland	77.70	−0.40	123.60	3.80	1.46	27.40	35.85	13.40	29.04	1.00	46.60	9317.00	13870.00	0.70	70.60	3.30
Portugal	81.50	1.90	113.00	3.30	1.42	29.80	23.38	9.50	35.62	1.00	36.20	12962.00	20740.00	2.80	75.50	6.50
Romania	75.30	−4.40	82.70	6.00	1.76	26.70	26.84	17.30	37.60	1.00	25.80	6196.00	11510.00	1.70	68.60	3.90
Slovenia	81.50	7.20	103.70	1.70	1.60	28.80	29.15	9.00	26.21	1.00	44.90	16048.00	23170.00	1.90	75.20	4.50
Slovakia	77.40	1.40	112.00	5.00	1.54	27.10	30.51	17.20	46.64	1.00	40.10	9869.00	17210.00	3.40	72.70	5.80
Finland	81.80	1.30	18.20	2.10	1.41	29.20	37.76	10.30	32.89	1.00	47.30	30065.00	43570.00	1.20	78.30	6.70
Sweden	82.60	9.50	25.20	2.00	1.76	29.30	45.26	6.40	37.37	1.00	52.50	27419.00	46160.00	0.90	82.90	6.80

Source: own study based on [8].

Table A2. Data on the standard of living—part 2.

	S17	S18	S19	S20	S21	S22	S23	S24	S25	S26	S27	S28	S29	S30	S31	S32
Belgium	562.24	950.00	312.96	26.10	15.70	1860.06	760.71	46.00	0.84	511.00	57.75	5.10	2.10	22.58	5902.24	15.00
Bulgaria	756.91	735.00	421.71	21.60	32.00	1382.13	588.61	42.10	0.52	396.00	6.86	6.40	1.20	35.15	18535.87	13.10
Czechia	661.82	369.00	403.76	36.20	5.60	2802.33	1043.01	50.90	0.88	540.00	15.87	2.40	1.50	33.92	2621.03	11.10
Denmark	242.97	442.00	419.44	31.30	2.80	3677.02	894.40	48.00	0.21	447.00	30.96	4.60	1.90	14.62	3693.70	8.40
Germany	800.23	6203.00	431.09	43.20	7.10	4188.11	1754.77	45.71	0.61	567.00	36.77	3.50	1.80	31.95	4884.70	25.20
Estonia	457.35	183.00	348.34	44.20	2.40	1956.05	1159.80	48.00	1.07	563.00	3.41	2.20	1.70	53.91	17500.93	10.20
Ireland	297.39	324.00	327.94	27.70	8.10	3322.74	1761.88	54.00	0.53	445.00	13.12	2.20	2.10	11.15	2851.97	6.50
Greece	419.77	1064.00	500.80	23.70	15.60	2202.70	854.44	49.50	3.53	487.00	0.00	5.60	1.20	29.55	4251.22	20.20
Spain	297.15	5374.00	402.08	29.20	7.80	3637.26	1433.42	61.48	1.13	513.00	30.80	2.20	1.90	36.70	2936.34	9.90
France	590.85	4851.00	317.08	38.00	3.00	4623.02	1831.68	50.00	0.44	478.00	18.43	4.20	1.90	27.12	5096.75	14.90
Croatia	561.25	210.00	344.06	36.60	12.40	1730.50	541.05	60.30	27.91	409.00	23.15	5.90	1.10	34.22	1359.91	5.90
Italy	314.05	5901.00	397.71	15.90	28.90	3579.82	1099.60	49.00	3.62	646.00	22.98	5.40	1.40	31.49	2857.92	12.40
Cyprus	330.09	79.00	407.32	38.80	39.40	1156.03	632.05	71.80	0.93	629.00	27.78	0.90	2.00	0.13	2628.32	8.40
Latvia	549.35	191.00	330.38	42.10	27.90	863.75	472.62	43.30	0.64	369.00	0.00	12.40	1.20	52.76	923.83	18.30
Lithuania	643.40	241.00	459.78	36.80	12.30	1719.37	751.48	44.00	1.34	512.00	4.96	5.40	1.60	33.70	2534.03	17.10
Luxembourg	450.70	22.00	298.49	25.10	16.60	566.15	202.22	30.87	0.69	676.00	63.81	3.20	2.00	34.30	14683.96	15.20
Hungary	701.29	470.00	338.37	39.70	12.10	1785.31	745.49	41.90	0.45	373.00	21.31	6.80	1.50	22.09	1879.67	12.40
Malta	430.84	88.00	397.21	30.50	14.40	962.30	407.40	66.20	0.49	608.00	0.00	1.40	2.20	1.46	5079.58	33.90
Netherlands	316.55	1868.00	366.96	32.20	6.20	4148.33	1492.09	50.20	0.51	494.00	66.35	1.70	2.00	8.87	8404.10	14.90
Austria	727.16	701.00	524.14	37.40	5.40	4120.70	1508.01	48.00	2.48	562.00	20.78	5.80	1.60	46.44	7412.55	10.50
Poland	653.69	1677.00	237.75	39.20	23.40	1966.12	742.57	41.70	0.30	617.00	5.24	8.10	1.10	30.29	4612.34	13.80
Portugal	344.51	934.00	442.42	41.20	14.80	2530.21	1137.99	51.05	0.70	514.00	33.23	5.40	1.70	35.91	1546.70	13.50
Romania	696.83	814.00	304.70	18.90	13.90	1268.17	546.22	39.72	0.42	332.00	3.45	4.10	1.10	29.07	10466.60	13.50
Slovenia	442.79	67.00	317.81	35.80	5.10	2115.91	733.83	43.99	4.60	549.00	30.73	4.50	1.60	61.16	3950.52	16.20
Slovakia	569.62	124.00	397.34	31.80	13.50	2050.79	707.96	36.16	0.63	426.00	9.83	1.40	1.20	39.28	2275.40	9.50
Finland	361.18	347.00	320.63	49.50	4.50	2906.83	1655.78	41.99	0.25	629.00	2.74	1.00	1.90	66.21	23232.55	9.40
Sweden	213.79	738.00	426.52	36.90	3.70	4613.31	2393.94	45.00	0.43	476.00	4.86	3.60	1.80	63.80	13554.75	6.60

Source: own study based on [8].

Table A3. Data on the level of development of renewable energy sources.

	R1	R2	R3	R4	R5	R6	R7	R8	R9	R10	R11
Belgium	0.01	0.07	0.03	0.01	0.01	20.83	8.31	6.81	0.00	66.85	1.18
Bulgaria	0.04	0.02	0.02	0.00	0.00	23.51	35.51	7.89	0.00	60.78	0.09
Czechia	0.03	0.01	0.02	0.02	0.00	14.05	22.65	7.83	0.16	52.11	0.74
Denmark	0.00	0.24	0.01	0.01	0.01	65.35	48.02	7.17	0.00	329.85	0.55
Germany	0.03	0.13	0.05	0.03	0.01	40.82	14.55	7.68	0.08	232.79	0.59
Estonia	0.00	0.04	0.00	0.00	0.00	22.00	52.28	5.15	0.00	0.00	0.06
Ireland	0.02	0.18	0.00	0.00	0.01	36.49	6.32	8.93	0.00	68.71	0.42
Greece	0.03	0.06	0.04	0.00	0.00	31.30	30.19	4.05	0.79	453.86	0.21
Spain	0.05	0.10	0.02	0.00	0.00	36.93	18.87	7.61	0.59	93.95	0.79
France	0.08	0.04	0.02	0.00	0.00	22.38	22.46	9.25	0.87	49.16	0.32
Croatia	0.13	0.03	0.00	0.01	0.00	49.78	36.79	5.86	0.00	66.78	0.09
Italy	0.07	0.03	0.03	0.01	0.00	34.77	19.67	9.05	1.97	71.96	0.36
Cyprus	0.00	0.02	0.02	0.01	0.00	9.76	35.10	3.32	0.00	1237.71	0.16
Latvia	0.09	0.01	0.00	0.02	0.00	53.42	57.76	5.11	0.00	11.29	0.09

Table A3. *Cont.*

	R1	R2	R3	R4	R5	R6	R7	R8	R9	R10	R11
Lithuania	0.03	0.05	0.00	0.00	0.00	18.79	47.36	4.05	0.02	0.00	0.28
Luxembourg	0.13	0.04	0.02	0.01	0.01	10.86	8.71	7.66	0.02	112.77	0.46
Hungary	0.00	0.01	0.01	0.00	0.00	9.99	18.12	8.03	0.01	35.81	0.19
Malta	0.00	0.00	0.04	0.00	0.00	8.04	25.70	8.69	0.00	148.94	0.44
Netherlands	0.00	0.06	0.03	0.00	0.01	18.22	7.08	12.51	0.00	39.98	1.59
Austria	0.43	0.07	0.02	0.01	0.00	75.14	33.80	9.77	0.16	570.10	0.86
Poland	0.01	0.03	0.00	0.00	0.00	14.36	15.98	6.12	0.03	71.00	0.44
Portugal	0.09	0.11	0.01	0.00	0.00	53.77	41.65	9.09	1.02	131.17	0.84
Romania	0.07	0.03	0.01	0.00	0.00	41.71	25.74	7.85	0.00	10.53	0.32
Slovenia	0.19	0.00	0.01	0.00	0.00	32.63	32.16	7.98	0.00	107.80	0.36
Slovakia	0.07	0.00	0.01	0.01	0.00	21.95	19.70	8.31	0.13	0.00	0.48
Finland	0.19	0.09	0.00	0.01	0.01	38.07	57.49	21.29	0.00	13.23	1.86
Sweden	0.55	0.17	0.01	0.00	0.01	71.19	66.12	30.31	0.47	44.87	0.66

Source: own study based on [8].

References

1. Bp p.l.c. Energy Outlook, 2020 Edition. Bp p.l.c. 2020. Available online: https://www.bp.com/content/dam/bp/business-sites/en/global/corporate/pdfs/energy-economics/energy-outlook/bp-energy-outlook-2020.pdf?utm_source=newsletter&utm_medium=email&utm_campaign=newsletter_axiosgenerate&stream=top (accessed on 2 March 2021).
2. Fraunhofer ISE. Public Net Electricity Generation in Germany 2019: Share from Renewables Exceeds Fossil Fuels, FRAUNHOFER INSTITUTE FOR SOLAR ENERGY SYSTEMS ISE. 2020. Available online: https://www.ise.fraunhofer.de/content/dam/ise/en/documents/News/0120_e_ISE_News_Electricity%20Generation_2019.pdf (accessed on 28 March 2021).
3. European Commission. *Reflection Paper towards A Sustainable Europe by 2030*; European Commission: Brussels, Belgium, 2019. [CrossRef]
4. European Commission. *Communication from the Commission to the European Parliament, the European Council, the Council, the European Economic and Social Committee and the Committee of the Regions the European Green Deal*; European Commission: Brussels, Belgium, 2019.
5. Njiru, C.W.; Letema, S.C. Energy Poverty and Its Implication on Standard of Living in Kirinyaga, Kenya. *J. Energy* **2018**, *2018*, 3196567. [CrossRef]
6. Getie, E.M. Poverty of Energy and Its Impact on Living Standards in Ethiopia. *J. Electr. Comput. Eng.* **2020**, *2020*, 7502583. [CrossRef]
7. Hussein, M.A.; Filho, W.L. Analysis of energy as a precondition for improvement of living conditions and poverty reduction in sub-Saharan Africa. *Sci. Res. Essays* **2012**, *7*, 2656–2666.
8. Eurostat. Available online: https://ec.europa.eu/eurostat/web/main/data/database (accessed on 20 February 2021).
9. Zeliaś, A. *Poziom życia w Polsce i krajach Unii Europejskie*; PWE: Warszawa, Poland, 2004.
10. Luszniewicz, A. *Statsytyka Społeczna: Podstawowe Problemy i Metody*; PWE: Warszawa, Poland, 1982.
11. Grzega, U. *Poziom życia Ludności w Polsce—Determinanty i Zróżnicowania*; Wydawnictwo Uniwersytetu Ekonomicznego w Katowicach: Katowice, Poland, 2012.
12. Mourad, M.; Perez, A.; Richardson, C. *Digital Inclusion Social Impact Evaluation*; Final Report; One Global Economy: Washington DC, USA, 2014.
13. Słaby, T. Poziom i jakość życia. In *Statystyka Społeczna*; Panek, T., Ed.; PWE: Warszawa, Poland, 2007; pp. 99–130.
14. Chabior, A.; Fabiś, A.; Wawrzyniak, J.K. *Starzenie się i Starość w Perspektywie Pracy Socjalnej*; Centrum Rozwoju Zasobów Ludzkich: Warszawa, Poland, 2014.
15. Frieske, K.W.; Popławski, P. *Opieka i Kontrola*; Interart: Warszawa, Poland, 1996.
16. Tomaszewski, T. Aktywność człowieka. In *Kształtowanie Motywacji i Postaw. Socjotechniczne Problemy Działalności Ideowo-Wychowawczej*; Czapów, C., Ed.; Książka i Wiedza: Warszawa, Poland, 1976; pp. 88–89.
17. Max-Neef, M.A. Human Scale Development. In *Conception, Application and Further Reflections*; The Apex Press: New York, NY, USA; London, UK, 1991.
18. Murray, M.; Pauw, C.; Holm, D. The House as a Satisfier for Human Needs: A Framework for Analysis, Impact Measurement and Design. In Proceedings of the XXXIII IAHS World Congress on Housing Transforming Housing Environments through Design, Pretoria, South Africa, 27–30 September 2005; pp. 1–9.
19. Ding, H.; Jiang, T.; Riloff, E. Why is an Event Affective? Classifying Affective Events based on Human Needs. In Proceedings of the Workshops at the Thirty-Second AAAI Conference on Artificial Intelligence, New Orleans, LA, USA, 2–7 February 2018; pp. 8–15.
20. Słaby, T. *Konsumpcja. Eseje Statystyczne*; Difin: Warszawa, Poland, 2006.

21. Sompolska-Rzechuła, A.; Oleńczuk-Paszel, A. Poziom życia ludności na obszarach wiejskich i miejskich w Polsce. *Wieś Rol.* **2017**, *4*, 77–96. [CrossRef]
22. Piasny, J. Problem jakości życia ludności oraz źródła i mierniki ich określania. *Ruch Prawniczy. Ekon. Socjol.* **1993**, *2*, 73–92.
23. Zeliaś, A. Dobór zmiennych diagnostycznych. In *Taksonomiczna Analiza Przestrzennego Zróżnicowania Poziomu życia w Polsce W ujęciu Dynamicznym*; Zeliaś, A., Ed.; Wydawnictwo Akademii Ekonomicznej w Krakowie: Kraków, Poland, 2000; pp. 35–55.
24. Gotowska, M. Problemy definiowania poziomu i jakości życia. In *Poziom i Jakość życia w Dobie Kryzysu*; Gotowska, M., Wyszkowska, Z., Gotowska, M., Eds.; Wydawnictwo UTP: Bydgoszcz, Poland, 2013; pp. 11–24.
25. Kryk, B. Jakość życia a społeczna odpowiedzialność przedsiębiorstwa za korzystanie ze środowiska. In *Jakość życia w Perspektywie Nauk Humanistycznych, Ekonomicznych i Ekologii*, Tomczyk-Tołkacz, J., Ed.; Akademia Ekonomiczna we Wrocławiu: Jelenia Góra, Poland, 2003; pp. 260–268.
26. Sulich, A.; Grudziński, A.; Kulhánek, L. Zielony wzrost gospodarczy—Analiza porównawcza Czech i Polski. *Pr. Nauk. Uniw. Ekon. Wrocławiu* **2020**, *64*, 192–207. [CrossRef]
27. Visser, W.; Pearce, D.; Markandya, A.; Barbier, E.B. *Blueprint for a Green Economy*; Greenleaf Publishing Limited: South Yorkshire, UK, 2013; pp. 70–76.
28. UNEP. *Towards a Green Economy: Pathways to Sustainable Development and Poverty Eradication—A Synthesis for Policy Makers*; United Nations Environment Programme: Saint-Martin-Bellevue, France, 2011.
29. Krakowińska, E. "Zielone" miejsca pracy w energetyce odnawialnej w Polsce i na Mazowszu. In *Rola Odnawialnych źródeł Energii w Rozwoju Społeczno-Ekonomicznym Kraju i Regionu*; Nowak, A.Z., Ed.; Wydawnictwo Naukowe Wydziału Zarządzania Uniwersytetu Warszawskiego: Warszawa, Poland, 2016; pp. 69–96.
30. Maśloch, G. Construction Autonomous Energy Regions in Poland—Utopia or Necessity? *Stud. Prawno-Ekon.* **2018**, *106*, 251–264. [CrossRef]
31. Nowak, A.Z.; Zaborowska, W. Wprowadzenie. In *Rola Odnawialnych źródeł Energii w Rozwoju Społeczno-Ekonomicznym Kraju i Regionu*; Nowak, A.Z., Ed.; Wydawnictwo Naukowe Wydziału Zarządzania Uniwersytetu Warszawskiego: Warszawa, Poland, 2016; pp. 10–13.
32. Kochańska, E. *Determinanty Rozwoju Odnawialnych źródeł Energii*; Centrum Badań i Innowacji Pro-Akademia: Łódź, Poland, 2014.
33. WWF. *Demaskowanie Mitów: Obalanie Mitów o Energii Odnawialnej*; WWF International: Gland, Switzerland; Warsaw, Poland, 2014.
34. Gnutek, Z. Wybrane aspekty użytkowania energii w rolnictwie. *Inżynieria Rol.* **2011**, *9*, 49–56.
35. Graczyk, A.M. Koordynacja polityki energetycznej—Regionalne ujęcie wsparcia odnawialnych źródeł energii w świetle Strategii rozwoju energetyki na Dolnym Śląsku. *Barom. Reg.* **2016**, *14*, 113–120.
36. McLuhan, M.; Nevitt, B. *Take Today—The Executive as Dropout*; Harcourt Brace Jovanovich: San Diego, CA, USA, 1972.
37. Zalega, T. Wykorzystanie Odnawialnych źródeł Energii w Gospodarstwach Domowych seniorów w Polsce w świetle wyników badań własnych. In *Rola Odnawialnych źródeł Energii w Rozwoju Społeczno-Ekonomicznym Kraju i Region*; Nowak, A.Z., Ed.; Wydawnictwo Naukowe Wydziału Zarządzania Uniwersytetu Warszawskiego: Warszawa, Poland, 2016; pp. 48–68.
38. Borys, G. System wsparcia energetyki prosumenckiej w Polsce. Studia Ekonomiczne. *Zesz. Nauk. Uniw. Ekon. Katowicach* **2014**, *198*, 35–43.
39. Woźniak, M.; Saj, B. Odnawialne źródła energii w badaniach ankietowych mieszkańców województwa podkarpackiego. *Zesz. Nauk. Inst. Gospod. Surowcami Miner. Energią PAN* **2018**, 61–80. [CrossRef]
40. Crețan, R.; Vesalon, L. The Political Economy of Hydropower in the Communist Space: Iron Gates Revisited. *Tijdschr. Econ. Soc. Geogr.* **2017**, *108*, 688–701. [CrossRef]
41. Văran, C.; Crețan, R. Place and the spatial politics of intergenerational remembrance of the Iron Gates displacements in Romania, 1966–1972. *Area* **2018**, *50*, 509–519. [CrossRef]
42. Świątek, M.; Cedro, A. *Odnawialne źródła Energii w Polsce ze Szczególnym Uwzględnieniem Województwa Zachodniopomorskiego*; Uniwersytet Szczeciński: Szczecin, Poland, 2017.
43. European Commission. *Communication from the Commission. A Clean Planet for All. A European Strategic Long-Term Vision for a Prosperous, Modern, Competitive and Climate Neutral Economy*; COM(2018) 773 Final; European Commission: Brussels, Belgium, 2018.
44. Jędrak, J.; Konduracka, E.; Badyda, A.J.; Dąbrowiecki, P. *Wpływ Zanieczyszczeń Powietrza na Zdrowie [Impact of Air Pollution on Health]*; Krakowski Alarm Smogowy: Kraków, Poland, 2017.
45. Szyszkowicz, M.; Willey, J.B.; Grafstein, E.; Rowe, B.H.; Colman, I. Air Pollution and Emergency Department Visits for Suicide Attempts in Vancouver, Canada. *Environ. Health Insights* **2010**, *4*, EHI.S5662. [CrossRef] [PubMed]
46. Rzędowska, A.; Rzędowski, J. *Smog. Czas na Decydujące Starcie. Raport*; Instytut Jagielloński: Warszawa, Poland, 2018.
47. WHO Regional Office for Europe—OECD. *Economic Cost of the Health Impact of Air Pollution in Europe: Clean Air, Health and Wealth*; WHO Regional Office for Europe: Copenhagen, Denmark, 2015.
48. Tomala, M. Energia odnawialna jako kluczowy element bezpieczeństwa zaopatrzenia energetycznego i środowiskowego państw nordyckich, Bezpieczeństwo. *Teor. Prakt.* **2016**, *1*, 105–117.
49. Ignarska, M. Odnawialne źródła energii w Polsce. *Poliarchia* **2013**, *1*, 57–72. [CrossRef]
50. Skubiak, B. Rola Samorządów w Przezwyciężaniu Barier Rozwoju Odnawialnych źródeł Energii. In *Trendy i Wyzwania ZrównOważonego Rozwoju*; Kryk, B., Ed.; Uniwersytet Szczeciński: Szczecin, Poland, 2011; pp. 275–294.

51. Ligus, M.; Poskrobko, T.; Sidorczuk-Pietraszko, E. *Pozaśrodowiskowe Efekty Zewnętrzne w Lokalnych Systemach Energetycznych*; Fundacja Ekonomistów Środowiska i Zasobów Naturalnych: Białystok, Poland, 2015.
52. REN21. *Renewables 2020 Global Status Report*; REN21 Secretariat: Paris, France, 2020.
53. BP p.l.c. *BP Statistical Review of World Energy 2020*, 69th ed.; Statistical Review of World Energy: London, UK, 2020.
54. IRENA. *Renewable Capacity Statistics 2019*; International Renewable Energy Agency (IRENA): Abu Dhabi, United Arab Emirates, 2019.
55. Simionescu, M.; Bilan, Y.; Krajňáková, E.; Streimikiene, D.; Gędek, S. Renewable Energy in the Electricity Sector and GDP per Capita in the European Union. *Energies* **2019**, *12*, 2520. [CrossRef]
56. Grabara, J.; Tleppayev, A.; Dabylova, M.; Mihardjo, L.W.W.; Dacko-Pikiewicz, Z. Empirical Research on the Relationship Amongst Renewable Energy Consumption, Economic Growth and Foreign Direct Investment in Kazakhstan and Uzbekistan. *Energies* **2021**, *14*, 332. [CrossRef]
57. Llamosas, C.; Sovacool, B.K. The future of hydropower? A systematic review of the drivers, benefits and governance dynamics of transboundary dams. *Renew. Sustain. Energy Rev.* **2021**, *137*, 110495. [CrossRef]
58. Walesiak, M. *Uogólniona Miara Odległości GDM w Statystycznej Analizie Wielowymiarowej z Wykorzystaniem Programu R*; Wydawnictwo Uniwersytetu Ekonomicznego we Wrocławiu: Wrocław, Poland, 2011.
59. Redclif, M. *Critical Concepts in the Social Sciences*; Routledge Taylor & Francis: London, UK; New York, NY, USA, 2005.
60. Bleys, B. Alternative Welfare Measures. *MOSI Work. Pap.* **2005**, *12*, 1–15.
61. Lawn, P.A. A theoretical foundation to support the Index of Sustainable Economic Welfare (ISEW), Genuine Progress Indicator (GPI), and other related indexes. *Ecol. Econ.* **2003**, *44*, 105–118. [CrossRef]
62. Gasparatos, A.; El-Haram, M.; Horner, M. A critical review of reductionist approaches for assessing the progress towards sustainability. *Environ. Impact Assess. Rev.* **2008**, *28*, 286–311. [CrossRef]
63. UNDP. *Human Development Indices and Indicators 2018 Statistical Update*; United Nations Development Programme: Washington, DC, USA, 2018.
64. UNDP. Human Development Report 2015. Work for Human Development—TECHNICAL NOTES. 2015, pp. 1–10. Available online: http://hdr.undp.org/sites/default/files/hdr2015_technical_notes.pdf (accessed on 6 March 2021).
65. The Economist Intelligence Unit's Quality-of-Life Index. 2005. Available online: https://www.economist.com/media/pdf/quality_of_life.pdf (accessed on 6 March 2021).
66. OECD. Better Life Index: Definitions and Metadata. 2019. Available online: https://www.oecd.org/statistics/OECD-Better-Life-Index_definitions_2019.pdf (accessed on 6 March 2021).
67. NEF. Happy Planet Index 2016 Methods Paper. Available online: https://static1.squarespace.com/static/5735c421e321402778ee0ce9/t/578cc52b2994ca114a67d81c/1468843308642/Methods+paper_2016.pdf (accessed on 28 March 2021).
68. Green, M.; Harmacek, J.; Krylova, P. *2020 Social Progress Index*; Executive Summary; Social Progress Imperative: Washington, DC, USA, 2020; Available online: https://www.socialprogress.org/static/37348b3ecb088518a945fa4c83d9b9f4/2020-social-progress-index-executive-summary.pdf (accessed on 28 March 2021).
69. Almond, R.E.A.; Grooten, M.; Petersen, T. *Living Planet Report—2020: Ratunek dla Różnorodności Biologicznej*; WWF: Gland, Switzerland, 2020.
70. Lazarus, E.; Zokai, G.; Borucke, M.; Panda, D.; Iha, K.; Morales, J.C.; Wackernagel, M.; Galli, A.; Gupta, N. *Working Guidebook to the National Footprint Accounts*; Global Footprint Network: Oakland, CA, USA, 2014.
71. The Global Green Economy Index. Available online: https://dualcitizeninc.com/global-green-economy-index/ (accessed on 28 March 2021).
72. Karpińska-Mizielińska, W. *Instytucje Infrastruktury Rynkowej w Kreowaniu Przedsiębiorczości Lokalnej*; Fundacja Promocji Rozwoju im. E. Lipińskiego: Warszawa, Poland, 1999.
73. Nermend, K. *Metody Analizy Wielokryterialnej i Wielowymiarowej We Wspomaganiu Decyzji*; PWN: Warszawa, Poland, 2017.
74. Kuc, M. Konwergencja społeczna i ekonomiczna regionów państw nordyckich. *Coll. Econ. Anal. Ann.* **2016**, *41*, 175–188.
75. Malinowski, M.; Smoluk-Sikorska, J. Spatial Relations between the Standards of Living and the Financial Capacity of Polish District-Level Local Government. *Sustainability* **2020**, *12*, 1825. [CrossRef]
76. Wysocki, F. *Metody Taksonomiczne w Rozpoznawaniu Typów Ekonomicznych Rolnictwa i Obszarów Wiejskich*; Wydawnictwo Uniwerystetu Przyrodncizego w Poznaniu: Poznań, Poland, 2010.
77. Młodak, A. *Analiza Taksonomiczna w Statystyce Regionalnej*; Difin: Warszawa, Poland, 2006.
78. Hwang, C.L.; Yoon, K. *Multiple Attribute Decision Making: Methods and Applications*; Springer: New York, NY, USA, 1981.
79. Korol, J.; Szczuciński, P. *Ekonometryczne Modelowanie Procesów Gospodarki Regionalnej Opartej na Wiedzy*; Wydawnictwo Adam Marszałek: Toruń, Poland, 2009.
80. Kaufman, L.; Rousseeuw, P.J. *Finding Groups in Data: An Introduction to Cluster Analysis*; John Wiley & Sons Inc.: Hoboken, NJ, USA, 2005.
81. UNESCO. *IDAMS. Internationally Developed Data Analysis and Management Software Package*; UNESCO: Paris, France, 2008.
82. Ter Braak, C.J.F. Interpreting canonical correlation analysis through biplots of structure correlations and weights. *Psychometrika* **1990**, *55*, 519–531. [CrossRef]
83. Cavadias, G.S.; Ouarda, T.B.M.J.; Bobée, B.; Girard, C. A canonical correlation approach to the determination of homogeneous regions for regional flood estimation of ungauged basins. *Hydrol. Sci. J.* **2001**, *46*, 499–512. [CrossRef]

84. Hardoon, D.R.; Szedmak, S.; Shawe-Taylor, J. Canonical Correlation Analysis. In *An Overview with Application to Learning Methods*; University of London: London, UK, 2003.
85. Weenink, D. Canonical Correlation Analysis. *IFA Proc.* **2003**, *25*, 81–99.
86. Naylor, M.G.; Lin, X.; Weiss, S.T.; Raby, B.A.; Lange, C. Using Canonical Correlation Analysis to Discover Genetic Regulatory Variants. *PLoS ONE* **2010**, *5*, e10395. [CrossRef]
87. Krzyśko, M.; Łukaszonek, W.; Wołyński, W. Canonical Correlation Analysis in the Case of Multivariate Repeated Measures Data. *Stat. Transit. New Ser.* **2018**, *19*, 75–85. [CrossRef]
88. Ebenezer, O.R. Canonical Correlation Analysis of Poverty and Literacy Levels in Ekiti State, NIGERIA. *Math. Theory Model.* **2012**, *2*, 31–39.
89. Chin-Tsai, K. Life Quality and Job Satisfaction: A Case Study on Job Satisfaction of Bike Participants in Chiayi County Area. *J. Int. Manag. Stud.* **2011**, *6*, 1–9.
90. Saeed, T.; Tularam, G.A. Relations between fossil fuel returns and climate change variables using canonical correlation analysis. *Energy Sources Part B Econ. Plan. Policy* **2017**, *12*, 675–684. [CrossRef]
91. Santos-Alamillos, F.J.; Pozo-Vázquez, D.; Ruiz-Arias, J.A.; Lara-Fanego, V.; Tovar-Pescador, J. Analysis of Spatiotemporal Balancing between Wind and Solar Energy Resources in the Southern Iberian Peninsula. *J. Appl. Meteorol. Clim.* **2012**, *51*, 2005–2024. [CrossRef]
92. Panek, T.; Zwierzchowski, J. *Statystyczne Metody Wielowymiarowej Analizy Porównawczej: Teoria i Zastosowania*; Oficyna Wydawnicza SGH: Warszawa, Poland, 2013.
93. Jaba, E. The "3 sigma" rule used for the identification of the regional disparities. Yearbook of the "Gheorghe Zane". *Inst. Econ. Res.* **2007**, *16*, 47–57.
94. Kasunic, M.; McCurley, J.; Goldenson, D.; Zubrow, D. *An Investigation of Techniques for Detecting Data Anomalies in Earned Value Management Data*; Software Engineering Institute: Pittsburgh, PA, USA, 2011.
95. Noble, R.B., Jr. *Multivariate Applications of Bayesian Model Averaging*; Virginia Polytechnic Institute: Blacksburg, VA, USA, 2000.
96. Tadeusiewicz, R.; Izworski, A.; Majewski, J. *Biometria*; AGH: Kraków, Poland, 1993.
97. Thompson, B. *Fundamentals of Canonical Correlation Analysis: Basics and Three Common Fallacies in Interpretation*; University of New Orleans and Louisiana State University Medical Center: New Orleans, LA, USA, 1987.
98. World Economic Forum. *Fostering Effective Energy Transition*, 2020 ed.; WWF: Cologny, Germany; Geneva, Switzerland, 2020.
99. GUS. *Energia ze źródeł Odnawialnych w 2019 r*; Główny Urząd Statystyczny: Warszawa, Poland, 2020.
100. Kopczewska, K.; Kopczewski, T.; Wójcik, P. *Metody Ilościowe w R. Aplikacje Ekonomiczne i Finansowe*; CeDeWu: Warszawa, Poland, 2009.
101. Więcek, G.; Sękowski, A. Powodzenie w kształceniu integracyjnym a wybrane zmienne psychospołeczne—Weryfikacja modelu teoretycznego. *Rocz. Psychol.* **2007**, *10*, 89–111.

 *energies*

Article

Characterization of Customized Encapsulant Polyvinyl Butyral Used in the Solar Industry and Its Impact on the Environment

Samer Khouri [1], Marcel Behun [1], Lucia Knapcikova [2], Annamaria Behunova [1], Marian Sofranko [1] and Andrea Rosova [1,*]

[1] Institute of Earth Resources, Faculty of Mining, Ecology, Process Control and Geotechnologies, Technical University of Kosice, 9, 04200 Kosice, Slovakia; samer.khouri@tuke.sk (S.K.); marcel.behun@tuke.sk (M.B.); annamaria.behunova@tuke.sk (A.B.); marian.sofranko@tuke.sk (M.S.)

[2] Department of Industrial Engineering and Informatics, Faculty of Manufacturing Technologies with the seat in Presov, Technical University of Kosice, 1, 08001 Presov, Slovakia; lucia.knapcikova@tuke.sk

* Correspondence: andrea.rosova@tuke.sk

Received: 4 September 2020; Accepted: 12 October 2020; Published: 15 October 2020

Abstract: Taking climate and geopolitical issues into account, we must shift our thinking towards "eco" and focus on renewable energy. The accessible solar energy represents 400 times the amount of consumption, while its potential represents 10,000 times the amount of demand. The paper aims to analyze recycled, customized polyvinyl butyral (PVB) with high purity (more than 98%) concerning its physicochemical and mechanical properties and its possible applicability in the photovoltaic industry as an encapsulating material. The detailed investigation on polyvinyl butyral starting from characterizations, homogenization, and moulding process to tensile tests and used exposure testing in laboratory apparatus are performed. Samples of recycled polyvinyl butyral were exposed to ultraviolet (UV) radiation of the value 0.76 $W.m^{-2}.nm^{-1}$ at 340 nm, water spray, drying at 50 °C and condensation for 320 h when the radiation was turned off. The results obtained were more controlled in a laboratory environment than those found in external, uncontrolled environments. These conditions subsequently accelerate any degradation of polyvinyl butyral as a material and subsequent degradation of the final product.

Keywords: polyvinyl butyral; solar panels; environment

1. Introduction

Energy consumption grows yearly worldwide by 2.6%. [1]. It is expected by the year 2100 that energy consumption will increase more than 3 times. [1,2]. Europe is committed to significantly reducing CO_2 emissions, with about a 20% share of renewable energy sources. This is supported by extensive research programs and cutting-edge technology and involvement in the solar industry. Nowadays, the installation of solar panels is becoming more efficient and cost-effective [2]. It is assumed that sufficient solar energy is available in densely populated areas.

One of the possibilities is to obtain electricity from sunlight, which is a source of life on our planet as we know it. Solar collectors can meet 60–70% of the annual hot water demand in the household, in summer almost wholly, and in the transition period and winter it will be pre-heated. Energy obtained from the sun is primarily a virtually inexhaustible, safe and renewable energy source accessible for most of the year. The use of the sun's energy contributes to a sustainable way of life and does not burden future generations [3,4]. The use of solar energy alone does not have any adverse environmental effects during the entire life of the technological equipment, which in our conditions is around 20 to 30 years [5]. The economic goal of the development of a new financial sector of producers and suppliers of technologies for renewable energy sources is also significant. According to a study by the European Federation for the Thermal Utilization of Solar Energy, the use of solar energy creates incomparably more jobs compared to fossil and nuclear power. For 1000 GWh of primary point supplied, there are 90 jobs created in the coal sector, 72 jobs in the nuclear industry, and up to 3960 jobs in the solar industry [5,6]. It is mainly the production, design, installation and maintenance of solar systems (Figure 1), which, unlike large energy sources, are not centralized in one place.

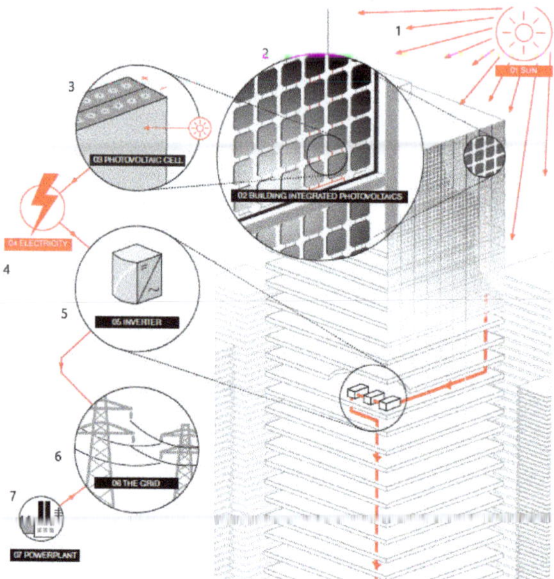

Figure 1. Description of the solar energy process.

Legend to Figure 1: 1-Sun; 2-Building Integrated Photovoltaics; 3-Photovoltaic Cell; 4-Electricity; 5-Inverter; 6-The Grid; 7-Powerplant.

A photovoltaic cell [7], otherwise called a solar cell, is a semiconductor device [8,9]. The structure of this semiconductor device is similar to the form of a photodiode. Generally, a photovoltaic cell transforms solar energy into electrical energy, using a photoelectric effect [8]. The term photoelectric effect means the release of electrons from a substance exposed to electromagnetic radiation [9]. The number of emitted electrons depends on the intensity of radiation [9]. Monocrystalline silicon was the first material to be used in practice for the production of photovoltaic cells. In its beginning in 1954, the efficiency of converting solar energy into electrical energy in such a section was around 6%. Later, as science became more concerned with this simple source of electricity, the efficiency value was increased to 17%, which was considered unsurpassed in 1980 [7]. The overall development of photovoltaic cells was hampered by the fact that prices were high and scientists sought to reduce the unit price per panel, thus using less pure silicon, which was not sufficiently efficient [4]. In the year 1992, the maximum efficiency was achieved. The laboratory measurement of the cell showed an efficiency value of up to 35.2%. Monocrystalline cells are composed of monocrystalline silicon crystals, measuring about 10 cm [8]. They are produced from ingots or, in other words, polycrystalline silicon rods by the Czochralski method by the slow drawing of pure molten silicon [10]. These bars are further cut with a particular saw for sheets with a thickness of approximately 0.25 mm to 0.35 mm. Figure 2 shows a chronological overview of the crucial milestones of technologies and materials used in the photovoltaic industry [10]. The year 1958 was essential for the use of the first solar cell made of silicon by Bell Laboratories (New Jersey, USA) [10].

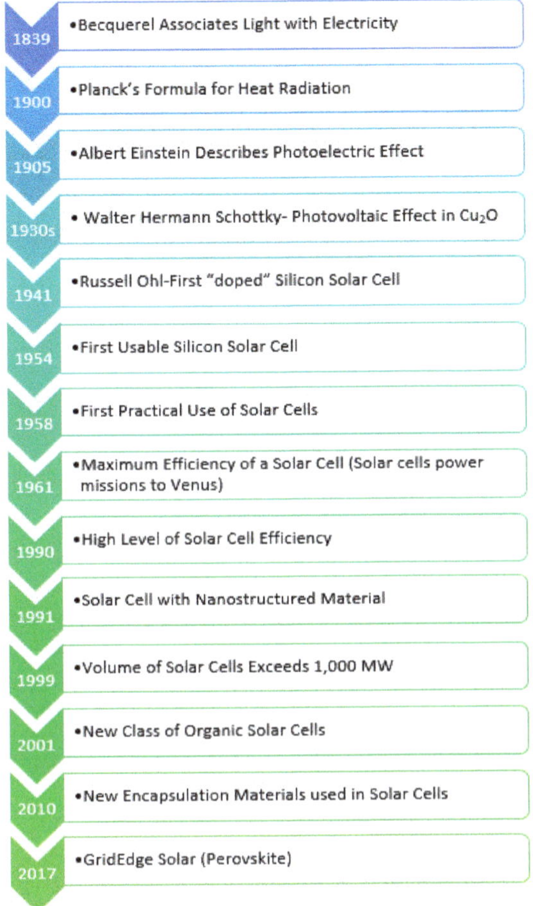

Figure 2. Chronology review of photovoltaics.

The latest research is from 2017 and continues to the present day. The Massachusetts Institute of Technology (MIT) is working on research and development to reduce the technological and market risks of lightweight and flexible photovoltaic modules through proof of testing analyses by which we can significantly reduce these impacts on the environment.

Nowadays, it is also possible to produce cells with a thickness of 0.1 mm, with the lowest possible waste that is generated when cutting sections. Polycrystalline silicon cells are today one of the most common types of solar cell [4,11]. Their production consists of the extraction of monocrystalline silicon, or by casting pure silicon into moulds, where they are subsequently cut to the required thickness [8]. This method of casting silicon is a much simpler way than drawing the single silicon crystal itself because the manufacturer can cast the shape he needs [9]. The disadvantage of casting is more low efficiency, but their advantage is a better use of this material, lower production cost and the possibility of larger size [7]. The manufacturing of amorphous solar cells is much simpler than monocrystalline and polycrystalline cells. Generally consumes much less material. This material saving, of course, means that when buying an amorphous cell, we save significantly compared to buying a single crystal and polycrystalline cell [12]. A 1.0 mm thickness of the amorphous cell absorbs 90% of the sun's radiation, making it possible to produce much thinner modules that can be used to cover the roofs of buildings [13]. According to research, the polymer ethylene-vinyl acetate (EVA) belongs to key material for solar cells as encapsulation. EVA with its excellent processing properties offers a wide range of applications possibilities. The company Kuraray Co., Ltd. (Tokyo, Japan) [7], through cooperation with the National Institute of Advanced Industrial Science and Technology (AIST)), has developed high-strength encapsulating polyvinyl butyral (PVB) films [12,14] (Figure 3). PVB films produce photovoltaic (PV) modules easier and cheaper, and the project has started to take them into the test markets.

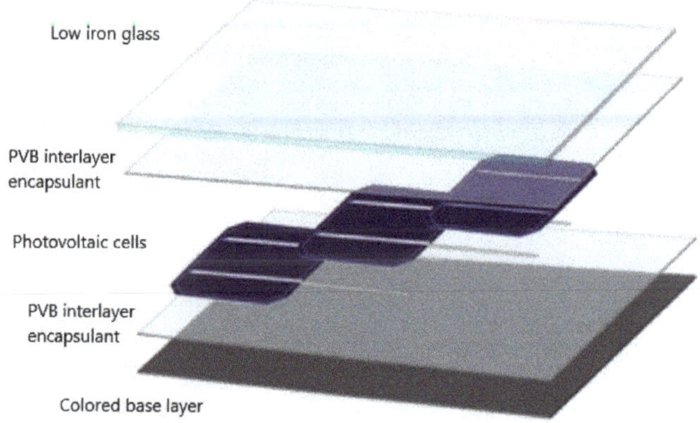

Figure 3. Solar panel composition used polyvinyl butyral (PVB) interlayer encapsulate.

For durability of the module and the long-term production of energy, the cell encapsulation materials must have specific essential properties, e.g.,

- Mechanical cell protection;
- Weather protection;
- Electrical insulation;
- Resistance to external impact;
- Protection against oxygen and water vapor;
- Minimization of cell corrosion;
- High adhesion to other components of the module (glass, chamber, base foil, contacts, etc.);
- Maximum guarantee of transparency and high protection against ultraviolet (UV) radiation.

The solar industry is one of the fastest growing areas of polyvinyl butyral end-use on the market. In terms of offer and demand for PVB interlayers in the photovoltaic industry, we can say that this material is a leader. Studies clearly show a shift to renewable energy sources, which is why the consumption of PVB interlayers is expected to increase [8]. The current state is that the intermediate layers of PVB film are mainly used in 70% of laminated glass applications. The basic function is to increase safety and improve the acoustic and UV protection of glass [11,12]. In terms of construction, these interlayers are used in traditional quadrilateral supported glass and 1–3 side and minimally supported glass in windows, façade systems, ceilings, floors and railings [12]. These interlayers also have limitations, like a low strength and higher weight with comparison to functional glass.

The most important representatives of PVB manufacturers are Eastman Chemical Company (US), Kuraray (Germany), Sekisui Chemicals (Japan), Everlam (Belgium), Genau Manufacturing Company (India), KB PVB (China), Chang Chun Group (China), DuLite (China), Huakai Plastic (China), Willing Lamiglass Materials (China), Jiangsu Darui Hengte Technology (China), and Tiantai Kanglai Industrial (China) [13].

PVB is a polymeric material with excellent mechanical properties and is mainly used as an interlayer for laminated glass [11–13]. PVB is used in the solar industry, in the production of PV panels, as well as in the automotive and construction industries. Recently, the demand for the use of PVB as an intermediate layer in sandwich laminated glass has increased significantly due to its use for safety and security purposes [14]. Acoustic insulation and protection against UV radiation are the main properties of PVB, which increase their use for the above applications. PVB interlayers also absorb more than 99% of UV light rays [11]. In buildings, PVB interlayers are used in facades, windows, ceiling glazing, partitions, and floors.

Demand for polyvinyl butyral (PVB) is increasing due to its increased use in the production of films and foils, which are further used in the automotive and construction industries for the production of laminated glass [14]. With the growing solar industry, the demand for excellent optical clarity, strong binding ability and exceptional adhesion properties increases. It is estimated that the photovoltaic end-use sector recorded the highest value between 2016 and 2021 in terms of the market share of use. The growing use of PVB films and films in the manufacturing of solar components is driving the market for the photovoltaic industry forward. The transparent polyvinyl buyral (PVB) film is placed on the front and back of the photovoltaic cells. When the melting point is reached, the polyvinyl butyral film melts and subsequently fills all gaps and removes air between the layers. [15].

The photovoltaic market is very dynamic and has developed rapidly in recent years. Due to the growing interest in renewable energy sources, production processes have recently improved considerably. The main goal of all manufacturers is to increase the efficiency of cells while reducing production costs. Price per unit of output is still a significant factor in this area. Therefore, in order to obtain that objective, it is necessary to lower the price of the input material, simplify and reduce the production technology, and increase the conversion efficiency of PV cells. Other no less important factors are, of course, the lifespan of the cells and the decline in performance over time.

It is known that the European Union (EU) as a whole is more than 50% dependent on imports of primary energy sources. In most cases, this is from politically or economically unstable regions [8]. Another impact on the EU's energy strategy is the commitments made in the field of air protection. Therefore, the EU's energy efforts focus mainly on energy efficiency and the use of renewable energy sources, the potential of which is not negligible in the individual member states [7,9].

2. Material and Methods

2.1. Material Characterization

PVB is produced from polyvinyl acetate (PVA) by reaction with butyraldehyde ($CH_3CH_2CH_2CHO$) and formaldehyde (CH_2O) [12]. PVB is a transparent, strong and flexible thermoplastic with high optical purity and good adhesion to a large number of substrates. Polyvinyl butyral belongs to the group of amorphous polymers with a very high VA (vinyl acetate) content higher than 63.3% by weight [16,17]. The melting point (Tm) is between 171 °C and 218 °C [18]. DSC 204 equipment from NETZSCH company was used to determine the Tg and Tm temperatures. The obtained data were evaluated by the NETZSCH Proteus program, which is a part of the equipment. Small weights of the test sample, in our case 9 mg, are used for analysis. The device was set in the temperature mode from –50 °C to 300 °C.

PVB is an essential example of the light, highly transparent, elastic and durable films. The polyvinyl butyral used in our research (Figure 4) came from the recycling of safety glasses with a purity of more than 98%.

Figure 4. Polyvinyl butyral used in this research.

With its properties such as toughness, good adhesion to other components, light fastness and excellent transparency, polyvinyl butyral is one of the key materials used in the solar industry [13,19]. PVB foil fulfils a safety function. Recycled polyvinyl butyral, which was used, has the following values (Table 1).

Table 1. Characterization of selected polyvinyl butyral.

Material	
Type	flakes form neutral
Coloring-matter	20–30 mm
Flakes parameter	>98%
Purity	<%
Impurity content	ca. 2%
Residual amount of humidity	Less as 2%
Ratio of glass particle content	
Fire point	none
Glasstransition temperature	130–170 °C
Viscosity (dynamic)	100–175 m Pa*s
MVR (Melt VolumeRate)	(DIN 53015)
MFR (Melt Flow Rate)	6–7 cm^3/10 min 5–6 g/10 min

2.2. Homogenization and Moulding Process

Thorough homogenization of the material was ensured by continuous mixing of polyvinyl butyral [18,19]. The material was mixed on a Brabender Plasti-Corder W 350 E, under laboratory conditions at 22 °C and 60% humidity. The kneading process took 15 min.

During this process, the removal of air bubbles, which are undesirable during the pressing process, was achieved.

After homogenization of the recycled polyvinyl butyral, the material was extruded and prepared for the tensile test. Moulding technology took place on a laboratory press Brabender W 350 with the following parameters (Table 2).

Table 2. Moulding process characteristics.

Moulding Parameters	Brabender W 350
Moulding Temperature	190 °C
Heating	20 min (±2 min)
Moulding Time	20 min (±2 min)
Cooling	20 min (±2 min)
Pressure	10 MPa (±0.1 MPa)

The moulding cycle consisted of the following operations [18]:

- opening the mould,
- material filling,
- moulding process,
- moulding process,
- opening the mould,
- removing process,
- cooling and cleaning process.

2.3. Polyvinyl Butyral Testing Used Methods of Exposure to Laboratory Light Sources

The application possibilities of polyvinyl butyral are undoubtedly an advantage in the solar industry. It is always crucial assign the intensity of sunlight, heat, humidity and other climatic influences on the behavior of the material [13,15]. Exposures in laboratory apparatus (Figure 5) is performed under more monitored aspects than those found in natural environments and aimed to accelerate any polymer degradation and product failure. In preparing the material for laboratory testing, we followed the standard EN ISO 4892 (Plastics–Methods of exposure to laboratory light sources) [15].

Figure 5. The ThermoTec Laboratory equipment.

According to EN ISO 4892 [15], the test conditions were determined, which are listed in Table 3. The test specimens were subjected to water, a temperature in the range from –50 °C to +120 °C, UV radiation with a value of 0.76 W.m^{-2}.Nm^{-1}, and relative humidity. The whole cycle lasted 10 h, and in our case, the samples were tested first for seven cycles i.e., 70 h and the total duration of action in the climate chamber was 32 cycles, i.e., 320 h. After taking samples from the climate chamber, we subjected the material to mechanical and physical-mechanical tests.

Table 3. Measurement characteristics.

No. Cycle	Load Cycle	The Lamp Type	Radiation Intensity	Prescribed Temperature	Relative Humidity
	8 h Drying		0.76 W.m^{-2}.Nm^{-1} By 340 nm	(50 ± 3) °C	Unregulated
	0.15 h Water spray	Type 1 A	-	Unregulated	Unregulated
2	0.15 h Water spray	(Ultraviolet	-	Unregulated	Unregulated
	0.15 h Water spray	(UV-A-340))	-	Unregulated	Unregulated
	0.15 h Water spray		-	Unregulated	Unregulated
	2 h Condensation		Radiation off	(50 ± 3) °C	Unregulated

UV stability is very important for long-term exposure of the material, such as the exposure required for the encapsulation of a photovoltaic module [19,20]. High transparency is required to achieve an effective wavelength for PVB components. By using benzotriazole (0.2–0.35% by weight) we achieve the mitigation of the effects of UV radiation. [13].

When testing the effect of radiation, heat and humidity on the physical, chemical and optical properties of polyvinyl butyral using artificially accelerated irradiation of specific laboratory light sources, we achieved simulation of real climatic conditions in the laboratory [21]. These controlled conditions are used to determine the rate of degradation of the material in the exposed environment.

3. Results and Discussion

Obtaining results from a real environment where the material is exposed to extreme conditions for a short time is difficult. Therefore, it is important to simulate real conditions in the laboratory and look for bottlenecks that would ultimately lead to material failure of the respectively finished product. Therefore, our primary goal in the presented paper was to test polyvinyl butyral and its physicochemical and mechanical properties, which were subsequently compared with the values of tests from the climate equipment ThermoTec.

3.1. Differential Scanning Calorimetry Testing

Differential scanning calorimetry (DSC) gives an overview of the glass transition temperatures (Tg) and melting points (Tm) for recycled polyvinyl butyral [22]. We obtained important information that helped us decide the right manufacturing process for the material. [18,23]. We can see from Figure 6 that the measurement of the recycled polyvinyl butyral seal took place as a function of mW/mg at a temperature of °C. If the measurement is set to mW/mg, it means that the calorimeter was calibrated before starting the measurement. The actual measured value (Tg) is the temperature deviation with respect to the reference sample. The first and most important information of the analyzed sample is the temperature at which we find the glass transition temperature Tg, with respect to melting point Tm. Peak is associated with the temperature at which the maximum reaction rate occurs (exothermic reaction). The glass transition (Tg) in materials is important information regarding the phase transition of materials. It is also used to quantify enthalpy characteristics that are associated with both physical and chemical transformations of substances in test systems. This method can be used to measure heat capacity, determine the purity of substances and for other particular purposes. The samples were analyzed under laboratory conditions, at a interior temperature of 22 °C and an interior humidity of 60%, following DIN ISO 113 57.

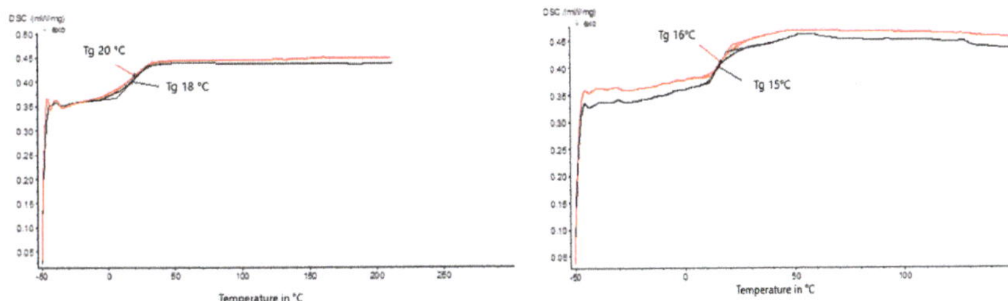

Figure 6. Differential scanning calorimetry (DSC) results for recycled polyvinyl butyral before (**left side**) and after (**right side**) exposure to laboratory light sources.

Legend to the Figure 6: Black line- 1st heating of recycled polyvinyl butyral before and after exposure to laboratory light sources; Red line-2nd heating of recycled polyvinyl butyral before and after exposure to laboratory light sources

By differential scanning calorimetry, the temperature of recycled polyvinyl butyral was maintained isothermally with the reference value. Such an amount of energy is needed to maintain isothermal conditions. The course is plotted as a function of time or temperature. The material is placed on metal pads, which reduce the thermal gradient to a minimum [24,25]. The condition was to achieve high heating values (in tens of K.min^{-1}, °C.min^{-1}) in the analysis, thus ensuring a high resolution of heating and, therefore, the obtained temperature scale [26].

3.2. Mechanical Testing (Tensile Test)

The tensile test was used to evaluate all requirement tensile strength characteristics of recycled polyvinyl material. The aim of testing was to stress the test specimen until the moment when the sample ruptured. The test specimen was clamped in the jaws of a tensile machine [27–29]. Regarding constant tensile speed, the material reached the maximum tensile strain. Along with the increasing deformation, the force required to maintain a constant speed of movement of the jaws of the tearing machine also increased. The tensile test of the tested samples from recycled polyvinyl butyral (Table 4) was performed at an interior temperature of 23 °C and humidity of 60%, according to standard DIN EN ISO 527.

Table 4. Tensile test characteristics.

Equipment	Zwick Z020 Universalpruefungsmaschine
Max. Force	20 kN
Testing Speed	100 mm/min
Standard	DIN EN ISO 527

The measured values of the tensile test were evaluated by statistical data processing evaluated by the software [30] Test Xpert (Table 5). We work with statistical constants with mean value, variance and deviation. The tested sample of material is clamped in the jaws of the tearing machine (Figure 7) and gradual stretching is performed at a constant speed [31,32], where max. the tensile strain value gives us a value of 145.9610%. The aim of our measurement was to get as close as possible to the correct value of the measured quantity for the required degree of reliability. However, using statistics, we can determine these values with limited accuracy from a set consisting of a finite number of measurements.

Table 5. Tensile test results according to DIN EN ISO 527 before exposure to laboratory light sources.

The Statistic Variables	The Thickness of the Test Sample [mm]	Width of the Test Sample [mm]	E-Module [MPa]	Tensile Stress at Break [MPa]	Tensile Strain at Break [%]	Max. Tensile Stress [MPa]	Max. Tensile Strain [%]
x	2.9701	6.0250	5.0000	17.2360	146.140	17.5182	145.9610
s	0.0294	0.0129	2.0000	1.4731	11.5940	1.6361	11.6471
ν	0.9900	0.2100	52.450	8.3401	7.9300	9.3400	7.9800

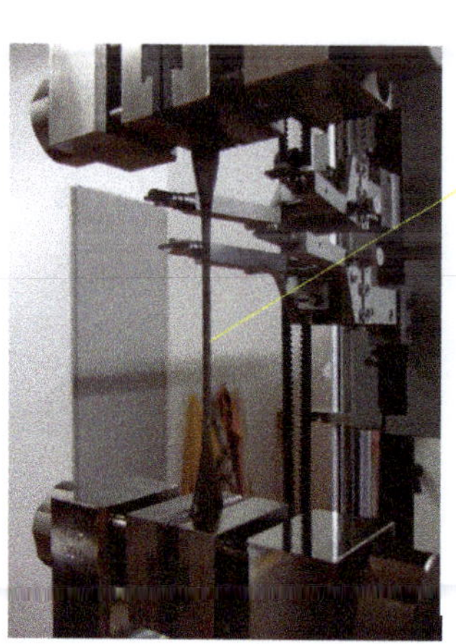

Figure 7. Tensile test of recycled polyvinyl butyral (**left side**—sample held, **right side**—PVB zoomed 50×).

With increasing deformation, the force increases. The force is required to maintain a constant feed speed of the jaw the tensile machine. The tensile test works on the principle of stressing the test specimen until the time [32], when the sample ruptures, which indicates the value of tensile stress at break σ_B (17.2360 MPa).

The test specimens were subject to a change of water, a temperature ranging from −50 °C to +120 °C, to UV radiation with a value of 0. 76 W.m^{-2}.Nm^{-1}, and relative humidity of 60%. One cycle lasted 10 h; the total time of action in the climate chamber was 32 cycles, i.e., 320 h. The value of the E-module after exposure to the set conditions was 4.9850 MPa, as can be seen from the previous measurement, and the decrease was minimal (5.000 MPa before exposure to laboratory light sources). At the values when the maximum effect of tensile characteristics σ_{max} and ε_{max} (maximum tensile stress and maximum tensile strain) occurs, the values after the action of climatic conditions were with minimal differences (Table 6). It was confirmed that polyvinyl butyral is UV stable and its mechanical properties do not change significantly under laboratory conditions.

Table 6. Tensile test results according to DIN EN ISO 527 after exposure to laboratory light sources.

The Statistic Variables	The Thickness of the Test Sample [mm]	Width of the Test Sample [mm]	E-Module [MPa]	Tensile Stress at Break [MPa]	Tensile Strain at Break [%]	Max. Tensile Stress [MPa]	Max. Tensile Strain [%]
x	2.8901	6.0050	4.9850	17.1890	145.940	17.4981	144.8700
s	0.0199	0.0021	2.0030	1.3931	10.6840	1.5931	11.2561
ν	0.8900	0.1990	51.650	7.9501	7.8700	9.2910	7.8500

The recycled polyvinyl butyral is exposed to many internal and external influences over time. Internal influences include thermodynamic imbalance, external include heat, solar radiation, atmospheric oxygen, ozone, humidity, rain, oxides of sulfur and nitrogen, dust gradient, aggressive media (gases, vapors, liquids), ionizing radiation, mechanical force (often variable) and microorganisms. These effects affect the structure, change it and thus also change the useful properties of the polymer. Each acts by a different mechanism and affects a different structural level of the polymer. Exposure to excessively high temperatures for some time severely disrupts the macromolecular chains of the polymers. Our priority was to present the possibility of using the polyvinyl butyral with our obtained results as an excellent alternative for encapsulation in the solar industry.

4. Conclusions

In the solar industry, in the production of photovoltaic modules, there has been an effort since 2005 to replace existing encapsulations with PVB films in order to increase substantially the safety of laminated glass in the solar industry [33]. Using solar energy saves the natural resources of our planet. The use of fossil fuels such as oil, coal or natural gas brings with it several severe problems with our environment. Nowadays, global warming and climate change have become a reality, so it is essential to strive for a wider use of "clean" technologies, which undoubtedly include the use of solar energy. Unlike the combustion of conventional fuels, the use of solar energy does not release pollutants or greenhouse gases released into the atmosphere, causing the atmosphere to warm up [34] gradually.

The presented research aimed to evaluate the physicochemical and mechanical properties of recycled polyvinyl butyral, which came from the recycling of car windows. The tests took place in a climatic chamber, under the prescribed conditions. The UV radiation had a value of 0.76 $W.m^{-2}.Nm^{-1}$ at 340 nm. Samples of extruded polyvinyl butyral were exposed to UV radiation, water spray, drying at 50 °C, and condensation for 320 h when the radiation was turned off. It is known from the results that the values of recycled material after the tests differ only exceptionally. The value of the E-module after exposure to the set conditions was 4.9850 MPa, as can be seen from the previous measurement, and the decrease was minimal (5000 MPa before exposure to exposure to laboratory light sources). At the values when the maximum effect of tensile characteristics σmax 17.4981 MPa and εmax 144.8700% (maximum tensile stress and maximum tensile strain) occurs, the values after the action of climatic conditions were minimal.

The applicability and application of polyvinyl butyral varies with respect to location, mainly due to differences in UV radiation, atmospheric humidity, temperature changes, etc.

Author Contributions: Conceptualization, A.B. and L.K.; methodology, L.K. and M.B.; software and measurements, L.K. and A.B.; validation, L.K. and A.R.; formal analysis, S.K. and A.B.; investigation, L.K. and M.S.; resources, L.K. and M.S.; data curation, L.K. and S.K.; writing—original draft preparation, A.B. and L.K.; writing—review and editing, M.B. and A.R.; funding acquisition S.K. and A.R.; All authors have read and agreed to the published version of the manuscript.

Funding: This work was supported by the projects of the Scientific Grant Agency of the Ministry of Education, Science, Research and Sport of the Slovak Republic and the Slovak Academy of Sciences project No. VEGA 1/0317/19, project No. VEGA 1/0797/20 and was supported by the projects of the Cultural and Educational Grant Agency of the Ministry of Education, Science, Research and Sport of the Slovak Republic and the Slovak Academy of Sciences project No. KEGA 006TUKE-4/2019, project No. KEGA 016TUKE-4/2020.

Acknowledgments: The authors would like to thank the anonymous reviewers for their helpful and constructive comments and suggestions that greatly contributed to improving the final version of this paper.

Conflicts of Interest: The authors declare no conflict of interest. The funders had no role in the design of the study; in the collection, analyses, or interpretation of data; in the writing of the manuscript; or in the decision to publish the results.

References

1. Sunshine Energy. Available online: http://www.inforse.org/europe/fae/OEZ/slnko/slnko.html#TOP (accessed on 30 June 2020).

2.   Average Sunshine a Year at Cities in Europe—Current Results. Available online: https://www.currentresults.com/Weather/Europe/Cities/sunshine-annual-average.php (accessed on 6 August 2020).

3.   Possibilities of Using Solar Energy. Available online: http://www.aee-now.at/cms/fileadmin/downloads/projekte/Homepageuploads/Infosolar_ECB_Moznosti_vyuzivania_slnecnej_energie.pdf (accessed on 11 August 2020).

4.   Urtinger, K.; Beranovsky, J.; Tomeš, M. *Fotovoltaika: Energies from the Sun*; Vyd. 1; EkoWATT: Praha, Slovakia, 2007; p. 81.

5.   Sapa Solar Building. Available online: http://www.sapabuildingsystems.com/globalassets/sapa-building-systems-ab/pictures/brochures/solar_bipv_low.pdf?ts=636316455108630000 (accessed on 12 August 2020).

6.   GridEdge Solar. Available online: https://gridedgesolar.org/ (accessed on 8 July 2020).

7.   A PVB Film Helping to Make Photovoltaic Modules Lighter and Less Expensive. Available online: https://www.kuraray.com/news/2013/131002 (accessed on 13 August 2020).

8.   Transparency Market Research. Available online: http://www.transparencymarketresearch.com/polyvinyl-butyral-films-sheets.html (accessed on 10 August 2020).

9.   Solar Panels. Available online: https://www.quest.sk/solarne-kolektory/uspory-navratnost/ (accessed on 15 August 2020).

10.   Cutler, J.C.; Morris, C. Available online: https://www.elsevier.com/books/handbook-of-energy/cleveland/978-0-12-417013-1 (accessed on 10 August 2020).

11.   Polyvinyl Butyral. Available online: https://www.ihs.com/products/polyvinyl-butyral-chemical-economics-handbook.html (accessed on 28 July 2020).

12.   Polyvinyl Butyral. Available online: https://www.britannica.com/science/polyvinyl-butyral (accessed on 26 July 2020).

13.   Global Polyvinyl Butyral (PVB) Films Market Status (2015–2019) and Forecast (2020–2024) by Region, Product Type and End-Use. Available online: https://www.marketintellica.com/report/MI3819-global-polyvinyl-butyral-pvb-films-market (accessed on 10 August 2020).

14.   Kolcun, M.; Medveď, D.; Petráš, J.; Stolárik, R.; Vaško, Š. Research of Characteristics of Photovoltaic Possibilities for Potential Design of Solar Systems. Available online: http://people.tuke.sk/dusan.medved/Subory/VADIUM-Kniha.pdf (accessed on 21 July 2020).

15.   Standard EN ISO 4892 Plastics—Methods of Exposure to Laboratory Light Sources. Available online: https://www.iso.org/standard/60048.html (accessed on 21 August 2020).

16.   Agroui, K.; Koll, B.; Collins, G.; Salama, M.; Arab, A.H.; Belghachi, A.; Doulache, N.; Khemici, M.W. Available online: https://www.spiedigitallibrary.org/conference-proceedings-of-spie/7048/70480G/Characterization-of-encapsulant-materials-for-photovoltaic-solar-energy-conversion/10.1117/12.794012.short?SSO=1. (accessed on 11 August 2020).

17.   Eisentraeger, J.; Naumenko, K.; Altenbach, H.; Meenen, J. A user-defined finite element for laminated glass panels and photovoltaic modules based on a layer-wise theory. *Compos. Struct.* **2015**, *133*, 265–277. [CrossRef]

18.   Drabczyk, K.; Sobik, P.; Starowicz, Z.; Gawlińska, K.; Pluta, A.; Drabczyk, B. Study of lamination of solar modules with PMMA front layer. *Microelectron. Int.* **2019**, *36*, 100–103. [CrossRef]

19.   Wang, L.; Zhou, H.; Li, N. Carrier transport composites with suppressed glass-transition for stable planar perovskite solar cells. *J. Mater. Chem. A* **2020**, *8*, 14106–14113. [CrossRef]

20.   Heath, G.; Silverman, T.J.; Kempe, M.; Deceglie, M.G.; Ravikumar, D.; Remo, T.; Cui, H.; Sinha, P.; Libby, C.; Shaw, S.; et al. Research and development priorities for silicon photovoltaic module recycling to support a circular economy. *Nat. Energy* **2020**, *5*, 502–510. [CrossRef]

21.   Straka, M.; Khouri, S.; Rosova, A.; Cagáňová, D.; Culkova, K. Utilization of Computer Simulation for Waste Separation Design as a Logistics System. *Int. J. Simul. Model.* **2018**, *17*, 583–596. [CrossRef]

22.   Pujadas-Gispert, E.; Korevaar, C.; Alsailani, M.; Moonen, S. Linking constructive and energy innovations for a net zero-energy building. *J. Green Build.* **2020**, *15*, 153–184. [CrossRef]

23.   Tartakowski, Z.; Cimander, K.; Bursa, J. New polymer construction materials for applications in electrical engineering. In Proceedings of the Conference on Innovative Materials and Technologies in Electrical Engineering (i-MITEL), IEEE, Sulecin, Poland, 18–20 April 2018.

24.   Tanyildizi, H.; Asilturk, E. Performance of Phosphazene-Containing Polymer-Strengthened Concrete after Exposure to High Temperatures. *J. Mater. Civ. Eng.* **2018**, *30*, 04018329. [CrossRef]

25. Mandičák, T.; Mesároš, P.; Tkáč, M. Impact of management decisions based on managerial competencies and skills developed trough BIM technology on performance of construction enterprises. *Pollack Period.* **2018**, *13*, 131–140. [CrossRef]

26. Chen, S.; Zang, M.; Wang, D.; Yoshimura, S.; Yamada, T. Numerical analysis of impact failure of automotive laminated glass: A review. *Compos. Part B Eng.* **2017**, *122*, 47–60. [CrossRef]

27. Liu, B.; Xu, J.; Li, Y. Constitutive Investigation on Viscoelasticity of PolyVinyl Butyral: Experiments Based on Dynamic Mechanical Analysis Method. *Adv. Mater. Sci. Eng.* **2014**, *2014*, 1–10. [CrossRef]

28. Zajac, J.; Dupláková, D.; Hatala, M.; Goldyniak, D.; Poklemba, R.; Šoltés, P. The Effect of Used Fillers on the Strength Characteristics of Polymer Concrete Test Bodies. *TEM J.* **2019**, *8*, 795–800.

29. Moseley, J.; Miller, D.; Shah, Q.; Sakurai, K. Available online: https://www.nrel.gov/docs/fy12osti/52586.pdf (accessed on 17 July 2020).

30. Straka, M.; Rosova, A.; Lenort, R.; Šaderová, J.; Besta, P. Principles of computer simulation designfor the needs of improvement of the raw materials combined transport system. *Acta Montan. Slov.* **2018**, *23*, 163–174.

31. Mouritz, A.P.; Mathys, Z. Mechanical properties of fire-damaged glass-reinforced phenolic composites. *Fire Mater.* **2000**, *24*, 67–75. [CrossRef]

32. Dodds, N.; Gibson, A.G.; Dewhurst, D.; Davies, J.M. Fire behavior of composite laminates. *Compos. Part A* **2000**, *31*, 689. [CrossRef]

33. Niaki, M.H.; Fereidoon, A.; Ahangari, M.G. Experimental study on the mechanical and thermal properties of basalt fiber and nanoclay reinforced polymer concrete. *Compos. Struct.* **2018**, *191*, 231–238. [CrossRef]

34. Weiheng, F.; Weirong, H.; Tianze, X.; Kim, Y.; Lyu, C.; Anqi, Z. Exploring the Interconnection of Greenhouse Gas Emission and Socio-economic Factors. *SAR J.* **2019**, *2*, 115–120. [CrossRef]

Publisher's Note: MDPI stays neutral with regard to jurisdictional claims in published maps and institutional affiliations.

MDPI
St. Alban-Anlage 66
4052 Basel
Switzerland
Tel. +41 61 683 77 34
Fax +41 61 302 89 18
www.mdpi.com

Energies Editorial Office
E-mail: energies@mdpi.com
www.mdpi.com/journal/energies